同济大学研究生教材

景观生态规划设计案例评析

Reviews on Landscape Ecological Planning & Design Cases

王云才 著

同济大学出版社
TONGJI UNIVERSITY PRESS

图书在版编目（CIP）数据
景观生态规划设计案例评析：汉英对照 / 王云才著. -- 上海：同济大学出版社, 2013.5
ISBN 978-7-5608-5112-9

Ⅰ. ①景… Ⅱ. ①王… Ⅲ. ①景观生态环境 – 生态规划 – 案例 – 世界 – 双语教学 – 高等学校 – 教材 – 汉、英
Ⅳ. ①X32

中国版本图书馆CIP数据核字(2013)第041481号

景观生态规划设计案例评析
Reviews on Landscape Ecological Planning & Design Cases

王云才　著
Yuncai WANG

责任编辑　孟旭彦
责任校对　徐春莲
装帧设计　张　微
　　　　　董春洁 [右序设计工作室]

出版发行　同济大学出版社（地址：上海四平路1239号
　　　　　邮编：200092　电话：021-65985622）
经　　销　全国各地新华书店
印　　刷　上海安枫印务有限公司
开　　本　787mm×1092mm　1/16
印　　张　24.5
字　　数　612 000
版　　次　2013年5月第1版　2019年5月第2次印刷
书　　号　ISBN 978-7-5608-5112-9
定　　价　75.00元

同济大学景观规划设计专业研究生教育系列教材
同济大学研究生精品课程与研究生教材出版基金资助
同济大学研究生专业学位教育综合改革试点专项资助

THE PROFESSIONAL DEGREE GRADUATE EDUCATION TEXTBOOK SERIES FOR LANDSCAPE, ARCHITECTURE AND URBAN PLANNING OF TONGJI UNIVERSITY

SUPPORTED BY TOP-QUALITY COURSE & TEXTBOOK FOR DEGREE GRADUATE EDUCATION OF TONGJI UNIVERSITY FOUNDATION

SUPPORTED BY DEGREE EDUCATION COMPREHENSIVE REFORM EXPERIMENTAL UNIT OF TONGJI UNIVERSITY FOUNDATION

前言
PREFACE

景观规划的过程就是帮助居住在自然系统中或利用系统中有限资源的人们找到最适宜的生活与生产途径（麦克哈格，1969年），通过土地和自然资源的保护与利用规划，实现可持续的景观或生态系统。景观是生态系统，一个好的或是可持续的景观规划，必须是一个基于生态学理论和技术的规划。在20世纪80年代后，生态规划设计已经形成了综合自然生态和人文生态为一体的整体系统规划。生态教学、科研和实践不仅仅是指对自然生态系统特有的生态关系的揭示。同时，文化作为人类适应和改造自然的有效工具，人文生态成为生态规划发展的另一个潮流，它以不同尺度规划空间内的自然与人文生态系统形成的有机整体——“整体人文生态系统”作为景观规划设计的对象，为现代景观生态规划设计指明了发展方向。无论是自然生态还是人文生态，景观作为客体，具有完整的视域范围，人在其中形成独有的认知和体验。同时，景观具有自己独特的语言，记录和述说着人与环境相互作用的关系。因此景观生态理论、人文生态理论和景观的语言成为景观生态规划设计的三大理论基础。在此基础上，经历长期的学科发展，景观生态规划设计将自己的适用范围从花园、场地、道路、广场、公园扩展到城市、风景名胜区、自然保护区、资源保

The process of landscape planning is to help those who live in natural or take advantage of limited resources in system, in order to find the most suitable approach for living and production (McHarg, 1969), and to realize sustainable landscape or ecological system through the land and natural resources protection and utilization planning. Landscape is the ecological system, a good or sustainable landscape planning must be a planning based on ecology theory and technology. In the 1980 s, ecological planning and design has already formed the total system planning integrated natural ecology and human ecology together. It not only refers to the special natural ecological system of the ecological relationship revealed in ecological education, scientific research and practice; but also refers to culture, a effective tool to adapt and transform the natural for human, the human ecology becomes another developing trend for ecological planning development. Whatever the natural ecology or human ecology, landscape are objective and with a whole visional scale, and in which people can got the different cognition and experience. At the same time, landscape has an unique language, records and tells the interaction relationship between people and the environment. So the theory of landscape ecology, the theory of human ecology and the language of landscape become the three theoretical basis of landscape ecological planning and design. On this basis and through long-term development of the discipline, the landscape ecological planning and design extend its field, from the garden, site, road, square, park to city, scenic spot, nature reserve area, resources protection area, land use, greenway system, river basin, region and territory land, it becomes the bridge for landscape planning and design

护、土地利用、绿道系统、流域、区域与国土等广泛的空间，成为景观规划设计积极融于国际发展潮流和参与国家重大发展方向建设的桥梁。

景观生态规划设计的理论教材《景观生态规划原理》（中国建筑工业出版社，2007年第一版），是在同济大学建筑与城市规划学院景观学专业教学探索的基础上，广泛吸收和借鉴国内外经验的基础上完成的。在其后的教学应用中，结合景观生态规划设计研究的新进展和新成果，在第一版的基础上又进一步完成了《景观生态规划原理》（第二版，2013年）的出版。本版仍立足景观生态规划的基本理论与方法、空间类型、关键切入点和效果评价四个层面展开，形成清晰的四大板块教学体系。《景观生态规划原理》建立起了理论教学的体系和框架，是中华人民共和国住房和城乡建设部（住建部）普通高等教育土建学科专业“十一五”和“十二五”规划教材和教育部“十二五”普通高等教育本科国家级规划教材。

《景观生态规划设计案例评析》是与《景观生态规划原理》配套的进阶学习教材。在系统理论学习的基础上，重点立足理论的实际应用，从成功的实践中学习和总结景观生态规划设计的知识和技能。本教材精选河流流域与滨水景观设计、湖泊景观规划设计、绿道网络规划设计、绿色基础设施规划设计、生物多样性设计、城市公园景观生态设计、生态技术应用、景观的再生与转型、适应全球气候变化等景观生态规划设计九大领域共三十个国内外经典案例，从项目背景、条件分析、规划设计、规划总结与评价、思想拓展5个环节进行系统的研究和评析，初步构建出景观生态规划设计技能、方法和经验体系和框架。本书旨在抛砖引玉，以期更多的学子能够加入到生态规划设计的实践中，推动可持续景观的设计和健康环境的营造。

2012年8月同济大学

to cater for the international development trend and to participate actively in the construction of country's important development.

The theory textbook of Landscape ecological planning and design The Principle of Landscape Ecological Planning (the first edition, 2007) was published first by China building industry publishing house, which based on the professional teaching exploration in College of Architecture and Urban Planning, Tongji University, and absorbed and drew lessons widely from the domestic and foreign experiences. In the course of later teaching and application, combined with new progress and new achievements of landscape ecological planning and design, The Principle of Landscape Ecological Planning (the second edition, 2013) was published on the base of first edition. The new edition set up the theoretical teaching system and framework which includes basic theory and method, space types, key points and effect evaluation. The Principle of Landscape Ecological Planning is the civil engineer education "11th five-year plan" and "12th five-year plan" planning textbook of house and construction ministry of China and high education "12th five-year plan" planning textbook of the education ministry of China.

Reviews on Landscape Ecological Planning & Design Cases is the advanced and matching teaching materials of The Principle of Landscape Ecological Planning. Based on theory study systematically and aiming at practical application of the theory, at the same time learned and summarized knowledge and skills from the successful practices. The book selected 30 cases in 9 fields which are river basin and waterfront landscape design, lake landscape planning and design, greenways network planning and design, green infrastructure planning and design, biological diversity design, urban park landscape ecological design, ecological technology application, landscape regeneration and transformation, landscape ecological planning and design adapted to the global climate change. Each case was researched and evaluated from five aspects which are project background, condition analysis, planning and design, planning summary and evaluation, thoughts develop. This book set up preliminarily the skills, methods, experience system and framework of landscape ecological planning and design. The purpose of the book is to throw out a minnow to catch a whale, and expect more students take part in the practice of landscape ecological planning & design, so can promote landscape sustainable design and the construction of healthy environment.

August, 2012 in Tongji University

目录
CONTENT

003	前言		Preface	
006	第一章	河流流域与滨水景观生态规划设计	Chapter Ⅰ	River Basin Waterfront ecological Planning & Design
008	第一节	波托马克河流域规划	Section Ⅰ	Potomac River Basin Planning
021	第二节	圣佩德罗河上游流域多解规划	Section Ⅱ	Alternative Futures for the Upper San Pedro River
038	第三节	沃辛顿河谷地区研究	Section Ⅲ	The Plans For Worthington Valleys
051	第四节	明尼阿波利斯滨河项目	Section Ⅳ	Minneapolis Riverfront Project
063	第五节	布法罗河口区规划	Section Ⅴ	Buffalo Bayou Planning
074	第六节	洛杉矶河复兴总体规划	Section Ⅵ	Los Angeles River Revitalization Master Plan
090	第二章	湖泊景观生态规划设计	Chapter Ⅱ	Lake Landscape Ecological Planning & Design
092	第一节	圣地亚哥潟湖规划	Section Ⅰ	The Plans for San Dieguito Lagoon
108	第二节	圣埃利荷潟湖规划	Section Ⅱ	San Elijo Lagoon Planning
120	第三章	绿道网络生态规划设计	Chapter Ⅲ	Greenways Network Planning & Design
122	第一节	佛罗里达绿道网络	Section Ⅰ	Florida Statewide Greenways Plan
133	第二节	迈阿密河流绿道概念规划	Section Ⅱ	The Concept Plan of Miami River Greenway
150	第四章	绿色基础设施规划设计	Chapter Ⅳ	Green Infrastructure Planning & Design
152	第一节	马里兰绿色基础设施评估	Section Ⅰ	Maryland's Green Infrastructure Assessment
166	第二节	哈罗新城绿色基础设施规划	Section Ⅱ	The Green Infrastructure Plan (GIP) for the Harlow Area
176	第三节	环城翡翠项链：亚特兰大新公共领域	Section Ⅲ	The Beltline Emerald Necklace: Atlanta's New Public Realm
190	第五章	生物多样性规划设计	Chapter Ⅴ	Bio-Diversity Planning & Design
192	第一节	梅诺莫尼河谷项目	Section Ⅰ	Menomonee River Valley Project
203	第二节	甘尼特报业集团总部景观设计	Section Ⅱ	Landscape Design for Gannett

209 第三节 克罗斯温茨湿地 Section Ⅲ Crosswinds Marsh
220 第四节 乌兰野生动物园 Section Ⅳ Woodland Park Zoo

232 **第六章 城市公园景观生态规划设计** **Chapter Ⅵ City Park Ecological Planning & Design**
239 第一节 清水湾公园 Section Ⅰ The Fresh Kill Park
251 第二节 安大略公园规划 Section Ⅱ Lake Ontario Park Planning
268 第三节 橘子郡大公园总体规划 Section Ⅲ The Planning of Orange County Great Park
277 第四节 伊利西安公园总体规划 Section Ⅳ Section Ⅳ Elysian Park Master Plan

292 **第七章 生态技术的应用** **Chapter Ⅶ Application of Ecological Technology**
294 第一节 纽约总督岛公园及公共空间总体规划 Section Ⅰ Governor Island Park and Public Space Master Plan
306 第二节 布鲁克林桥公园规划 Section Ⅱ Brooklyn Bridge Park Planning
318 第三节 碳技术、浮桥藻类公园案例 Section Ⅲ Carbon T.A.P.//Tunnel Algae Park
326 第四节 生态城——阿科桑底 Section Ⅳ Arcology-Arcosanti

338 **第八章 景观的再生与转型** **Chapter Ⅷ Landscape Regeneration & Transformation**
340 第一节 北杜伊斯堡景观公园 Section Ⅰ North Duisburg Landscape Park
350 第二节 郭瓦纳斯运河海绵公园规划 Section Ⅱ Gowanus Sponge Park Design
359 第三节 纽约东河滨水景观的转型 Section Ⅲ Transforming the East River Waterfront, The City of New York

370 **第九章 适应全球气候变化的景观生态规划设计** **Chapter Ⅸ Landscape Ecological Planning & Design Adaptive to Climate Change**
372 第一节 格罗宁根：适应气候变化的设计 Section Ⅰ Groningen: Adaptation to Climate Change
383 第二节 重庆龙湖项目 Section Ⅱ The Longhu Project, Chongqing

391 **后记** **Epilogue**

第一章　河流流域与滨水景观生态规划设计

CHAPTER I RIVER BASIN WATERFRONT ECOLOGICAL PLANNING & DESIGN

第一节 波托马克河流域规划 Section Ⅰ Potomac River Basin Planning
第二节 圣佩德罗河上游流域多解规划 Section Ⅱ Alternative Futures for the Upper San Pedro River
第三节 沃辛顿河谷地区研究 Section Ⅲ The Plans For Worthington Valleys
第四节 明尼阿波利斯滨河项目 Section Ⅳ Minneapolis Riverfront Project
第五节 布法罗河口区规划 Section Ⅴ Buffalo Bayou Planning
第六节 洛杉矶河复兴总体规划 Section Ⅵ Los Angeles River Revitalization Master Plan

第一节 波托马克河流域规划

Section I Potomac River Basin Planning

一、 项目概述

波托马克河地区的范围是确定的，其流域是可以清晰描述勾勒出来的——由水体统一起来的固定单元。显然，流域是个水文单元，而不是一个地理单元，人们很难寻求一个更确切的土地划分，因为地理区域会呈现不同特点，其变化剧烈，而且不连续。波托马克河流域横跨一系列地理区域，规划的重点是区域内未来的土地利用。这项规划是一次关于适宜性生态规划方法的实验，是要探索并挖掘出这一流域内所有土地最好最大的使用潜力，但在不同的情况下，还要厘清这些土地利用方式结合在一起的最大可能，而大自然就是一个具有相互作用及储藏功能的价值体系。

二、 区域环境介绍

1. 自然条件

在考虑气候时，最值得注意的因素是气候和自然地理的相互关系。阿巴拉契亚山脉对阿勒格尼高原造成影响，在这个区域的东部形成“雨影区”。这里夏秋两季多雾多云，邻近的山脊和山谷气温变化很大，山谷中时常雾气缭绕。强烈的暴风雨和短

I.Overview

In this case there was no problem of defining the area to be studied it was the Potomac River Basin. Thus, we were spared the agonies that rack the socio-economic planners whose regions are ephemeral and transitory. At least the river basin is describable it is united by water; and it is permanent. Yet, it is clear that, while the river basin is a hydrologic unit, it is not a physiographic one; and, if one seeks a more finite division of land, the physiographic region offers this character to an unequaled degree. Physiographic regions vary dramatically. They are discrete. In the Potomac River Basin there is a single hydrologic unit that transects a number of physiographic regions in which the preoccupation is with all prospective land uses. This is a fitting test for the ecological planning method. Now we have a program; we seek to find the highest and best uses of all the land in the basin, but in every case we will try to identify the maximum conjunction of these. This, then, is the image of nature as an interacting and riving storehouse a value system.

II. Regional Form Natural Pattern

1. Natural Conditions

The most notable factor in considering climate in the basin is the marked correlation with physiography. The Appalachian Mountains affect the Allegheny Plateau and result in a "rain shadow" in the east of this region. Summer and autumn fogs and cloudy

暂的生长季节成为这一区域显著的特点。皮得蒙高原和海岸平原气候是相似的，只是海岸平原常有飓风，夏季温暖炎热潮湿，冬天温度适中，在整个流域中生长季节最长。因而这里的气候变化也仅仅是区域性的。

波托马克河流域是大西洋和海湾海岸体系中的次区域，由寒武纪之后的地质活动所形成，约有五亿年的历史。从地质学上讲，整个区域由三大地带组成：第一，皮得蒙高原的老结晶岩地区；第二，阿勒格尼高原，由较新的沉积物形成；最后，在东面是最新的海岸平原疏松沉积层。从波托马克河源头至海洋，该河共横向跨越了六个自然地理区域：从阿勒格尼高原至岭谷地区，然后到达大河谷、蓝岭和皮得蒙高原，最后到达海岸平原的河口湾。

波托马克河的流域面积大约为15 000平方英里(约38 850平方公里)：其主要支流为北支流和南支流，还包括大谢南多厄、卡卡朋、康纳康奇和莫诺卡西等河溪。和那些高原地区一样，岭谷地区的土壤层非常薄，贫瘠且易遭侵蚀；而大河谷的石灰岩土壤和岭谷地区中那些罕见的谷底滩地一样肥沃，这里是整个流域的农业核心区。流域内的矿物资源包括：煤、石灰岩、砂砾石和漂白土。

在这一望无际的流域中，由于其自然地理特点的多样性，因此能看到众多的植被种类和群落。特定的环境生长特定的植物，因此生态学家通过对植物的外观、类型和分布的了解，就能从现有的关于气候、土壤、水分状况和其他因素的资料中，推断出更确切的关于植物环境的情况和信息。这种在最广泛的范围中进行的研究，揭示了分布在由东到西的广阔地带上的三大森林群落的划分情况：第一是橡树—松树群落；第二是橡树—栗树群落；第三是神奇地躲过更新世冰盖破坏的喜温湿的混合中生林，它的中心是阿巴拉契亚山脉。

在波托马克河流域横跨的一系列地形区域中，不同的坡度对于广泛而多种多样的使用因素来说，影响是很大的。这些坡度明显地体现了区域性特征，在岭谷地区坡度富于各种变化，在海岸平原则较少。同时，皮得蒙高原和大河谷地区的斜坡被河流所划分。

conditions are found here. The adjacent Ridge and Valley has great temperature variations and frequent valley fogs. Intense storms and a short growing season mark this province. Piedmont and Coastal Plain share a similar climate, save for the proclivity of the latter to hurricanes. Summers are warm to hot, humidity is high, winters mild, and the growing season longest in the basin. There is then, a marked regional climatic variability.

The Potomac basin is a subregion of the Atlantic and Gulf Coast system and results from geological activity since Precambrian time, some half billion years ago. Geologically, the region consists of three major zones: first, the area of very old crystalline rocks in the Piedmont Plateau; and, second, the Allegheny Plateau, which is of more recent sedimentary origin; and, finally, on the east, the very recent series of generally unconsolidated sedimentary strata of the Coastal Plain.From source to ocean, the Potomac transects six physiographic regions from the Allegheny Plateau to the Ridge and Valley Province, thence to the Great Valley, Blue Ridge and Piedmont, and, finally, to its estuary in the Coastal Plain.

The Potomac drains a basin of almost 15,000 square miles; its major tributaries are the North and South Branches, the great Shenandoah, the Cacapon, Conococheague Creek and the Monocacy. Like those of the Plateau, the soils of the Ridge and Valley are thin, erodible and in fertile, except in certain limestone uplands and in the valley bottomlands, which are as fertile as can be found in the entire country. Yet, with the exception of shale formations, the limestone soils of the Great Valley are as fertile as those infrequent valley bottoms in the Ridge and Valley: here is located the great agricultural heartland of the basin. The mineral resources of the basin include coal, limestone, sands and gravels, fuller's earth.

In this enormous basin with its range of physiographic expression, it is to be expected that a wide range of vegetation types and communities will be observed and, indeed, this is so. As plants are very specific to environments, the ecologist who knows the presence, pattern and distribution of plants can infer more accurate information about their environments than is generally available from existing information on climate, soils, the water regimen and other factors. The broadest level of examination reveals the presence of three major divisions of forest associations distributed in broad bands from east to west. The first of these is the oak-pine association; the second, the oak-chestnut; and the third is the legendary mixed mesophytic forest that escaped the Pleistocene ice sheets and whose very center is the Appalachian Mountains.

2. 历史人文

波托马克河汇集了整个流域的各种文化，从上游的西弗吉尼亚州的煤矿工人到美国首都的城市居民，以及波托马克河下游的弗吉尼亚州最北端的船夫。由于河流位于美国历史和文化遗产比较丰富的地区，波托马克河被称为“民族之河”。据调查，美国第一任总统乔治·华盛顿，就出生于波托马克河流域，并在此居住了很久。美国首都华盛顿特区，也坐落于该流域内。而1859年在河流交汇处发生的战役就是美国南北战争的起始点。波托马克河位于马里兰州段内的切萨皮克–俄亥俄运河于1831年至1924年开通，连接了坎伯兰和华盛顿特区，实现了波托马克河大小急流险滩的货物运输。华盛顿渡口于1864年开放，从此华盛顿开始使用波托马克河河水作为饮用水的主要来源，饮用水的取水口建于大瀑布。1998年，美国总统比尔·克林顿将波托马克河定为美国河流遗产之一。

三、 波托马克河流域规划设计

1. 单项规划

1) 农业规划

地表地质、气候、土壤、坡度及朝向，可以作为确定整个流域中农业发展类型的衡量标准。在流域中，上述因素是变化多端的，但它们在自然地理区域和次区域中展现出一定的连续性，因此，我们能根据区域的特点来预示其适应性，并能够依据大河谷明显的特征变化预测其适宜性。皮得蒙高原呈现出广阔的农业生产区域，这是整个岭谷地区狭窄的山谷中所稀缺的；而阿勒格尼高原几乎没有农业用地。海岸平原的土壤虽然贫瘠，但施加足够的肥料，就能生长出有价值的蔬菜（图1–1–1）。

2) 森林规划

规划将森林分为三类用途：商业类森林、非商业类森林以及不可采伐的森林。适宜作为商业用途的森林分成两类：第一类，其地理条件需位于纸浆厂25英里(约40公里)半径内，坐落在一条五级河流或更大的河流旁，处在实行宽松的分区管理或非

2. Historical Context

The Potomac River brings together a variety of cultures throughout the watershed from the coal miners of upstream West Virginia to the urban residents of the nation's capital and, along the lower Potomac, the watermen of Virginia's Northern Neck. Being situated in an area rich in American history and American heritage has led to the Potomac being nicknamed "the Nation's River." George Washington, the first President of the United States, was born in, surveyed, and spent most of his life within the Potomac basin. All of Washington, D.C., the nation's capital city, also lies within the watershed. The 1859 siege of Harper's Ferry at the river's confluence with the Shenandoah was a precursor to numerous epic battles of the American Civil War in and around the Potomac and its tributaries. The Chesapeake and Ohio Canal operated along the banks of the Potomac in Maryland from 1831 to 1924 and also connected Cumberland to Washington, D.C.. This allowed freight to be transported around the rapids known as the Great Falls of the Potomac River, as well as many other, smaller rapids. Washington, D.C. began using the Potomac as its principal source of drinking water with the opening of the Washington Aqueduct in 1864, using a water intake constructed at Great Falls. President Bill Clinton designated the Potomac as one of the American Heritage Rivers in 1998.

III. Potomac River Basin Planning and Design

1. Special Planning

1)Agriculture Planning

Subsurface geology, climate, soils, slope and thus drainage together with exposure, determine the appropriate types of agriculture that should, or can, be practiced in the entire basin. These factors are variable in the basin but exhibit some consistency within the physiographic regions and subregions, so we can predict suitabilities according to characteristics Immediately the primacy of the Great Valley is apparent. The Piedmont reveals extensive productive areas; these are sparse in the narrow valleys of the Ridge and Valley Province and all but absent in the Allegheny Plateau. While the soils of the Coastal Plain are poor and infertile, with abundant fertilizer these can be made to produce valuable vegetable crops(Fig.1-1-1).

2)Forestry Planning

Planning divided forestry into three categories: commercial forestry, noncommercial forestry and

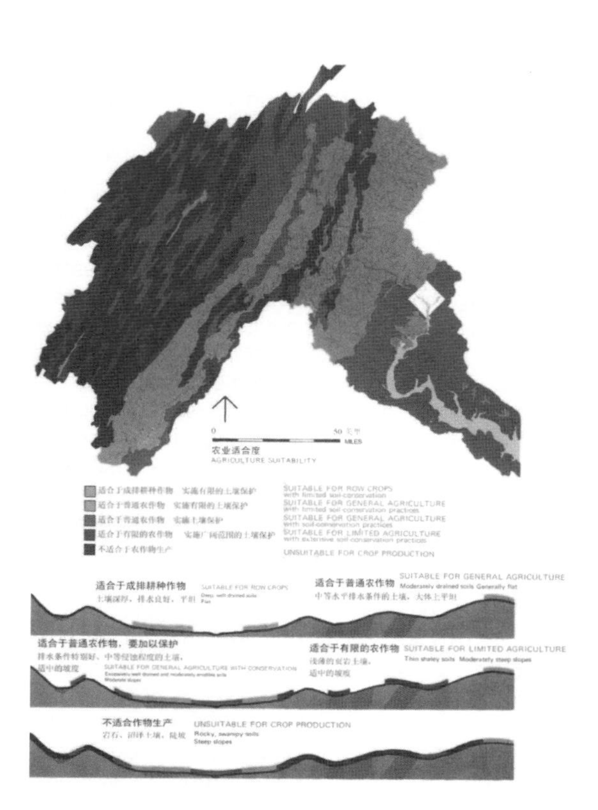

图1–1–1 农业适合度

Fig.1-1-1 Agriculture Suitability

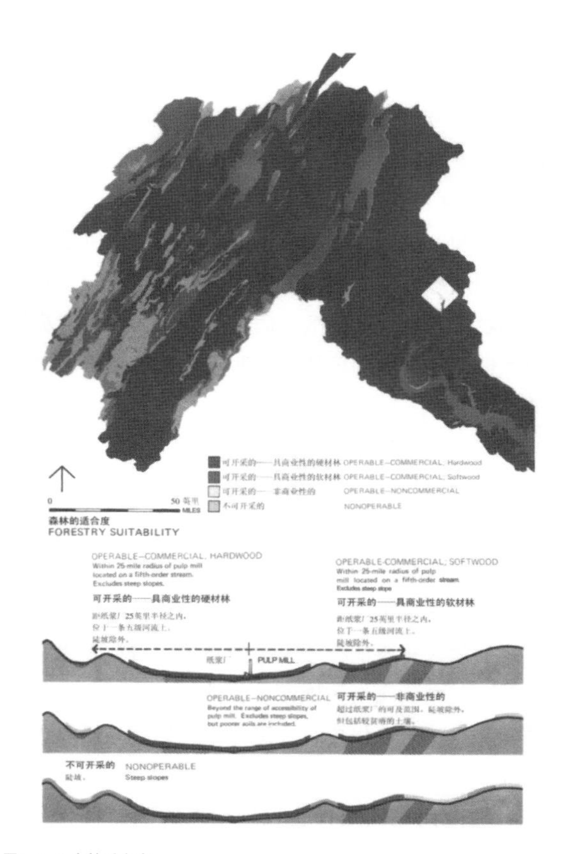

图1–1–2 森林适合度

Fig.1-1-2 Forestry Suitability

分区管理地区，森林应在小于坡度25%的坡地上；第二类，以生长在海岸平原上的针叶林、软材林为主，非商业用途森林：可以采伐该地区的林木，但森林会因此遭到破坏，短时间内难以再生。最后一类是不可开采的森林，主要是指由于交通不能到达、陡坡、离工厂或河流太远，因此不具经济性的林木。除此之外，有两类最不适宜农耕的土地——陡坡和易遭侵蚀的土壤，建议将它们作为森林覆盖之用（图1–1–2）。

3）游憩规划

根据各地区资源的分析进行规划：海岸平原的游憩资源主要是以水为中心的游憩活动；阿勒格尼高原和岭谷地区提供了大量的陆上游憩活动。大河谷地带和皮得蒙高原很少能提供独特的游憩机会，而蓝岭地区却能提供游憩活动场所。虽然阿勒格尼高原提供了如同宽广的阿巴拉契亚森林一样最高质量的天然林地，但是这一地区已经被滥用，掠夺性

nonoperable forestry. Commercial forestry are divided into two categories. The locational determinants for the first category are a radius of 25 miles from an existing pulp mill, on a fifth-order stream or larger, with lax or nonexistent zoning restrictions, and with forests on slopes of less than twenty-five per cent. A second category of commercial forestry, based on softwoods, occurs in the Coastal Plain. A further category of forestry is operable noncommercial: the areas that may be logged but will be so devastated that their regeneration is not in the foreseeable future. The final category is nonoperable as a result of inaccessibility, steep slope, distance from mill or stream, there is no present possibility of economic lumbering. In addition to these categories the two lowest classes of agricultural suitability, associated with steep slopes and erodible soils, are recommended for forest cover (Fig.1-1-2).

3)Recreation Planning

Clearly, there are regional resources: the Coastal Plain is the major resource for water-based recreation; the Allegheny Plateau and the Ridge and Valley Province offer the maximum opportunities

的露天采煤留下条条巨大的沟壑，森林被大量砍伐，河流水质酸化等，目前基本丧失了发展潜力。岭谷地区拥有该流域最大的游憩活动潜力，凉爽的夏季，加上适当的降雨和优美的风景，提供了独一无二的游憩场所。在大河谷地区，探索岩洞或驾车领略农场景色活动受限于周边山丘。在皮得蒙高原也能够感受同样的乐趣，同时还增加了调研历史遗迹的机会（图1–1–3）。

4) 城市规划

为了确定适合于城市化建设的用地，人们提出了一系列标准——土地的坡度不大于5%，必须不能位于50年一遇的洪泛平原之上；不能位于重要的地下水回灌区；不能位于大雾笼罩的山谷或曝露在日晒和风雨中的高海拔地区。另外城市用地必须能够获得充足的水源供应，公路建设用地的坡度不大于15%。

城市用地的选择也呈现明显的区域特点。在整个阿勒格尼高原都不可能找到如此大规模的城市用地，这里的小块土地只能供小村庄使用。在岭谷地区这类用地在山谷中变成了带状，但仍能找到一些。大河谷地区提供的城市用地要更少，因为这一地区的许多土地是在地下水回灌区之上。而皮得蒙高原，有整个流域内最适合城市化建设的土地，有适合城市功能的自然地形，不过，许多适合于城市化的土地也同时是优良的农业用地（图1–1–4）。

海岸平原由于存在着地下水回灌、地下水位高和森林火灾的危险，因此作为城市用地开发的机会受到限制。阿勒格尼高原只能获得最小的村庄用地，岭谷地区能够找出一些较大的城市用地，再大一些的能在大河谷地区找到，皮得蒙高原和海岸平原都有建设几个新城镇的可能性。

2. 功能兼容

之前的单项适宜性规划揭示了每一个区域在单一方面的价值，同时也研究了流域内每一地区具体的土地利用。但是我们寻找的不是单一的土地适宜性利用，而是寻求多种功能兼容的土地利用。为了实现这一目标，我们规划设计了一个矩阵，把所

for terrestrial recreation; the Great Valley and the Piedmont offer little that is unique, while the Blue Ridge does to an exceptional degree. While the Allegheny Plateau intrinsically offers the maximum quality as the site of the great Appalachian forest, it has been so abused with great swaths of open cast mining spoils, mauled forests and acid rivers that this potential is incapable of capture at the present. In the Ridge and Valley Province exists the greatest recreational potential in the basin. The cool summer climate, combined with low rainfall and the lovely and dramatic landscape, offer unrivaled opportunity. In the Great Valley, this is limited to the surrounding hills and the opportunities for exploring caves or driving through a farm landscape. In the Piedmont much of the same experience is possible, but increased by the opportunities to examine the emblems of history(Fig.1-1-3).

4) Urban Planning

In order to determine the sites that qualified as suitable for urbanization, a number of criteria were developed. The land should have slopes of no greater than five per cent incline; it must not be in the 50 year floodplain, nor in an important recharge area, nor in fog pockets or high and exposed elevations, Adequate water supplies must be available, and the required highways must not need to be constructed through slopes over fifteen per cent.

There is again a marked regionality. It is all but impossible to find such sites in the Allegheny Plateau, and where they exist they could support only hamlets or small villages, in the Ridge and Valley Province the sites are attenuated in the valleys, but a number can be found. The Great Valley offers fewer opportunities than might have been expected, as much of the area overlies aquifer recharge. In the Piedmont there is confirmation of the present location of urbanization in the basin: this is the most suitable of the physiographic regions for this function. However, much of the suitable land is also prime farmland (Fig.1-1-4).

The Coastal Plain offers opportunities that are restricted by the presence of aquifer recharges, a high water table and the forest's tendency to burn. Only the smallest of sites are available in the Plateau, rather larger ones are discerned in the Ridge and Valleys, still larger in the Great Valley, with opportunities for several new towns revealed in both Piedmont and Coastal Plain.

2. Multiple Land Uses

The preceding studies of intrinsic suit abilities for agriculture, forestry, recreation and urbanization reveal the relative values for each region and for the basin within each of the specified land uses. But

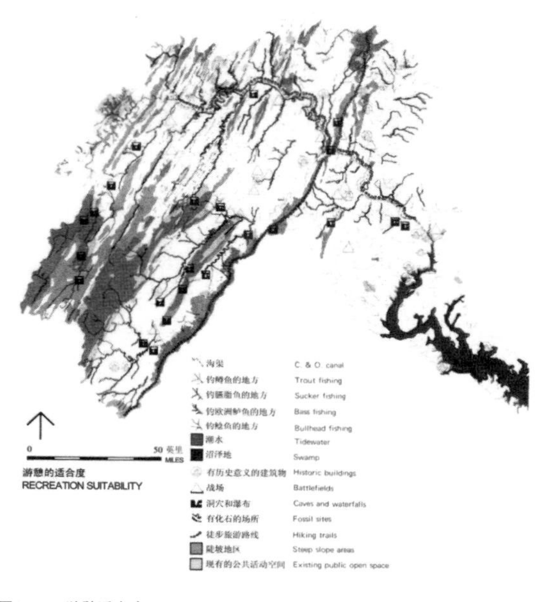

图1–1–3 游憩适合度
Fig.1-1-3 Recreation Suitability

图1–1–4 城市规划适合度
Fig.1-1-4 Urban Planning Suitability

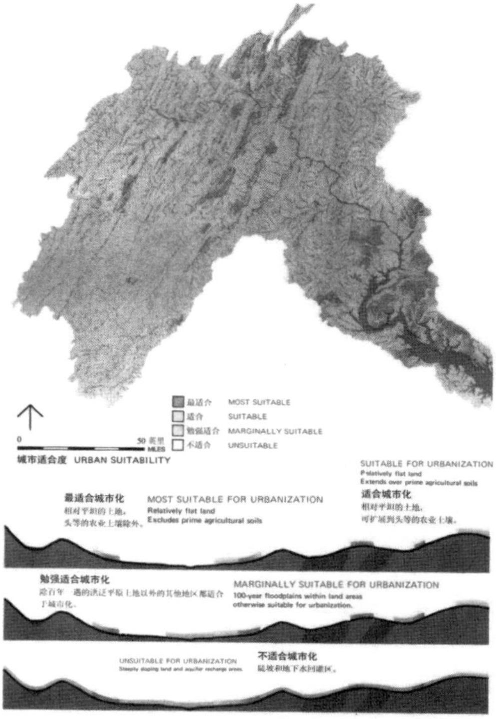

有未来的土地利用都标注在坐标上，从而将每种土地利用与其他所有的土地利用进行对比检验是否兼容，及两种土地利用相互干扰的程度。借助于这个矩阵，研究单一的土地利用适宜性并确定它与未来的土地利用之间的兼容程度就成为可能。例如，某一地区已经显示出具有高度的林业潜力，也可以同时与游憩和野生动物资源管理兼容。在这一地区还有机会发展小规模的农业，特别是畜牧业，而整个地区可以水资源为目标加以经营管理。而在另一个区域中，将提供一个农业为主的土地利用的机会，同时有可能支持游憩活动和一定的城市化建设以及有限的矿产开采（图1–1–5）。

矩阵中紧挨着关于土地之间相互兼容程度的栏目是识别未来的土地利用的必要资源：农业所需的肥沃土壤，采矿所需的煤和石灰石，城市所需要的平坦土地及水资源等等。矩阵最后一栏显示了这些土地利用实施后的结果。在采煤的地区，会产生酸性污水问题，农业常与水土流失和沉积相联系，城市化会产生污水，工业会造成大气污染。把这些因

we seek not to optimize for single, but for multiple compatible land uses. Towards this end a matrix was developed with all prospective land uses on each coordinate. Each land use was then tested against all others to determine compatibility, incompatibility and two intervening degrees. From this it was possible to reexamine the single Optimum and determine the degree of compatibility with other prospective land uses. Thus, for example, an area that had been shown to have a high potential for forestry would also be compatible with recreation, including wildlife management. Within it there might well be opportunities for limited agriculture-pasture in particular while the whole area could be managed for water objectives. Yet, in another example, an area that proffered an opportunity for agriculture as dominant land use could also support recreation, some urbanization and limited exploitation of minerals (Fig.1-1-5).

Adjacent to the matrix on inter-compatibility is another that seeks to identify the resources necessary for prospective land uses productive soils for agriculture, coal and limestone for mining, flat land and water for urban locations, and so on. The final matrix is devoted to the consequences of the operation of these land uses. Where there is coal

素结合在一起，人们就能大体上评估土地利用相互兼容的程度，决定兼容程度的自然因素和土地利用造成的结果（图1–1–6）。

当对这些土地利用的结果进行研究之后，显然，以采矿、煤和水资源为基础的工业用地规划在阿勒格尼岛原的机会最大，同时林业和游憩可以成为这里次要的用途。在岭谷地区，游憩活动具备主要的发展潜力，以林业、农业和城市化为次要用途。在大河谷地区，农业是最大的优势资源，游憩和城市化只能占用较少的土地。蓝岭地区显示出相对单一的游憩潜力，但品质却是最高的。皮得蒙高原主要适合于城市化用地，附带可作为农业和一般游憩活动使用。海岸平原显出具有以水资塬为基础的和与水有关的游憩活动潜力，以及林业的潜力，而作为城市化和农业用地可能性不大。

3. 分区规划

整个流域地区的研究是在1:250 000比例尺的图纸上进行的，许多细节人们看不到，因而不能十分完整地识别自然地理、土壤、气候和植被等因果关系。因此，在对每个自然地理区域更为详细研究

图1–1–5 综合、多种用途交替的适合度
Fig.1-1-5 Synthesis Alternative Suitability

mining, there will be acid mine drainage; agriculture is associated with sedimentation, urbanization with sewage, industry with atmospheric pollution. The sum of these, in principle, allows one to consider the inter-compatibility of land uses, the natural determinants for their occurrence and the consequences of their operation (1-1-6).

When the results are examined, it is clear that mining, coal and water-based industry offer the maximum opportunity in the Allegheny Plateau, with forestry and recreation as subordinate uses. In the Ridge and Valley Province, the recreational potential is dominant, with forestry, agriculture and urbanization subordinate. In the Great Valley, agriculture is the overwhelming resource, with recreation and urbanization as lesser land uses. The Blue Ridge exhibits only a recreational potential, but of the highest quality. The Piedmont is primarily suitable for urbanization with attendant agriculture and non-differentiated recreation. The Coastal Plain exhibits the highest potential for water based and related recreation and forestry, and a lesser prospect for urbanization and agriculture.

3. District Planning

The studies of the entire basin were conducted at a scale of 1:250,000, and thus there were many details that escaped attention, and indeed, the causal relations of physiography, soils, climate and vegetation could only be imperfectly discerned. For that reason, areas were selected in each of the physiographic regions for more detailed study where this causality could be seen. These were undertaken at the scale of 1:24,000. Each of the areas selected was thought to be typical of its region.

1)Allegheny Plateau

This great province has been savaged-forests felled and burned, coal carelessly mined, wastes widespread and streams acid. The land was rich but the wealth was removed, a degraded land and impoverished people remain. Yet there are resources still, abundant coal, latent forests, a wildlife and recreation potential of the highest value. But to capture this requires a knowledge of resources as revealed here, plans and management policies which reflect this understanding. It will also require time and man to heal this land and its people (Fig.1-1-7).

2)Ridge And Valley

The absence of coal may explain the lesser depredations in this region. It offers the greatest resource of terrestrial recreation in the Basin. Although valleys are narrow they are remarkably fertile. The forests are not of high commercial value but have great value for recreation the prime

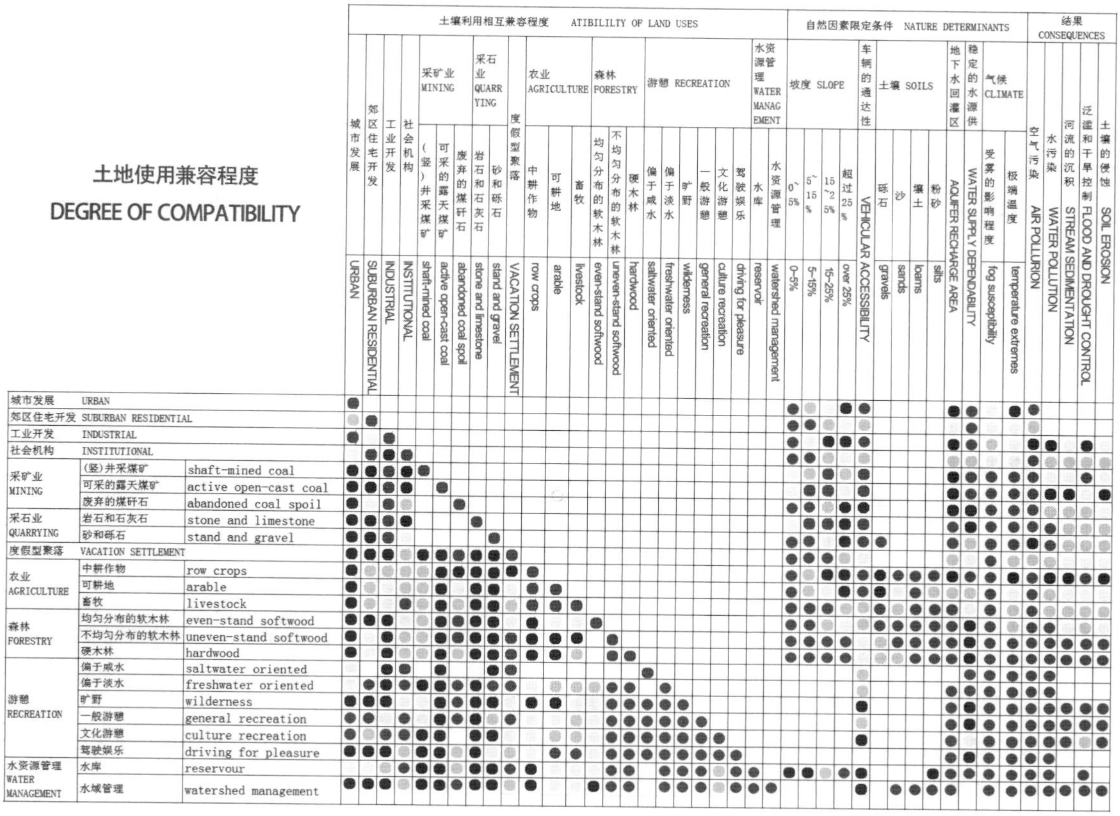

土地使用兼容程度

DEGREE OF COMPATIBILITY

图1-1-6 土地使用兼容程度

Fig.1-1-6 Degree of Compatibility

的基础上，我们选择某些地区作进一步研究，从而使因果关系清晰地呈现出来。这些地区的研究是在1:24 000比例尺的图纸上进行的。每一个挑选出的地区都是该区域内的典型。

1) 阿勒格尼高原

这一广阔地区已遭野蛮掠夺：森林被砍伐和焚烧，煤矿被缺乏计划地采掘，废物到处堆积，河流酸化。这里曾是富饶的土地，但其价值已被破坏，留下的是损毁的土地和贫困的居民。然而这里仍有许多资源——丰富的煤矿、潜藏的森林、野生动物和最具潜在价值的游憩活动资源条件。但是要实现这些目标，需要掌握这里显示出来的资源条件，并根据已有资料，制订出开发计划和管理政策。还需要时间和人力去恢复这块土地，造福这里的人民（图1-1-7）。

resource of this region. Urbanization has too often been located on flood plains. Better sites on higher ground with medium slopes, good orientation, above frost pockets can be located. Recreation is an important regional resource, constituting a high social value (Fig.1-1-8).

3)The Great Valley

The Great Valley is one great agricultural region east of the Rockies a broad, generally fiat valley with predominantly rich limestone soils. There are, however, three subdivisions the western hills on sandstone, shale, limestone and quartzite, the wide belt of Martinsburg shale and the valley proper of limestone and dolomite. In brief the hills provide the maximum recreational potential, the limestone the agricultural resource, and the shale the best locations for urbanization. This last is important as it ensures that urbanization does not occur over the aquifer. The resources and their distribution are most felicitous wooded hills, a fertile valley, a swath of shale suited for

2) 岭谷地区

由于没有煤矿，这一区域较少遭到破坏。本地区提供了该流域最大的陆上游憩资源，虽然谷地狭窄，但土壤十分肥沃。森林虽没有很高的商业价值，但有很高的游憩价值，成为这一地区的重要资源。城市化过程过去常常发生在洪泛平原区，较好的城市应当地势较高、坡度适中、朝向好、高于降霜带。游憩是一项重要的区域性资源，可以形成很高的社会价值（图1–1–8）。

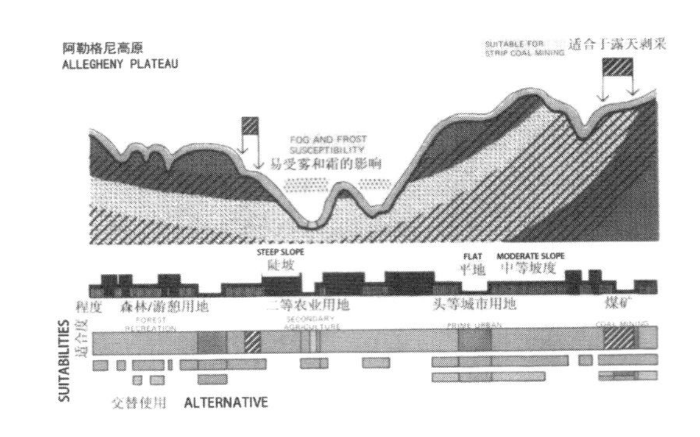

图1–1–7 阿勒格尼高原
Fig.1-1-7 Allegheny Plateau

3) 大河谷地区

大河谷地区是洛基山脉以东最大的一块农业区域。谷地幅员宽广、地势平坦，主要蕴含丰富的石灰岩土壤。整个区域可再分为三个分区：由砂岩、页岩、石灰岩和石英岩构成的西部山丘，宽广的马丁斯堡页岩带，石灰岩和白云岩构成的河谷。简言之，这些山丘提供了最大的游憩潜力，石灰岩能成为农业资源，页岩地带是最佳的城市建设位置。重要的是，这保证了城市用地不会建在地下含水层上面。这一地区内的资源及其分布是适宜的——林木覆盖的山丘，肥沃的河谷，一条适合于城市化的页岩带，而且页岩带紧靠一条美丽的河流，展现出非常高的景观品质（图1–1–9）。

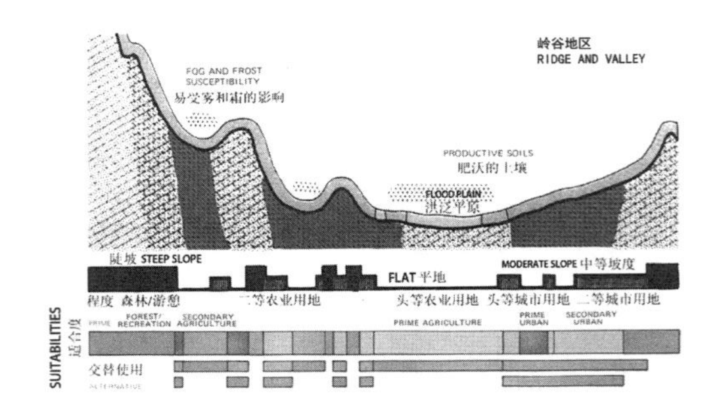

图1–1–8 岭谷高原
Fig.1-1-8 Ridge and Valley

4) 皮得蒙高原

皮得蒙高原的剖面图显示出地层结构的高度复杂性——一条石灰岩和白云岩的河谷，一块前寒武纪的结晶岩高地被侵入岩插入，一条宽广的石英岩带，还有一条页岩带。内在因素反映出地质构成和相应的地形地貌、水文和土壤情况。石灰岩和石英岩的谷地最适合于农业，页岩的河谷作为畜牧和非商业性的森林最为合适，某些农作物、畜牧和森林在结晶岩地区的谷地和洪泛平原上也是合适的。最适合于城市的用地是在结晶岩地区的平坦高原或山岭上。在石灰岩上没有适宜的城市用地，在页岩上也很少。这是一个在城市化边缘的地区。城市化的机遇大量存在，但是规划必须反映这个地区特定的可能性和限制条件（图1–1–10）

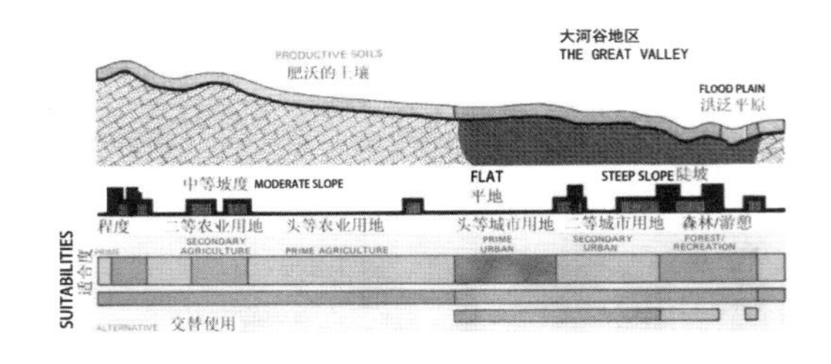

图1–1–9 大河谷高原
Fig.1-1-9 The Great Valley

urbanization, the latter bordered by a fine river and exhibiting considerable scenic quality (Fig.1-1-9).

4)The Piedmont

The section of the Piedmont illustrated reveals a great complexity a limestone and dolomite valley, a

5) 海岸平原

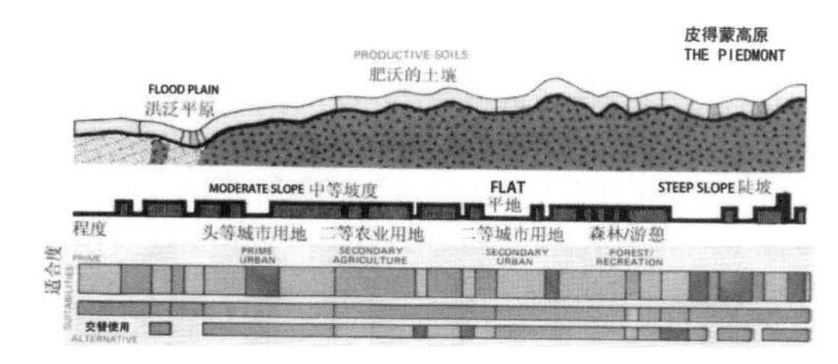

图1–1–10 皮得蒙高原

Fig.1-1-10 The Piedmont

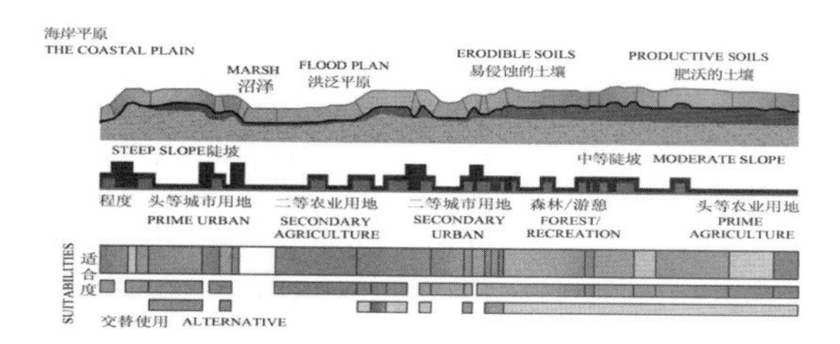

图1–1–11 海岸平原

Fig.1-1-10 The Coastal Plain

波托马克河下游地区显示出海岸平原的自然地理特点：河口，海湾、海口、弯曲的溪流和沼泽等。这里森林茂密，地下水位很高，土壤主要是石英砂。这一区域提供的水上游憩活动的机会是独一无二的；但农业用地受到限制，城市用地也是如此；森林和渔业目前已有很高的价值。水资源在这一区域具有重要的价值，是内在适合度的主要决定因素。森林易发生火灾，地表水易被污染，地下水位对地市来说普遍过高，土壤对大面积的发展农业来说又太贫瘠。不过，有些高起的土地适合于建筑，冲积土壤适合于农业，陆上的和水生的野生动物栖息地均很丰富(图1–1–11)。

四、 总结

波托马克河流域横跨一系列地形区域，地貌变化丰富，并汇集了各种文化，是美国河流遗产之一。规划范畴包括农业规划、森林规划、游憩规划和城市规划建设等内容，从宏观的综合规划到重点地区的详细规划，确保了方案的可操作性。规划因地制宜，对阿勒格尼高原、岭谷地区、大河谷地区、皮得蒙高原、海岸平原等不同地域环境采取不

Precambrian upland of crystalline rocks fissured with intrusions, a broad band of quartzite, yet another of shales. Intrinsic suitabilities respond to geology and the consequential physiography, hydrology and soils. The limestone and dolomite valley is most suited for agriculture, the shales for pasture and non-commercial forests, some crops, pasture and forests are appropriate to valleys and flood plains in the crystalline area. The most suitable urban sites fall in the crystalline region on flat plateaus and ridges. They are absent on limestone, rare on the shales. This is an area on the edge of urbanization. Opportunities abound but planning must respond to the specific opportunities and constraints afforded by the region (Fig.1-1-10).

5)The Coastal Plain

The area of the lower Potomac reveals the characteristic physiography of the Coastal Plain-estuary, bays, inlets, meandering streams and marshes. Forests are abundant, the water table is high, and soils are mainly quartzite sand. This region offers water based recreational opportunity to a unique degree. Agriculture is limited, as are sites for urbanization. Forestry constitutes a present value as do fisheries.

Water is the major value here and the main determinant of intrinsic suitabilities. Forests are fire prone, ground water can be polluted, the water table is generally too high for urbanization, soils too infertile for wide- spread agriculture. Yet there are elevated sites suitable for building, alluvial soils suited for agriculture, both land and water wildlife habitats are abundant (Fig.1-1-11).

IV. Summary

The Potomac River Basin stretches over a series of terrain area, landform changes in abundance, and brings together a variety of culture, is one of heritages of the United States. The culture includes agricultural planning, forest planning, tourism planning and city planning construction and so on, from the macroscopic comprehensive planning to key sections of the detailed planning, and which guarantees the maneuverability of projects. The planning fits itself to local conditions, and different measures are carried out to specific areas, like Allegheny plateau, ridge valley area, big valley region, Pidemeng plateau, coastal plain, and so on. Synthetically considering the project system and integrity, and ensure the strategy being targeted and effective.

The planning takes analysis of the mono land use suitability, and uses the matrix analysis method, to seek a variety of compatible land use plan.

同的规划措施。既保障项目的系统性和完整性，又确保实施策略的有效性和针对性。

规划分析了单一的土地利用适宜性，并采用矩阵分析法，寻求多种兼容的土地利用方案。图1–1–5展示的土地使用兼容程度，类似我国城市规划体系中控制性详细规划的“用地兼容性表”，该表的主要功能是分析同一土地使用性质的多种选择与置换的可能性，表现为土地使用性质的“弹性”、“灵活性”与“适建性”。这就使城市规划管理部门在维护规划的原则性与严肃性的同时，还能对规划可能存在的缺陷进行合理修正，并符合市场调节的需求。用地兼容性分析在明确每块土地使用性质的同时，更提供一种灵活性与宽容度。同时，我们也应认识到，土地使用兼容性是相对于一定时期的规划目标与环境标准而言的，随着这些外在条件的改变，土地兼容性的指标也会相应发生变化，没有一成不变、脱离既定目标约束的兼容。

Fig.1-1-5, the land use compatibility degree, is similar to China's city planning system of the regulatory detailed planning "for the compatibility table", the main function of that is the analysis of the multiple selection and replacement of possibilities for the nature of the same land use, and which is being comprehended as land "elastic", "flexibility" and "constructability". This makes the city planning and management department can plan with shortcomings being rationally corrected, basing on maintaining planning principle and seriousness, and accords with market adjustment needs. When land use compatibility clears with each nature of the use of land, it also provides a more flexibility and tolerance for that at the same time. Also, we should be aware of that, land use compatibility is relative to a certain period the planning objectives and environmental standards, as the external conditions change, standards for land compatibility will change correspondingly, there is no compatibility that is away from a predetermined target.

思考题：

1. **本项目采用适宜性生态规划方法，它包含了哪些指标？**
2. **每一个单项规划的指标是什么？**
3. **土地利用综合规划的方法是什么？有什么优点与缺点？**
4. **尺度是规划过程中一项重要的因素，本项目规划在不同尺度下研究的内容有哪些？**
5. **波托马克河流域规划带来哪些启示？它所采用的研究方法具有怎样的优点？**

Questions:

1. The project applies a fitting test for the ecological planning method, what element does it contain?
2. What are the standards of every special planning?
3. What is the method of multiple land use? What are advantages and disadvantages?
4. Scale is an important factor in the plan process, what are studies at different scales in the project?
5. What does the Potomac River Basin plan tell us? What are the advantages of the method which are applied in the plan?

注释/Note

图片来源/Image：本案例图片均来自《设计结合自然》/ Design With Nature

参考文献/Reference：本案例引自《设计结合自然》/ Design With Nature

推荐网站/Website：http://www.potomacriver.org/2012/

第二节 圣佩德罗河上游流域多解规划

Section II Alternative Futures for The Upper San Pedro River

一、 项目简介

1. 项目简介

位于亚历桑那州和索诺拉州的圣佩德罗河上游流域的未来发展，是一个反映保护与发展间矛盾的典型例子，而主要军事设施的存在使这里情况更为复杂。研究区域包括圣佩德罗河上游流域，从索诺拉的卡纳内阿附近的发源地到亚利桑那的雷丁顿。调查也包括了位于流域附近，对整个区域生物多样性具有不可分割作用的地区。总的看来，该研究区域共10 660平方公里。亚利桑那占研究面积的74%，其余26%都在索诺拉。

圣佩德罗河发源于墨西哥的索诺拉州，向北流经美国的亚利桑那州，与流入科罗拉多河的吉拉河汇合，最后在加利福尼亚湾入海（图1–2–1）。位于亚利桑那州和索诺拉的圣佩德罗河上游流域（图1–2–2）反映了一系列有争议的复杂问题：该地区作为鸟类栖息地的国际重要性；它的开发吸引力；景观改变的脆弱性，其直接因素是开发，间接因素是地下水位持续下降。这些问题亟待解决。

1996年，国防部派出代表参加生物多样性研究协会。该协会是由政府机构与大学合作创办的组织，其目标是开发数据库和分析方法以进行生物多

I. Overview

1. Introduction

The future of the Upper San Pedro River Basin in Arizona and Sonora is just one example of the tensions between conservation and development, and it is further complicated by the presence of a major military installation. The research area includes the Upper San Pedro River Basin from its headwaters near Cananea, Sonora, to Redington, Arizona. Areas adjacent to the basin that are integral for the maintenance of regional biodiversity are also included in the investigation. In total, the study includes 10,660 sqkm (nearly 4100 sq mi). Arizona encompasses 74 percent of the study area, and the remaining 26 percent is in Sonora.

The Upper San Pedro River begins in Sonora, Mexico, and flows northward through Arizona, United States, before joining the Gila River, which flows into the Colorado, and finally empties into the Gulf of California (Fig.1-2-1). The Upper San Pedro River Basin (Fig.1-2-2) in Sonora and Arizona is the focus of the a number of urgent, complex, interrelated, and controversial issues, including its international importance as a bird habitat, its attractiveness to development, and the vulnerability of its landscape to changes caused directly by development and indirectly via continued lowering of the groundwater table.

In 1996, the Department of Defense sent representatives to the Biodiversity Research Consortium, a partnership of government agencies and universities. BRC's goal is to develop

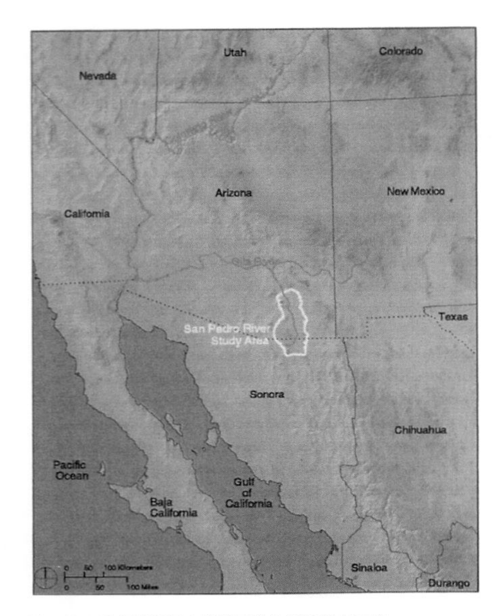

图1-2-1 圣佩德罗河上游地区的位置及研究区域
Fig.1-2-1 Location and study areas of the Upper San Pedro River

图 圣佩德罗河走廊
Photo Upper San Pedro River

图 比斯比德拉维德·皮特铜矿
Photo Lavender Pit Copper Mine, Bisbee

图 草地沙漠矮树
Photo Mesquite Chihuahua desert scrub in grassland

图 圣佩德罗河三角杨柳栖息地
Photo Upper San Pedro River cottonwood-willow habitat

图 圣大卫
Photo Saint David

图 卡纳内阿的铜矿
Photo Copper mine at Cananea

图 花园河谷入口
Photo Mouth of Garden Canyon

图 本森南部的农业用地
Photo Agricultural fields south of Benson

图 索诺拉德合作农场
Photo Ejido in Sonora

图1-2-2 圣佩德罗河上游地区的景色
Fig.1-2-2 Landscape of the Upper San Pedro River

样性的风险评估和管理。为进一步支持变化景观的多解规划研究，环境保护署于2000年10月发起了联邦净水行动计划。根据该计划的指导，各联邦机构在环境管理中应注意保护流域，并应通过增加联邦属地流域管理中的公众参与进一步改进自然资源管理工作。其中，流域计划包括对流域条件的评估和

databases and analytical methods for assessing and managing risks to biodiversity. In further support of a study of alternative futures for this changing landscape, the Environmental Protection Agency initiated the Federal Clean Water Action Plan in October 2000. The plan directs federal agencies to assume a watershed perspective for environmental management and improve natural

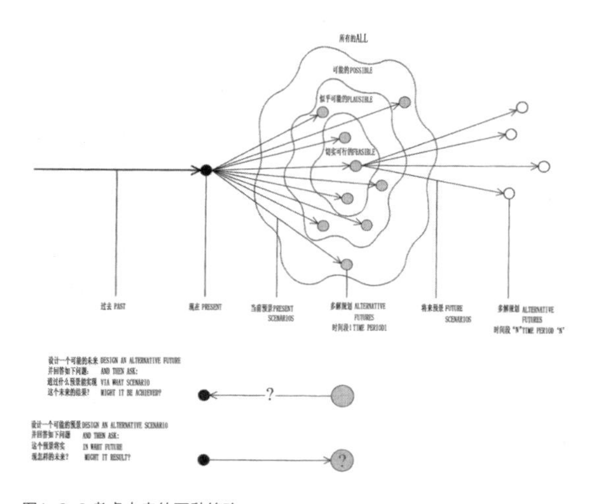

图1-2-3 考虑未来的两种策略

Fig.1-2-3 Two Strategies for Considering the Future

监控，以及对优先发展的流域的确定以使预算及其他资源得到有重点的使用。卡尔·斯坦尼茨的多解规划框架正是流域管理方案的主要组成部分。

可以通过两种主要方法来研究未来的可能情况（图1-2-3）。该研究旨在为圣佩德罗上游流域、亚利桑那州和索诺拉地区制定一系列未来可供选择的用地模式，并对这些模式可能对生物多元化和植被、水文与视觉感受等相关环境因素产生的影响进行评价。该研究的一个基本前提是从区域的角度充分理解用地和生态系统规划的相关问题。其目的并不在于指导社区向某一特定的方向发展，而是为当地规划者提供了一种方法，帮助他们预见当地可能的几种发展模式带来的不同结果，提高他们对当地未来发展模式的决策能力。

二、 卡尔·斯坦尼茨的变化区域的多解规划

圣佩德罗河上游流域项目的基本研究方法是多数政府、组织和个人制定地区未来发展方案的决策过程。可以设计一个预景来反映各种相互联系的政策的多种选择方案。在多解规划的预景研究中，每个独立的政策方案要么改变了既可吸引又可排斥未来发展的空间多样化特点，要么改变了未来变化影响评价流程模型的某一个参数。作出选择之后，应

resources stewardship through an increase in public involvement in watershed management on federal lands. Watershed planning includes assessment and monitoring of watershed conditions and identification of priority watersheds on which to focus budget and other resources. Carl Steinitz's alternative futures framework is a major component of this approach.

There are two main ways of thinking about alternative futures (Fig.1-2-3). This study is designed to develop an array of possible alternative future patterns of land uses for the region of the Upper San Pedro River Basin, Arizona, and Sonora, and to assess the resultant impacts that these alternative futures might have on patterns of biodiversity and related environmental factors, including vegetation, hydrology, and visual preference. A basic premise of the research is that issues related to land use and ecosystem planning can best be understood on a regional basis. This study is not an attempt to steer the community in a particular direction. It is, rather, a means to help local planners predict the consequences of the region's potential alternative futures, and therefore improve their foresight in choosing among them.

II. Alternative Futures for Changing Landscapes by Carl Steinitz

The approach of the Upper San Pedro River Basin is that decision-making processes of the many governmental, organizational, and individual choices that shape the future for a region. A scenario is then created to reflect choices among the possible options for each policy in the set. In a scenario-based study of alternative futures, each single policy option either alters a spatially varied characteristic that can attract or repel future development or alters a parameter in one of the several process models that assess the impacts of future change. Choices are made, and the resulting scenarios are used to direct the allocation of future land uses using a model of the process of development. The alternatives are then assessed for their consequences.

1. The Research Framework

This study of the Upper San Pedro River Basin is organized according to the framework for alternative futures studies developed by Carl Steinitz (1990,1993) and shown in Fig. 1-2-4 The framework consists of six questions that are asked several times during the course of the study. In designing a study of alternative future for an area, the answers-the models and their applications-are particular to the case study.

The six questions are:

通过发展过程模型，利用相应的预景来指导未来用地的分配，并对各种方案的效果进行评价。

1. 多解规划研究框架

如图1-2-4所示，圣佩德罗河上游流域的研究项目采用卡尔·斯坦尼茨开发的多解规划研究框架（1990年，1993年）。该框架包含了研究过程中多次提及的6个问题。某个地区的多解规划研究所得的答案——模型及其应用——对该案例具有特殊性。

这六个问题和三个阶段如下所述：

（1）如何以内容、空间和时间来描述景观的状态？这个问题应由表述模型和研究所依赖的有关数据和信息来回答。

（2）景观是如何运作的？景观各要素之间的功能和结构关系如何？这个问题应由过程模型来回答，过程模型为研究内容的几项分析提供了信息。

（3）当前的景观是否运行良好？这个问题应由评价模型来回答，评价模型取决于决策者的文化知识水平。

（4）可以通过哪些政策或行为，何时何地对景观做出怎样的改变？这个问题应由改变模型来回答，研究将对改变模型进行检验。这也是假设的未来数据。

（5）这些变化可能产生什么样的影响？这个问题应由影响模型来回答，影响模型是过程模型在改变条件下所产生的信息。

（6）景观应作何种改变？这个问题应由决策模型来回答，与评价模型一样，决策模型取决于决策者的文化知识水平。

第一阶段：范围描述。研究中，首先应对研究的主要问题和自然环境作广泛的调查。

第二阶段：明确方法。为多解规划研究设计一套方法，涉及许多决定，尤其是基于经验和判断的复杂决定。

第三阶段：执行方法。表述、过程、评价、改变、影响、决策。

最终归结为两种决定：是和否。

三个阶段，六个问题框架贯穿始终。

图1-2-5中展示了研究组与决策者之间的关

(1)How should the state of the landscape be described in content, space, and time? This question is answered by representation models, the data upon which the study relies.

(2)How does the landscape operate? What are the functional and structural relationships among its elements? This question is answered by process models that provide information for the several analyses that are the content of the study.

(3)Is the current landscape working well? This question is answered by evaluation models, which are dependent on cultural knowledge of the design-making stakeholders.

(4)How might the landscape be altered, by what policies and actions, where and when? This question is answered by the change models that will be tested in the research. They are also data, as assumed for the future.

(5)What difference might the changes cause? This question is answered by impact models, which are information produced by the process models under changed conditions.

(6)How should the landscape be changed? This question is answered by decision models, which, like the evaluation models, are dependent on the cultural knowledge of the responsible design-makers.

First Iteration:Describe Scope. The study process begins with a broad survey of major issues and the physical setting of the study.

Second Iteration:Define Method. Designing the methodology for a study of alternative futures involves designs that are especially complex and that are most often based on the experience and judgment.

Third Iteration:Implement, MethodRepresentation,Process,Evaluation,Change,Impact,Design.

At the extreme, two decision choices present themselves: no and yes.

The framework is used from top to bottom in carrying out the study.

Fig. 1-2-5 shows the relationship between the research team and stakeholders. In order to make decisions, questions must be asked and answered, and options for choice must be framed and deliberated.

2. The Organization of the Research

This study of alternative futures for the Upper San Pedro River Basin applies the six-question framework in the organization of the research and in carrying it forward.

(1)How should the state of the landscape be described in content, space, and time?

The research area includes the Upper San Pedro River Basin from the headwaters near Cananea, Sonora, to Redington, Arizona. In total, the analysis covers 10,660sq km (nearly 4100sq mi). To describe the geography and the dynamic processes at

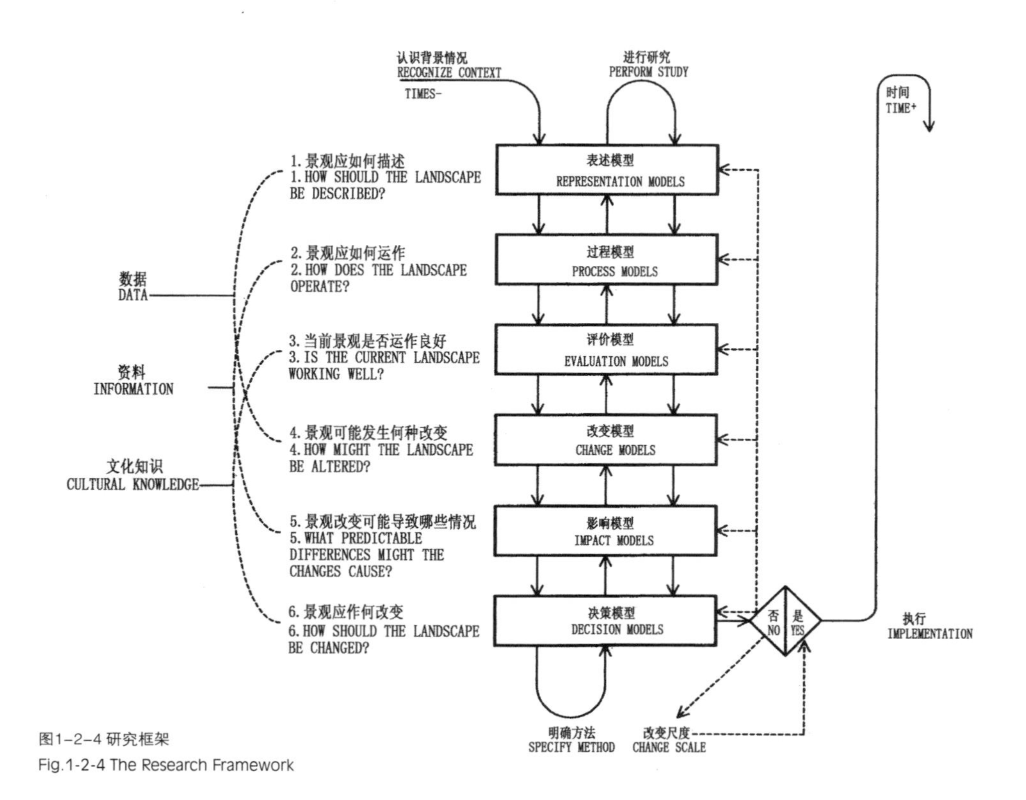

图1-2-4 研究框架

Fig.1-2-4 The Research Framework

系。为了制定决策，必须提出问题并给出答案，必须制定和考虑各种选择方案。

2. 研究的组织

本次对圣佩德罗河上游流域多解规划研究在组织和执行时采用了6个问题框架。

（1）如何从内容、时间和空间上描述景观状态?

本研究区域包括圣佩德罗河上游流域，从索诺拉的卡纳内阿附近的发源地到亚利桑那的雷丁顿。调查区域10 660平方公里。为了描述如此大面积研究区域的地理和动态工作过程，研究组建立了一个计算机地理信息系统以储存该地区已公开的空间信息。

（2）景观如何运作? 景观元素在功能及结构上的关系如何?

有关景观如何运作的问题应由过程模型来回答，该模型为研究中的几项分析提供信息（图1–2–6）：发展模型、水文模型、植被模型、景观生态格局模型、单一物种潜在栖息地模型、视觉偏

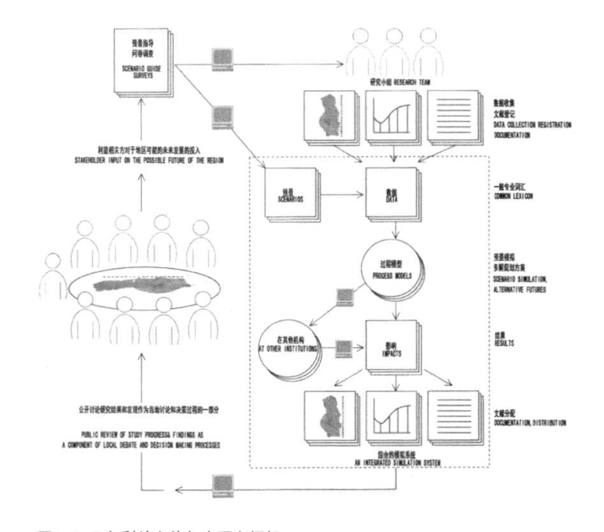

图1–2–5 各利益主体与本研究框架

Fig.1-2-5 The Stakeholders and the Research

work in this large study area, a computer-based geographic information system was organized to contain spatially explicit and publicly available data on the region.

(2)How does the landscape operate? What are the functional and structural relationships among its

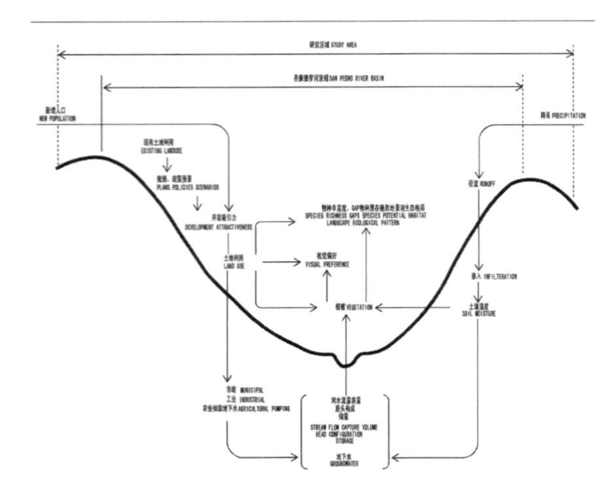

图1-2-6 过程模型
Fig.1-2-6 Process Models

好模型等。研究组使用一套计算机过程模型来描述及评价当前的景观，并评定每种多解规划方案作用于2000年的情况可能带来的影响。

发展模型评价了可获取的土地对5种开发模式的吸引力：商业、工业、城市、郊区、农村和城市远郊，其中后四种为住宅开发。随后，根据不同的改变方案模拟该区域的发展模式。每种方案都包括对未来人口设定、对不同房屋类型的需求以及相关政策和规划，这些政策和规划对私人的城市开发活动可能起到制约作用，也可能起到促进作用。

该地区的生物多样性评价分为三个部分：景观生态格局、一组单一物种的潜在栖息地模型和脊椎动物的物种丰富度，评价以预测中的未来植被格局为基础。

景观生态格局关注于景观结构和功能要素之间的空间联系。任何形式或规模的景观都可以比喻成马赛克——由廊带连接起多个斑块地以及背景基质。景观生态格局的评价利用了假设，即有些空间景观格局可以保持所有景观的主要自然演变。研究中选择了6种陆地脊椎动物，以单一物种模型评价了未来的变化对这些动物潜在栖息地的影响。

（3）当前的景观是否运行良好?

每个过程模型都适用于研究区域在1997～2000年之间的情况，这就设立了参考年限，以此为基础来衡量未来变化的影响。

（4）可以通过哪些政策或行为，何时何地对景

elements?

Questions about how the landscape operates are answered by process models that provide information for the several analyses that are the content of the study (Fig. 1-2-6): The Development Model, The Hydrological Model, The Vegetation Model, The landscape Ecological Pattern Model, Single Species Potential Habitat Models, The Visual Preference Model ect.

A suit of computer-based process models is used to describe and evaluate the current landscape and also to assess the potential impacts of each of the alternative futures relative to the conditions in 2000.

The development model evaluates the attractiveness of the available land for five kinds of development: commerce-industry and four residential types: urban, suburban, rural, and exurban. It then simulates the development pattern of the region in accordance with several different scenarios for change. Each scenario includes assumptions about future population, the demand for different housing types, and policies and plans that might constrain or attract new private urban development.

The predicted new vegetation patterns form the basis for a three-part assessment of regional biodiversity: the landscape ecological pattern, a group of single species potential habitat models, and vertebrate species richness.

The focus of the landscape ecological pattern model is the spatial relationships between structural and functional elements of the landscape. Any type of landscape at any scale can be described as a mosaic—a background matrix and patches connected by corridors. Evaluation of the landscape ecological pattern presents the hypothesis that there are spatial landscape patterns that will conserve the majority of natural processes in any landscape.

Single-species models of six terrestrial vertebrates evaluate the impacts of future changes on their potential habitat.

(3)Is the current landscape working well?

Each of the process models is applied to data on conditions in the study area during the period 1997-2000, referred to here as 2000.

(4)How might the landscape be altered—— by what policies and actions, where and when?

Three groups of scenarios are projected from 2000 to 2020 in 5-year increments by the allocation portion of the development model. The first , called PLANS, is based on interpretation of the existing planning documents and land use practices of the region. The second, CONSTRAINED, investigates lower-than-forecast population growth and tightly controlled development zones. The third, OPEN,

观作出怎样的改变?

研究组制定了2000～2020年期间的三组发展方案，以每5年作为一个分阶段建立发展模型。第一组为计划方案，以该区域现有的规划文件和用地实践为依据。第二组为限制方案，研究了在实际人口增长低于预计水平的情况下严格控制发展地区的模型。第三组为开放方案，研究了在实际人口增长高于预计水平的情况下低密度开发区域的模型。每组方案根据主要政策形势的差异细化为10种不同的模型，由此产生10种关于未来发展的预景。通过比较主要政策的改变而导致的不同效果为更好地理解何种行为可以产生最好的效果提供了基础。

（5）这些改变可能带来何种影响?

使用过程模型研究2020年多解规划方案，并将不同方案产生的各种效果与参考年份2000年的情况进行比较，由此产生影响评价。

（6）景观应做何种改变?

每个决策过程都包含着决策模型，并建立在决策者的文化水平基础上。本研究旨在为决策过程提供更多的参考信息，而不是推荐某种解决方法或政策。索诺拉、亚利桑那和瓦丘卡堡的有关人士可以利用多解规划方案的影响分析，以及影响的各种政策的合理性进行评价。图1-2-7展示了本研究的全部构成。

3. 变化预景

研究组在对圣佩德罗地区面临的环境问题进行调查后，制定了一份情景调查问卷，供亚利桑那的研究区域使用。调查问卷提供了一系列有关该地区未来规划问题和优先发展方案，被调查者可以作出选择或者提出自己的意见。这些问题涉及土地开发、水资源利用及土地管理。研究和预景调查问卷的时间范围都是20年，即2000年至2020年。调查结果显示，按调查对象的反馈意见分成几种模式。第一种方案为规划方案，以亚利桑那现有的规划和人口预测为基础。第二种方案为限制方案，将亚利桑那的发展限制于现在的开发区，并假定未来人口增长将比现在的预测水平要少。第三种方案为开放方案，取消对于土地发展的大部分限制，并假定未来

anticipates greater-than-forecast population growth and low-density development across the region. Each of these is expanded by variations that alter key policy positions, resulting in a total of ten different scenarios, each of which results in an alternative future. By comparing the effects caused by changing these key policies, the variations provide a basis for better understanding which actions will produce the greatest effects.

(5)What difference might the changes cause?

Applying the process models to the alternative futures for 2020 and comparing the results with the reference year 2000 yields impact assessments.

(6)How should the landscape be changed?

Every decision-making process implies the existence of a decision model, based on the cultural knowledge of those bearing the responsibility for making choices. This research is intended to inform these decision-making process, rather than to recommend specific solutions or policies. The projected impacts of the alternative futures can be used by stakeholders in Sonora and Arizona, including Fort Huachuca, to assess the desirability of the various policies that generated them. Fig. 1-2-7 shows the overall organization of the research.

3. The Scenarios for Change

After investigation of the environmental issues facing the San Pedro region, a questionnaire called the Scenario Guide was developed for use in the Arizona part of the study area. The Scenario Guide gave respondents the opportunity to select from a set of options (or to suggest their own strategies) related to planning issues and priorities for the future of the region. It contained questions relating to land development, water use, and land management. A 20-year period, 2000 to 2020, was chosen for the study and for the Scenario Guide. Analysis of the Scenario Guides revealed several recognizable patterns of response that represented the observed interests of the respondents. One scenario, referred to as PLANS, is based on current plans in Arizona and accepts current population forecasts. Another, called CONSTRAINED, directs development in Arizona into currently developed areas and reduces the forecast population. The third, OPEN, removes most constraints on land development and assumes higher population than forecast.

Ten scenarios guide the model of how development could take place, creating the ten alternative futures. The ten scenarios result from changes to important components of the three main scenarios, PLANS, CONSTRAINED, and OPEN, made in order to examine the changes in the alternative futures when various policy choices are changed.

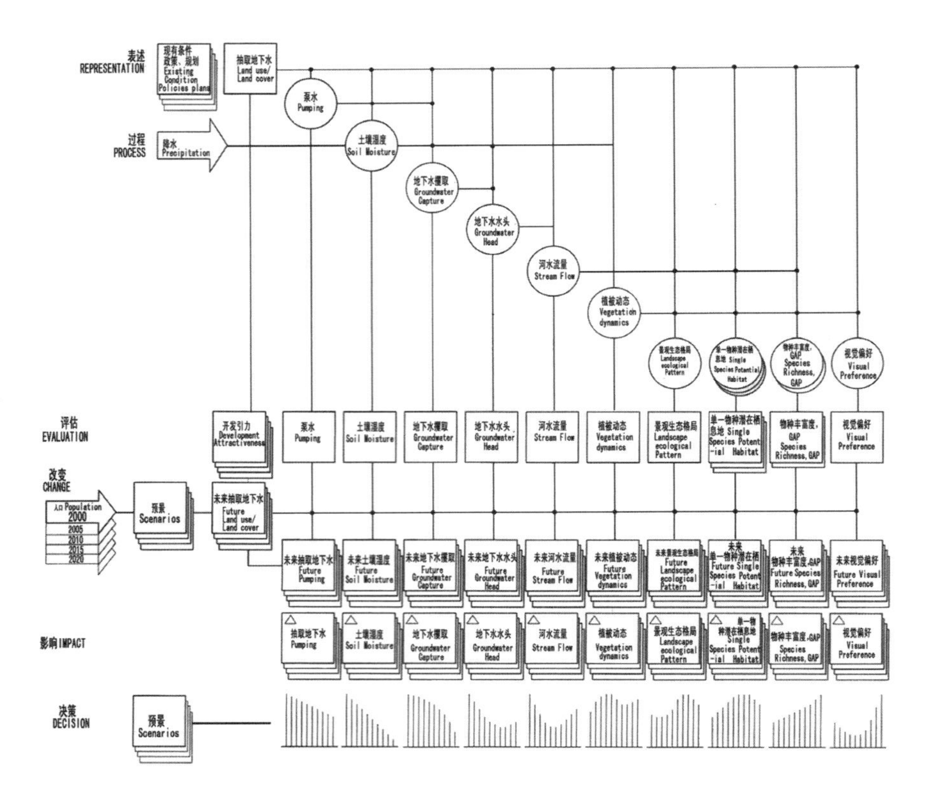

图1-2-7 本研究的组织

Fig. 1-2-7 The Organization of the Research

人口增长比现在的预测水平要高。

研究组制定了10种方案来模拟10种未来发展的预景。这10种方案源于对3种主要方案重要组成要素的改变，即规划方案、限制方案和开放方案。由此，可以判断如果政策发生改变，未来的发展将随之发生何种改变。

三、圣佩德罗河上游流域多解规划研究

1. 开发模型

开发模型对住宅项目、商业项目和工业项目进行选址，这是开发商在各种情况下可能做出的选择。开发模型用于分配五种类型的开发，一种是工商业项目，其他四种都是住宅项目：城市住宅、近郊住宅、农村住宅和远郊住宅。

开发模型首先分配工商业用地，接下来分配住宅用地。

三组发展方案——规划方案、限制方案和开放方案，组合了新发展的各种不同模式，如图

III. The Research of Alternative Futures for the Upper San Pedro River Basin

1. The Development Model

The development model selects the locations for residential and commercial/industrial projects that developers would likely make under the several scenarios. The development model is used to allocate the five kinds of development-one commercial/ industrial and four residential types: urban, suburban, rural, and exurban.

The model proceeds by first allocating commercial/ industrial land use.Next, the development model allocates residential land uses.

The three groups of scenarios-PLANS, CONSTRAINED, and OPEN –create very different patterns of new development, as shown in Fig. 1-2-8-Fig. 1-2-10. Differences in the patterns of development are largely attributable to the combination of constrains on developable land, the attractiveness of the available land for each housing types, and the differing housing demands under each scenario. Among the impacts caused by the alternative futures is change in overall attractiveness for residential development.

Fig. 1-2-11, Fig. 1-2-12 shows the 5-year time

1-2-8～图1-2-10所示。开发模型的不同主要是通过对可开发土地的限制条件、可开发土地对每种住宅类型的吸引力和每种发展方案之下不同的住房需求组合而得出的。多解规划方案产生的影响之一就是住宅开发吸引力的改变。

图1-2-11、图1-2-12显示了规划方案、限制方案和开放方案以每5年为一个阶段的制定情况。

图1-2-13、图1-2-14显示了由2020年规划方案、2020年限制方案和2020年开放方案得出的土地利用和土地覆盖的模型。

2020年多解规划方案、限制方案和开放方案的比较显示，在开放方案模式下，远郊区开发将得到

stages in which the PLANS,CONSTRAINED, and OPEN alternative futures are developed

Fig. 1-2-13, Fig. 1-2-14 shows the land use/land cover patterns resulting from the alternative futures PLANS2020,CONSTRAINED 2020 , and OPEN2020. Comparison of the alternative futures in 2020 for PLANS, CONSTRAINED, and OPEN shows that the greatest expansion of exurban development comes under the OPEN scenario. It creates much more development in and near Sierra Vista, the Palominas /Bisbee area, and in the lower slopes of all the mountain areas in Cochise County. The CONSTRAINED scenario produces more exurban development than the PLANS scenario, because the CONSTRAINED scenario does not allow rural residential development.

2. The Landscape Ecological Pattern Model

Landscape ecology is based on the concept that spatial landscape patterns of land use and land

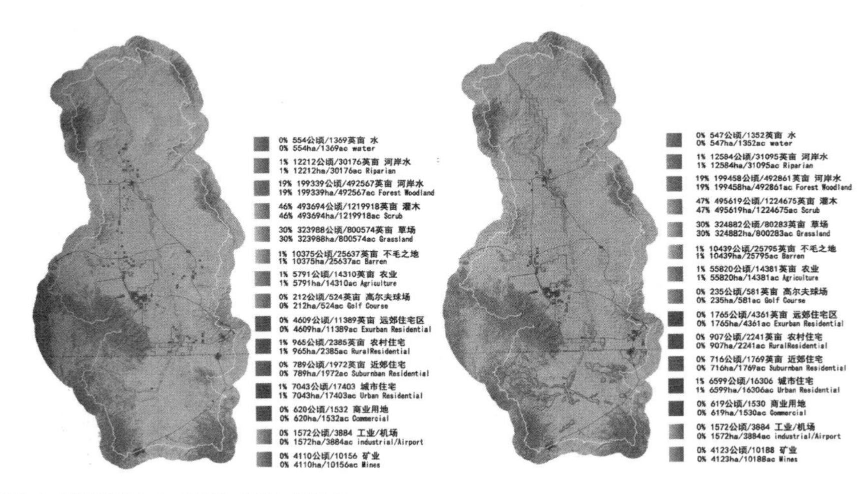

图1-2-8 规划方案2-3，新开发，2000—2020年
Fig.1-2-8 Plans 2-3:New Development,2000-2020

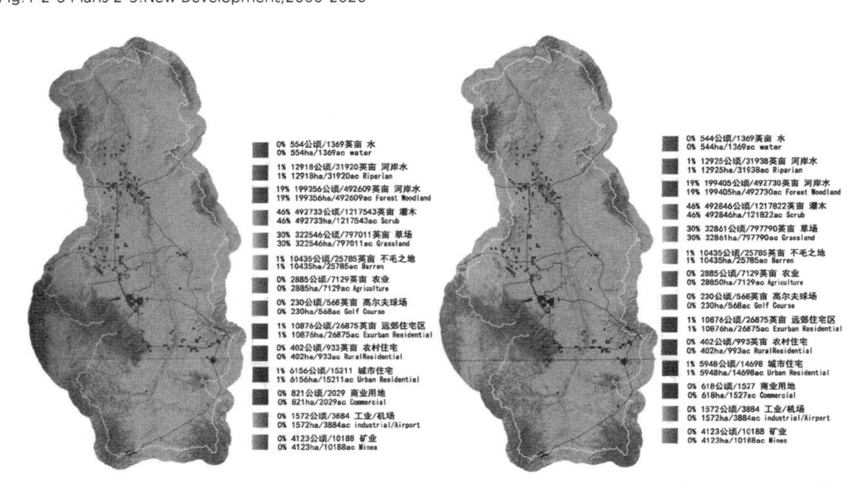

图1-2-9 限制方案1-2，新开发，2000—2020年
Fig.1-2-9 Constrained 1-2, New Development,2000-2020

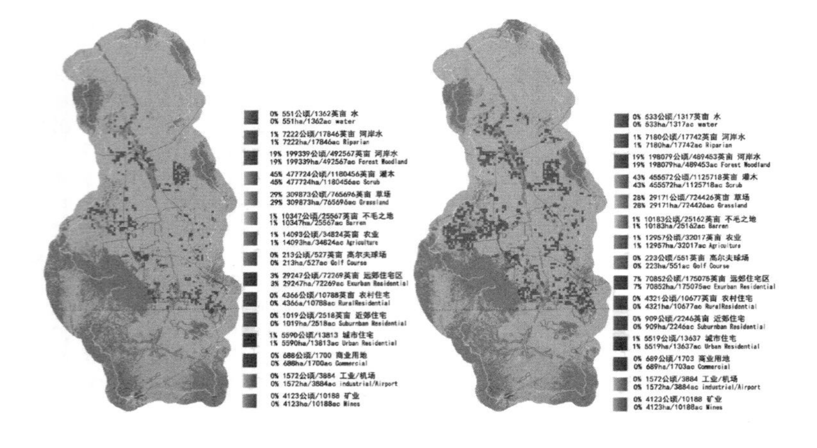

图1-2-10 开放方案1-2，新开发，2000—2020年
Fig.1-2-10 Open 1-2, New Development,2000-2020

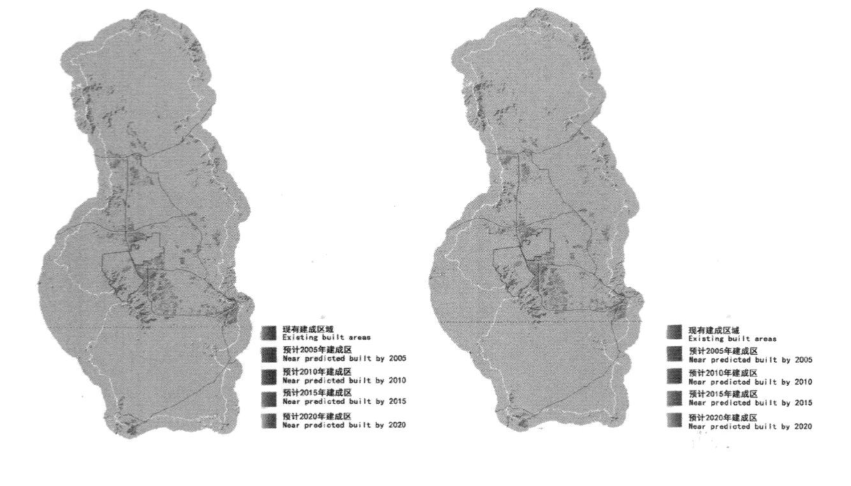

图1-2-11 规划方案限制方案，开发各时间段
Fig.1-2-11 Plans Constrained, Time Stages of Development

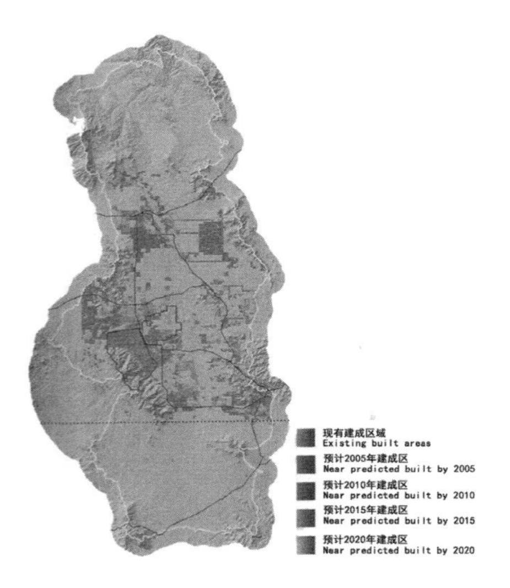

图1-2-12 开放方案，开发各时间段
Fig.1-2-12 Open, Time Stages of Development

cover define the relationship among structural and functional elements in any landscape. These patterns can be described as a mosaic made up of a background matrix, and natural and disturbed patches, either isolated or linked by corridors. The usefulness of landscape ecology is not for the preservation of any one species, but rather for the maintenance of the current mosaic to preserve the ecological diversity present in a given local (Dramstad et al.1996;Forman 1995).

The landscape ecological pattern model identifies large natural patches by excluding development, agriculture, and roads, and their effects. All of the scenarios create alternative futures that have negative impacts on the landscape ecological pattern (Fig. 1-2-15, Fig. 1-2-16). The impacts are caused be the future development pattern resulting from each scenario. In all the alternative futures, the ability of species to move among the existing large patches is reduced.

The most dramatic impacts result from the OPEN

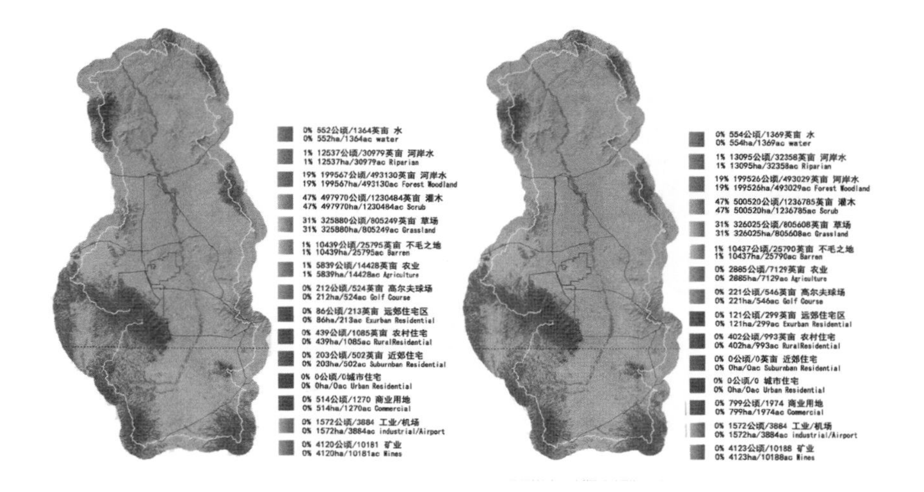

图1-2-13 规划方案、限制方案，土地利用土地覆盖，2002年

Fig.1-2-13 Plans Constrained, Land UseLand Cover,2002

最大限度地扩展，谢拉维斯塔、普罗米纳斯-比斯比地区和科奇斯县的所有山地矮坡以及上述地点附近地区都呈现了更多的发展机会。限制方案比规划方案创造了更多的远郊区开发机会，因为限制方案不允许农村住宅的开发。

2. 景观生态格局模型

土地利用与土地覆盖的空间景观格局界定了所有景观结构和功能要素之间的关系，这一点正是景观生态学的依据。这些格局可以用一个土地镶嵌体来表示，由背景基质、天然和人工斑块构成，相互间或独立或通过廊道相连。景观生态学的用途并不在于对单一物种的保护，而是维持当前特定区域内的土地镶嵌模式和生态多样性。

景观生态格局通过限制开发行为、农业活动和公路建设及其影响，确定了对大型天然斑块的保护。所有的多解规划方案都会对景观生态格局产生负面影响（图1-2-15，图1-2-16），影响是由每种方案的未来开发模式造成的。在所有的多解规划预景中，物种在原有大面积天然斑块的活动能力都减弱了。

开发方案下的开发模型会对景观生态格局产生最严重的影响。规划方案和限制方案下大面积天然斑块的减少程度差不多，比其开放方案相对要少。减少的斑块主要位于丘陵和山谷地区。

3. 单一物种的潜在栖息地模型

研究中选择了6种陆地脊椎动物，以单一物种

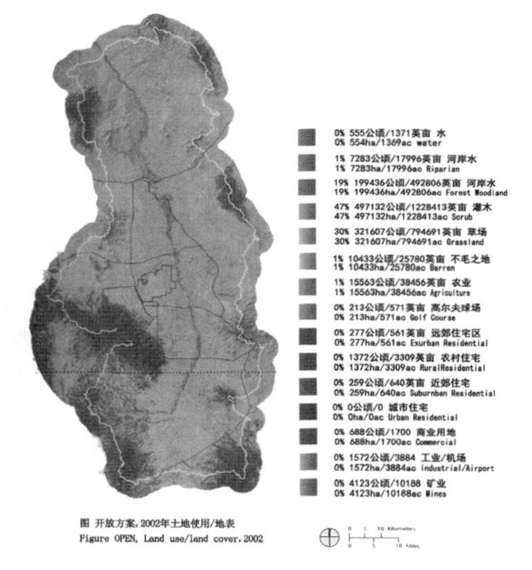

图1-2-14 开放方案，土地利用地表，2002年

Fig.1-2-14 Open, Land UseLand Cover,2002

scenarios. The PLANS and CONSTRAINED scenarios result in generally similar and smaller reductions in large patch size. Loss occurs mainly in foothill and valley locations.

3. Single Species Potential Habitat Model

Single-species models of six terrestrial vertebrates evaluate the impacts of future changes on their potential habitat. The species were chosen so that each of the vegetation communities within the study area was represented by at least one animal inhabitant. Additional criteria for selection included threatened or endangered species status (Empidonax traillii extimus, Sonoran pronghorn), existing or proposed reintroduction programs(beaver, jaguar),and species requiring large areas or unique features(northern

图I 大面积天然地块（栎木林） PhotoI Large natural patch(oak woodland)

图II 波波康玛莉牧场的旱谷 PhotoII Arroyo on Bobocomari Ranch

大面积天然斑块的变化/公顷	Change in area of large patches/ha
Baseline 2000	687,983
Plans	-5,466
Plans 1	9,632
Plans 2	-9,302
Plans 3	-1,601
	-4,232
	15,503
	-15,503
Open	-67,486
Open 1	-101,455
Open 2	-49,991

相关影响
Relative impacts

影响比例
Percent impacts

图1-2-15 景观生态格局：影响，2000—2020年
Fig.1-2-15 Landscape Ecological Pattern : Impacts,2000-2020

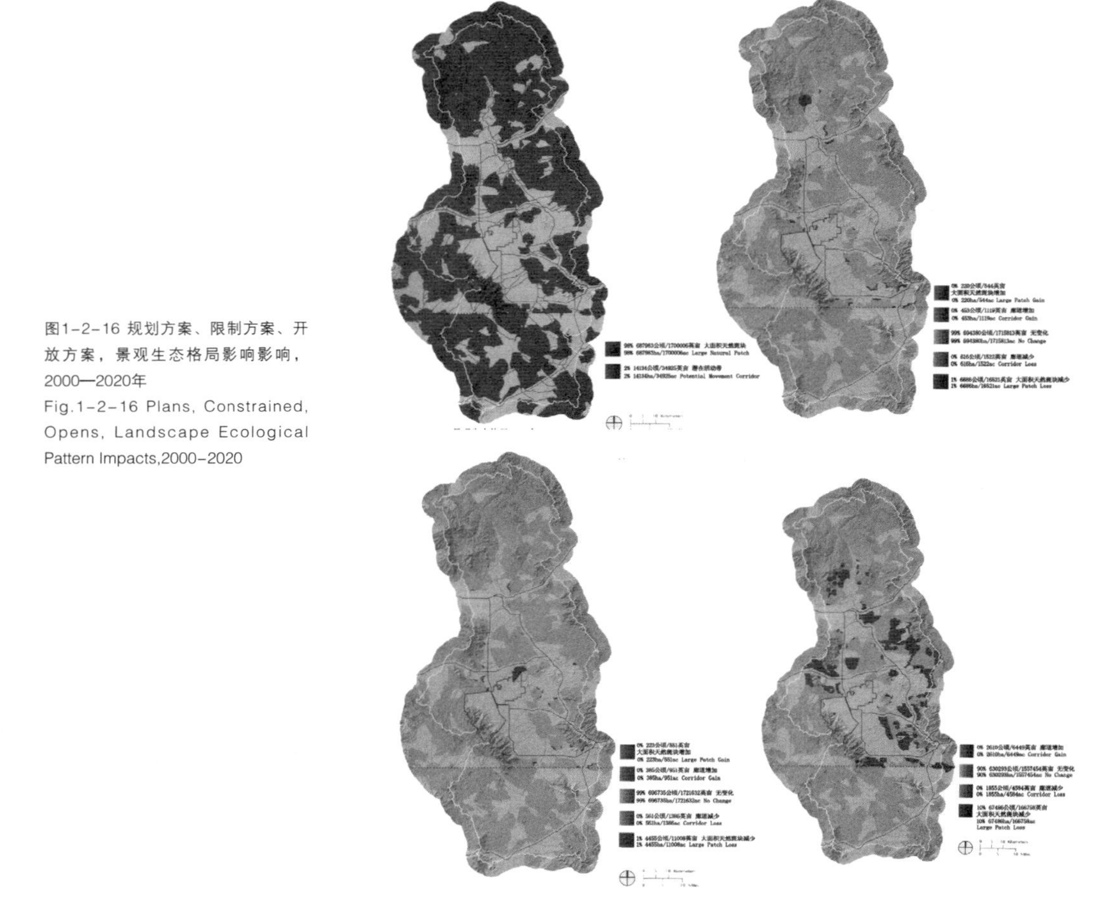

图1-2-16 规划方案、限制方案、开放方案，景观生态格局影响影响，2000—2020年
Fig.1-2-16 Plans, Constrained, Opens, Landscape Ecological Pattern Impacts,2000-2020

模型评价了未来的变化对这些动物潜在栖息地的影响。这样，研究区域内的每个植被区域都至少有一种栖息动物作为代表。另一个选择的标准包括受威胁或濒危物种的状态（西南柳木鹟，索诺拉叉角羚羊）、现有的或建议实行的物种回迁方案（河狸、美洲豹），以及需要生活在面积广大以及特色鲜明环境中的物种（北方苍鹰，大毒蜥）。

而6种陆地脊椎动物详细内容繁多，因案例篇幅限制，在此仅介绍一处，供读者参考。

美洲豹及其潜在栖息地：

美洲豹是西半球最大的猫科动物，体重68公斤到100公斤（150～225磅），是一种凶猛的食肉动物。

尽管1997年8月起，美洲豹就被列为濒危物种，在亚利桑那州却几乎看不见美洲豹。相信通过适当的关注和管理策略，保证大型食肉动物在区域内的活动空间，美洲豹有可能会重现。

各种多解规划方案对美洲豹潜在栖息地造成了不同的影响效果（图1-2-17，图1-2-18）。在规划方案下，美洲豹潜在栖息地有少量增加，这主要是由圣大卫北部地下水水位升高引起的。规划方案下各种子方案对美洲豹栖息地产生的影响差别极小。在限制方案下，美洲豹潜在栖息地有也少量增

goshawk, Gila monster).

Because of various contents of the detailed description for six terrestrial vertebrates, as well as limitation of length of the case, there is only one description in the introduction, to be a reference for readers.

Jaguar Potential Habitat:The jaguar, Panthera onca, is the largest cat native to the Western Hemisphere, ranging in weight from 68 to 100 kg (150 to 225lb). The jaguar is a formidable predator.

Although the jaguar was placed on the list of endangered species in August 1997, there is little or no sign of its presence in Arizona. It is believed that with proper care and a management strategy that allows large carnivores to move throughout the region, the jaguar may return.

The various scenarios create a range of effects on jaguar potential habitat (Fig. 1-2-17, Fig. 1-2-18). The PLANS scenarios cause small increases in potential jaguar habitat. Most of the expansion is due to localized increases in groundwater levels north of Saint David. Very little difference exists among the PLANS alternative futures. The CONSTRAINED scenarios all create small increases in potential jaguar habitat, similar to those under PLANS scenarios. The only difference between PLANS and CONSTRAINED appears to be in amount. The greatest loss of potential jaguar habitat occurs under the OPEN scenarios, with minor differences among the variations. Under the OPEN alternative future, vegetation loss fragments cover habitat, eliminating much of the potential jaguar habitat.

图片I 美洲豹 PhotoI Jaguar

图片II 美洲豹栖息地 PhotoII Jaguar habitat

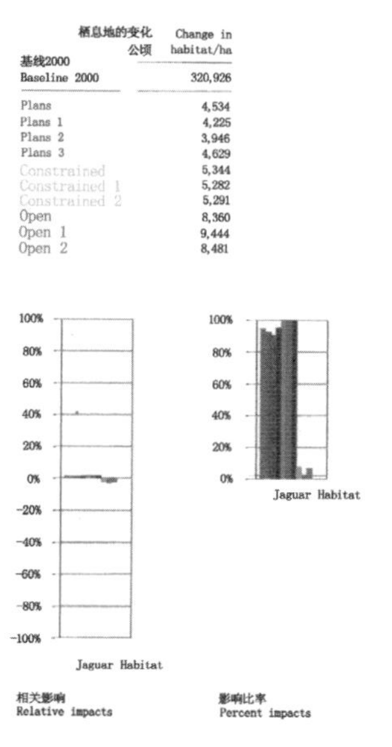

图1-2-17 美洲豹栖息地影响，2000—2020年
Fig.1-2-17 Jaguar habitat：Impacts,2000-2020

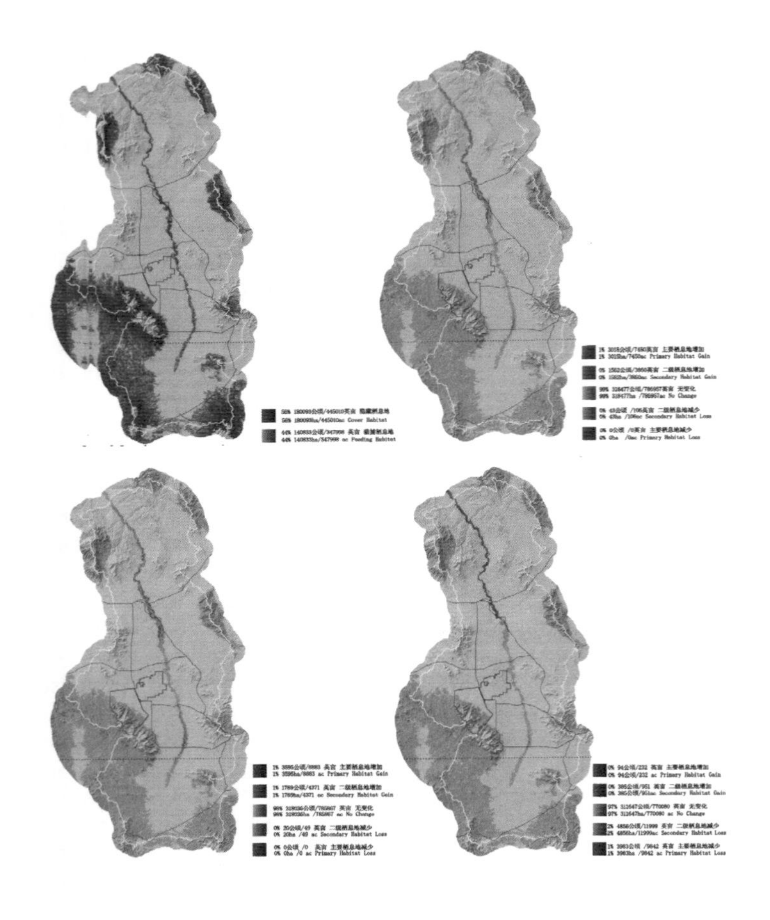

图1-2-18 美洲豹潜在栖息地&多解规划方案

Fig.1-2-18 Jaguar potential habitat & the alternative futures

加，这与规划方案下的情况类似。规划方案与限制方案的唯一区别是增加的程度有所不同。在开放方案下，美洲豹栖息地大范围减少，各种子方案对美洲豹栖息地产生的影响差别极小。从开放方案制订的未来情景模拟中，可以看到植被的流失使得隐藏栖息地变得支离破碎，导致大量潜在栖息地消失。

4. 影响总结

将过程模型应用于2020年多解规划方案中，并将其结果与2000年的数据相比，由此产生影响评价。图1-2-19显示了10种多解规划方案对2020年产生的影响。这些影响通过两种不同的图示法进行总结：与基本条件相比得出的变化比例和10种多解规划方案范围内的等级比例。除了居住吸引力等级外，所有与2000年基本条件相比较得出的变化比例都得以列出。

在所有的多解规划方案下，圣佩德罗河上游流域地区的用地模型和土地政策对植被和栖息地造成的

4. Summary of Impacts

Applying the process models to the alternative futures for 2020 and comparing the results with the reference period 2000 yields impact assessments. Table XX shows the impacts of the ten alternative futures to 2020. Impacts are summarized graphically in two different ways: by percent change relative to baseline conditions, and by percent ranking within the range of the ten alternative futures. For all but the residential attractiveness ratings, percent change measures are expressed relative to baseline conditions in 2000.

In all the alternative futures, the effects of changes in patterns of land use and land policies on vegetation and on habitat in the region of the Upper San Pedro River Basin will be significant and dramatic. Increasing urbanization within the uplands, especially near Sierra Vista and Cananea, will result in a decrease in the extent of grassland. However, changes in development policies, fire management practice, retirement of agricultural water rights, and management of riparian vegetation can result in dramatic improvement to both upland and riparian vegetation.

The comparative pattern of impacts (Fig. 1-2-19)in which the OPEN alternative futures have considerably

影响是巨大且惊人的。高度城市化的蔓延将造成草地的减少，这在谢拉维斯塔和卡纳尼亚附近尤为突出。然而，开发政策的改变、消防管理、农业用水权的取消和河岸植被管理等一系列措施可以促使高地与河岸植被得到巨大改善。

从影响比较模型中（图1-2-19）可以看到，开放方案造成的不利影响比规划方案与限制方案要严重得多，这在所有与栖息地相关的衡量标准中都可以看出。

5. 变化景观的多解规划研究总结

10种方案及其产生的多解规划结果囊括了一系列预计会影响圣佩德罗河上游流域变化的选择。该区域具有很高的开发及环保价值。总结性开发吸引力以10种方案下的分配模型为基础。4种住宅开发类型的每一个吸引力模型都是通过简单取均值后按以10种方案的比例分配进行加权。这种公平的加权计算结果显示，未来近郊开发是一个主要的影响因素。谢拉维斯塔南部及其附近区域尤其具有开发吸引力。

优先保护地图显示了保护吸引力总结性评价。

more harmful impacts than do the PLANS and CONSTRAINED alternative futures is seen throughout all the habitat-related measures.

5. Summary of Alternative Futures for Changing Landscape

The ten scenarios and their resulting alternative futures encompass a likely range of choices that are expected to influence change in the Upper San Pedro River Basin. The region is highly valued for both development and conservation. Summary development attractiveness is based on the allocation patterns resulting from the ten scenarios. Each of the attractiveness patterns for the four residential development types is weighted by its proportional allocations across all ten scenarios using simple averaging. As a result of the equal weighting, future suburban development is a major influence. Areas adjacent to and south of Sierra Vista are particularly attractive for development.

The summary evaluation of attractiveness for conservation is shown in the conservation priority map. This evaluation is the result of an unweighted sum of the evaluations for all large patches and corridors from the landscape ecological pattern model, The assessment clearly shows the great importance of the San Pedro riparian corridor, the

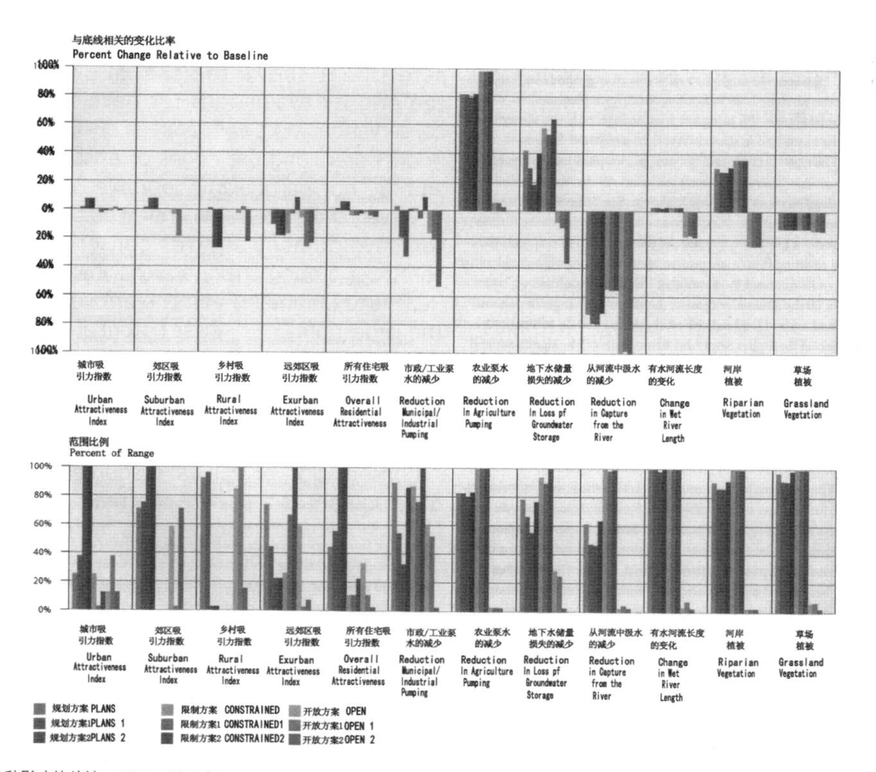

图1-2-19a 各种影响的总结，2000—2020年

Fig.1-2-19a Summary of impacts,2000-2020

图1-2-19b 各种影响的总结，2000—2020年
Fig.1-2-19b Summary of impacts,2000-2020

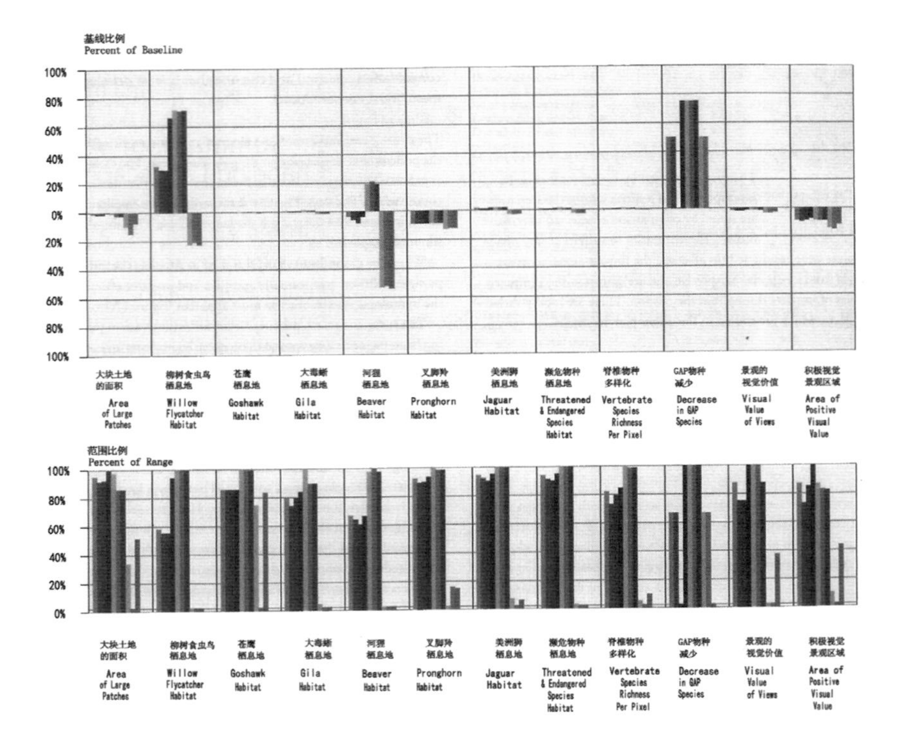

这项评价是一系列评价的未加权总和，涉及的地带包括景观生态类型格局中所有大块斑块和廊道，该评价清楚地标明了圣佩德罗河岸廊道、矮坡山区以及与其相连的小溪的重要性。

开发、保护竞争评价显示了开发与保护优先权评价在空间上的重叠性。大多数具有高度开发优先意义的区域位于现有居住区边缘地带，没有特殊的保护意义。然而，仍有一些重点地区在开发和保护上存在着冲突，其中矛盾最突出的是瓦丘卡山脉的矮坡地带和谢拉维斯塔的旱谷地带。瓦丘卡堡是潜在矛盾冲突突出地带面积最大的一块。

四、 总结

自然资源保护与经济发展是一对矛盾，找到其中的平衡点一直是规划师苦苦追寻的目标，本项目是解决保护与发展问题的典型案例。处理好保护与发展的关系，不仅是一个理论问题，更是一个实践问题。规划中积极开展生物多样性保护工作，对有效缓解生物多样性压力，改善生态环境起到了积极作用。

多解规划是本案例最成功之处，“它是一种基于

lower slopes of the mountain zones, and the pattern of arroyos that connect them.

The development/conservation competition assessment shows the spatial overlap between the priority evaluations for development and for conservation. Most of the areas of high development priority are along the edges of existing settled areas and are not particularly desirable for conservation. However, there are some critical areas where demand for development and for conservation conflict. Foremost among these are the lower slopes of the Huachuca Mountains and the arroyos in Sierra Vista. The largest area of highest potential conflict is within Fort Huachuca.

IV. Summary

Conservation of natural resources and economic development is a pair of contradiction, to find the balance between them is the goals that planners always searching for. It is a typical case study to solve the problems between protection and development. Dealing with the relationship between the them correctly, is not only a theory problem, but also a practical one. The active protection of biodiversity in planning work, can relieve the biodiversity pressure effectively, and play an active role in improving the ecological environment.

Alternative futures are the most successful in this

GIS的模拟模型策略的规划方法，该策略考虑了一个地区的人口、经济、自然和环境过程，同时预测了不同的土地利用规划和管理决策所带来的后果。”多解规划并非简单的几个可选方案的罗列，更强调的是规划或设计过程的多种可能性、框架性和弹性，研究组制定了10种方案来模拟10种未来发展的预景，这10种方案源于对3种主要方案组成要素的改变，即规划方案、限制方案和开放方案。

综上所述，不确定性是城市规划和开发中所面临的一个普遍问题，城市规划就是要对这些不确定性因素做出合理恰当的预先安排，而多解规划正是解决这一问题的成功之路。

“Scenario”这个词通常被理解为事件的情况，或故事、戏剧或电影的情节。同样，在本研究中，“Scenario”指框架或情节，据此可以预测圣佩德罗河上游流域的未来。

case study. "It is a planning method based on the strategy of GIS-based simulation modeling strategy that considers the demographic, economic, physical, and environmental process of an area and projects the consequences of various land-use planning and management decisions." Different alternatives superpose simplely are not the method of Alternative futures, on the contrary, many kinds of possibility, framework and elasticity of the planning or design process are more emphasized in Alternative futures. Ten scenarios guide the model of how development could take place, creating the ten alternative futures. The ten scenarios result from changes to important components of the three main scenarios, PLANS, CONSTRAINED, and OPEN.

To sum up, uncertainty is a common problem during city planning and development, Urban planning is to make reasonable arrangement in advance aiming at some uncertainty factors, however, alternative futures is the road to success to solve the problem.

思考题

1. **什么是“多解规划”？**
2. **什么是“Scenario”？**
3. **卡尔·斯坦尼茨的多解规划框架内容?**
4. **简要叙述圣佩德罗河上游流域多解规划研究在组织和执行时采用的组织框架?**
5. **圣佩德罗河上游流域多解规划研究如何运用过程模型体现景观的运作过程?**

Questions

1. What is “Alternative futures”?
2. What does the word scenario mean?
3. What are the contents of the research framework of the Alternative futures?
4. What does the framework applies in the organization of the research and in carrying it forward in the study of alternative futures for the Upper San Pedro River Basin briefly?
5. How does the landscape operate in this study of alternative futures for the Upper

注释/Note

图片来源/Image: 本案例图片均引自郑冰，李劼译，朱强，俞孔坚校，变化景观的多解规划,中国建筑工业出版社，北京；Carl Steinitz ;Hector Manuel Arias Rojo; Scott Bassett; Michael Flaxman; Tomas Goode; Thomas MaddockIII; David Mouat; Richard Peiser; Allan Shearer, Alternative Futures for Changing Landscapes——THE UPPER SAN PEDRO RIVER BASIN IN ARIZONA AND SONORA

参考文献/Reference: 郑冰，李劼译，朱强，俞孔坚校，变化景观的多解规划,中国建筑工业出版社，北京；Carl Steinitz ;Hector Manuel Arias Rojo; Scott Bassett; Michael Flaxman; Tomas Goode; Thomas MaddockIII; David Mouat; Richard Peiser; Allan Shearer, Alternative Futures for Changing Landscapes——THE UPPER SAN PEDRO RIVER BASIN IN ARIZONA AND SONORA

推荐网站/Website: http://www.usppartnership.com/ ；http://www.nature.org/ourinitiatives/regions/northamerica/ unitedstates/arizona/placesweprotect/san-pedro-river.xml

第三节 沃辛顿河谷地区规划
Section III The Plans For Worthington Valleys

一、 项目综述

1. 项目简介

一个大都市地区郊区发展的方案，通常也是早期郊区化发展失败的案例。其关键是应用生态规划的原则并试用这些原则来抵制大都市的发展和市场机制的要求。沃辛顿河谷地区是一个能够很好的说明私人参与，通过市民和政府行动相结合的例子。

马里兰州巴尔的摩县在20世纪50～60年代经历了早期的城市化。格林奈尔·洛克是居住在格林斯普林河谷当地的建筑师，他意识到这样的开发会对如此美丽独特的农业开放空间带来威胁。为了保护河谷，他提出新的规划设想，既满足河谷未来发展，又可以控制公共设施的无限扩张，保护现有的公共开放空间。

洛克首先针对河谷问题与宾夕法尼亚大学的华莱士和麦克哈格进行探讨，他们是美国公认首屈一指的大尺度景观规划设计公司。

这次讨论的结果是在县规划和分区管理办公室的支持下，于1962年建立格林斯普林及沃辛顿河谷规划委员会。规划委员会筹集了125 000美元支付华莱士与麦克哈格联合事务所，希望他们能够为面积达70平方英里（约181平方公里）的3个河谷中的代表区域进行总平面规划。

I. Overview

1. Introduction

It is a case study of suburban growth in a metropolitan region-an area that usually becomes the victim of inchoate suburbanization. The problem then is to apply ecological planning principles and test them against the demands of metropolitan growth and the market mechanism. The Green Spring Worthington Valleys represent a unique opportunity to demonstrate the conjunction of private concern and action through civic and governmental process.

The late 1950s and early 1960s witnessed the virtually untrammeled expansion of residential real estate development in Baltimore County. Grinnell W. Locke, a local architect who resided in the Green Spring Valley, recognized the threat posed by such development to the unique and beautiful agricultural open space, to protect the valleys, he conceived of a plan which would accommodate growth and limit random extension of public utilities, while preserving existing open space.

He initiated discussions with David Wallace and Ian McHarg of Philadelphia, their firm being acknowledged as the premier large-scale landscape planning firm in the United States.

The result of these discussions led to the establishment of the Green spring and Worthington Valleys Planning Council, Inc.– predecessor of today's Valleys Planning Council, Inc-the latter here in after being referred to as "Council" or "VPC".A committee was established to raise the necessary $125,000 to pay Wallace-McHarg Associates for the development of a comprehensive plan covering the 70 square miles,

2. 河谷定义

研究的区域范围自贝尔特路延伸到小西河的北坡，自赖斯特敦路和西马里兰铁路到巴尔的摩-哈里斯堡高速公路（图1-3-1）。遍布着郁郁葱葱的山川谷地、广阔的平原以及一系列用地类型。这是一份美丽的遗产，使设计师肩负严肃的责任；这也是一个受到威胁的地区，既充满挑战又存有机遇（图1-3-2，图1-3-3）。

二、 沃辛顿河谷地区规划与设计

1. 河谷地区规划

1) 规划问题实质

规划将决定河谷开发的适宜程度，规划结果要与

and roughly 45,000 acres of land represented by the three valley area.

2. Definition of Valley

The Valleys have the advantage of being defined by natural planning boundaries. The area of study extends from the Beltway to the northern slope of the western Run, from Reisterstown Road and the Western Maryland Railroad to the Baltimore-Harrisburg Expressway (Fig.1-3-1). This area contained widespread valleys, plateaus, wooded ridges, and an intricate array of many land uses. It is a beautiful inheritance, a serious responsibility, an area threatened, a challenge and opportunity (Fig.1-3-2, Fig. 1-3-3).

II. The Plan for Worthington Valleys

1. The Plan for Valleys

1)The Nature of the Problem

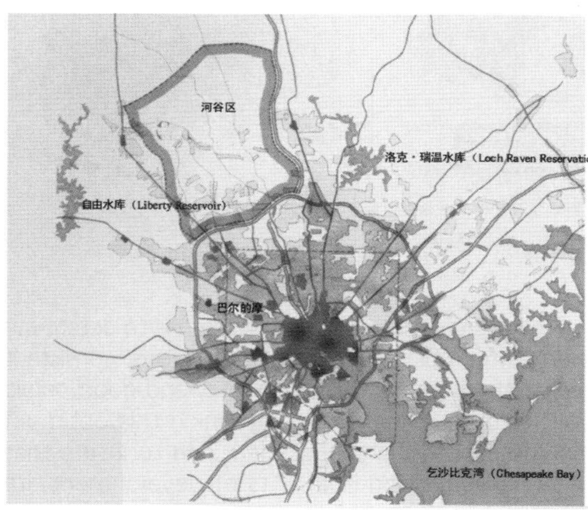

图1-3-1 巴尔的摩地区
Fig.1-3-1 Baltimore

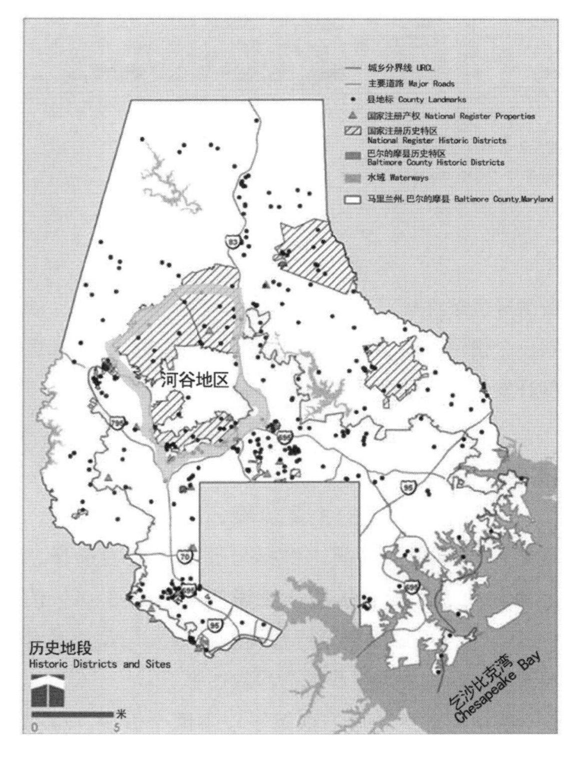

图1-3-3 沃辛顿河谷历史区域位置
Fig.1-3-3 The position of Worthington Valley Historic District

图1-3-2 鸟瞰图
Fig.1-3-2 Bird's Eye Perspective

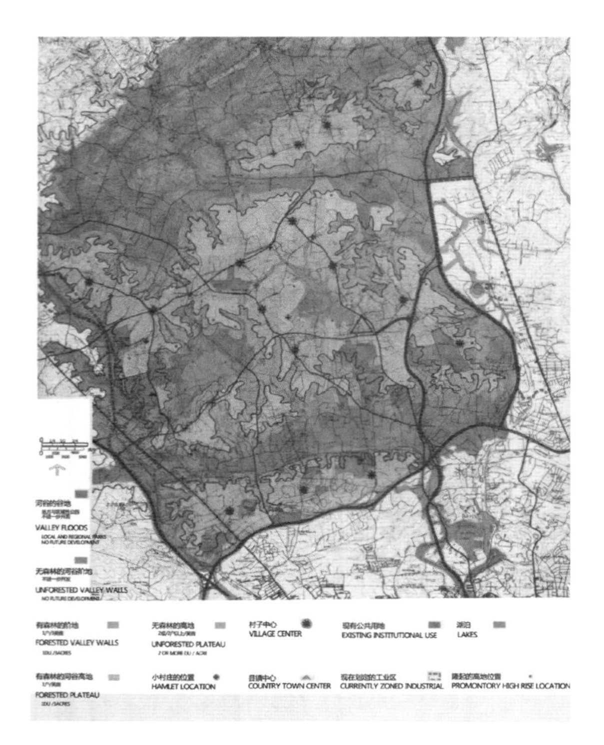

图1-3-4 最优土地利用

Fig.1-3-4 Optimum Land Use

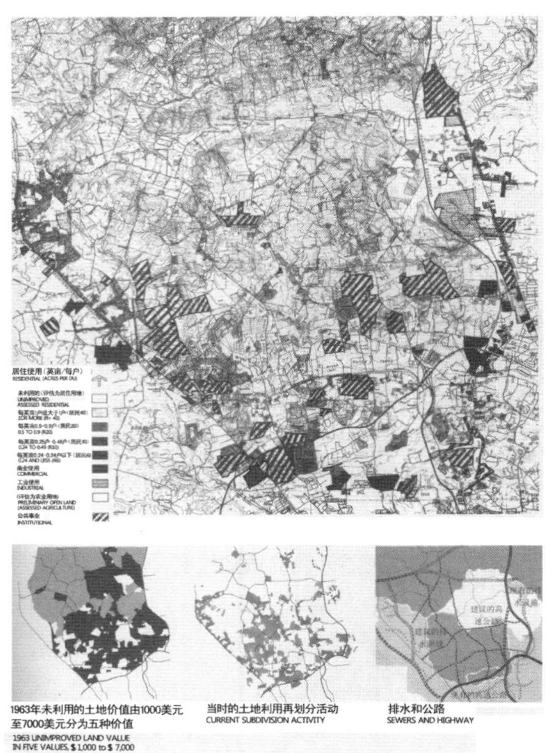

图1-3-5 1963年土地使用情况

Fig.1-3-5 1963 Land Use

自身环境条件相符合，与良好的公共政策相一致；规划一个使河谷达到最佳可持续的状态的过程；保证适度开发使得长期整体利益能够平等分配。

2) 解决问题方法

解决问题方法的主要步骤包括：现状分析，可能性增长模型预测，目标与形式可能性选择，土地利用适度发展模型详细说明（图1-3-4，图1-3-5），规划成果可行性检测，私人与政府联合控制与执行、发展过程的详尽叙述等。

解决问题的哲学方法基于以下两个假设：存在主导未来农场发展方向的现实景观；将规划看作是一种过程，在此过程中，设计师提出备选方案，权衡各个方案的利弊，以及需要解决与社区居民基本价值观相博弈所产生各种问题。

3) 规划建议

在规划中，就提出问题与解决措施两方面同时

The nature of the planning problem is to determine the optimum development of The Valleys, consistent with a high level of natural amenity and good public policy; to devise a process by which this optimum may be achieved and sustained; and to insure that the total long range benefits of development are equitably distributed.

2)Approach to the Problem

The major steps used in carrying out this approach included analysis of the existing situation, forecast of probable growth, development of goal and form alternatives, development of detailed optimum land use patter(Fig. 1-3-4, Fig. 1-3-5), testing for consequences and feasibility, the elaboration of possible private and government controls and the development of implementation process.

The philosophic approach to this problem is based upon two assumptions. The first is that there exists an essential landscape quality which must be recognized and understood as a guide to the farm and location of future growth. The second assumption conceives planning as a process of posing alternatives, weighing them, against each other and against a basic value system shared by the community.

提出建议：这是一个美丽而又容易遭到破坏的地区；发展是不可避免的，必须容纳增加的人口；缺乏控制的发展必然是破坏性的；发展必须和区域目标相一致；遵守保护的原则能避免破坏并保证提高环境质量；这个地区能接受所有预期的发展而不受破坏性掠夺；有规划的发展较无规划的发展更理想，可以获得更多的利润；公共和私人的力量能联合起来，共同参加到实现规划的过程中去。

这是一个美丽而又容易遭到破坏的地区。这些广阔的河谷地区极容易遭受破坏。不像别的景观拥有绿茵如画的广阔河谷那样，一旦侵入一些小小的建设项目，它就会立刻遭到破坏。河谷由两侧的林木台地和河谷的谷地形成。要是树木被砍伐或被建设项目所取代，美丽和平静的景色就会消失。只是因为河谷地区缺少排水设施，才使得它们至今免受开发的劫掠。

发展是不可避免的，必须容纳增加的人口。今天，这个地区还没有开发，但土地价格已经很高，而且还在上升，这证明这个地区的开发已经迫在眉睫。由于它优美舒适的环境、大量可供利用的建设用地以及交通上的便利等优点，在巴尔的摩整个区域发展中所分担的发展份额会不断增长。

无控制的发展必然是破坏性的。无控制发展模型（图1-3-6）提供了从1963年到1980年，甚至到2000年，如果没有规划及建议指导控制河谷发展，

图1-3-6 无控制发展的混乱局面

Fig.1-3-6 Specter of Uncontrolled Growth

3)The Proposition

In the planning, the problem and solution are presented as a proposition: The area is beautiful and vulnerable; Development is inevitable and must be accommodated; Uncontrolled Growth is inevitably destructive; Development must conform to Regional Goals; Observance of conservation Principles can avert destruction and ensure enhancement; The Area can absorb all prospective growth without despoliation. Planned growth is more desirable than uncontrolled growth; and more profitable; Public and Private Powers Can be joined in partnership in a process to realize the plan

The area is beautiful and vulnerable.These broad valleys are vulnerable. No landscape can be so quickly destroyed by small intrusions of development as can the broad valley in pasture. Their character is as dependent upon their wooded walls as upon the valley floor. Should the woods be felled and replaced by development, the beauty and serenity of scene will vanish. Only the absence of sewers in the valley has protected them from development and spoliation.

Development is inevitable and must be accommodate.Today the area is undeveloped, but high and rising land values are testimony to the imminence of development. Its advantages of amenity, the availability of developable land and accessibility, make inevitable a growing share of regional growth.

Uncontrolled growth is inevitably destructive. The uncontrolled Growth Model(Fig. 1-3-6) provides a picture of the sequence of change in the valleys from 1963 to 1980, and to the year 2000, if current trends continue with no other devices for guidance and control than now exist. The uncontrolled Growth Model shows a spectre of future, sufficient to convince even the most insensitive as to the likely despoliation of The Valleys. It also provides an analytical device to trace the implications of current development policies and programs of the Country and State, particularly in relation to highways and utilities.

The third purpose of the Uncontrolled Growth Model is to determine the ultimate developed land value (in aggregate) that the valleys can expect in the foreseeable future. This aggregate value can then be set as a bench mark against which the gains from development to the current property owners, developers.

Development must conform to regional goals. Regional growth to be accommodated by the study area has been identified. Regional and county planning agencies, deploring uncontrolled growth, have recommended four major concentrations bordering the study area: at Pikesville , Reisterstown, Towson, Hereford.

Observance of conservation principles can avert destruction and endure enhancement. These

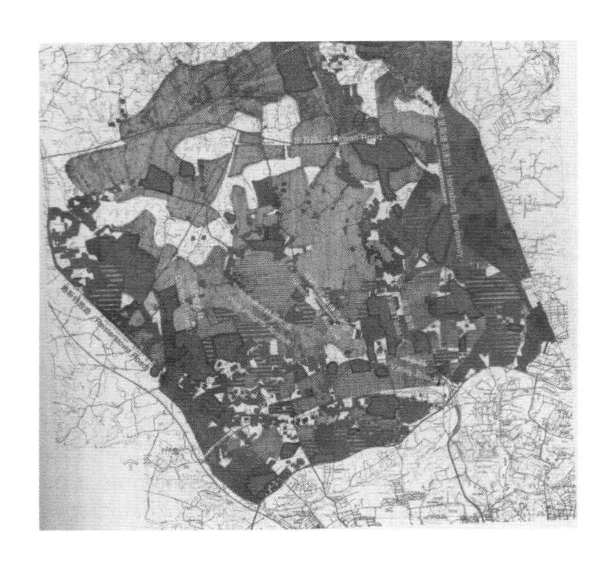

图1-3-7 1963—2000年扩展图

Fig.1-3-7 Expanded Growth 1963-2000

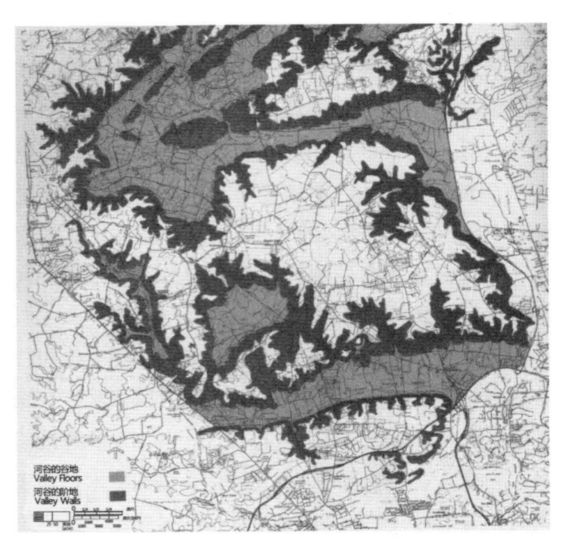

图1-3-9 基本环境优美地区

Fig.1-3-9 The Basic Amenity

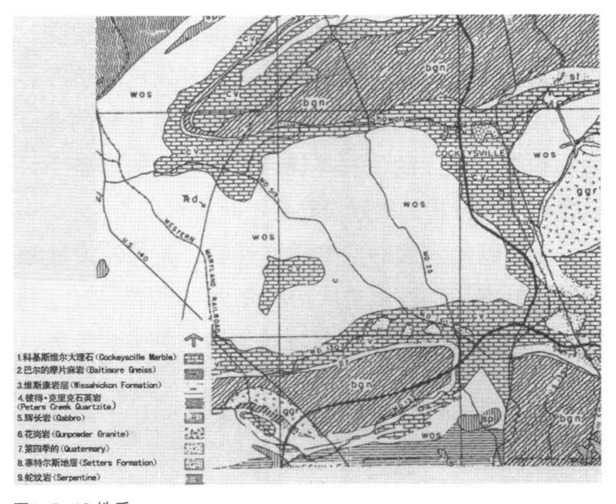

图1-3-10 地质

Fig.1-3-10 Geology

◀ 图1-3-8 裸露的高原、植被覆盖的高原和植被覆盖的岩壁

Fig.1-3-8 Unforested Plateau, Forested Plateau，Forested Valley Walls Baltimore

现有趋势任其发展的一系列未来变化图景。无控制发展模型展示了噩梦般的未来，充分证明即使是最不敏感的行动也会将河谷毁灭。它还提供了一种分析方法，分析当前国家及州的发展政策及规划所产生的意义，尤其是高速公路及公共事业设施。

无控制发展模型的第三个目标是确定土地的最终发展价值（总体价值），而且这是河谷地区在未

principles indicate the types of development and densities appropriate to the various physiographic characteristics. Conservation principles are: The valleys should be prohibited to development save by such land uses as compatible with the present pastoralscene. These would include agriculture, large estates, low intensity use institutional open space, parks and recreation, public and private; Development should be prohibited overall formations of Cockeysville Marble and the aquifers they contain; 50 year Flood

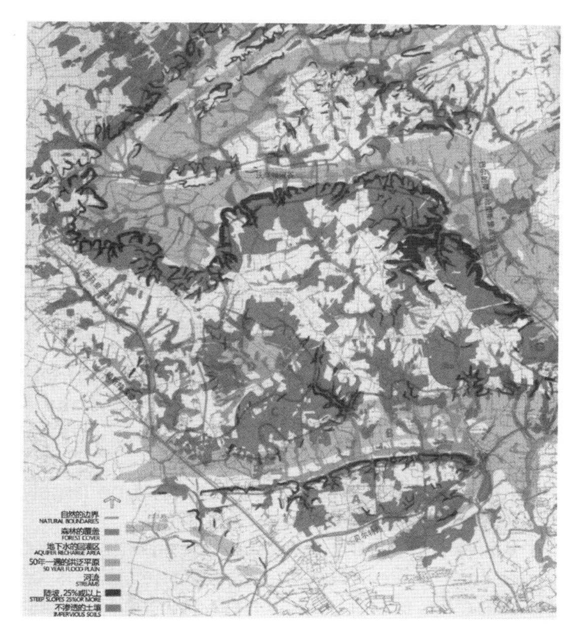

图1-3-11 自然地理特征

Fig.1-3-11 Physiographic Features

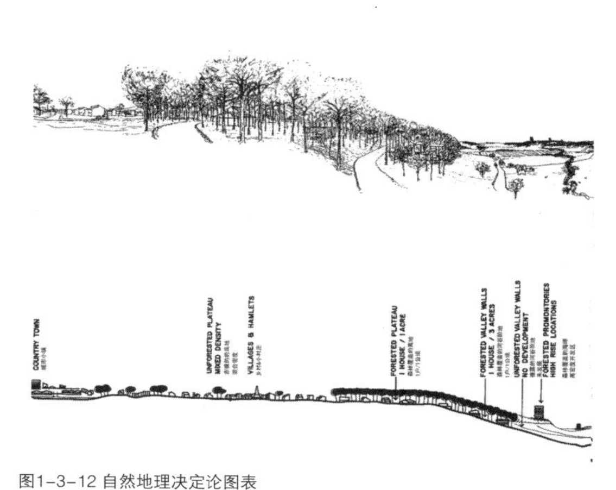

图1-3-12 自然地理决定论图表

Fig.1-3-12 Physiographic Determinants of Form

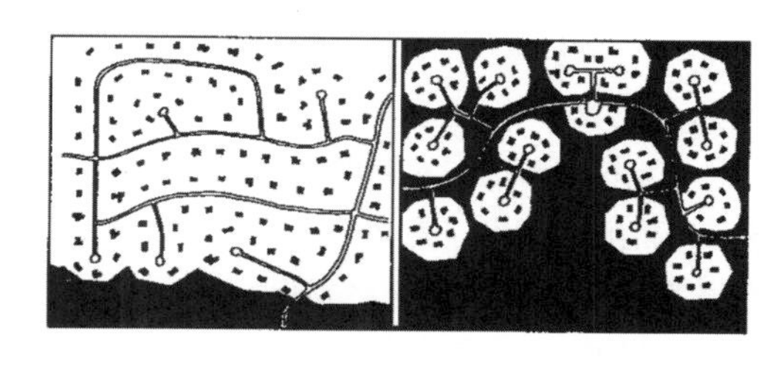

图1-3-13 集群化发展副本

Fig. 1-3-13 Cluster Development

来能够预测到的价值。总体价值预测不仅可以作为衡量其他价值的基准点，也可作为反对目前财产所有者、开发商的依据。

发展必须和区域目标相一致。这个研究区域可容纳的区域发展规模已明确了。区域和县的规划机构对无控制的发展已感到后悔，因此提议在研究区域的边缘建立四个大的集中开发地区，即派克斯维尔、赖斯特顿、陶森和赫里福德。

遵守保护的原则能避免破坏并保证提高环境质量。自然地理的保护和发展原则表明了，与各种各样自然地理特点相适应的建设类型和建设密度：如河谷地区应该禁止开发建设，将其保留起来，用作与现有的田园景色相协调的土地。其中包括：农业用地、大庄园、低密度建设、公共事业开放空间、公共和私人花园及游憩场所；所有科基斯威尔大理石含水层上应禁止建设；在50年一遇的洪泛平原上应禁止一切建设，留作农田、公共开放空间和游憩使用；目前的州卫生标准规定，禁止在不适于建化粪池的土壤上修建化粪池，应该严格执行这一规定。在允许施建的土壤上进行建设时，建设密度应该根据土壤的渗透性和对地下含水层的影响作出规定等。

开发原则如下：

Plains should be exempted from all development save agriculture, institutional open space and recreation; Current State Health Regulations prohibiting development on all soils unsuitable for septic tanks should be rigidly enforced. On other soils, density of development using septictanks should be regulated in relation to soil permeability and with reference to aquifers.

Development principles are as follows:

(1)Valleys without forest cover. Such lands should be prohibited to development and should be planted to forest cover when they are covered with the appropriate distribution of mixed hardwoods to an average height of 25feet they may be considered as below;

(2)Valley walls in forest cover. Valley walls in forest

（1）无森林覆盖的河谷台地。这种土地应禁止建设，应种植树林。当在这些土地上适当混植硬木，树木平均高度达到25英尺（约7.6米）时，则按以下内容考虑；

（2）森林覆盖的河谷阶地。这些台地，坡度在25%或更大的时候，只有在永久保持现有森林面貌的情况下，才可进行建设。建设的最大许可密度为每3英亩（约12 140平方米）建一户住宅；

（3）河谷台地和坡度25%或更大的坡地禁止建设而应植树覆盖；

（4）森林高地。在高地的森林和林地上，建设密度应该不大于每英亩（约4 047平方米）1户；

（5）岬地。在特定的种有林木的岬地，密度限制可以放宽，允许建造低密度的塔式公寓住宅；

（6）空旷的高地。建设应大量地集中在空旷的高地上。

这个地区能承受所有预期的发展而不受破坏性掠夺。人们已经目睹了宽阔的河谷及森林繁茂的河谷台地等景观，这些地区必须保持原状。相比之下，高地根据自身地理特征，尤其适合非破坏性开发。如果开发过程中智慧与艺术并存，河谷开发也会提高其景观价值。然而，一旦河谷免除开发的命运，建筑被限制只能建造在有森林的台地，结果势必会降低整个地块的建造价值。

社区可以分为城镇、村子、小村庄（图

cover, exclusive of slopes of 25% or greater, should be developed at such a density arid in such a manner to perpetuate their present wooded aspect. The maximum density permitted for development should be 1house per 3 acres.

(3)Walls and slopes of 25% or greater. Valley walls, and all slopes of 25% or greater should be prohibited to development and should be planted to forest cover;

(4)Wooded Plateau. Forest and woodland sites on the plateau should not be developed at densities in excess of 1 house per acre;

(5)Promontory sites. On specific promontories, in wooded locations, the density limitations can be waived to permit tower apartment buildings with low coverage;

(6)Open plateau. Development should be largely concentrated on the open Plateau.

The area can absorb all prospective growth without despoliation.We have seen that the genius of landscape lies in the broad valleys and their wooded slopes. These are the areas which must be sustained unchanged. In contrast, the Plateau is particularly suited, by its physiography, to absorb development without despoliation. If wisdom and art prevail, development may enhance this landscape. However, if the valleys are exempted from development, and building limited upon wooded slopes, we will have significantly reduced the total land area available to building.

Country Town, Villages and Hamlets. These communities are identified as Country Town, Villages and Hamlet (Fig. 1-3-14, Fig. 1-3-15). The Country Town is the major community with a population of 20,000 and is located at Falls Road and Broadway Road at the head of a large lake, a number of villages

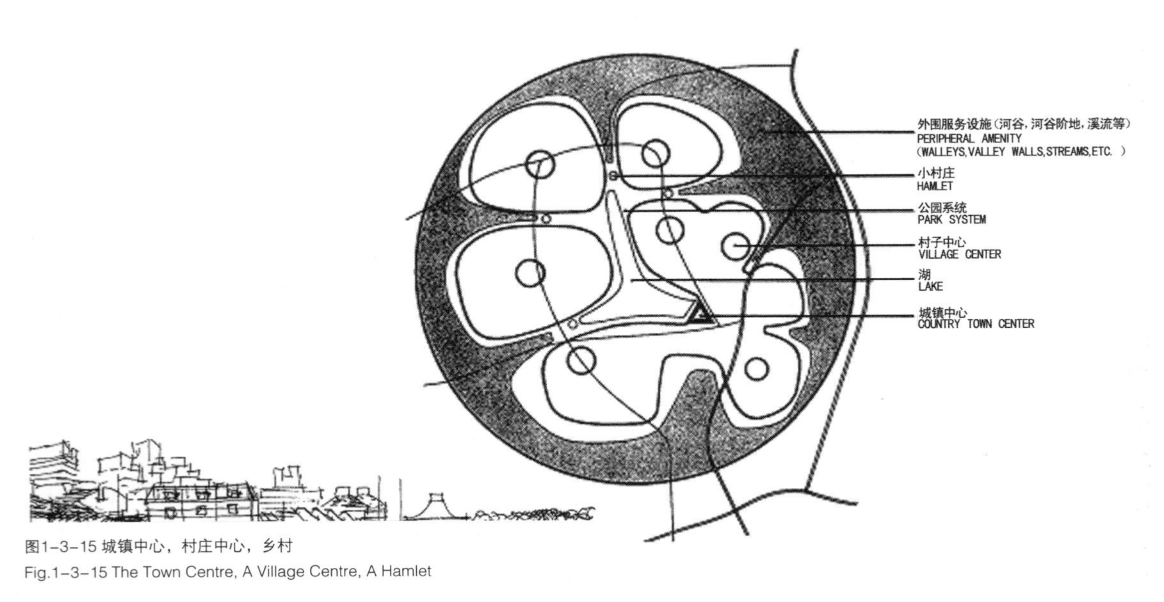

图1-3-15 城镇中心，村庄中心，乡村
Fig.1-3-15 The Town Centre, A Village Centre, A Hamlet

城镇中心
Town Center

村庄中心
A Village Centre

乡村
A Hamlet

图1-3-14 城镇及大村落概念图
Fig.1-3-14 County Town and Village Concept

1-3-14，图1-3-15）。位于大湖的前端福尔斯路与百老汇路的2万人口城镇，有6 000～10 000人的大村落，也有500人的小村庄。小村庄毗邻公园，位于各个大村落之间。城镇，村子，小村庄发展用地要求与高原的自然地理特征相对应，且以不破坏河谷为前提扩大景观面积。

有规划的发展较无规划的发展更富有吸引力，并可获得更多的盈利。有规划的发展与无控制的发展相比较，它的优越性几乎毋庸置疑。我们称赞的发展较好的城市、城镇、市郊，一般都与规划之初设想的一样。这种情况同样适用于该研究区域。这样，有规划的发展必然优于无控制的发展。

公共和私人的力量能联合起来，共同参加到实现规划的过程中去。检验规划方案的最终方法是：如果不能全部达到土地的最优利用，规划也应大规模达到最优利用。无论沃辛顿河谷或者巴尔的摩县的居民如何相信理想的规划可以达到预期目标，如果不能实现，则所有的分析与规划都是徒劳。因此在指导、控制和实施建设这一连续过程中，必须把积累各种力量视为重要的部分。

县规划政策。县规划委员会及规划区域办公室最重要的首次行动是检查河谷地区规划的基本原理，查看是否满足县及社区居民的要求，是否是正确的规划原理。

开放空间政策。大部分土地规划为“开放空间”之后用途是：农场、庄园、低密度使用机构、

of 6,000~10,000population and as many hamlets of 500 population are also proposed. Hamlets are located adjacent to parks, between villages. The location of development in Country Town, Village, and Hamlet, in response to the physiography of the Plateau, allows for an enhancement of the landscape without despoliation of the valleys.

Planned growth is more desirable and as profitable as unplanned growth.The superiority of planned versus uncontrolled growth seems scarcely worth argument. Those cities, towns and suburbs which we admire are generally as good as they are well planned. This would hold true for the study area. Certainly planned growth as proposed is superior to uncontrolled development.

Public and private powers can be joined in partnership to realize the plan.The final test of The Plan is whether the Optimum Land Use can be achieved in large part if not in entirety. No matter how ideal the residents of The Valleys or the County may believe the Plan to be, if it cannot be realized then all of the analysis and planning will have served no useful purse. Powers must be accumulated as part of a continuous process of guidance, control and implementation.

County planning policy. The most important first action of the County planning Board and Office of Planning and Zoning is to examine the basis of the Plan for The Valleys, and see where it does——or perhaps does not truly meet with the goals of both the County and the residents, and with sound planning principles.

Policy open space. Much of the land planned for"open space"will be in use: as farms, estates, institutions at low densities of uses, game preserves, golf clubs, and a wide variety of uses compatible with retaining the great open vistas of the valleys.A most important obligation of the area is the provision of

禁猎区、高尔夫球场等能与河谷原有开放空间相协调的土地利用方式。该地区最重要的职责是提供区域开放空间作为各种休闲游憩活动场地。

4）已达预期效果的规划

（1）绿春河谷：绿春河谷历史区域位于巴尔的摩县郊区，18、19世纪和20世纪早期历史建筑，以及19世纪晚期20世纪早期公园似的风景使其成为重要的河谷地区（图1-3-16）。这些建筑的主体都是沿着起伏的群山，森林高地以及浓荫夹道的宽阔大道而建，风格从美国较流行的18世纪新古典主义到20世纪30年代以前的英国乔治王朝时代以及其他众多朝代，建筑设计特征具有独特魅力。作为巴尔的摩县郊区富足的居住地区，在20世纪早期，绿春河谷典型的郊区发展，变成十分重要的历史区。

（2）沃辛顿河谷历史区域：沃辛顿河谷如此重要在于它延续了自身200多年的独特乡村景观（图1-3-17）。人们将这片土地划分成无数农场，有的面积甚至达到了200或者300公顷，目前的农场主一直努力按照前人的方式，经营这片美丽的土地，保留着以往的公共开放空间，重建或者恢复分散在各处的家园。在19世纪末期曾经有大批人迁离这片土

Regional open space for many kinds of recreational activities.

4)Planning of Achieved Expected Effects

(1)The Green Spring Valley. Significance: The Green Spring Valley Historic District is a suburban area of Baltimore that acquires significance from the collection of 18th, 19th, and early 20th century buildings and for its park-like setting that retains a late 19th-early 20th century atmosphere (Fig. 1-3-16). The buildings, primarily houses set among rolling hills, forested highlands, and tree-lined drives and roads, embody the distinctive design characteristics of the major architectural styles popular in the United States from the Neoclassical of the 1700s to the Georgian and other revivals of the pre-1930 period. As an affluent suburban residential region near Baltimore, the Green Spring Valley Historic District is also important historically for its association with typical patterns of suburban development in the early 20th century.

(2)Worthington Valley Historic District. Significance: The significance of the Worthington Valley lies in its unaltered, rural atmosphere which has not changed appreciably in over 200 years (Fig. 1-3-17). The land is divided into numerous farms, some up to 200 or 300 acres in extent. Many properties have been inherited by the present owners who are endeavoring to run them as in the past, retaining the open spaces and restoring and occupying the substantial homes which dot the countryside. It has not always been thus: following a period of settlement and rapid growth in the 18th and 19th centuries, many of the large family

图1-3-16 格林斯普林河谷历史区域

Fig.1-3-16 The Green Spring Valley Historic District

图1-3-17 沃辛顿河谷历史区域

Fig.1-3-17 Worthington Valley Historic District

地，20世纪三四十年代还是有这样的情况发生。而这一时期选择留下来居住的人们，主要认为这里的土地十分适合饲养马匹。现在饲养马匹与竞赛已经是盈利丰厚的大型贸易活动，当地已有16家注册的纯种马马场。最有名的或许就是萨加莫尔农场。自从1922年，斯诺希尔与沃辛顿农场举办的马里兰狩猎俱乐部越野障碍赛，就是享有国家及国际盛誉的独特运动项目之一。

（3）凯夫斯：沿着公园高地林荫道向南到高地，穿过沃辛顿小镇，就到达凯夫斯。凯夫斯湖长一英里（约1 609米），宽半英里，基本覆盖了整个河谷。高尔夫球场围绕整个湖水修建。房屋半隐在林中，而公寓建筑却点缀在周围的山脊中。

（4）高地：两大主要河谷都保留着历史乡村景观。而高地居住着许多和睦的居民，拥有村落、小村庄以及沿着湖水的城市小镇。河谷公园以及森林将他们彼此分开，使其拥有各自不同的秉性却共同享受河谷带给他们生活上的舒适便利。

2. 今天的河谷

1）特殊项目

（1）乡村道路设计标准研究。2008年10月，巴尔的摩县委员会采取乡村道路设计标准。新标准来自一份河谷规划委员会推荐的报告，现在是市政工程部设计指南。整个研究作为巴尔的摩县为了保护城乡结合处外部的乡村道路特质，而采用的一项标准。它借鉴了2010年巴尔的摩县总体规划中乡村道路发展标准。整个研究建议维持现存乡村道路规模、桥梁、山体等资源原貌，以便于不同的使用者欣赏不同的景色。

正如河谷规划委员会的董事长彼得·芬威克所说，“乡村道路是整个保护中关键部分。巴尔的摩县的居民为了土地保护，已经做出大量的投资，并放弃发展数以千亩的土地。我们愿意以一种最大限度维持乡村特质原貌的形式保护那些项目。宽直的快速道路已经太多见，这正是许多人想要颁布真正的乡村道路标准的原因”。

（2）溪流缓冲区。河谷规划委员会与土壤保持区、漂亮男孩流域同盟、斯帕克斯-格兰克社区规划

estates fell into disrepair and were sold to others. During the 1930s and 1940s, still more change took place. Those who came were people appreciative of the qualities of the soil, especially for raising horses. Horse breeding and racing is a very large and lucrative business in the valley. There are 16 registered thoroughbred horse farms in this district. Probably the best known is Sagamore Farm. Since 1922, Snow Hill and Worthington Farms have been the scene of the Maryland Hunt Club Steeplechase, one of the most famous and unique sporting events of national and international fame.

(3)The Caves. We follow Park Heights Avenue south up onto the Plateau, through the village of Worthington, and towards the Caves. Caves Lake is a mile long and half as wide, almost filling the bottom of the valley. A golf course borders the lake. Hidden among the trees are houses, while apartment buildings rise from the surrounding ridges.

(4)The Plateau. The two major valleys are unchanged. The Plateau contains a number of and friendly settlements, villages, hamlets and finally the country town by the lake. These communities are divided from each other by valley parks and forests, they sustain independent identities and yet all share the enclosing amenity of the great valleys.

2. Valley Today

1)Special Projects

(1)Rural road design standards study. In October 2008, the Baltimore County Council adopted Rural Road Design Standards. The new standards stemmed from a report and recommendations by the Valleys Planning Council and are now part of the Department of Public Works Design Manual. The study called for Baltimore County to adopt rural road standards aimed at preserving the rural character of county roads located outside the urban rural demarcation line (URDL). The study echoed recommendations in the county's Master Plan 2010 which called for the development of rural road standards in several sections. The study recommendations centered on maintaining existing dimensions of rural county roads and bridges as well as the hills and curves that make them scenic routes for a variety of users.

"Rural roads are a critical part of the conservation equation," stated VPC President Peter Fenwick. "Baltimore County residents have made a substantial investment in land conservation by surrendering development rights on thousands of acres of land. We would like to protect that investment by maintaining as much rural character as possible. Wider, straighter, faster roads are seen as counterproductive to those efforts and that is why many want to see rural road standards adopted

(2)Stream Buffer Participation. The VPC is working with the Soil Conservation District, Pretty boy

委员会以及其他社会团体合作，联合调查土地所有者不积极利用免费或者资助的缓冲区种植项目的原因。溪流缓冲区提供溪流系统急需的过滤及保护措施，鉴于此，政府及非盈利组织提供土地所有者不同种类的鼓励措施，支持他们沿着自己土地中贫瘠裸露的溪流栽植林木。为了更好地了解植被恢复力的状况，由农民、住宅业主组成的三个观察小组，合作得出相关反馈结论，在这个过程中，会雇佣咨询公司指导工作。由观察小组得到研究数据，并进一步以电话采访的形式核实准确度。

2) 文化项目

艺术保护、土地保护是河谷规划委员会保护巴尔的摩县文化遗产的重要手段。描绘巴尔的摩县美丽风光的艺术作品是该县尊重土地、提高民众合理使用土地意识以及保护土地巨大价值的另一种重要方法。“河谷中的艺术家”项目的目标是给当地艺术家提供更多近距离接近巴尔的摩县自然风光的机会。参加该项目的河谷规划委员会委员邀请各类艺术家在自己的土地中作画、速写及拍照。河谷规划委员会每年组织特殊展览活动，将邀请大约50位著名的艺术家参加。并称之为“为了土地的艺术”。

3) 地役权计划

“保护我们的河谷”是马里兰环境托管局与土地保护托管局创立的合作项目，河谷规划委员会因此增加了巴尔的摩县中的绿春河谷、沃辛顿及贝尔法斯特河谷永久性保护的土地面积。保护地役权是土地所有者与土地信托之间自愿、合法的协议，通过限制发展来保护特殊产权。土地托管局确保合约条款生效之前土地仍然归私人所有。在满足土地所有者财政及保护目标的要求时，性质与结构都是极其灵活的。为了保护共有的公共开放空间，严重削减公众财政以强调捐赠的土地的重要性。

4) 乡村遗产计划

马里兰乡村遗产计划是国家仅有几项计划中，通过保护土地役权方式，保留大量乡村土地作为永

Watershed Alliance, Sparks-Glencoe Community Planning Council and other groups to evaluate why land owners are reluctant to take advantage of free or subsidized buffer planting programs. Stream buffers provide much-needed filtering and protection for stream systems, and recognizing this, government agencies and non-profits offer incentives to get land owners to plant trees along naked streams on their properties. To better understand the resistance, a consulting firm (Opinion Works) was hired to conduct three focus groups made up of farmers and residential property owners to elicit feedback. The focus group research will be augmented by phone interviews.

2)Cultural Programs

Maintaining Baltimore County's cultural heritage through the arts and land preservation is central to the VPC's efforts. Works of art which depict the many scenic landscapes of Baltimore County is another way to honor these areas and raise awareness of land use and the enormous value of land preservation. The goal of the Artists in the Valleys program is to provide opportunities for local artists to gain access to scenic landscapes in Baltimore County. VPC members who participate in the program permit artists to visit their properties to paint, sketch or photograph. The VPC hosts a special event every other year, Art for Land's Sake. This is a well-attended event that features the work of approximately 50 artists.

3)Easement Programs

"Conserving Our Valleys" is a collaborative project initiated by the Maryland Environmental Trust, the Land Preservation Trust, and the VPC to increase the land base of permanently protected acreage in the Green Spring, Worthington, and Belfast Valleys of Baltimore County. A conservation easement is a voluntary, legal agreement between a landowner and a land trust which protects the special features of a property by restricting development. The land remains in private ownership while the land trust assures that the terms of the agreement are followed in perpetuity. It is flexible in nature and structured to meet a landowner's financial and preservation objectives. The severe cutbacks in public funding for land conservation programs underscore the importance of donated conservation easements to conserve our remaining open spaces.

4)The Rural Legacy Program

Maryland's Rural Legacy Program is among the few such programs in the nation that design a large blocks of rural lands for permanent protection through conservation easements. Baltimore County has five such designated areas-more than any other county in Maryland.

图1-3-18 沃辛顿河谷历史区域农业景观
Fig.1-3-18 Landscape of Farm in Worthington Valley Historic District

久性保留土地的计划之一。巴尔的摩县已经有五个指定区域——是马里兰州数目最多的县。

乡村遗产计划是保护农田与森林地区结合体，以及湿地和沿海区域等环境敏感地区。它的主要目标是保护那些可以为野生动物提供栖息地、保护环境及未来农业经济提供最高价值的大幅连续地块。

5）农场计划

农场是巴尔的摩县乡村景观的主要组成部分，也是生活方式的一部分。尽管农场的数量一直在稳定减少，至今仍然有750家。总面积达78 000公顷，平均面积104公顷。巴尔的摩县种植的农作物包括玉米、大豆、小麦、干草及大麦；饲养的牲畜包括牛、羊等。当地的农业收益性研究及实施规划中，明确分析了该县农场发展的优势、劣势、机遇与挑战，以及针对该县农业的总体发展与市场分析、训练与教育、公共政策与调控三方面提出各项建议（图1-3-18）。

三、总结

此案例位于马里兰州巴尔的摩县，是关于大都市市郊发展的方案。规划运用生态原则来抑制大都市的无序蔓延，保护市郊珍贵的自然资源。该地区在19世纪50年代已经历了早期的城市化，70平方英里（约181.3平方公里）的绿春河谷地区是受到城市化蔓延威胁的郊区，河谷开发的适宜程度需要准确测算和严格控制，才能确保这个美丽而又容易遭到破坏的地区实现可持续发展。

The Rural Legacy Program is designed to protect a combination of farm and forest lands, as well as environmentally sensitive lands such as wetlands and coastal areas. Its primary objective is to protect lands in large, contiguous blocks that will provide the highest value for wildlife and environmental protection as well as agricultural economic opportunity for the future.

5)Farming Programs

Farms are a major part of the landscape and way of life in rural Baltimore County. Number of farms has been steadily declining, there are still about 750 farms in the county, totaling over 78,000 acres, with an average size of 104 acres. Although the include corn, soybeans, wheat hay and barley. Livestock raised include cattle, sheep. Rural Baltimore County Agricultural Profitability Study and Action Plan made analysis of the strengths, weaknesses, opportunities and challenges of the development of the farming, in addition, put forward to the suggestion in three factors, that is AG Development and Marketing; Training and Education; Public Policy and Regulation(Fig. 1-3-18).

III. Summary

The case study is located in Baltimore County, Maryland, It is a case study of suburban growth in a metropolitan region. The plan is to apply ecological planning principles against the disorder of metropolitan growth, and protect valuable natural suburbs resources. In the late 1950s witnessed the virtually urbanization early in Baltimore County, 70 square miles Green Spring Valley is suburb that posed threat to by such development. Only have the suitability of Valley growth require accurate measurement and strict control, then can the beautiful and vulnerable area achieve sustainable development.

The plans For Valleys analysis a picture of the sequence of change that uncontrolled growth might cause in the valleys from 1963 to 2000, public and private powers both realized the gravity of the

规划分析了1963年至2000年期间无控制发展给这一地区带来的破坏程度，政府、公众都意识到问题的严重性，并对合理的保护与开发原则达成一致。规划中，执行乡村道路设计标准，设计溪流缓冲区，启动乡村遗产计划以及农场计划，保持郊区固有的优美和舒适，实现了生态适宜性规划的目标。

本案例中提出的控制与发展策略与我国“城市空间增长边界的划定”、“四区控制”（禁止建设区、限制建设区、适宜建设区、已经建设区）以及“非建设用地规划控制”有相通之处，目的都是为城市未来的潜在发展提供合理的疏导，将增长引向最适合开发的地区。对非建设用地进行控制是防止建设用地无序蔓延的一种有效途径，其控制不是阻止城市发展而是谋求更好的发展。

problem, and principled consensus on the the principle of reasonable protection and development. They adopted Rural Road Design Standards, designed Stream Buffer Participation, launched into its plans for The Rural Legacy Program and FARMING PROGRAMS, and keep the inherent beautiful and comfortable suburb and achieve the goals of the Suitability Method.

Control and development strategy proposed in the case study, has some common points with some strategy of China, such as,"Delimitation of Urban Growth Boundary ", "Four area control", ("Constructive expansion prohibited area", "construction limited area", "constructable area", "build-up area",)"Urban Non-construction Land Control", The purposes are all to provide reasonable guidance for the potential future development of the city, lead up to the most suitable development for the region. Non-construction area Control is an efficient way to prevent construction area from sprawling, it is not to stop city growth but seeking a better development.

思考题

1. 河谷地区的规划建议有哪些?
2. 河谷地区规划开发的原则?
3. 地方各个部门采取了哪些措施保护河谷?
4. 概括性叙述麦克哈格的适宜性评价方法在沃辛顿河谷规划中的应用。
5. 在我国景观规划中如何应用麦克哈格的适宜性评价方法?请简要叙述。

Questions

1. What is the Proposition of the Plans for the Valleys?
2. What are the Development Principles of the Plans for the Valleys?
3. What do the measures adopt by the local departments?
4. How does the suitability method operate in the Plans for the Valleys by McHarg briefly?
5. How does the suitability method operate by McHarg in the landscape Architecture of China? Please give a general overview of the subject.

注释/Note

图片来源/Image：本案例图片均引自IAN L. McHarg, Design with nature[M].New York:1992 ；Forster Ndubisi, Ecological Planning: A Historical and Comparative Synthesis [M]. Santa Fe, New Mexico, and Harrisonburg, Virginia:2002； Information on http://www.thevpc.org/; 伊恩•论诺克斯•麦克哈咯著，芮经纬译. 李哲审校，设计结合自然，天津:天津出版社，2008：97-113.

参考文献/Reference：IAN L. McHarg, Design with nature[M].New York:1992 ；Forster Ndubisi, Ecological Planning: A Historical and Comparative Synthesis [M]. Santa Fe, New Mexico, and Harrisonburg, Virginia:2002；Information on http://www.thevpc.org/; 伊恩•论诺克斯•麦克哈咯著，芮经纬译. 李哲审校，设计结合自然，天津:天津出版社，2008：97-113.

推荐网站/Website：http://worthingtonvalleylandscaping.com/

第四节 明尼阿波利斯滨河项目
Section IV Minneapolis Riverfront Project

一、 项目简介

明尼阿波利斯滨河项目为大家提供了这样的一个机会：思考公园系统如何保护明尼阿波利斯滨河区的国家生态遗产和密西西比河；如何保障社区健康与富足；如何带动明尼阿波利斯滨河区构建适应力强、多功能的可持续生态基础设施。由于能源价格不断攀升、气候变暖而导致冬季闭港期缩短等原因，水路航运的价值不断提升。

二、 总体规划

工业仍然是密西西比州的发展基础，工业生产促进了城市经济繁荣发展。然而城市与河流之间距离较远的现状限制了经济发展。“河流优先”的观点意味着将河流视为一个具有生命的生态系统，重现它的活力及多样性。在这个理念下，规划人员也把城市的自然、社会和经济三方面的健康发展视为优先考虑的对象。

“河流优先”这个项目的设计灵感来自河流动态特征。在河水冲击和侵蚀发生的地方，规划方案基于相同原理，通过生态阶梯的建立实现河水与社区的联系。方案将新的沉积产生地，改造成新的公

I. Project Introduction

The Minneapolis Riverfront project provides a chance to consider how a Park system can protect the national ecological heritage of the Minneapolis Riverfront and the Mississippi River, provide for community health and prosperity and prepare Minneapolis with resilient, multi-tasking and sustainable eco-infrastructure. River access may increase in value as energy costs rise and warming climate shortens winter port closures.

II. The Master Plan

The Mississippi is still a working place for industry, and the city needs to work in order to prosper. But the current distance between the river and the city is not a component of prosperity. When we say “put the river first”, we mean to renew its strength and diversity as a living ecosystem. In doing so, we also put the health of the city first--physical, societal and economic.

The RIVERFIRST design is guided by the dynamics of the river. Where water carves and erodes, we employ the same principles to create ecological stairways between water and communities. Where it deposits new material, we mold this into new Park lands. RIVERFIRST means renewing and using the natural land topography to solve problems. Topographical design in section reconnects Northside’s historic Farview Park with the River and new skilled jobs in the River City Innovation District. The Northside Perkins Hill land form enables us to “discover” available land in section for clean industry and a future Green Port. We create moments of vertical separation between Park Trail

园用地。“河流优先”的内涵就是场地问题的解决依托于自然地形的重建及利用。按照区间进行的地形设计，实现了河流北岸历史悠久的远景公园、明尼阿波利斯河、城市创新区新技术工作三者间的联系。河流北岸的帕金斯山的地形让规划人员找到了一块适于建设清洁产业及未来绿色港口的理想场地。在公园小道与驳船地点之间，创建了许多垂直分隔空间，以利于公园、航运及工业的共同发展。市政雨水治理系统设计及公共公园创建均是基于河岸地形进行的，它们缓解了21世纪时期城市的洪涝灾害并提供了生态过滤等生态服务。

“河流优先”认为应该建立并强化现有基础设施、社区及生态资产之间的协同合作。结之桥实现了公园北部和东北部道路的连接，同时也把东北艺术区的创新力量与城市河流创新区及中心区连接起来。美国地质探测局沿结之桥布置的实时河水监测系统，将结合照明设备向公众展示监测结果。以桥墩锚定的佰黑文群岛为迁徙鸟类及濒危野生生物提供了超过8公顷的河滨保护区。河流通讯及太阳能动力的公园无线网络将向公众提供密苏里大学罗拉分校关于本地及国内生态的公众教育，吸引世界一流的机构、公司及组织机构参与到“河流优先”的提议中来（图1–4–1～图4–4–4）。

and barge operations, allowing Park, navigation, and industry to coexist. Riverbank topography organizes the remediation design of municipal storm water and creates a public Park which provides Minneapolis with flood mitigation and bio-filtration eco-services for the 21st century.

RIVERFIRST initiatives create and enhance synergies among existing infrastructures, community and ecological assets. Knot Bridges tie together North and North East Park trails and link the vibrant creative energy of the NE Arts District with the River City Innovation District and Downtown. Real time stream water monitoring from the USGS website is made public with smart illumination along Knot Bridges. Biohaven Islands anchored to bridge piers provide more than 8 acres of protected riparian habitat for migrating birds and endangered wildlife. The River Talk iPhone app and solar powered Park WiFi network provide unprecedented opportunities for local and national public education about the ecology of the UMR, attracting world class institutional, corporate, and organizational partners to the RIVERFIRST Park initiative (Fig.1-4-1~Fig.1-4-4).

III. Major Areas Planning and Design

1. River Bank North

The RIVERFIRST Park system begins with a multidimensional idea of public access. This includes physical access to the river, access to education and information about river ecology and access to jobs, and the full spectrum of tangible community benefits that flow from the 21st century green economy along

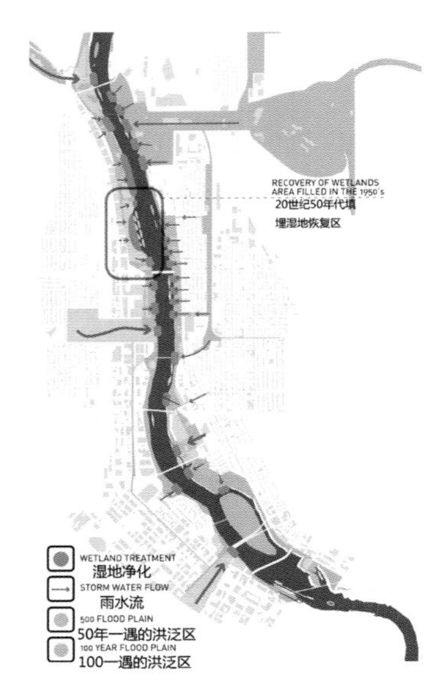

图1–4–1 水体
Fig1–4–1 Water

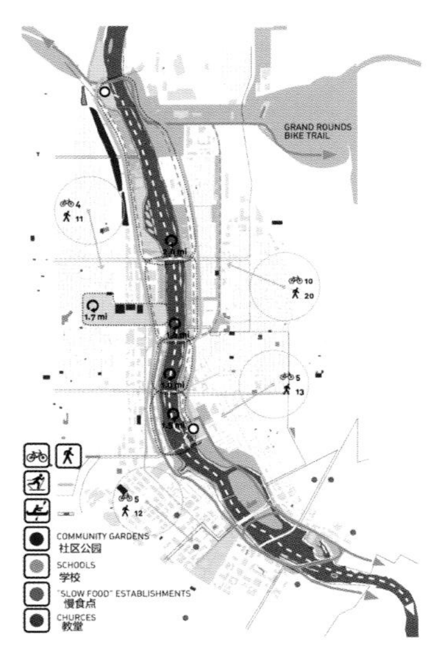

图1–4–2 健康
Fig.1–4–2 Health

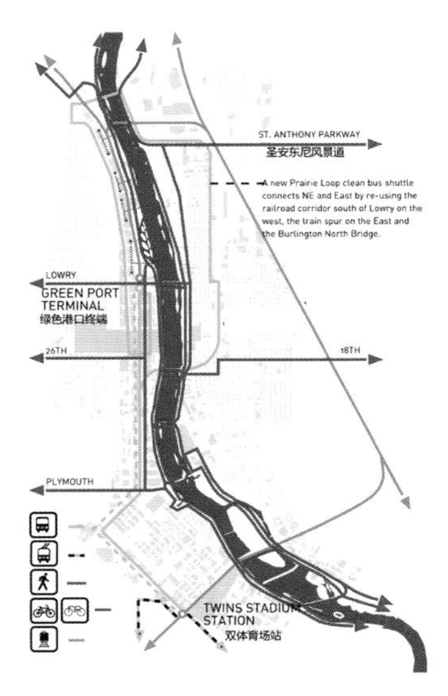

图1–4–3 交通
Fig.1–4–3 Mobility

三、主要区域规划设计

1. 河流北岸

“河流优先”公园系统源于公众可达性的多方面思考，包括河流自然环境可达性、公众获取河流生态教育及信息的便利性、工作地点的可达性以及21世纪沿河流域绿色经济所带来的多方面社区利益。不连续的市属土地及工业用地曾经限制了场地的可达性。北部河岸景观设计源于以下想法：将航运业的废弃物视为河流原始材料，用这些材料建设新公园。由于河流挖掘出的淤积物目前常常被堆放在沿河地带或者被贱卖作建筑填充物，因而规划人员建议赋予这些淤积物对城市友好的新用途。这些河流淤积物将成为公园道路及公共滨河道路的基本建造材料。为公众提供亲近河流的机会，使河流能够与工业和谐共存，这是规划设计的第一步，具有成本低、可行性大、效果显著的特点。河流淤积物被塑造成一条护堤形式的连续步行及自行车道，这条道路凌空穿过现有驳船停靠点及运营区。根据这种想法，在河流北岸建造了一条波动起伏、安全可靠的公共公园小道，同时将对现有工业场地、设备以及运作的影响降至最低。

同样的地形塑造方法被用于建设一条联邦政府资助的标准高速公路天桥建设，实现与北部社区的联系。规划设想是将城市历史制高点——远景公园与新建的沿河道路系统连接起来，形成一片凌空跨越94号州际公路的绿色“平原”，在不必拆除现有建筑的前提下，为市民、动物、雨水提供到达河流的通道（图1–4–5）。

2. 远景公园

远景陆桥是城市河流创新区的核心部分，它带动了明尼阿波利斯在创新材料、医学、自然科学及智能技术等领域的研究。四层办公楼及轻工业分区成为公园重要的税源。更新的工业建筑近期可以改造成工作室、居住房。未来的教育和职业培训集群及住宿将沿劳里桥分布（图1–4–6）。

远景公园是城市历史制高点，公园将跨越94号州际公路，将明尼阿波利斯北部区域与河流重新连

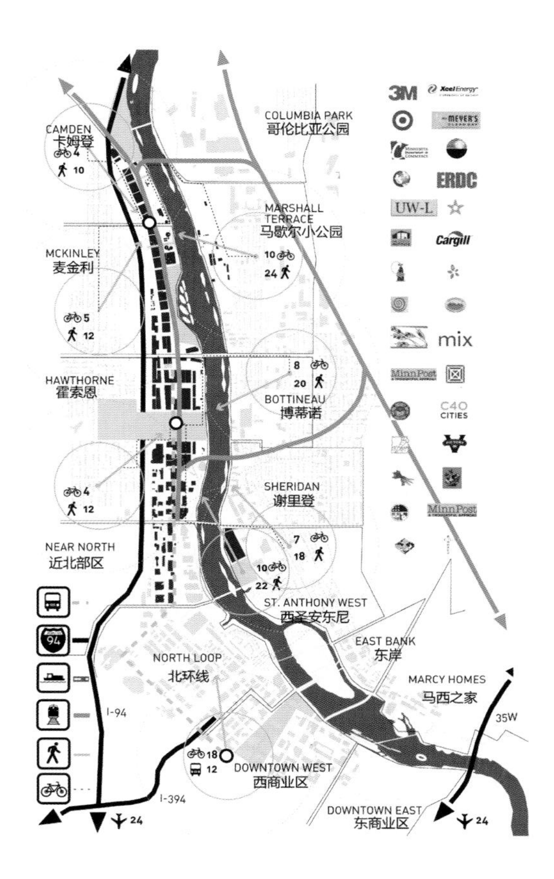

图1–4–4 绿色经济

Fig.1–4–4 Green Economy

the river. Physical access has been limited due to non-continuous parcels of city-owned land and industry operations. The Riverbank North landscape design springs from seeing navigation's 'waste' products as raw river material with which to build the new Park. We propose a new beneficial and civic use of dredged sediments, which are currently stockpiled on riverfront land and/or sold at very low cost for construction infill. The river sediments become the molded materials from which the RIVERFIRST park and public riverfront trails are made. The first step is affordable, practical and impactful: provide public access to the River which can coexist with the existing industries. River sediment is molded in berms enabling a continuous pedestrian and bike trail to rise and bridge over the locations of existing barge terminals and operations. This creates an undulating and safe public Park trail along the Riverbank North which can be built today with minimal impacts on industrial property, equipment, and operations.

The same process of molding land berms is used with a standard federally funded highway overpass

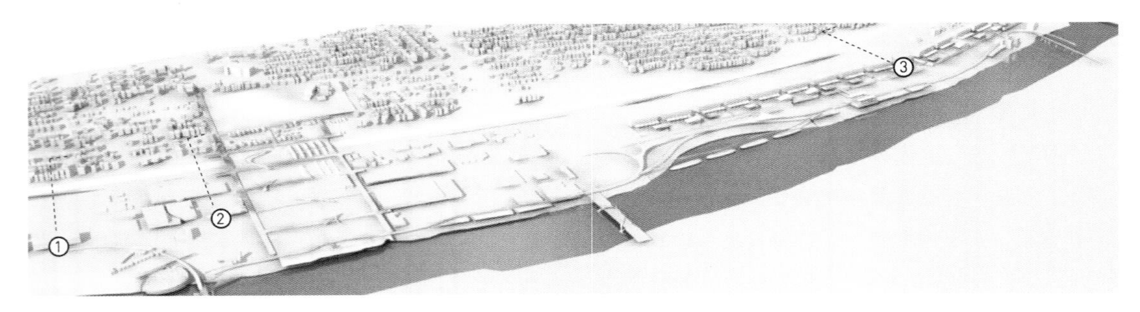

图1-4-5 河流北岸模型
Fig.1-4-5 Model of River Bank North

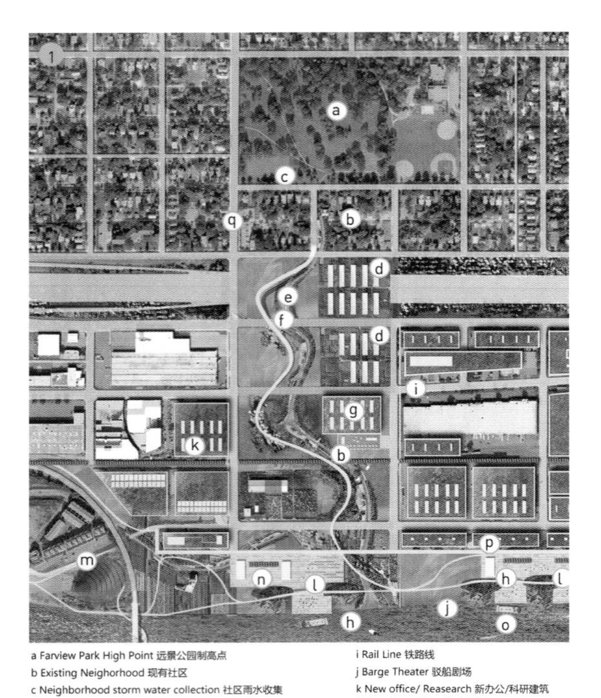

a Farview Park High Point 远景公园制高点
b Existing Neighorhood 现有社区
c Neighborhood storm water collection 社区雨水收集
d Urban Farm and Greenhouses 城市农场及温室
e Daylighted Stream and Landform Park Daylighted 溪流及地貌公园
f Pedestrain Bridge步行桥
g Farmer's Market 农夫市场
h Barge Restaurant 驳船餐馆
i Rail Line 铁路线
j Barge Theater 驳船剧场
k New office/ Reasearch 新办公/科研建筑
l Waterfront Plaza 滨水广场
m Riverfront Amphitheater 滨河圆形剧场
n Park Pavillion 公园展览亭
o Barge Restaurant驳船餐馆
p Nice Ride Bike Station 莱斯莱德自行车站点
q 26th Street Bike Path 第26大街自行车道

图1-4-6 远景公园总体规划图
Fig.1-4-6 The Master Plan of Farview Park

接起来，提供了一条到达城市河流创新区中心区的通道。一座新建的高速路桥包括了社区形式的城市农场及桥下供给植物并吸收二氧化碳的拱形温室。一系列公共公园空间通过拥有步行天桥以及护堤形式的地形，将降低季节性的洪峰流量。

通过与北部社区管理人员及组织的合作，“河流优先”公园农场提供了一种新型的、适于本地的、可

construction to connect to the North neighbourhoods. Our vision is to reconnect the City's historic high point, Farview Park, with the new Riverfront trail system, extending a great plain of green that bridges over I-94, and providing a direct land access to the River for people, animals, and storm water, without demolition of any existing homes (Fig.1-4-5).

2. Farview Park

The Bridge is the centerpiece for the new River City Innovation District, which draws on Minneapolis expertise in innovative materials, medical, scientific and smart technologies. A zone of four story office and light industrial buildings offset the Parkland tax base. Renovated industrial buildings are adapted for work/live housing near term. A future education and job training cluster and hotel will be developed along the Lowry Bridge (Fig.1-4-6).

Farview Park marks the historic high land point in the City. This park will now extend across I-94 to reconnect North Minneapolis to the river, providing access to the center of the River City Innovation district. A new highway bridge includes a community-based urban farm and hoop houses with CO_2 harvesting for plants from the highway below. A seasonal storm water stream will descend through a series of public park "rooms" with pedestrian bridges and land berms.

In cooperation with North side community leaders and organizations, RIVERFIRST Park farming offers a new model for local, sustainable urban food production that celebrates diverse cultures and culinary traditions while building local communities. As more people and businesses in the City adopt local Park-grown foods, the more they may question mega-scaled mono agriculture policies which must rely on chemicals and fertilizers that inevitably leek into the river.

The River City district will attract corporations such as Coloplast who can recruit employees who value a life style with proximity to Park trails, locally grown and healthy food, and a sustainable commute by bike or shuttle bus. Corporations create a job base that

持续的城市农业生产，这种新模式在社区建设的过程中促进了多样社区文化的发展，保存了当地的饮食文化传统。随着越来越多的市民和企业接受本地公园生产的农产品，他们也就越来越质疑依赖化学药剂及肥料的大规模的单一农业政策，因为这些化学物质最终会不可避免地流入河流。城市河流区还将会吸引诸如康乐宝之类的企业落户，他们的员工享受着这样一种生活方式：临居于公园道路，食用当地产的健康食物，骑自行车或者乘坐班车等可持续交通方式上下班。企业也提供了工作岗位，包括企业员工及培训生、诊所、食品加工、餐饮、银行、健身中心、制造业、3D打印等先进制造领域的小企业。“河流优先”策略还包括一个会议中心，用于员工的继续教育、高等教育及职业培训，它们为员工提供了全球化经济下所需要的技能。

3. 湿地

受全球气候变化影响，密西西比河及明尼阿波利斯市气候变化不定。根据国际气候变化委员会2007年的模拟气候平均数据表明，明尼苏达州到2050年气温将上升8℉（13.3℃）。气温升高将加大夏季河水蒸发量，容易出现旱灾或者水位下降的现象。河流低水位时期，河水停留时间延长，细菌爆发的可能性加大，将对明尼阿波利斯河的水质及市民健康产生有害影响，河流水生生物及野生生物也会受到影响。蒸发到大气中的水分也会增大夏季风暴产生的机率；这些风暴会造成巨大破坏，诸如降雨量骤增、洪灾爆发，同时也出现河流泥土干涸、吸水能力弱的现象。

这些2050年将出现的气候现象表明城市管理者及普通市民需要重新思考城市的基础优先事项并做出相应调整，同时制定城市应对气候变化的战略。明尼阿波利斯市及城市公园局拥有极大的机遇来建立一个重要的城市生态基础设施，它为市民提供了游憩、社区活动及亲近自然的场地。“河流优先”指导原则是发展规模经济：多经济部门合作投资每一个可能的基础设施建设；将分散方案整合成整体的、综合性的东西河岸系统；基于三重底线经济原理构建替代方案，这些替代方法更大程度上基于统一战略，促进城市全方位发展并提供生态服务（图1-4-7～图1-4-9）。

includes corporate employees and trainees, clinics, food, dining and local catering services, banks, health clinics, fabricators, and small businesses associated with rapid prototyping and advanced manufacturing. The RIVERFIRST strategy includes a conference center for continuing and higher education and job training to develop the skills needed by workers and employers in a dynamic global economy.

3. Wetlands

Climate change impacts on the river and Minneapolis are characterized by extreme uncertainty and variability. Averaged climate projection data from the International Panel on Climate Change 2007 simulation models suggest that 2050 temperatures in Minnesota could increase by as much as 8˚F. Hotter weather will create extensive river water evaporation in summer causing droughts and/or low water events. As water residence times increase in low flow events, there is a greater risk for bacteria blooms, impacting Minneapolis water quality and human health, as well as river biology and wildlife. Evaporated water in the atmosphere also creates the chance for many more damaging summer storms, with heavy precipitation and floods as river soils dry and loose capacity to absorb moisture.

These 2050 climate scenarios point to the need for City leaders and citizens to rethink fundamental priorities and make changes, while planning for municipal climate change adaption strategies. The City of Minneapolis and the Parks Board have the opportunity to create a significant municipal eco-infrastructure that is also a place for leisure, community activities, and close daily contacts with nature. RIVERFIRST guiding principles are to develop economies of scope: where every possible infrastructure investment works across multiple lines of business; nest discrete decentralized solutions into larger synthetic Western and eastern riverbank systems and; use triple-bottom-line economics to compare and generate alternatives - where alternatives are more broadly generated based on an integrated strategy to deliver the full range of city and eco-services (Fig.1-4-7～Fig.1-4-9).

4. Sustainable Infrastructure: Green Port

A 21st century Park offers Minneapolis a chance to step back and reassess how the city's recycling, heavy industry and port businesses are conducted, what tools and infrastructure are used, and how green house gas emissions can be reduced. If smokestack industries "go away" or are "relocated" somewhere else it only shifts the site of the problem since we "walk on water" in Minneapolis as aquifers, groundwater, storm water and river are inevitably linked (Fig.1-4-10).

RIVERFIRST proposes to use the Park as a catalyst

图1-4-7 湿地规划图

Fig.1-4-7 Wetlands Plan

图1-4-8 湿地剖面图

Fig.1-4-8 Section Through The Wetlands

图1-4-9 堰州岛效果图

Fig.1-4-9 View of Barrier Islands

图1-4-10 绿色港口规划图

Fig.1-4-10 Green Port Plan

4. 可持续基础设施：绿色港口

作为一座21世纪的公园，它促使明尼阿波利斯市重新审视和评估城市循环回收、重工业和港口业务的运作情况，某些设备及基础设施使用情况，以及温室气体减排现状。如果排放废气的企业搬离城市或者重新选址，这些做法仅仅转移了问题发生的地点，而并未切实解决问题。由于市民要“穿行”于明尼阿波利斯的水环境中，因而蓄水层、地下水、雨水及河流必然相互连通（图1-4-10）。

“河流优先”提出将公园作为城市催化剂，促进城市由重工业城市向21世纪绿色经济转变。项目提出了绿色港口概念，它将逐步成为城市绿色经济门户，为明尼苏达州提供可再生风能，为船舶提供清洁岸电（当轮船停泊时通过大型电缆与岸边的发电网连接以获取电力）、为火车和工业等提供清

to accelerate the transition from smokestack industry to a 21st century green economy. We imagine a green port, which evolves to a green economy portal connecting the city with Minnesota wind renewable energy, providing clean on shore power OSP (cold ironing) for barges, trains and industry and reducing future energy cost risks, noise and emissions. By engaging the land in section, our proposal meets the current Port space needs in half the footprint, densifying operations and increasing efficiencies. Five green economy corporate headquarters would increase port job density from less than 1 job per acre today to over 1,000 skilled jobs per acre.

The plan provides Minneapolis with a Green Port that is also a working Parkland with needed civic scaled storm water remediation ponds, 2 acres of permeable ground surface that provides future climate change floodplain carrying capacity, and clusters of diverse floodplain forest. RIVERFIRST begins by providing public access to the river, initiating a comprehensive Park trails system and returning the river shore to the

洁能源，降低未来能源消耗成本、噪音及气体排放。通过土地分区处理，规划建议满足了当今港口在碳足迹减半、运作频度增大、效率提高等方面的要求。五个绿色经济企业总部将增加港口工作岗位数量，每英亩低于一个工作岗位增加到每英亩超过1000个熟练工作岗位。

规划所设计的绿色港口还是一片公园，它包含了一个大型雨水处理池、近两公顷的可渗透地表以及数处泄洪平原森林。渗透地表增加了洪水承载力，这也是应对气候变化的一项措施。“河流优先”项目开始于为公众提供河流可达性，提供一个全面的公园道路系统，河岸公共化以及市民成为变化的催化剂。

5. 佰黑文岛栖息地

密西西比河总长2 320英里(约3 733公里)，其中60%的区域是北美鸟类重要的迁徙通道。佰黑文保护网将恢复对候鸟极为重要的河滨栖息地，为它们提供筑巢、活动场地，为濒危水生及陆生动植物提供栖息地。这些生物包括本土贻贝、布兰丁龟、鱼鹰、伯劳鸟和卡纳蓝蝴蝶。佰黑文岛将建立8公顷多的恢复栖息地，同时也会强化泛舟的娱乐性及探险性（图1-4-11）。刚性结构物不仅成本昂贵、易破裂而且耗能高，因而基于水的浮力及弹性软质工程原则，佰黑文群岛创造了一个由坚固耐用材料、轻质材料及100%再循环PET材料组成的新型浮岛系统。岛屿表层覆盖了土壤，为植物、灌木丛及小树木提供持久的营养媒介。佰黑文岛还可以提供野生稻谷及本土的莓果，这些农业生产可以交付给当地机构组织和企业管理。现有码头用于固定浮岛，这些浮岛栖息地的治理及恢复与ACE目标相一致。群岛形态设计顺应了河流的走向，其边缘由太阳能船用水泵保护，以避免冬天的冰层碰撞。下游浮岛由弹性绳索固定，浮岛一旦受到船舶的撞击就会滑向一侧。桥上铰链将会转移下游水流及风所形成的负荷。锚提供了浮岛侧面的稳定性。在水位变化期间，所有连接在一定程度上提供了安全支持（图1-4-12，图1-4-13）。

图1-4-11 佰黑文岛规划模型
Fig.1-4-11 Model of Biohaven Islands Plan

public, allowing citizens to become the catalyst for change.

5. Biohaven Island Habitats

At 2,320 miles, the Mississippi River is a vital migration corridor for 60 percent of North America's bird species. A protected network of floating Biohavens will restore riparian in-river habitat essential for migratory birds, provide nesting and staging areas and support endangered aquatic and land animals and plants. These include native mussels, blanding's turtle, osprey, loggerhead strike and karner blue butterfly. The Biohaven islands will create over 8 acres of restored habitat and enhance the enjoyment and exploration of the River by kayak (Fig.1-4-11). Instead of costly, disruptive hard structures with high embodied energy construction, Biohavens are based on water buoyancy and resilient soft engineering principles that utilize an innovative flotation system made of robust, lightweight geo-textiles and 100% re-cycled PET materials. Soil placed on top of the island provides an enduring nutrient medium for plants and shrubs, and small trees. Biohaven Islands can also support harvests of wild rice and native berries, and can be 'adopted' and supported by local organizations or corporations. Existing bridge piers are used to tether the floating islands and their habitat remediation and restoration mission is consistent with ACE objectives. The shape of the islands is designed to accommodate

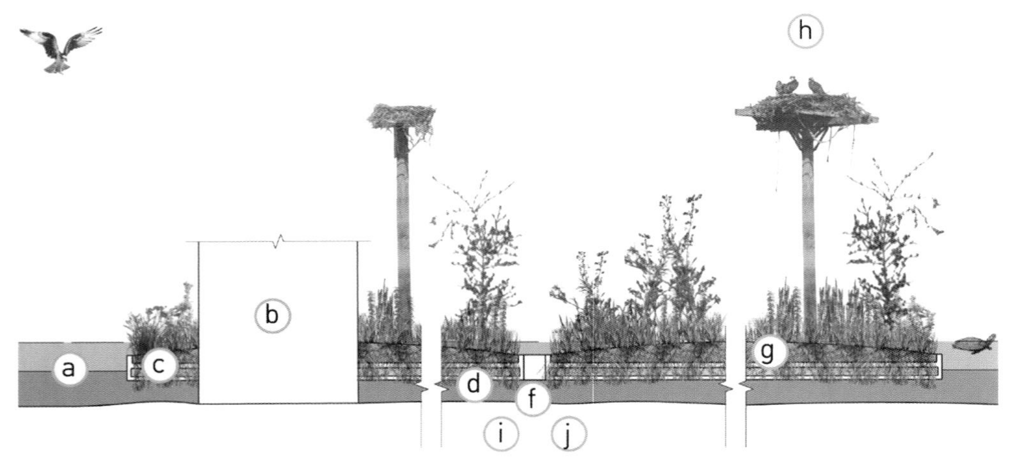

a Average Water Level 平均水平面　b Bridge Support 桥墩　c Float (styrene foam) 浮板（泡沫聚苯乙烯）
d Reclaimed Plastics,PVC PVC再生塑料　e Lightweight Soil Mix 轻重土壤混合物　f Osprey Nest Habitat 鱼鹰巢聚集地
g Fixed Island 固定的岛屿　f Floating Island 浮岛

图1-4-12 佰黑文岛剖面图
Fig.1-4-12 Section Through of Biohaven Island

图1-4-13 佰黑文岛效果图
Fig.1-4-13 View of Biohaven Island

6. 东北峭壁

“河流优先”项目增加了公园所有绿化区域的生物多样性：高地森林、河滨树林、湿地及草地。陡峭的河滨林地延伸至可以俯瞰下游河流的峭壁处。河岸由巨石及大型废弃木材等加固，而这些材料是在河岸地势突兀之处营造新栖息地必不可少的条件。现存的柳树及黄槿树等植被加固了悬崖斜坡。规划方案在东北部峭壁区制定了特定保护方案，以吸引那些濒临灭绝的动物，这些动物包括本土贻贝、布兰丁龟、鱼鹰、游隼、 越冬鹰类、伯劳鸟和卡纳蓝蝴蝶（图1-4-14）。“河流优先”公园在东北区域的设计顺应了河流冲蚀石灰石峭壁的现象。当河流遇到向外弯曲的河岸线时，河水便会不断地侵蚀河岸。河流对土

a river placement and the edges of the islands are protected with solar powered marine pumps to avoid ice stacking in winter. A downstream island is anchored on springing ropes allowing it to simply move aside if bumped by barges. A hinge connection at the bridge bearing acts as a load transfer for downstream running water and wind. Side stabilization is provided by anchors. All connections would be constructed in a way to allow a safe support during moving water levels (Fig.1-4-12, Fig.1-4-13).

6. Northeast Bluffs

The RIVERFIRST proposal increases biodiversity in all of the Park's vegetated zones: upland forest, riparian forest, wetlands and prairies. Steep riparian forest areas move from the bluff overlook down to the river. The banks are stabilized with boulders and large woody debris which are essential to create habitat components at the toe. Established vegetation of willows and cottonwoods stabilize the bluff slopes. The North east bluffs support a targeted approach to conservation biology to attract animals that are endangered or at risk. These include native mussels, Blanding's turtle, osprey, peregrine falcon, wintering eagles, loggerhead shrike and the Karner blue butterfly (Fig.1-4-14). The RIVERFIRST park design for the North East side follows principals of carving produced by the action of river water on the northeast side limestone bluffs. When the river current meets a shoreline with an "outside curve", water erodes the bank material over time. It continually carves into the land and creates steep slopes or vertical faces. The North

地的不断冲刷形成了陡峭的斜坡或者竖直的立面。河流东北部河岸由抗性材料组成，这段河岸包含有断裂的砂岩及密实的冰碛层。通过规划人员对这些耐腐蚀材料的雕琢，此处形成峡谷和梯田景观。峡谷的存在使得雨水可以在此停留而不是直接排入河中。峡谷中的生态阶梯水渠不仅为生物提供了栖息地，也为公众提供了到达河边的通道。经过“雕琢”的土地将会就地再次利用，营造出高起的梯田景观，同时人们可以在此处近距离观察树梢上越冬的老鹰及河流远景（图1-4-15，图1-4-16）。

7. 谢勒公园

沿着西南朝向的曲线河岸，新谢勒公园复原了霍尔斯岛，为人们提供了一处可以远眺市区优美天际线的河湾。谢勒公园让公众再次意识到一个核心观点：密西西比河是一个充满活力、动态的生态系统，它拥有自己的独特景观，如沙丘和浅水塘的形态随着积雪融化、泥沙淤积以及河水流量的不同而产生变化。这是一个人们可以踏水嬉戏的地方，任何力量都无法勾勒出如此生机勃勃的景观，因为它

east River bank is composed of resistant material – fractured sandstone and tightly packed glacial till. We carve into this resistant material to create ravines and terraces. The ravines allow city storm water to be remediated instead of falling directly into the river. Stepped ravine ecostairs channel water, provide habitat and access to the public to the river's edge. Carved land material is reused on site to form raised terraces with intimate views of treetop habitats, wintering eagles, and broad overlooks to the river (Fig.1-4-15, Fig.1-4-16).

7. Scherer Park

Located along the only river bend shore with south/west exposure, the new Scherer Park restores Hall's Island and offers a river beach cove with a an amazing skyline view of Downtown. Scherer Park re-introduces the public to the core idea of the river as a living, dynamic eco-system which produces its own landscape of sand bars and shallow pools that shift shape according to winter melts, patterns of sediment deposition and river flows. This is a place where people can get their feet wet, a living landscape which can't be "mapped" because it is different each day (Fig.1-4-17).

As a demonstration site for the new Park system, market rate housing and artist work/live studios define the edges of the park, enjoy spectacular views up and down river, and overlook Scherer Park

图1-4-15 峭壁区剖面图
Fig.1-4-15 Section Through Bluff

图1-4-14 东北峭壁规划模型
Fig.1-4-14 Model of Northeast Bluffs Plan

图1-4-16 峭壁区效果图
Fig.1-4-16 View of Bluffs

图1-4-17 谢勒公园鸟瞰
Fig.1-4-17 Aerial View of Scherer Park

图1-4-18 谢勒公园皮艇区
Fig.1-4-18 Scherer Park Kayak District

每时每刻都在变化（图1-4-17）。

作为新公园系统的示范点，商品住宅及艺术家生活加工作室界定了公园边界。在这里人们可以欣赏到河流各段的秀美景观，也可俯瞰谢勒公园——保证了公园廊道的人气、安全及活力，同时税收收入也比当初增加了6倍。西布力大街上的艺术中心将为工薪族、青年人及退休人员提供接触艺术、培养自己特长的机会。职业艺术家也可以在这里相互交流、学习甚至展览自己的作品。这也使得公园系统将东北艺术区的创新力量与河流北部城市创新区的创新能力联结起来。

木质步道曲折蜿蜒，向游客充分展示了滨河景致，形成了一个充满乐趣的大型城市“甲板椅”，人们在这里可以晒太阳和户外活动。公园维护费主要来源于：当地餐馆租赁，这些餐馆的食材来自北部公园所生产的有机食品及当地食品；霍尔斯岛上的一处游泳、溜冰场地；一个向公众开放的游艇餐厅。新公园和水生景观区是中心出发点，道路系统将引导游客通过泛舟、步行、骑车或者滑雪等方式体验“河流优先”公园。明尼阿波利斯的居民，还有什么地方可以让你在清晨上班之前，泛舟于密西西比河之上享受那宁静的美景（图1-4-18，图1-4-19）？

8. 明尼阿波利斯市中心

这块开放空间成为人们热议的绿色“城市客厅”，而这也是目前明尼阿波利斯市中心一直缺少的场所。公园创造了水景和视觉中心，它形成了一条通向市中心的绿色通道，沿着亨内平大街将市中心和河流联系起来。可以获得的开放空间及路面停车场可以改建成林地区域的地表排水道。公园中设

as watchful neighbors — bringing people, safety and activity to the park corridor and generating 6 times more in property taxes than the current facilities. A community art center on Sibley Street will provide an opportunity for working adults, youth and retirees to access studio space and develop their skills. Working artists from the area will gain an opportunity to network, teach, and exhibit at the Center, allowing the Park system to link the creative energy of the NE arts district with the innovation capacity of the Northside River City innovation district.

A public promenade of timber strips lifts and dips to reveal riparian landscape, creating playful large urban “deck chairs” for sunning and working outdoors. Park maintenance revenue is generated through local restaurant tenants making use of Northside Park-grown organic produce

and locally-sourced food, a swim/skate facility on Hall’s Island and a public kayak launch. This new park and aquatic landscape space is the central launching point and trail head for all visitors who wish to explore the RIVERFIRST trails by kayak, foot, bike or on skis. Minneapolis residents: Where else can you kayak on the Mississippi in the still of the morning before you go to work (Fig.1-4-18, Fig.1-4-19)?

8. Downtown Minneapolis

This open space creates the much-discussed green “urban living room” that has been missing from downtown Minneapolis. The park creates the water source and visual center piece of a great forested gateway to the downtown connecting to the river along Hennepin Avenue. Available open spaces and surface parking lots can be converted to surface drainage ways within this native forest. The park is designed as great sloping planes of lawn, emerging from the forest which receive rainfall and gently tilt toward a wetland water course flanking Nicollet Mall and flowing toward the River (Fig.1-4-20, Fig.1-4-21).

图1-4-19 游泳池效果图
Fig.1-4-19 Pool View

图1-4-21 图书馆广场鸟瞰图
Fig.1-4-21 Aerial View of Library Square

a Niocellet Mall/Plaza 尼科莱特商场/广场
b 3rd st. Plaza 第三大街　c Forest　森林
d Green Vegetation 绿色植被
e Storm Water Collection Stream 雨水收集所形成的溪流
f Wetland Garden 湿地公园
g Cancer Survivors Garden 癌症幸存者公园
h Roof Water Reservoir 屋顶雨水储存池
i Riparian Forest Gateway 河岸森林门户
j Daylighted Storm Water Collection 露天雨水通
k Nice Ride Bike Station 莱斯莱德自行车站点

图1-4-20 图书馆广场规划图
Fig.1-4-20 Library Square Plan

计了众多的斜坡场地，从可以吸收雨水的林地一直延伸到一条湿地溪流，溪流流经尼科莱特商场侧面，并最终汇入河流（图1-4-20，图1-4-21）。

四、总结

“河流优先”项目聚焦于更大尺度的社会生态学、自然生态学、城市生态学以及经济生态学等方面的战略关系设计。“河流优先”公园提出了一系列创造性的、相互联系的设计策略，这些策略在各个尺度上发挥了作用。项目的提议均能够在4～5年内实施且具有远见性，也会对区域未来40年（至2050年）的发展产生重要影响。公园区域为城市创造了新的重要机遇：城市农业发展、提供安全食品以及扩大社区居民享有健康食品及营养品的通道，建设社区以及发展有机生产、快速生长等方面的地方产业可以促进这些机遇的实现。

公园将可持续多模型与相互联系的公共交通系统结合起来，为城市提供通勤、游憩以及社会流动

IV. Summary

RIVERFIRST project focuses on the design of strategic relationships between larger social, natural, civic and economic ecologies. The RIVERFIRST Park proposes a set of creative, inter-related design initiatives that function at multiple scales. Its proposal is visionary and practical in the 4-5 year time frame and impactful in the 40 year frame of 2050. Park lands create significant new opportunities for urban agriculture, provide food security, and expand neighborhood access to healthy food and nutrition in ways that build community and build local businesses in organic produce, the fastest growing.

RIVERFIRST Park incorporates a sustainable multi-model and interconnected public transportation system for commuting, recreation, and mobility in the city area. It proposes to use the Park as a catalyst to accelerate the transition from smokestack industry to a 21st century green economy.

How to balance the public interest and the private interest is the biggest difficult problem of waterfront space of development. But this project figures out the problem. The main idea of it is "the river first", that means maximum protection of waterfront space

等。规划建议将公园作为催化媒介加速由重工业向21世纪绿色经济的转型。

如何平衡公共利益和私人利益是滨水空间开发的最大难题，本项目中，“河流优先”的主题表达了将河流放在第一位的规划思想，即最大化保障滨水空间的公共性、景观性和生态性。案例中，公众可达性是设计重点之一，“可达性是人们从空间中的任意一点克服空间阻力到达该目的地的相对或绝对难易程度，反映了公众到达目的地的水平运动过程所克服的空间阻力大小。”城市规划及景观设计中常以可达性作为原则之一，目的在于衡量游览区给市民提供服务的可能性或潜力，也可以评价不同群体接近公园的程度是否公平。本项目中，可达性并非简单指交通可达性，更包括河流自然环境的可达性、公众获取河流生态教育及信息的便利性、工作地点的可达性以及享受多方面社区利益的便利性等内容，将可达性的内涵进一步深化。

of public nature and landscape and ecology. In this project, public accessibility is the key of the design, "accessibility is the ability that people from any point in space to overcome space resistance to the destination, which reflects the degree of space resistance for people overcoming the horizontal movement." City planning and landscape design considers accessibility as one of principle, in order to measure the potential possibility of the service provided by the resort; also accessibility can evaluate the fairness for different groups getting close to the park. In this project, accessibility is not that simply which refers to the traffic accessibility, actually it includes reachability to natural river environment, the convenience for public to get access to the river ecological education and information, accessibility to work location and that of enjoying the many community benefit, etc. And all of this shows the development of the meaning of accessibility.

思考题

1. **“河流优先”如何互动人类活动与河流之间的关系？**
2. **“河流优先”公园系统如何满足公众可达性？**
3. **城市农业是远景公园的重要组成部分，你如何看待城市农业？它在未来城市发展中有着什么样的作用？**
4. **项目采取了哪些措施保护公园中的各类栖息地？采取哪些措施营造新的栖息地？**
5. **“河流优先”提出了哪些设计策略来平衡社会、生态、自然、经济等方面的关系？**

Questions

1. How does the RIVERFIRST strengthen the relationship between human activities and the river?
2. How does the RIVERFIRST park system meet the needs of public access?
3. Urban agriculture is an important part of Farview Park, what is your opinion on it? What role does it play in the urban development in the future?
4. What measures are taken to protect different habitats and to create new ones?
5. What are strategic design guidelines made which focuses on the balance of society, ecology, nature, and economy?

注释/Note

图片来源/Image：图1-4-1～图1-4-20来自于http://minneapolisriverfrontdesigncompetition.com/

参考文献/Reference：http://minneapolisriverfrontdesigncompetition.com/ ;http://www.archdaily.com/112055/riverfirst-kennedy-violich-architecture-and-tom-leader-studio/

推荐网站/Website：http://minneapolisriverfrontdesigncompetition.com/

第五节 布法罗河口区规划
Section V Buffalo Bayou Planning

一、 项目综述

布法罗河口是流经美国得克萨斯州哈里斯郡休斯敦市的一条主要水道。它源自得克萨斯州凯蒂市的水牛湾，流域全长53英里（约85公里），东至休斯敦航道，流入加尔维斯顿湾和墨西哥湾。河口水道沿途纳入很多著名的河口支流，例如白橡河口，格林斯河口以及布瑞斯河口等，还流经了很多大型公园和小型社区公园。自艾伦兄弟1836年建立休斯敦以来，布法罗河口在休斯敦的历史上一直是一个焦点，其商业发展推动了休斯敦地区早期的繁荣。一个世纪前，当铁路取代了沿海航运，一些港口城市例如盖维斯顿已经渐渐衰落，而休斯敦预见到了这一趋势，将自己打造成20世纪最伟大的港口城市之一。今天，休斯敦凭借布法罗河口为城市发展和居住的核心，再一次成为21世纪的中心城市之一（图1–5–1）。

布法罗河口是休斯敦市乃至整个得克萨斯州墨西哥湾东南部的重要资源。它在休斯敦发展史上有着重要的地位，为休斯顿地区当前和未来的发展发挥着重要作用。布法罗河口作为洪水分流的主要通道为公共安全提供了服务。同时它作为加尔维斯顿湾的重要自然生态区，是当地一笔特殊的资源。但同时，它自身也存在着很多问题有待协调与解决（图1–5–2）。

布法罗河口总体规划是一个涵盖了保护与发展

I. Overview

Buffalo Bayou is a main waterway flowing through Houston, in Harris County, Texas, USA. It begins in Katy, Fort Bend County, Texas and flows approximately 53 miles (85 km) east to the Houston Ship Channel and then into Galveston Bay and the Gulf of Mexico. Along the way the bayou receives several significant tributary bayous, such as White Oak Bayou, Greens Bayou, and Brays Bayou and passes by several major parks and numerous smaller neighborhood parks. Buffalo Bayou has been a focal point in Houston's history since the Allen brothers founded the City in 1836.Houston's early success was driven by the Bayou's commerce. When coastal shipping was eclipsed by the railroads a century ago, port cities such as Galveston were bypassed. Anticipating this trend, Houston reinvented itself as one of the world's great 20th century ports of call. Today Houston is once again poised to reinvent itself as a great 21st century world center-with Buffalo Bayou as the focal point for its development and livability(Fig.1–5–1).

Buffalo Bayou is a significant resource for Houston,and the southeast Texas Gulf Coast. It is significant in the historical development of Houston, and in the region's current and future vitality. The Bayou serves public safety as a major channel of floodwater conveyance. Within the natural ecology of Galveston Bay, it is a special resource for the region. It also has many problems for which coordinated solutions must be found(Fig.1–5–2).

The Master Plan for Buffalo Bayou is a multifaceted plan integrating aspects of conservation and

图1-5-1 布法罗河口现状卫星图
Fig.1-5-1 The Map of Buffalo Bayou

图1-5-3 布法罗河口一角改造后现状
Fig.1-5-3 Buffalo Bayou After Planning

的多层次综合性规划，它综合考虑了众多学科，使总体规划能够兼顾到各方面并加以实施（图1-5-3）。这其中包括城市发展和设计、环境质量、生态区域、洪水管理、景观设计、河口交通运输以及水上活动等，它们都在总体规划中得到了很好的处理并给出了行动框架。

二、 场地和文脉

布法罗河口流经休斯敦市，是当地著名的地标。它将城市分为北岸和南岸，以及东部和西部地区。在第二次世界大战后的几十年间，它已经成为一个被忽视和隐蔽的垃圾场，城市中最不堪的一角。现在，作为一个亟待振兴的城市中心地带以及计划中的重要综合区，它又重新进入到人们的视野中，成为连接市中心各区的重要门户。

休斯敦市是在河口的河岸上建立起来的，城市的街道、建筑以及基础设施都是依河口而生。河口地区也是19世纪休斯敦工业化发展的交通大动脉，中央大街从艾伦兄弟着陆的地方如雨后春笋般向陆地延

图1-5-2 布法罗河口一角改造前
Fig.1-5-2 Buffalo Bayou Before Planning

development. Leadership from disciplines with complementary expertise must come together to detail and implement the Plan's many aspects(Fig.1-5-3). These include urban development and design, environmental quality and the eco-region, flood management, landscape, Bayou access and transportation, and water-based activities; all are addressed and outlined here in a framework for action.

II. Introduction of the Site and Context

The Buffalo Bayou running through Houston has long been a significant landmark, slicing the city into north and south banks, into east and west zones. In the postwar decades it has become a neglected and well-hidden dumping ground, the back door of downtown. Now, envisioned as a vital integrated District defined by a continuous central spine with revitalized edges, it again becomes the front door to all parts of the city.

Houston was founded on the banks of the Bayou and its urban form the streets, buildings, and infrastructure were developed in response to it. The Bayou was the transportation artery of 19th century Houston engendering industrial uses on its banks, with Main Street springing landward from the original Allen's Landing. Major streets to the water (such as Main Street) housed important civic and commercial uses. West and East of Downtown, historic neighborhoods, set back from the flood plain of the Bayou, were organized in a similar manner. When the ship channel was dredged, it spurred heavier industrial use of Bayou lands east of Downtown. Some of Houston's most distinctive residential neighborhoods were established on the Bayou at about the same time, including great civic amenities like Sam

图1-5-4 布法罗河口规划平面图
Fig.1-5-4 The Plan of Buffalo Bayou

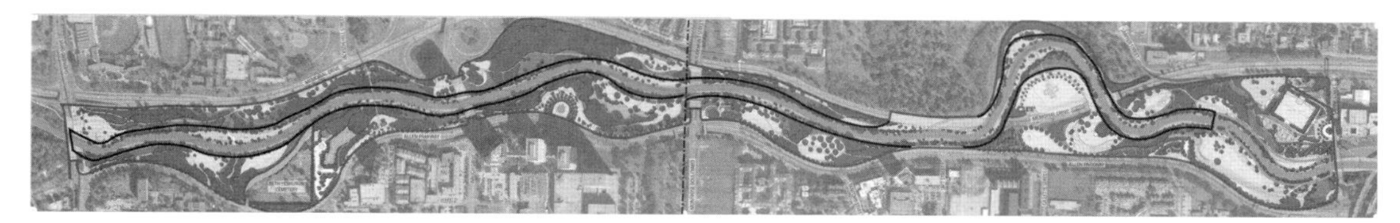

图1-5-5 布法罗河口规划范围
Fig.1-5-5 The area of the Buffalo Bayou Planning

伸。一些临水的主要街道（如中央大街）都作为重要的市政和商业用途而建造。市中心的西部和东部的历史街区，都是以相同的形式在洪水冲击平原上建造而成的。随着航道的开发，重工业在市中心东部逐渐发展起来。休斯敦的一些具有特色的居民区基本上是在同一时期在河口地区兴建的，包括一些著名的市政设施，例如休斯顿山姆公园和纪念公园。随着市中心经济的发展对河口的依赖越来越少，河口地区发展越来越滞后，甚至成为城市发展难以逾越的障碍。

河口地区大量的工厂影响了这里的环境质量。河口流域80%的面积都已经城市化。城市化使河口流域大部分面积都成为不透水表面（例如沥青、混凝土等），从而使流域内的降雨到达河口的时间比自然状况下要快。由于降雨到达河口区过快，导致水位不稳定并且变化过快。而休斯敦市仍然继续毫无节制地扩张，使河口区的不透水城市化面积迅速增长，问题进一步升级。

布法罗河口总体规划主要聚焦于由西至东约10公里左右的范围，沿着布法罗河口的航道从谢泼德大道延伸至特宁贝辛集装箱码头。廊道沿河岸向南北向延伸约1.5英里（约2.4公里），并且包括大量的现有社区。河口地区的流域范围广，占地约500平方英里（约1 295平方公里）（图1-5-4，图1-5-5）。由于总体规划涉及相当大的一片区域，它的作用，特别是在环境质量方面，只能建立适用于河口等较大流域规划的原则和政策。

Houston Park and Memorial Park. As the Downtown economy became less dependent upon the Bayou, development turned its back on it, leaving the Bayou as an impediment to be crossed over but seldom visited.

A number of factors impact the environmental quality of Buffalo Bayou. Eighty percent of the Buffalo Bayou watershed is urbanized. Urbanization that introduces large areas of impervious surfaces (asphalt, concrete, etc.) causes storm water from rainfall in the watershed to reach the Bayou faster than under normal or natural conditions. Waterthat reaches the Bayou too quickly results in erratic and rapid water level changes. Houston continues to sprawl with little restraint, rapidly adding impervious urbanizedareas to Bayou's watershed, so the problem escalates.

The Master Plan's primary focus extends 10 miles, west to east, along the course of Buffalo Bayou, reaching from Shepherd Drive to the Turning Basin. The corridorstretches approximately one half-mile into each north and south bank, and includesa number of existing neighborhoods.The Bayou watershed is large, covering nearly 500 square miles(Fig.1-5-4,Fig.1-5-5). While the Plan addresses a large area, its efficacy, especially with regard toenvironmental quality can only be established if planning principles and policiesare applied to the Bayou's larger watershed.

III. Plan Strategies

Plan strategies are:

1)Rehabilitate the Bayou as an Ecologically FunctionalSystem

Restore to Buffalo Bayou its natural functions, includingwater retention and drainage, sustenance of wildlife habitat, cleansing of storm water, and ability to maintain water quality.

三、 规划战略

布法罗河口区的规划战略如下：

1) 将河口区恢复为一个具有生态化功能的系统

布法罗河口恢复至其原生状态下的功能，包括水体涵养和排水系统、野生动物栖息地、雨水自净以及水质保持的能力。

2) 增强洪水的排放能力

开挖新的泄洪渠道，治理河岸并清理市中心阻碍泄洪的相关设施，从而有效地提高河口支流的洪水运送能力，以解决规划区域内的洪水影响。

3) 鼓励低冲击负荷开发

鼓励新型、密集化地开发，以节约型规划和工程建设来积极响应河口区生态保护。

4) 增强河口区的可视性

重塑河口区河岸，以保护自然栖息地以及市民的休憩地，并方便市民亲近土地，并使之能够从周围街道、行人进行观赏。

5) 增加市中心的住宅率

扩展滨水住宅区的建设，从而保护河口和市中心的独特资源。

6) 确保河口区的活动开放度

通过规划和设计，确保公众能够沿着河口区河岸和滨水区活动。

7) 确保河口区的承载力

未来河口区的发展会吸引越来越多的人来投资，应使规划能够承载未来河口区发展增长，不超出现有社区发展的承载力。

8) 鼓励混合利用增长模式

利用滨海地区的吸引力，吸引混合发展模式和创新的发展项目来增加市中心的就业机会和收入，例如旅游业、零售业和越来越多公众参与的滨水运动。

2)Increase Floodwater Conveyance Capacity

Address regionalflooding impacts by effectively increasing the Bayou's floodwaterconveyance capacity—with new supplementary channels, banktreatments, and removal of obstructing structures in Downtown.

3)Promote Low-impact Development

Encourage new and infilldevelopment that responds to the ecology of the Bayou watershedwith conservation-minded planning and engineering.

4)Increase the Visibility of the Bayou

Reshape the Bayou banksto promote natural habitat and recreational use, to ease user access, and to make it visible from streets and sidewalks.

5)Increase Residential Opportunities Downtown

Expand thewaterfront residential community to build a dedicated constituencythat works to preserve the unique resources of the Bayou andDowntown.

6)Ensure Equity of Access along the Bayou

Ensure publicaccess along both Bayou banks and on the waterfront throughplanning and design standards.

7)Maintain Affordability

Anticipate future benefits of investmentin Bayou improvements to ensure that increasing appeal of theBayou waterfront does not cause displacement of residents of existing neighborhoods.

8)Encourage Mixed-use Redevelopment

Leverage the appealof the waterfront with mixed use development and innovativeprogramming to attract jobs and revenue to downtown, fromtourism, increased retail activity, and growing public participationin waterfront events.

9)Promote Joint Public-Private Development

Develop anintegrated implementation plan that combines the expertise of bothpublic and private sectors, and builds their continuous collaboration.

IV. Urban Development and Design

The proposed Buffalo Bayou District is a vital core for Houston: a "super"neighborhood that is structured and defined by the dynamics of Buffalo Bayou. Inan easy coexistence with nature, it brings the West End, Downtown, and the EastEnd to the Bayou, joining land and water.

The Plan highlights the following key actions and improvements:

A 25 million SF expansion of Downtown as a mixed-use Water View district focused on a grand waterfront promenade along Commerce Street.

A 2500-acre linked park centered on Buffalo Bayou

9) 促进公众-个人共同参与的发展模式

制定一个综合的实施计划，建立社会和个人共同贡献专业知识的平台，并促进他们的持续性合作。

四、主要设计内容

规划中的布法罗河口区是休斯敦市的一个重要核心区——一个以布法罗河口为依托形成的“超级社区”。在一个人与自然能够更和谐相处的环境中，以陆地和水系使城市的西部、市中心和东部与河口区紧密相连。

这一规划突出了以下的措施和改进项目：

将市中心拓展为一个沿商业街的2 500万平方英尺（约23.2公顷）的滨水混合商业区。沿布法罗河口2500英亩（约10.1平方公里）的公园链——这是布法罗河口区域生态保护的具有里程碑意义的第一步。

市中心防洪能力提升与滨水区整体保护：百年一遇的洪水线下降约8英尺（约2.44米）。

850英亩（约344公顷）新公园和20公里步道将市中心布法罗河口区连接成一个连续的公共空间。

新的投资使东部的交通流动性更强，同时将相近的社区更紧密的连接起来，并使其对河口区更加开放。

在滨水区主要节点开发了14个可以停泊船只的码头：高质量的码头综合服务区打造了区域性特色休闲场地。

在靠近修复的河口区兴建住宅和工作场所，沿水路的关键节点兴建新的住宅社区会为河口区带来新的生命力。规划预计在布法罗河口区未来20年内将增长12 000新居民以及15 000个新的工作岗位。

五、 城市设计

布法罗河口区的重建为将河口区重新塑造成为一个休憩场地提供了千载难逢的机会，这个机会将极大推动这一地区高水平的发展。沿河口地区的规划提出了新的旅游地和发展重点，这些场所将休斯敦市中心被忽视的滨水旅游资源转变为一个充满活力的旅游中心，丰富周围居民的生活并吸引周围的游客。

is the first step towards alandmark regional eco-preserve.

Flood capacity improvements integrated with waterfront growth protectDowntown: 100-year flood levels can be reduced by an estimated eight feet.

850 acres of new park and 20 miles of trails connect a continuous public realmalong both sides of Buffalo Bayou in the center City.

New investments improve mobility in the East End and connect nearbyneighborhoods to an accessible Bayou and to each other.

14 accessible boat landings at key destinations activate the waterfront: a high quality multi-service boat operation creates a regional recreational destination.

New residential neighborhoods on key sites along the waterway will bring life tothe Bayou, offering homes and workplaces in close proximity to the rehabilitatedBayou. The Plan anticipates a growth of up to 12,000 new residents in theBuffalo Bayou District, and 15,000 new jobs within 20 years.

V. Urban Development and Design

The revitalization of Buffalo Bayou presents a unique opportunity tore discover the Bayou as a potent amenity — a powerful impetus for high quality development. The Plan proposes new destinations and development sites along the Bayou,"places" that will transform central Houston's neglected waterfront into an active and vibrant center, enriching the lives of its residents, and attracting visitors from the region and beyond.

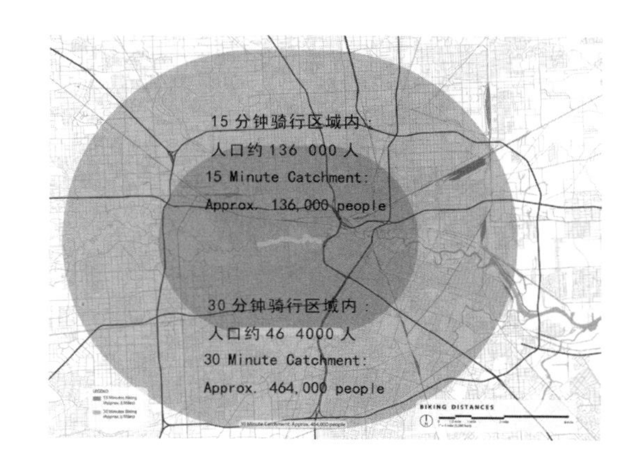

图1-5-6 规划骑行区范围

Fig.1-5-6 The Biking Distance

1）一个具有地方性的城市设计与发展规划的关键：

河口区的土地是一个混合型社区，需要深入了解其地方性背景和区域性人类活动。

应着力发展综合型项目，而不是单独发展某一单一作用的项目，从而建立起一种多用途场地，可以协同发展和多元化发展。

基于场地的思考模式可以关注使用者的需求和体验，而不是局限地提供实体基础设施。

2）城市发展和设计规划应遵循以下原则：

鼓励兴建新的住房，特别是在市中心和东部地区，以增加滨水地区的建设。

鼓励城市住宅的多元化发展——各种面积、户型、以及价位的住宅多元发展。

通过设计一系列的连续关键节点，激活滨水地区的活力。

通过设计穿过市中心的街道、人行道和步道使社区间实现无缝连接（图1-5-6，图1-5-7）。

通过新的高质量的城市基础设施建设来配合滨水区的发展（图1-5-8，图1-5-9）。

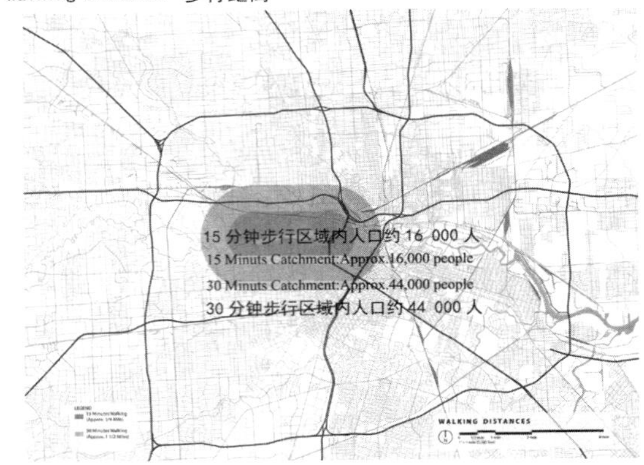

图1-5-7 规划步行区范围

Fig.1-5-7 The Walking Distance

1)The key of a place-based approach to Urban Development and Design is:

Bayou lands are a heterogeneous mix of neighborhoods, calling for a recognitionof context and area-specific actions.

Projects that are mixed-use, rather than isolated single-use developments, createa sense of “place” for different kinds of users, allowing synergistic and diversifiedmarkets to evolve.

Place-based thinking allows a focus on users'needs and experiences, ratherthan a narrow focus on providing physical infrastructure.

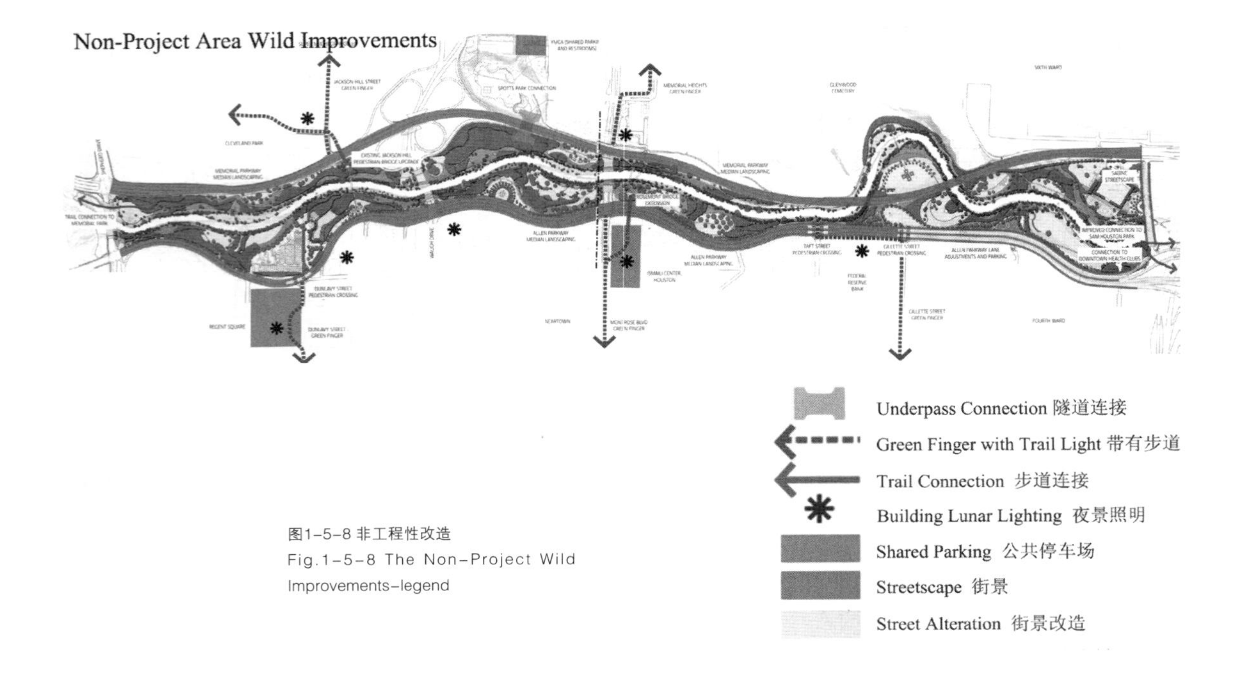

图1-5-8 非工程性改造

Fig.1-5-8 The Non-Project Wild Improvements-legend

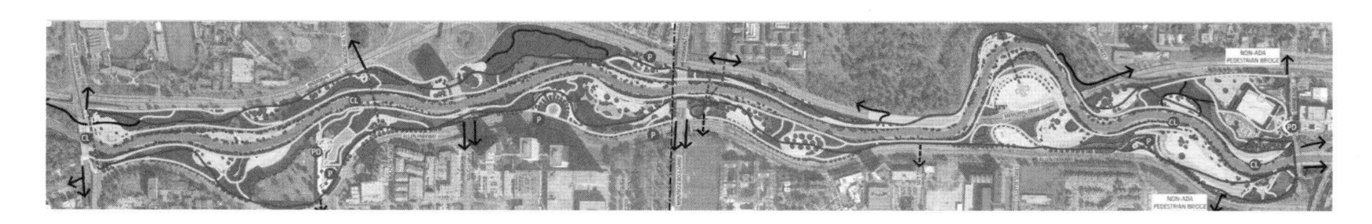

图1-5-9 步道系统规划

Fig.1-5-9 The Trail System Planning

通过激活和再利用老街区，保护历史遗产——因为这是一座城市的集体记忆。

六、 环境质量和生态区域保护

布法罗河口区是休斯敦独有的自然资源。作为加尔维斯顿湾自然生态系统的组成部分，它是休斯敦经济和休闲服务业的重要资源。对于现在河口地区的很多环境问题，总体规划提出了一套先进并行之有效的与新的发展相配套的环境保护解决方案。

布法罗河口地区目前的环境情况已经难以承受更多的环境负荷。在这里几乎没有野生动物的栖息地。其关键问题包括：

水力情况不稳定：布法罗地区80%的土地已经城市化，导致大部分城市表面被不透水铺装覆盖（沥青、混凝土等），从而使布法罗地区的暴雨冲刷负荷比正常情况下高，每年达到48～52英寸（约1.2～1.3米）。在此情况下，洪水频繁爆发。蔓延过河岸的洪水导致河岸侵蚀，增加了泥沙，改变了植被的栖息地，并危害到邻近的基础设施，这进一步增加了洪水发生几率，从而带来更多的河岸侵蚀。

水质环境差：雨水和非点源工业污水未经处理直接送入了河口。许多非点源污染流域聚集在河口。大量泥沙进入河口水域，降低了水体透明度。这影响了河口的可视度，同时也影响了动物物种对河水的利用率。

难以修复的棕地；流域河岸植被覆盖率低：入侵物种取代了更具生态性的本地物种；休闲娱乐设

2)The Urban Development and Design Plan is based on the following principles:

Populate the waterfront by encouraging new housing, especially inDowntown and the East End.

Encourage the diversity of urban residential opportunities – a variety of sizes, types and price points.

Activate the waterfront by programming a system of linked destinations andspecial events.

Create seamless connections to all neighborhoods and through Downtown viainterrelated streets, sidewalks and walking trails.

Coordinate waterfront development with new high quality urban infrastructuralinvestments and amenities.

Preserve historic assets--the City's collective memory --through adaptive reuse.

VI. Environmental Quality and the Eco-Region

Buffalo Bayou is a significant resource for Houston. As an integral component of the naturalecological system of Galveston Bay, it is a critical resource for Houston's economy and recreation. For numerous environmental problems nowapparent in all zones of the Bayou, The Master Plan proposes that a set of advanced and effective environmental solutions can come alongwith new growth and development.

The existing environmental conditions of Buffalo Bayou bear little resemblance to its pre-settlement or native condition. Very little native wildlife habitat has survived. The key concerns include:

Hydraulic destabilization:Eighty percent of the Buffalo Bayou watershed is urbanized. Urbanization that introduces large areas of impervious surfaces (asphalt, concrete, etc.) causes storm water from rainfall in the watershed to reach the Bayou faster than under normal or natural conditions. Rainfall in the Buffalo Bayou watershed is high, between 48~52

施的需求与野生动物栖息地需求相冲突；城市热岛效应产生的相关问题。

环境规划提出通过一系列循序渐进的方案显著改善河口地区的环境质量。规划不建议将河口地区完全恢复至其完全原生态的状态，而是建议通过建立一系列环境丰富的栖息地来构建区域生态系统。通过最大限度减少洪灾环境影响，改善水质以增加栖息地，加上不断增加的野生生物，河口地区的环境将成为吸引人们重返河口的关键因素。

环境规划将遵循以下的原则：

调节支流暴雨径流时间和径流量，调节河口的大小以适应预期的暴雨径流量，确定支流的非点源污染及其排向河口地区的排放点，确定支流地区的点源污染，减少支流的水土流失，改善水质；改善候鸟和保护物种的栖息地，增加河口流域的本地物种的生物多样性，建立与城市河口水文特质相匹配的生物栖息地。

七、 洪水治理规划

在休斯敦城市化进程中，频繁的洪涝灾害带来了巨大的财产损失和民众生命危险。热带风暴艾琳曾经造成近50亿美元的损失。城市化进程蚕食了自然形成的洪水冲击平原，减少了河口支流泄洪能力。洪水治理规划将针对上游地区和市中心进行洪水治理。

布法罗河口地区洪水治理规划遵循以下原则展开：

为河口区提供足够的防洪能力——即使遇到百年一遇的洪水，下游水位仍不致增长；在关键河段提高运载能力——从白橡树河口下游到麦基大街的交汇处；减少目前有问题的桥交汇处产生的洪水问题；限制百年一遇洪水区的发展。

这一规划旨在大幅减少洪水影响。洪水治理规划将使中央大街百年一遇的洪水线（例如艾琳热带风暴）降低8.5英尺（约2.6米）。这种安全系数的提高将为大城市带来极大的效益。依靠更高效的区域洪水排水系统，白橡支流上游额外的洪水可以得到分流而不致影响市中心，以减少洪水对上游地区

inches per year. Flooding is frequent under these conditions. As a result, over-bank flooding occurs along with bank erosion. Bank erosion in turn removes vegetation and habitats, adds sediment, and threatens adjacent infrastructure. The increased flood water is directed to the Bayou and this in turn brings more erosion.

Poor water quality:Storm water and non-point industrial discharges feed directly into the Bayou without treatment. Many non-point sources of pollution in the watershed ultimately accumulate in the Bayou. High levels of sediment released into the Bayou waters reduce water clarity. This affects the Bayou's visual character, as well as animal species utilization of the water.

Unremediated brownfields; Inadequate and inappropriate vegetated cover along Bayou banks: with invasive species displacing more ecologically valuable native species; Competing demands of recreation and wildlife habitat; Urban heat-island related issues.

The Environmental Plan proposes significant environmental quality improvementsthrough a set of sequenced steps. The Plan does not propose that the Bayou canbe restored to its native, pre-urbanized state. It does propose creating areas of environmentally rich habitat that are integrated into a regional system. By minimizing environmental impacts of flooding, improving water quality, andincreasing habitat, the Bayou – and the great increase in wildlife it will attract – willbe key factors in drawing people back to the Bayou environment.

The Environmental Plan is based on the following principles:

Moderating the time required for storm water to reach the Bayou, and the quantity of storm water put into the Bayou; Modifying the Bayou size to accommodate anticipated storm water quantities.

Addressing non-point sources of pollution throughout the watershed and at pointsof entry to the Bayou; Addressing point sources of pollution along the Bayou; Eliminating erosion along Bayou banks; Improving water quality; Providing habitat for migratory and resident species; Increasing diversity of native plant species along the Bayou; Creating habitats that are responsive to the urban hydrology of the Bayou.

VII. Flood Management Plan

Frequent flooding of the bayous in the urbanized core of Houston results in substantial property damage and even loss of life. Tropical StormAllison caused nearly $5 billion in damage regionally. These impacts result from urbanization that has encroached on the naturally absorbentflood plain, reducing the

社区的影响。

洪水治理规划是基于以下的三管齐下的治理方法：

（1）新建泄洪渠道，使布法罗河口和白橡树河口的洪水可以直接穿过市中心，绕开容易造成洪水的问题河段。

（2）巩固布法罗河口的桥梁交汇处，减少阻抗的洪水流量，必要时减少口岸和桥梁。

（3）在艾琳登陆处至麦基大街等关键节点通过拓宽和疏浚河岸的方法增加布法罗河口的运载能力。

八、景观设计

景观规划设计使布法罗河口重新成为一个主要的休憩场所。目前河口地区的公共形象非常的负面：这是一个被污染的、垃圾遍布、泥泞的地方，水道周期性洪水泛滥而引发灾难。由于在公众心中这些负面的印象以及在设施上难以靠近，河口地区鲜有人迹。景观规划将重塑河口地区形象。

景观规划遵循以下原则：

在河口两岸设置连续的行人通道和步道；将景观设施与防洪设施整合；建立一个连续的公园休憩系统，包括公园、海滨长廊和游戏区；通过建设绿色街道，增强学校与河口的联系并且建立社区绿道，将社区设置与滨水的开敞空间中；通过提供更具亲和力的连接处和休闲设施吸引公众更多地在河口区活动，包括河口上岸连接空间，步道以及照明设施等，在整个城市和区域范围内形成彼此相连的公园体系（图1–5–10）。

河口区的新形象建设有赖于水患、垃圾以及环境污染等问题的改善，也同样有赖于其外观设计。开放河口地区将使其具有更好的视觉效果。提供小径、步道、码头以及相关设施将使人们更安全地与水亲近。增加更多的池塘将使人们享有更多的水景和其中的乐趣。拥有更多湿地的浅滩将为野生动物创建更多的生境，从而为人类提供更多的观赏价值。大片开放河口地区将会为人们带来翻天覆地的变化。在不久的将来，整个流域的各类设施所形成的清洁水源将为河口地区带来更加可观的观赏效果。经过精心设计和适度建设，将使城市边缘形成一道别样的风景线。到那时，无论是河中泛舟还是

bayous' ability to drain storm water swiftly. The Plan addresses flood protection of both upstream and Downtown.

The Flood Management Plan for the Buffalo Bayou District is based on the following principles:

Provide adequate floodwater capacity in the Bayou District to address rainfallevents up to and including the 100-year frequency without causing increases indownstream flood levels;increase conveyance capacity of Buffalo Bayou in the critical reach, from theconfluence with White Oak downstream to McKee Street.; reduce present flood impacts caused by problematic bridge crossings; restrict development within the 100-year flood plain.

The Plan aims for a significant reduction in flood impacts. The Flood ManagementPlan will reduce the impact of a future 100-year storm event, similar to TropicalStorm Allison, by 8.5 feet at Main Street. This additionalsafety margin promises metropolitan as well as local flood benefits. Additional floodwater upstream on White Oak Bayou can be conveyed without impact throughthe downtown area to reduce flooding impacts in these upstream neighborhoods, thereby contributing to a more effective regional storm water drainage system.

The Flood Management Plan is based on a three-pronged approach:

(1)Direct floodwater flow carried by Buffalo Bayou and White Oak through Downtown by creating supplementary channels that bypass problematic flood-causing configurations of the two bayous.

(2)Consolidate bridge crossings of Buffalo Bayou to reduce impedance to the flow of floodwater, eliminating crossings and reconstructing bridges where necessary.

(3)Increase the conveyance capacity of Buffalo Bayou in the critical reach from Allen's Landing to McKee Street by widening and channeling the Bayou banks.

VIII. Landscape

The Landscape Plan reclaims Buffalo Bayou as a recreational spine. The current public image of the Bayou is often a negative one: a polluted, trash-filled, muddy, industrial channel that periodically causes disastrous floods. Due to this poor public image and lack of physical access to it sedges, the Bayou is infrequently visited or used. The Landscape Plan looks to the remarkable rewards of reclaiming Buffalo Bayou.

The Landscape Plan is based on the following principles:

Providing continuous and public pedestrian access and trail ways on both banks of Buffalo Bayou.Integrating landscape amenities with flood management infrastructure; creating a linked system of recreational destinations in a park setting,

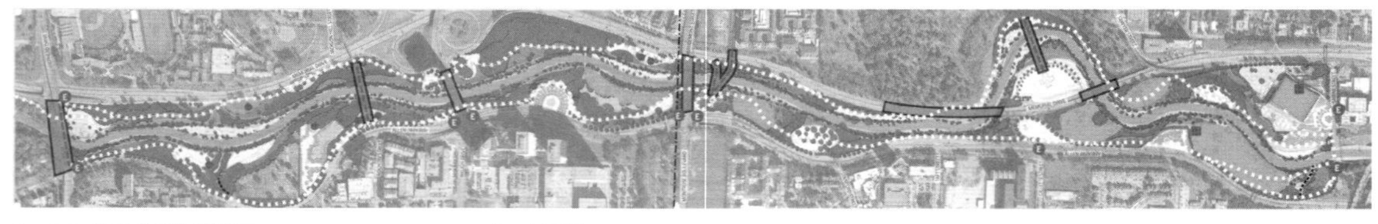

图1–5–10 环境设计规划

Fig.1–5–10 The Environmental Design

图1–5–11 桥下风景

Fig.1–5–11 The view under the Bridge

图1–5–12 河岸风景

Fig.1–5–12 The view along the band

徜徉在河岸上的人们，都会欣赏到一幅迷人而又安全，整洁而又美好的画面——改变河口地区的视觉效果也将最终改变河口地区在人们心中的形象（图1–5–11，图1–5–12）。

九、总结

规划力求解决该区域社会经济发展缓慢和工业污染严重两大问题。事实证明，布法罗河口区规划恢复

includingparks, promenades, and play areas; integrating Bayou neighborhoods into the waterfront open space system, usingGreen Streets,linking schools to the Bayou, and creating neighborhoodgreenways; encouraging public use of the Bayou by providing improved access andrecreational amenities, including Bayou landings, integrated trails, and streetlighting, extending the system of linked parks throughout the city and region(Fig.1-5-10).

The Bayou's new image will be partly based on addressing flooding, trash, and pollution problems. It will also be based on a change in its visual appearance. Opening up the banks of the Bayou will make it more visible. Providing paths, trails, landings and related amenities will allow people to safely come to the water's edge. Creating more ponds will create more surface water for people to see and enjoy. Shallower banks with more wetlands and riparian areas will create a greater diversity of wildlife to watch. Large expanses of open marsh will create dramatic vistas up and down the Bayou. Over time, clearer water resulting from basin-wide mitigation measures will create a more visually appealing Bayou. Construction of carefully designed and appropriately scaled buildings will provide an attractive urban edge. And people in boats and on the Bayou's banks will signal that it is an inviting, safe, clean place to be – and this in turn will generate more use. Changing the visual appearance of the Bayou can bring about a major change in people's perception and use of the Bayou and its banks(Fig.1-5-11,Fig.1-5-12).

IX. Summary

Buffalo Bayou Planning Strive to solve the two major problems of slow social economic development and serious industrial pollution in this area. Facts prove that Buffalo Bayou Planning has restored the advantages of original resource, made ecological system and regional development system joined closely , greatly improved the original environment quality, and put forward effective management plan for the city.Buffalo Bayou Planning has found the balance point of common development of the nature and cities.,and also laid a good foundation for the

了该区域原有的资源优势，生态系统与区域发展体系衔接紧密，极大的改善了原有的环境质量，并提出有效的治理方案，令自然与城市的共同发展找到了平衡点，也为顺利进行详细规划奠定了良好的基础。

本设计中的亮点之一就是洪水分流措施，洪水治理计划在最大程度上保护了城市的安全。规划着力于促进区域经济发展、吸引人口、促进经济复兴。科学的人口和用地发展规模能符合区域承载力要求，顺应未来经济发展趋势。另一处亮点是对于老城区的激活再利用，严格控制规划设计范围，保留原有城市特点。规划注重塑造艺术美学特色，水系两侧设置步行系统，基础设施完善，亲水平台安全性高且极具亲和力。

总体上，该规划令布法罗河口恢复了其作为休憩场所的本真面貌，极大程度上改善了之前环境污染的问题，为布法罗河口区域注入新生，使其再次成为令人向往的迷人风景线。

detailed planning.

One of the highlight of Buffalo Bayou Planning is the flood diversion measures, it has protected the city's safety in the greatest degree. Buffalo Bayou Planning Strive to promote regional economic development, attract people, promote economic revival.The scale of the population and land development accord with regional capacity requirements, conform with the future economic development trend. The other highlight of Buffalo Bayou Planning is the activation of the original city, which has strictly controlled the scope of planning and design, retained the characteristics of the original city. Buffalo Bayou Planning pay attention to create art esthetics characteristic. Foot set system are designed on both sides of the river. The infrastructure is perfect, and the Hydrophilic platforms are high safety and most affinity.

All in all, Buffalo Bayou Planning has restored the Buffalo Bayou as a Resting place, improved the environmental pollution problems in a great degree, and Infuse new vitality for Buffalo. It has made Buffalo become a charming scenery line once again.

思考题

1. 请简述布法罗河口区规划的区位特征。
2. 布法罗河口规划的规划战略是什么？
3. 布法罗河口规划的重点措施和改进项目是什么？你认为其中地方性的城市设计与发展规划的关键是什么？
4. 布法罗河口规划中需要解决的关键性环境问题是什么？环境规划如何通过一系列循序渐进的方案显著改善河口地区的环境质量？
5. 布法罗河口规划中洪水治理规划的原则是什么？是如何与景观设计有机融合的？

Questions

1. Please introduce the regional overview of Buffalo Bayou.
2. What are the plan strategies of Buffalo Bayou Planning?
3. What are the key actions and improvements of Buffalo Bayou Planning? What do you think about the key of place-based approach to urban development and design in Buffalo Bayou Planning?
4. What are the key concerns with the existing environmental conditions of Buffalo Bayou? How does the environmental plan propose significant environmental quality improvements through a set of sequenced steps?
5. What are the principles of the Flood Management Plan for the Buffalo Bayou? How does the flood management fuse with the landscape design?

注释/Note

图片来源/Image：图1-5-1来自Google Earth ;图1-5-2来自http://www.flickr.com/photos/91975704@N00/2885693635/ ;图1-5-11 来自http://houston.culturemap.com/newsdetail/04-16-10-a-symphony-for-the-bayou-brad-sayles-turns-to-the-trees/ ;图1-5-12来自http://www.houstondowntownalliance.com/en/cev/827/ 其余图片来自" Buffalo Bayou Park Master Plan"

参考文献/Reference：来自" Buffalo Bayou Master Plan" 以及" Buffalo Bayou Park Master Plan"

推荐网站/Website：http://www.buffalobayou.org/masterplan.html

第六节 洛杉矶河复兴总体规划

Section VI Los Angeles River Revitalization Master Plan

一、项目综述

1. 项目简介

洛杉矶河流经洛杉矶市的32英里（约51.5公里），覆盖着位于该市中心区750英亩（约303.5公顷）的地域（图1–6–1），将这些土地的一小部分转变为具有综合效益的新型用地便可以真正地复兴这条河流，包括恢复自然生态系统、处理雨水径流、建立河流绿道以及公园道路网络等。在长期的转变过程中，不仅要恢复该河的生态功能，还要将它恢复为一条彰显洛杉矶的过去和将来特征的河流；一条能有效提升洛杉矶人的生活品质的河流；一条将成为重要的旅游目的地的河流，一条象征着自然与城市共同复兴的河流（图1–6–2）。

2. 规划目标、理念与原则

1) 规划目标

洛杉矶河复兴的目标是规划出未来20年洛杉矶河发展和管理的蓝图。 为了达到这个目标，洛杉矶市下属的洛杉矶河特别委员会制订出以下条款来指导复兴进程：建立具有环境敏感度的城市设计导则，土地利用导则，流域导则；创造经济发展机会，通过提供更多的开敞空间、住宅、零售空间

I. Overview

1. Introduction

The 32 miles of the River that flow within the City of Los Angeles represent more than 750 contiguous acres of real estate(Fig.1-6-1), in the very heart of the City. Transforming even a small portion of that land for new, multiple-benefit uses, including natural system restoration, treatment of storm water runoff, establishment of a continuous River greenway, and an interconnected network of parks and trails, could indeed revive the River. These changes would go a long way in not only restoring the ecological function of the River, but also in restoring the River's identity to one that celebrates the past and the future of Los Angeles——one that significantly enhances the quality of life for Angelenos, that becomes an important destination for visitors, and that survives as a symbol of natural resilience and revival of the City itself (Fig.1-6-2).

2. Goals Ideas and Principles

1)Goals

The goal of the Los Angeles River Revitalization Plan is to create a 20-year blueprint for development and management of the LA River. To accomplish this, The City of Los Angeles Ad Hoc Committee on the Los Angeles River identified the following objectives to guide the revitalization process: Establish environmentally sensitive urban design guidelines, land use guidelines, and development guidelines for the River zone that will create economic development opportunities to enhance and improve River-adjacent communities by providing open space, housing, retail

图1-6-1 洛杉矶河流经洛杉矶市的32英里

Fig.1-6-1 The 32 miles of the River that flow within the City of Los Angeles

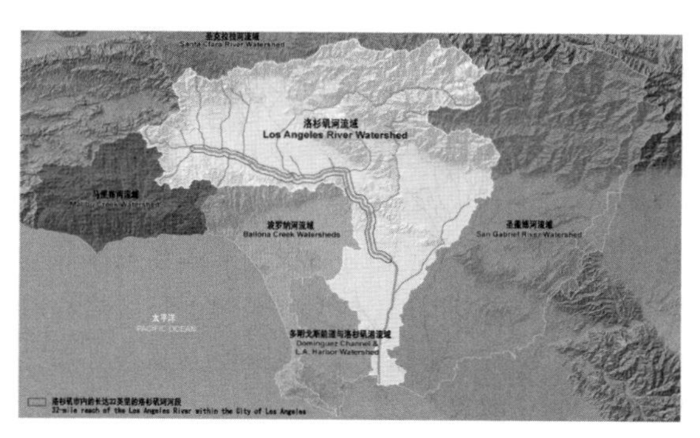

图1-6-2 洛杉矶河及其流域

Fig.1-6-2 Los Angeles River and its watershed

（如饭馆、咖啡厅、教育设施和其他公共机构）使沿河社区得到强化和改善；提升环境质量，提高水质及水源质量，强化河流的生态功能；使河流具有公众可达性；设立游憩和开敞空间，建设新的步道，给野生动物提供更好的自然栖息地；保持和提高河道的抗洪能力；在社会中培养保护洛杉矶河的社会意识和荣耀感。

在经历了18个月的规划过程后，洛杉矶市将会出台一系列文件，让洛杉矶河复兴的目标更明确，并更具可操作性。

2）规划理念

洛杉矶河将成为一条连接自然与社区的充满生机的绿色中枢，为居民提供游憩空间与场所，以进行人与人、人与自然环境之间的交往；通过了解自然界的美丽与神秘，激发青少年对环境与科学的兴趣，洛杉矶河将扎根并成长于当前和将来每一代人的心灵中；作为居住、成长和繁荣的绝佳场所，洛杉矶河将再次成为城市的心脏与灵魂。

3）规划原则

复兴河流；绿化毗邻社区；抓住社区机遇；创造价值。

二、 项目背景

1. 地理环境

洛杉矶河长51英里（约82公里），发源于洛

spaces such as restaurants and cafes, educational facilities, and places for other public institutions; Improve the environment, enhance water quality, improve water resources, and improve the ecological functioning of the River; Provide public access to the River; Provide significant recreation space and open space, new trails, and improve natural habitats to support wildlife; Preserve and enhance the flood control features of the River; Foster a growth in community awareness of the Los Angeles River, and pride in the Los Angeles River.

At the end of the 18-month planning process, the City will have a set of documents that provide a clear, implementable vision for the revitalization of the Los Angeles River.

2)Ideas

The River can become the living, green spine that connects nature and communities, providing space for active and passive recreation, and room for residents to interact with each other and their natural environment;

The River can grow in the minds of current and future generations by engaging children in the beauty and wonders of nature, inspiring interest in the environment and the sciences; the River can again become the heart and soul of the City as a great place to live, grow and prosper.

3)Principles

Revitalize the river; Green the neighborhoods; Capture community opportunities; Create value.

II. Context

1. Geographic Context

The Los Angeles River flows approximately 51 miles from its origin in the San Fernando Valley region of the City of Los Angeles, to Long Beach Harbor and the Pacific Ocean. The River runs east/southeastward through Los Angeles and along the cities of Burbank and Glendale in its northern reaches, and then heads

杉矶市的圣费尔南多山谷地区，于长滩港流入太平洋。该河以东-东南向流经洛杉矶市，接着其北沿伯班克和格伦代尔两城流过，然后转向朝南分别流经弗农、科默斯、梅坞、贝尔、贝尔加登斯、南盖特、林伍德、康普顿、派拉蒙、卡森和长滩等城市。该河的前32英里（约51.5公里）流经洛杉矶市，贯穿了10个行政区、20个毗邻社区以及10个社区规划区域（图1-6-3）。

两个水利因素影响了洛杉矶河的整治。第一，也是最重要的一条是保持河流已有的蓄洪能力。增加植被以营造栖息地和设置亲水平台的行为都会降低河流的蓄洪能力。除非通过现有的河道拓宽或加深来补偿，或者通过开挖一条地下涵洞将河水引流到现有的河道之外。第二个因素是河水的流速（图1-6-4）。在河床松软的地区，这里的水流速度能让植被生长。在流速较快的地区，植被将难以扎根，因为流水会将它们冲走。 这个问题不是仅仅通过例常的更换植物就能解决的，这关系到河道的结构整体性，因为河水还会冲走一部分的水泥和其他防护设施。

随着洛杉矶的城市化进程的加快，河流的水质有明显的下降。大多是由于城市里约2 200个出水口

图1-6-3 洛杉矶河地理背景

Fig.1-6-3 Geographic context of the Los Angeles River

southward, flowing through the cities of Vernon, Commerce, Maywood, Bell, Bell Gardens, South Gate, Lynwood, Compton, Paramount, Carson, and long Beach, respectively. The first 32 miles of the River flows through the City of Los Angeles, intersecting 10 Council Districts, 20 Neighborhood Councils and 10 community planning areas (Fig.1-6-3).

Two hydraulic considerations influence modifications to the River. The first and most important is the need to maintain existing flood capacity. Adding vegetation to create habitat, or providing terraced access to the water, are actions that would reduce flood capacity unless the current channel is widened or deepened to compensate, or unless new underground box culverts are installed to transport flood flows outside of the existing River channel.A second consideration

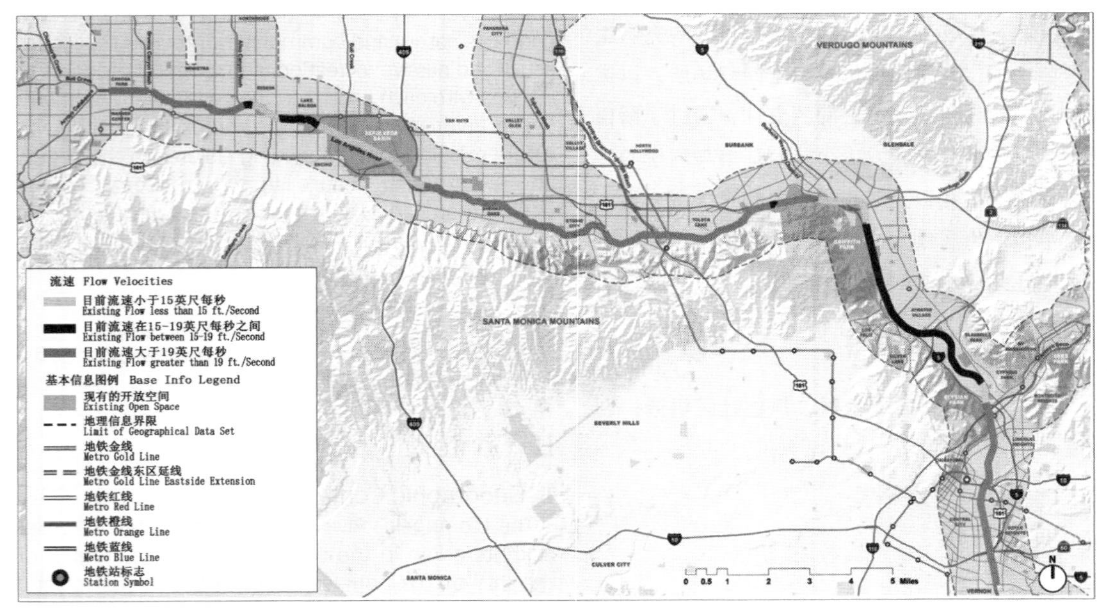

图1-6-4 河道内河水流速

Fig.1-6-4 Flow velocities within the channelr

将未经处理的雨水排入河流造成的（图1–6–5）。20世纪农业、工业、居住区的发展和杀虫剂、化肥、家用化学试剂的使用，都导致了该地区地表和地下水质的下降。

河流中漂浮的各种垃圾、碎片及其他漂浮物多是由于包装物，街道垃圾和植物残骸被随意丢弃而产生的。除了不美观之外，这些垃圾还会滋生细菌，抑制河中的含氧量，从而影响水生物的生长。2002年强制执行的日垃圾最大负荷量标准，要求每年向河中排放的垃圾减少10%，至2006年9月垃圾排放量已下降20%，2012年的排放量将减少100%。

2. 生态价值

根据加州海岸保护条例，洛杉矶河流域将近100%的原始湿地和90%～95%的滨水栖息地已经消失，这是由于城市化和河流和溪流渠道化造成的。在长达32英里的洛杉矶河项目流域里，拥有滨水栖息地的如今只有赛普维达盆地和格伦代尔峡谷。这些地段正在受到外来物种入侵、水文环境改变、垃圾倾倒和持续的侵蚀。

洛杉矶河流域包括了三个地区性的重要的生态

is the relative speed or velocity of water flows in the River(Fig.1-6-4). Within the River's soft-bottomed sections, flow velocities allow the existing vegetation to grow. At higher speeds, vegetation is difficult to sustain because flood flows wash it out. The issue is not simply one of maintenance -- having to replace plantings on a routine basis -- it also concerns the structural integrity of the channel itself, since flood flows can also wash out portions of the concrete or other armoring systems.

With the increasing urbanization of Los Angeles, the quality of the water in the River has declined significantly. Most of this is due to untreated stormwater runoff that enters the River through one of approximately 2,200 storm drain outlets (Fig.1-6-5). Agricultural, industrial, and residential development over the 20 century, coupled with the use of pesticides, fertilizers, and household chemicals, have resulted in degradation of both surface and groundwater within the region.

Trash, debris, and other floatables in the River result from careless disposal of packaging, street litter, and plant debris. In addition to negative aesthetic impacts, trash may harbor bacteria and inhibit dissolved oxygen levels, affecting aquatic life. A TMDL for trash was the first to be imposed, in 2002, and requires a 10 percent reduction per year in the discharge of trash into the River, with the first 20 percent reduction accomplished by September 2006 and a 100 percent reduction required by 2012.

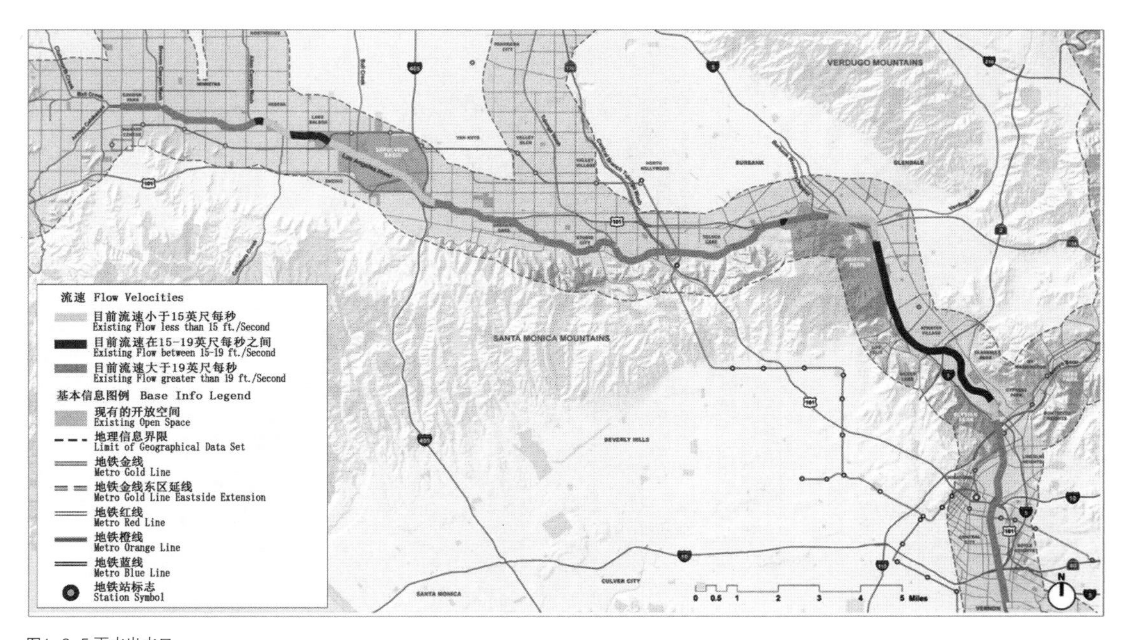

图1–6–5 雨水出水口

Fig.1–6–5 Storm drain outfalls

区域，分别是圣塔莫尼卡山，瓦多戈山和格里菲斯公园，它们和河道已经失去联系。据洛杉矶县1996年的洛杉矶河的总体规划记载，候鸟和留鸟会将洛杉矶河、三大生态区域和汉森大坝及塞普尔维达盆地的地表水系，作为主要的飞行和迁徙路线。洛杉矶水库、帕奎玛水库、恩西诺水库和图洪加水库等开阔的淡水水库也提供了筑巢空间和食物来源。据2006年南加州大学“绿色视觉”项目研究的结果，河道现有的功能是作为哺乳动物的迁徙廊道。

3. 更大的流域环境

本规划认为，我们不能，也不应该只关注降低洪水造成的损失和更大流域范围内的水质威胁。本规划明确关注这些问题的同时，更强化和信赖与其他城市和县的流域规划相配合，以强调整个流域面临的挑战，尽最大可能地在规划中关注这些挑战。

三、 规划设计介绍

1. 总体规划

1) 复兴河流

增强蓄洪能力，提高水质，强化安全的公共通道，恢复功能性生态系统（图1–6–6）。

2. Ecological Value

According to the California Coastal Conservancy, close to 100 percent of the original wetlands and 90 to 95 percent of in-stream riparian habitat within the Los Angeles River watershed have been lost, a consequence of urbanization and the channelization of rivers and creeks. Within the 32-mile Los Angeles River project area, the only areas that presently support riparian habitat are Sepulveda Basin and the Glendale Narrows. These areas are increasingly stressed by exotic species, hydrologic modifications, dumping of trash and debris, and encroaching development.

The Los Angeles River watershed includes three regionally significant ecological areas (SEA's) that are disconnected from the River corridor, these are: the Santa Monica Mountains, Verdugo Mountains, and Griffith Park. As noted in the Los Angeles County's 1996 Los Angeles River Master Plan, migratory and resident birds move along the major flyways between the River, the SEA's and other sites with surface water such as Hansen Dam, and the Sepulveda Basin. Open freshwater reservoirs such as the Los Angeles, Pacoima, Encino, and the Tujunga also offer feeding and nesting grounds. According to research by the University of Southern California's Green Visions program (2006), the channel and rights-of-way currently function as movement corridors for mammals.

3. The Larger Watershed Context

The perspective taken by this Plan is that the River cannot - and should not -be expected to be the sole means for addressing flood damage reduction and water quality challenges in the larger watershed.

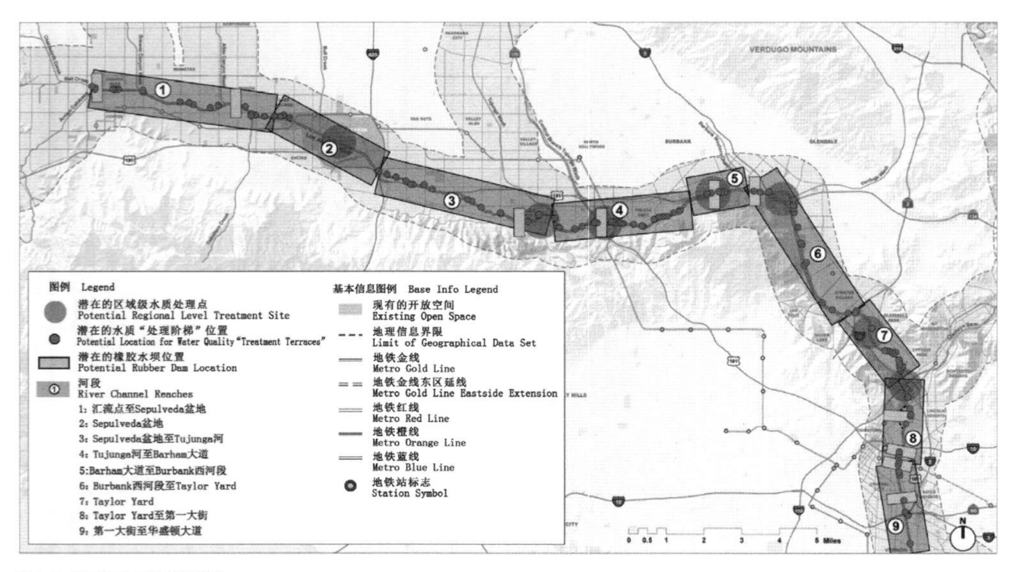

图1–6–6 可能的河道整治潜力

Fig.1–6–6 Potential in–channel river improvements

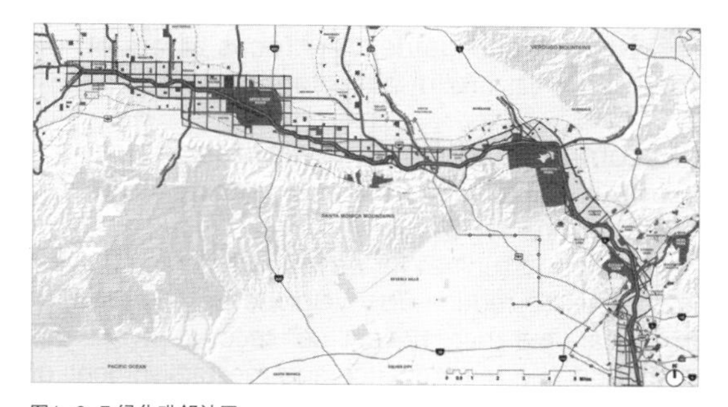

图1-6-7 绿化毗邻社区
Fig.1-6-7 Green the neighborhood

2）绿化社区

建立一条连续的河流绿道，将毗邻社区与河流连接起来，将开放空间、游憩及水质处理设施延伸至毗邻社区，提高河流的个性特征，将公共艺术融入沿河一带（图1-6-7）。

3）抓住社区发展机遇

使河流成为活动中心，培养市民的自豪感，让居民参与社区规划过程和决议，为教育和公共设施提供机会，呈现河流的文化遗产（图1-6-8）。

4）创造价值

改善生活品质，为就业者住房和零售空间增加机会，建立基于环境敏感性的城市设计和土地开发机会和

While the River can make a significant contribution to address these issues, the Plan reinforces and relies upon other City and County watershed planning initiatives in emphasizing that challenges generated within the watershed should be, to the maximum degree possible, addressed within the watershed.

III. Planning and designing

1. Master Plan

1)Revitalize the River

Enhance flood storage, enhance water quality, enable safe public access, and restore a functional ecosystem (Fig.1-6-6).

2)Green the Neighborhoods

Create a continuous river greenway, connect neighborhoods to the river, extend open space, recreation, and water quality features into neighborhoods, enhance river identity, incorporate public art along the river (Fig.1-6-7).

3)Capture Community Opportunities

Make the river the focus of activity, Foster civic pride, Engage residents in the community planning process and consensus building, Provide opportunities for education and public facilities, Celebrate the cultural heritage of the river (Fig.1-6-8).

4)Create Value

Improve the quality of life, increase employment housing and retail space opportunities, create environmentally-sensitive urban design and land use

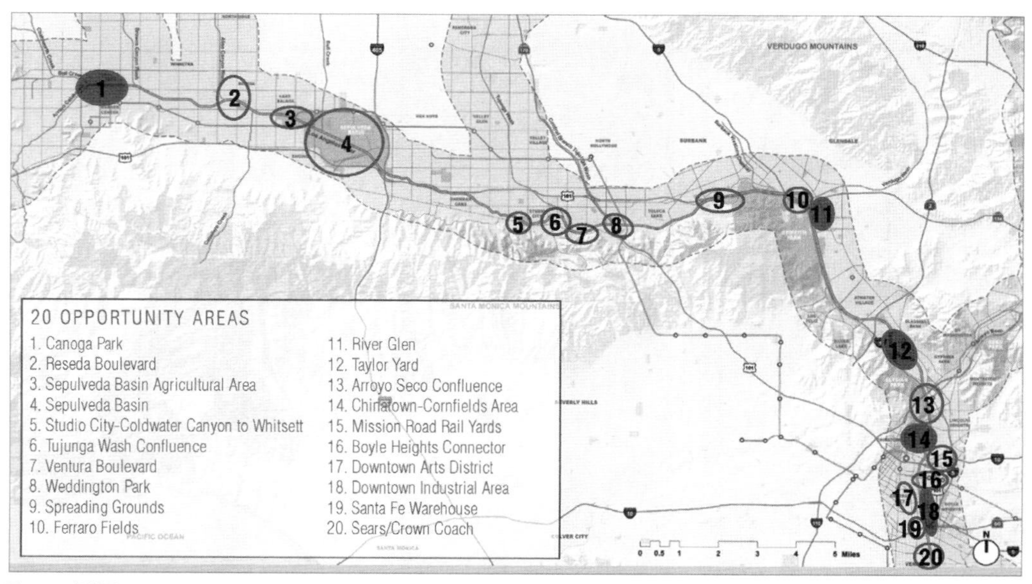

图1-6-8 机遇区
Fig.1-6-8 Opportunity areas

导则，集中精力于未充分开发地区和条件不佳的社区。

2. 河流复兴具体措施

1) 增强蓄洪能力(前提条件)

确定河道外部的洪峰蓄洪能力，以将河水流速减缓至临危数值内（小于12英尺每秒，约3.65米每秒）。这有利于维护和重建河流植被系统（图1-6-9）。

确定有选择性地获取让河流覆盖更多范围的机会，以扩大河流的冲积平原。

2) 提高水质(近期改进措施)

强调具有综合效益的景观处理和绿色基础设施；施行多尺度的水质处理以实现效率最大化（图1-6-10，图1-6-11）；在河流交汇处建造景观化的水质处理设施以净化来自支流的污染物（图1-6-12）；在河道内建造“理水阶梯”来处理浮于河流表面的雨水流（图1-6-13）；在河岸顶部、毗邻的带状公园和街道中设计景观化的“绿带”，以处理来自街道的雨水径流（图1-6-14，图1-6-15）。

3) 强化安全的公共通道(近期改进措施)

为安全亲水通道提供机会，确保人们能够快速地离开河道，并建立一个河水流速过高情况下

opportunities and guidelines, focus attention on underused areas and disadvantaged communities.

2. Recommendations of Revitalization

1)Enhance Flood Storage(Preconditions for Revitalization)

Identify opportunities for peak flood storage outside the channel to reduce flow velocities in the River to sub-critical (less than 12 feet per second) levels. This will support the maintenance and reestablishment of vegetation (Fig.1-6-9).

Identify opportunities for selective acquisition of additional rights-of-way to expand the River's floodplain.

2)Enhance Water Quality(Near-Term Improvement)

Emphasize multiple-benefit landscape treatments and "green infrastructure" improvements; Implement water quality treatment at multiple scales to maximize efficiency (Fig.1-6-10, Fig.1-6-11); Create landscape-based water quality treatment at major confluences of the River to treat pollutants carried by tributaries(Fig.1-6-12); Develop "treatment terraces" within the channel to treat stormwater flows that "daylight" or surface in the River(Fig.1-6-13); Create landscape-based "green strips" at the top of Riverbanks and in adjacent linear parkland and streets to treat stormwater runoff from streets(Fig.1-6-14, Fig.1-6-15).

3)Enable Safe Public Access(Near-Term Improvement)

Provide opportunities for safe access to the water, ensure that people can quickly exit the channel, and establish a flood warning system in the event of high flow conditions; Provide opportunities for temporary

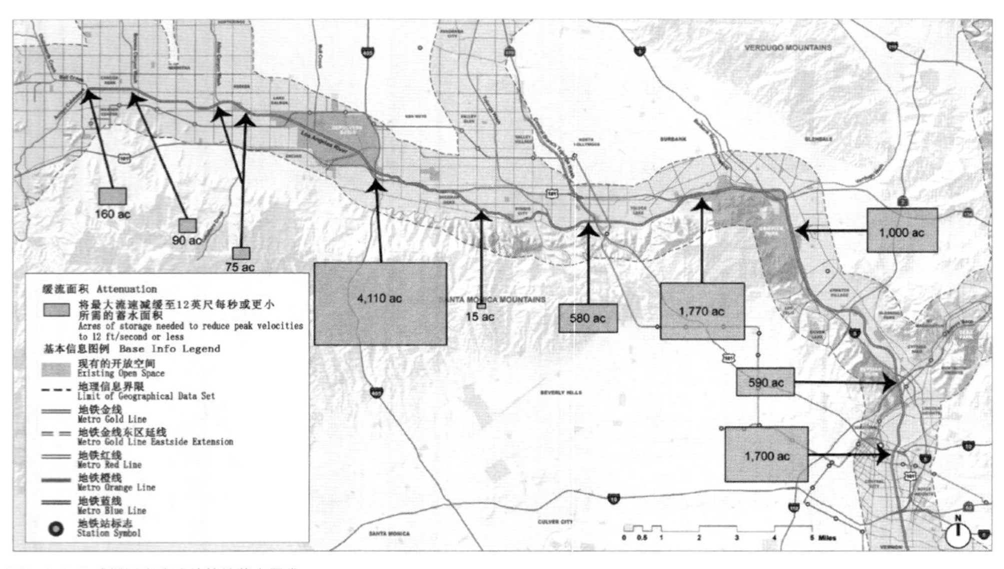

图1-6-9 为减缓河水流速估算的蓄水需求

Fig.1-6-9 Estimated water storage needs to reduce river flow velocities

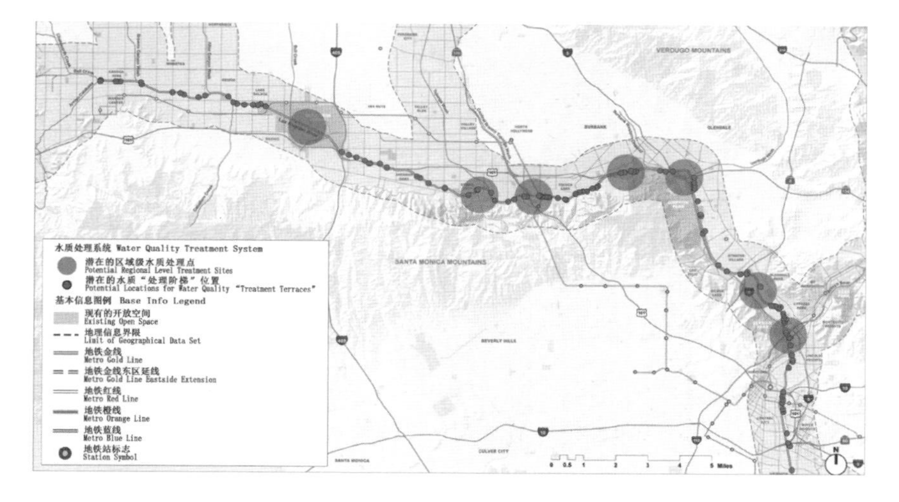

图1–6–10 可能的区域和河道内水质处理区
Fig.1–6–10 Potential regional and in-channel treatment areas

图1–6–11 位于泰勒角的潜在的大区域水质处理湿地区示意
Fig.1–6–11 An illustration of a potential large regional water quality wetlands area at Taylor Yard

图1–6–12 位于阿洛约・塞科河道和洛杉矶河交汇处的水质处理系统
Fig.1–6–12 Potential water quality system at the confluence of the Arroyo Seco and the River

图1–6–13 结合当地植物的河道内“理水阶梯”
Fig.1–6–13 An illustration of in–channel water quality treatment terraces that incorporate native plantings

图1–6–14 洛杉矶动物园停车场内的一个“绿带”，用于过滤雨水径流
Fig.1–6–14 A “green strip” in the Los Angeles Zoo parking lot filters stormwater runoff

图1–6–15 这种“雨花园”可用于公共或私有场所
Fig.1–6–15 This “rain garden” can be used on public and private sites

的洪水预警系统；安装充气式橡胶堤坝，为用于亲水游憩活动的临时性水池和湖泊提供机会（图1–6–16）；创造多样化的公共空间，包括小型袖珍

pools and lakes for water-based recreation by installing inflatable rubber dams (Fig.1-6-16); Create a variety of public spaces, including small pocket parks, natural areas, and urban plazas and civic spaces in “reclaimed” areas of the channel (Fig.1-6-17); Ensure public safety by using alternate “greening” techniques in areas where

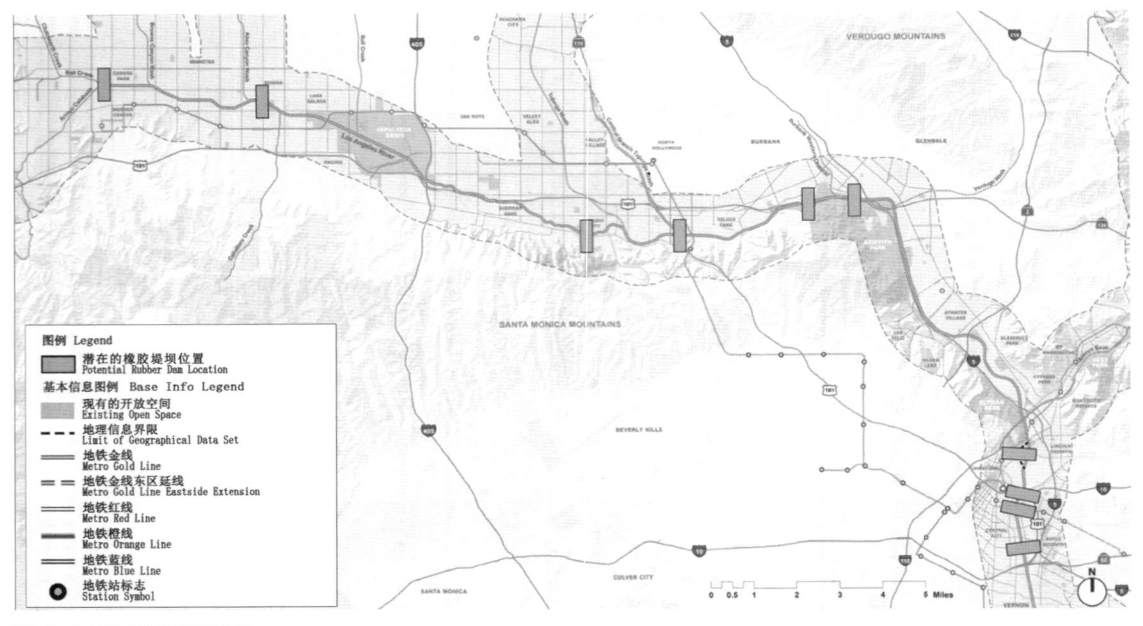

图1-6-16 可能的的橡胶堤坝位置

Fig.1-6-16 Potential rubber dam locations

图1-6-17a 图洪加河下游（现状）

Fig.1-6-17a Downstream of Tujunga wash (existing)

图1-6-17b 图洪加河下游（将来）

Fig.1-6-17b Downstream of Tujunga wash (future)

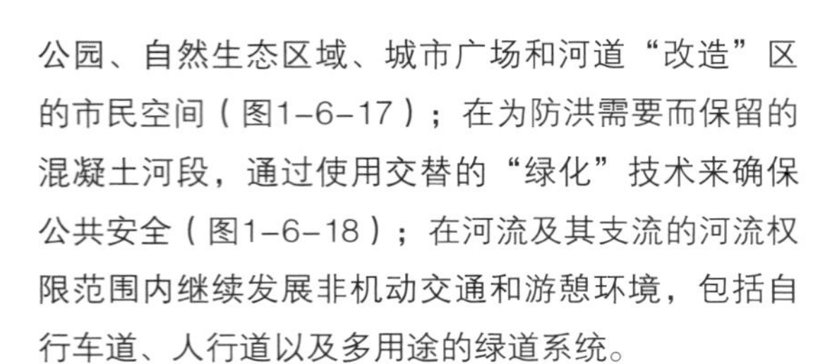

公园、自然生态区域、城市广场和河道“改造”区的市民空间（图1-6-17）；在为防洪需要而保留的混凝土河段，通过使用交替的“绿化”技术来确保公共安全（图1-6-18）；在河流及其支流的河流权限范围内继续发展非机动交通和游憩环境，包括自行车道、人行道以及多用途的绿道系统。

4）恢复功能性滨河生态系统（远期愿景）

建造一条连续的功能性滨河廊道，为鸟类、哺乳动物、两栖动物、爬行动物、无脊椎动物以及河道底部的鱼类提供栖息地（图1-6-19）；将这条廊道与其他重要栖息地以及沿支流进入山区的迁徙路线连接起来（图1-6-20）；改善水质，并提供能维

the concrete remains necessary for flood damage prevention(Fig.1-6-18); Continue development of non-motorized transportation and recreation elements including bike and pedestrian paths and multiuse trails in the River and tributary rights-of-way.

4)Restore a Functional Riparian Ecosystem(Long-Term Vision)

Create a continuous functional riparian corridor that provides habitat for birds, mammals, amphibians, reptiles, invertebrates, and fish within the channel bottom(Fig.1-6-19); Connect this corridor to other significant habitat and migration routes along the tributaries and into the mountains(Fig.1-6-20); Improve water quality and provide fish passages, ladders, and riffle pools that would support desirable fish species, including steelhead trout if feasible; Bio-engineer the River's edge where feasible to create and restore wildlife habitat along the upper reaches of the River.

图1-6-18a 由自行车道、本地藤植物和自然带来"绿化"混凝土河道示意图

Fig.1-6-18a An illustration of "greening" the concrete with a trail, native vines, and natural

图1-6-18b 由自行车道、本地藤植物和自然带来"绿化"混凝土河道示意图

Fig.1-6-18b An illustration of "greening" the concrete with a trail, native vines, and natural

图1-6-19a 河底用本地植物得以恢复的滨水廊道，洛杉矶市中心第一大街段（现状和将来）

Fig.1-6-19a The channel bottom with a restored riparian corridor native plants, section above 1st Street in Downtown Los Angeles

图1-6-19b 河底用本地植物得以恢复的滨水廊道，洛杉矶市中心第一大街段（现状和将来）

Fig.1-6-19b The channel bottom with a restored riparian corridor native plants, section above 1st Street in Downtown Los Angeles

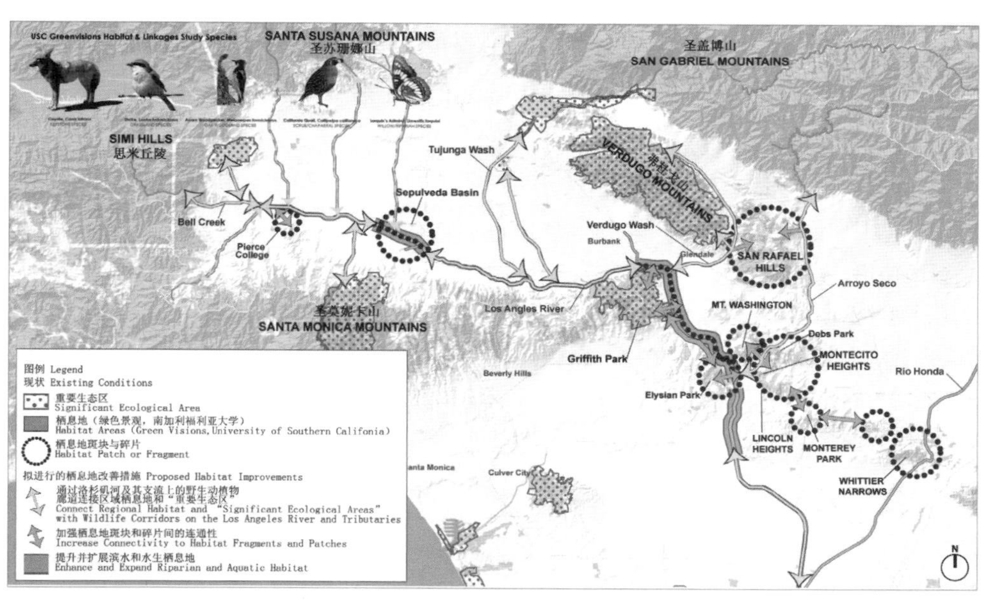

图1-6-20 经改造后的栖息地有希望建立连接的区域

Fig.1-6-20 Areas where improved habitat connectivity are desirable

持足够种群的鱼类（可能条件下应包括铁头鲑鱼）生存的鱼道、阶梯或水道和浅滩；沿河上游河段尽可能地采用生物工程来处理河流边缘，以营造和恢复野生动物栖息地。

3. 机遇区设计

1) 格伦河

格伦河机遇区的备选方案A的愿景，需要可回收

3. Opportunity Areas Designing

1)River Glen

The vision for the River Glen Opportunity Area's Alternative A, which calls for the acquisition of the metal-recycling facilities, allows for its transformation into almost 15 acres of new, functional, riparian habitat and water quality treatment wetlands that terrace gently from Doran Street down to Verdugo Wash (Fig.1-6-21).

The water quality wetlands are part of the City's comprehensive stormwater management strategy, which establishes regional-scale, landscape-based

金属的设施，以建设为15英亩（约6公顷）新型有效的滨水栖息地及从多伦街流向瓦多戈河的经过水处理的湿地（图1-6-21）。

湿地水质处理是城市综合雨水管理系统工程的一部分，在区域尺度上的每条支流的交汇处建设景观化的水处理设施。这15英亩的地区有能力过滤掉大量地表径流中含有的金属和化肥一类的营养物。在有效性这一点上，如此规模的基地非常有价值，同时，因为支流的水往往会夹杂污染物，截留和处理这些来自上游和河流交汇处的污染物会带来很大的生态效益（图1-6-22）。

在河流交汇处的这一大片湿地和滨水栖息地的恢复也使得上下游的栖息地重建联系，这样可以一直绵延至上游的瓦多戈河和瓦多戈山。

通过一系列的木栈道和瞭望台，人活动的区域和鸟类筑巢捕食的区域得以隔离缓冲，使干扰得以减小至最小。游客可以通过改造后的多伦街的十字路口到达这个自然生态公园（图1-6-23）。

2）泰勒角

在泰勒角地区，是基于以下情境规划的：G2地区因过去轨道交通的使用，已经遭受了很严重的污染。为发挥其作用，通过挖走沿河污染的土壤，拢成堆，营造了一个雕塑式的土地形态，与上游的草场式景观相呼应。被挖走泥土的地方将被覆盖，防止污染物渗透进地下水，一系列的湿地水质处理过程可以在

treatment at the River's confluence with each tributary. This 15-acre site has the potential for removing significant quantities of metals and nutrients, such as fertilizers, carried in stormwater runoff. From an efficiency standpoint, sites of this size are quite valuable. Also, because the tributaries carry significant pollutant loads, capturing and treating pollutants upstream in the tributary as well as at the confluence yields substantial ecological benefits (Fig.1-6-22).

The reestablishment of large wetland and riparian habitat zones at the confluence also reconnects upstream and downstream habitat within the River to very significant habitat zones further upstream in Verdugo Wash and the Verdugo Mountains.

A series of boardwalks and overlooks wind through the wetlands; these human-use areas are carefully buffered from shorebird nesting and feeding areas to minimize disturbance. Visitors are able to reach this new natural-area park from an improved Doran Street crossing (Fig.1-6-23).

2)Taylor Yard

The Plan's vision for Taylor Yard is based on the following scenario: the G2 site was heavily contaminated from past rail uses. To make it functional, contaminated soils can be excavated from the area next to the River's edge and used to create a series of mounded, sculptural landforms that support an upland, meadow-like landscape. The excavated area will be capped to prevent infiltration of contaminants into the groundwater, and a series of water quality treatment wetlands can be constructed on top of the cap, using imported soils that were stabilized to prevent erosion from the force of flood flows. This large site can be a very efficient regional water quality treatment facility, capturing runoff from very large box culverts that transport stormwater runoff that

图1-6-21a 格伦河首选方案（A方案）

Fig.1-6-21a River Glen preferred alternative (Alt. A)

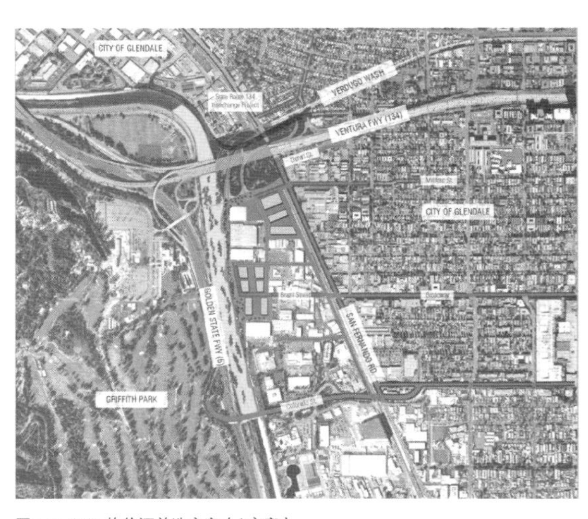

图1-6-21b 格伦河首选方案（A方案）

Fig.1-6-21b River Glen preferred alternative (Alt. A)

图1-6-22 瓦多戈河与洛杉矶河交汇处的水质处理湿地

Fig.1-6-22 Water quality

图1-6-23 在瓦多戈河畔游客可以享有一个美丽的开放空间公园

Fig.1-6-23 Visitors

图1-6-24a Taylor Yard首选方案

Fig.1-6-24a Taylor Yard preferred alternative1

图1-6-24b 泰勒角首选方案

Fig.1-6-24b Taylor Yard preferred alternative2

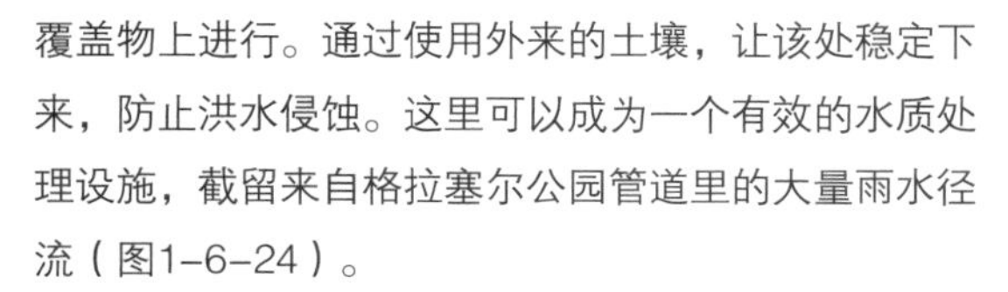

覆盖物上进行。通过使用外来的土壤，让该处稳定下来，防止洪水侵蚀。这里可以成为一个有效的水质处理设施，截留来自格拉塞尔公园管道里的大量雨水径流（图1–6–24）。

G2地区的处理将会还原河流的最初的自然生态。沿河的步道和小径联系了该地区的开敞空间，联系了阿洛约·塞科河道，乐土公园和“红色汽车走廊”。在与阿洛约·塞科的交汇处，一个区域性的通道可以用来联系阿洛约和其丰富的网络，步道和开敞空间。河流步道在两岸延续着机会区域，一直往南，利用了现有的19大街桥上的步行道，穿过110号高速公路桥下，以及其南都的铁路桥。该区域的健身和交通环道连结了河道、乐土公园、“红色汽车走廊”，以及阿洛约·塞科河道的交汇处（图1–6–25）。

3)唐人街—玉米地区域

这条河是周边地区的主要景观，通过改造，它的

emerges from pipes in Glassell Park (Fig.1-6-24).

The acquisition of the G2 parcel would allow for the creation of a premier natural River area. A River Promenade and River Trails can connect the area's open space, and link the Arroyo Seco, Elysian Park, and the Red Car Corridor trails. At the confluence with the Arroyo Seco, a Regional Gateway can celebrate the connection with the Arroyo and its rich network of trails and open space. River trails continue south of the Opportunity Area on both sides, using the designed pedestrian sidewalks on the Avenue 19 bridge and moving under the 110 Freeway and the railroad bridge just south of the Riverside Drive bridge. Fitness and transportation loops in this area can connect the River with Elysian Park, the Red Car Corridor, and with the Arroyo Seco Confluence (Fig.1-6-25).

3)Chinatown-Cornfields Area

The River is the main attraction in this neighborhood. Its volume and steepness through this reach has allowed for the creation of a secondary "diversion" channel that creates recreational access and use. On most weekends in good weather, kayakers in great numbers flock to this area for a chance to

图1–6–25 新建的人行天桥和可能的河道内下穿式桥将把塞科河与洛杉矶河绿道连接起来
Fig.1–6–25 A new pedestrian bridge and possible in–channel Bridge undercrossing connect the Arroyo Seco with the Los Angeles River Greenway

流量和湍急程度可以使其具有第二"游憩"的选择，开发了其游憩可达性和使用性。在大多数晴好的周末，大量的皮划艇爱好者会来到这里在河中荡舟。设计者为了这个项目，在下游建了一座充气橡皮水坝，使其成为一个更像池塘的区域（图1–6–26，图1–6–27）。

作为这类新型游憩的延伸补充，占地20英亩（约8公顷）的社区公园和剧场，公园和河岸成为这个有活力的，新生的混合功能的社区的核心。小剧场沿着河道的坡延伸，在水岸边建起一座舞台，供夏天夜晚音乐会演出用，遍布公园的步道延伸系统为早起和工作后的慢跑提供了可能（图1–6–28）。

4. 实施与管理

1）社区规划框架

本规划将在现行的城市规划和分区背景下实施。采纳本规划后的第一步，便是通过一个包括所有相关社区的参与过程，对洛杉矶河流经的区域进行社区规划的更新。先行的社区规划进程是规范河流复兴建议措施的最适宜方式，因为它为每个与众不同的社区提供了一个机会以调整河流发展方向，来满足地方股东所表达的不同意见。分区变化可能也要遵循这些规划更新。

为了完善社区规划程序，将划定一个河流改造覆盖区，并具备以下三个重要功能，在私人地产与河流之间建立一个高质量界面，增加开放空间并改善环境质量，建造充满活力的通往河流的步行街。

paddle in the River. Downstream of the kayak course, an inflatable rubber dam creates a small, ponded area (Fig.1-6-26, Fig.1-6-27).

Complementing this new recreational stretch of the River is a 20-acre community park and amphitheater. This parkland and River edge form the heart of a vibrant, emerging, mixed-use community. The small amphitheater along the diversion channel slopes to a stage at the water's edge, providing a venue for summer evening concerts, while an extensive system of trails through the park provide opportunities for early-morning or after-work jogs (Fig.1-6-28).

4. Implement and Management

1)A Community Planning Framework

Implementation of this Revitalization Master Plan would take place within the existing City planning and zoning context. An important next step, following adoption of this Plan, would be to update existing Community Plans in areas that include the River, through an inclusive community involvement process. The City's established community planning process is the most appropriate way to formalize revitalization proposals because it gives each unique neighborhood an opportunity to tailor River developments to the sentiments expressed by local stakeholders. Zoning changes may also follow these Plan updates. To complement the Community Plan process, a River Improvement Overlay (RIO) district would be created, with three important functions: Establish a high-quality interface between private property and the River; Increase open space and improve environment quality; Create active pedestrian streets leading to the River.

2)A Management Structure

A three-tiered structure is proposed for managing a revitalized Los Angeles River. Because of the multiple public entities with jurisdiction over various aspects of the River, the management structure must be

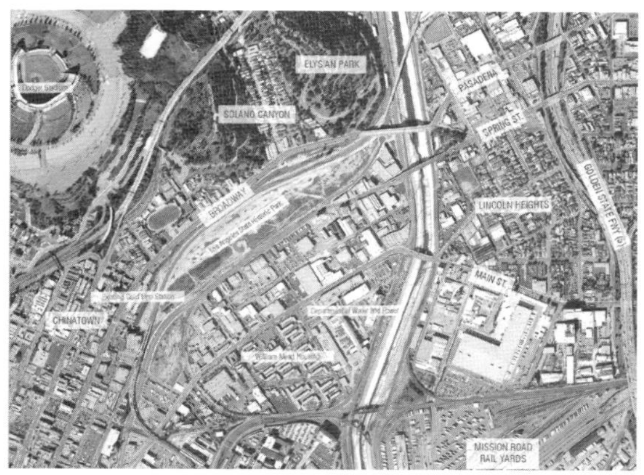

图1-6-26a 唐人街—玉米地区域首选方案（B方案）
Fig.1-6-26a Chinatown-Cornfields Area preferred alternative (ALT.B)

图1-6-26b 唐人街—玉米地区域首选方案（B方案）
F ig.1-6-26b Chinatown-Cornfields Area preferred alternative (ALT.B)

图1-6-27 因橡胶堤坝的安装而创造出的沉塘区的一条河道
Fig.1-6-27 A new channel with a ponded area created by installation of a rubber dam

图1-6-28 从教会路铁路站往北看
Fig.1-6-28 Looking north from the Mission Road Rail Yards

2) 管理构架

为管理复兴后的洛杉矶河，本规划提出了一个三级管理构架。由于多个公共机构拥有对洛杉矶河多方面的权限，管理构架必须足够综合和灵活才能使这些机构协同工作，并在必要时相互独立。

本规划所提出的管理构架包括三个方面：①政府：包括洛杉矶市、洛杉矶郡以及美国陆军工程公司在内的洛杉矶河管理局，作为一个联合权力机关，将作为管理洛杉矶河的主要机构。它拥有的权利和责任包括：洛杉矶河重建，路权管理和维护，承担公共义务、许可和土地开发的责任；②企业：由公共和私人资金直接支撑的洛杉矶河复兴公司将是一个非盈利公司，负责洛杉矶河相关及毗邻社区的复兴工程；③慈善：洛杉矶河基金会是一个由私人个体组建的非营利团体，主要支持本规划的河流复兴目标。

comprehensive, flexible enough to allow these entities to work in collaboration, and comprehensive enough to proceed independently when necessary.

The Plan's proposed River management structure includes three elements:1)Governmental: The Los Angeles River Authority, a joint powers authority (JPA) that includes the City of Los Angeles, the County of Los Angeles, and through a memorandum of understanding (MOU), the U.S. Army Corps of Engineers. The JPA would serve as the principal entity with authority and responsibility for River reconstruction, right-of-way management and maintenance, assuming responsibilities for public liability, permitting, and land development. 2) Entrepreneurial: The Los Angeles River Revitalization Corporation would be a not-for-profit entity charged with directing public and private financing for River-related and neighborhood revitalization projects. 3)Philanthropic: The Los Angeles River Foundation would be a not-for-profit body established by private individuals to support the Plan's revitalization goals.

This new management structure would enable the

这个管理构架将使得洛杉矶市及其合伙人能够长期有效地致力于洛杉矶河复兴，以确保洛杉矶河能为后世所享。

四、 总结

洛杉矶河流经洛杉矶市中心，该项目中，滨水区域生态功能与游憩功能的恢复对提升城市景观品质具有深远意义。城市问题学家芒福德在《城市发展史》中倡导保持城市的林木绿地，阻止城市无限制生长吞噬绿色植物、破坏城乡生态环境。不仅需要农业和园艺地、天然园地，还要增加活动场所。其中，特别强调休闲场所的邻近性，当邻近土地被侵占，除了保护外，所依赖的是更远处的休闲用地。但由于距离远，公共使用程度低，交通时间长，费用贵，休闲土地的价值丧失，将导致人们加倍涌入其他的自然风景区。因此，要创造性的利用景观，使城市变得自然而适于居住，就要发挥城市邻近休闲场所的最大功效。洛杉矶河滨水区正是城市中心的绿脉，绿色空间的营造不仅满足了游憩功能的需求，更保护了人们赖以生存的自然环境。

本规划提出了四项基本组织原则和一系列的支撑目标，以及无数达成这些目标所需的具体措施。随着本规划的实施，它将带来多方面的综合效益：

社会方面：洛杉矶河提供了开放空间、安全的公共通道以及非机动车路线等，这将使其成为洛杉矶市的活动中心以及康乐场所。而且，它将培养市民自豪感，并增强社区凝聚力。

环境方面：作为洛杉矶市的“绿色脊梁”，洛杉矶河将为居民和自然环境提供各种类型的交流和相互影响的空间，从而连接人类社会和自然界。作为一条连续的功能性滨水廊道，它将为各种类型的生物提供栖息地，并与其他重要栖息地以及迁徙路线相连。此外，通过户外课堂以及自然环境的体验，河流将会激发孩子们对环境和科学的热爱。

文化方面：通过彰显洛杉矶河的文化遗产、历史以及未来，以及可以增强其具有识别性特征的沿河公共艺术设计，它将成为洛杉矶市的心脏和灵魂。

City and its partner to maintain a long-term focus on River revitalization in order to ensure that the River remains a priority for future generations.

IV. Summary

Los Angeles River flows through downtown Los Angeles, so it has profound significance to enhance the quality of the urban landscape with the recovery of the ecological and recreational functions of waterfront area. Mumford, expert of urban problems, advocates in the book The City in History that it needs not only agricultural, horticultural and natural land, but also increased activity places, to keep the city's green land, prevent the growth of cities from unlimitedly swallowing the green plants and destroying ecological environment in urban and rural areas, which, with particular emphasis on the proximity of the leisure facilities, when the neighboring land was occupied, it has to rely on more distant leisure land. Because of the far distance, low level of public use, long traffic time, high costs, and the loss of the value of recreational land, more and more people will swarm into other natural scenic area. As a result, it is necessary to give full play to the functions of the neighboring urban recreational space to make creative use of the landscape, so that the city becomes more natural to live in. The waterfront area of Los Angeles River is the greenway of the central city, the creation of green space not only meet the need of recreational functions, but also protect the natural environment people live in.

This Plan includes four basic organizing principles, series of supporting goals, and numerous specific recommendations to accomplish the goals. And with the implement of this Plan, it will bring multi-benefits in several aspects:

Social: The River provides open spaces, safe public accesses, and non-motorized routes that would make the River the activity focus and civic amenity of the City. Further, it would foster civic pride, and thereby bringing communities together.

Environmental: To function as the "green spine" of the city, the River would connect nature and communities, providing space for kinds of communication and interaction between the residents and the natural environment. And as a restored continuous functional riparian corridor, the River would provide habitat for all kinds of living beings and connect to other significant habitat and migration routes. Furthermore, through the experience in the outdoor classroom and natural environment, it would inspire the children's interest in the environment and the sciences.

Cultural: With the celebration of the cultural heritage, history and future of the River, and the incorporation of public art along the River that will enhance its identity, the River would become the heart and soul of the City.

Economic: With the physical access and visual corridor it provides, and the enhancement of its attractiveness, the River would induce new residential, commercial development and thereby encourage

经济方面：凭借其提供的物理通道及视觉走廊，及其吸引力的增强，洛杉矶河将吸引新的房地产业和商业来此发展，从而增加就业、促进经济繁荣。

洛杉矶河的复兴规划延续了城市的发展脉络，并进行了独特的创新，昭示了洛杉矶市未来的发展。规划中强调了河流本身条件的局限和不足，并提供了切实的改进措施。增强蓄洪能力，改善环境水质以及强化安全通道都是重要的安全问题。不难想象，复兴后的洛杉矶河将会是绿树与河水交相辉映的一幅迷人画卷。在整个洛杉矶市的生态系统中，它为各种生物提供栖息地，削弱了整个城市工业化的印象，凸显了城市的人文气息。

employment and increase economic prosperity.

The revitalization panning of Los Angeles River continues the development history of the city, and carries out many unique innovations; it indicates the future development of the city of Los Angeles. This plan emphasizes the limitations and shortcomings of the river's conditions, and then provides practical improvement measures. Enhancing the flood storage capacity,improving the water quality and strengthening the safe access are all important security issues. It's not difficult to imagine that, the Los Angeles River will become a charming picture illuminated with green trees and clear water after the revitalization. In the ecosystem of the entire City of Los Angeles, the River provides habitat for a variety of organisms, and weakens the impression of industrialization, highlights the city's cultural atmosphere.

思考题

1. 本规划是否如篇首所述，有效彰显了洛杉矶河以及洛杉矶市的过去和将来的特性？若是，是如何彰显的？反之，你认为应从哪些方面来实现这一目的？

2. 结合本案例，思考项目背景的调查分析和研究对生态规划的价值和意义在于何处？

3. 在机遇区的具体设计中，本规划的规划目标、理念和原则是如何得以体现的？

4. 思考“增强蓄洪能力”、“提高水质”、“强化安全的公共通道”、“恢复功能性滨河生态系统”这四方面相互依存和影响的关系，以及在时序上为何进行规划中的先后安排？

5. 思考规划后的实施与管理的重要性，本案例中的实施与管理办法和措施对我国的景观生态规划实践有何借鉴意义？

Questions

1. Does this plan celebrate the past and the future of Los Angeles River and the City of Los Angeles effectively as the beginning of section described? If so, how does it celebrate that? If not, how to achieve this purpose in your opinion?

2. What is the value and significance of the project background investigation and research for ecological planning in this subject?

3. How the goals, ideas and principles of this plan are embodied in the specific design of opportunity areas?

4. Think about the interdependent and interactive relationship of the four recommendations of “Enhance flood storage ”, “Enhance water quality”, “Enable safe public access”, “Restore a functional riparian ecosystem”, and why does it make such timing arrangement of them in this plan?

5. Think about the importance of the implementation and management after planning, and what can we learn from the implementation and management methods and measures in this case to improve the landscape ecological planning practice in China?

注释/Note

图片来源/Image：本案例图片均引自《洛杉矶市洛杉矶河复兴总体规划文本》/LOS ANGELES RIVER REVITALIZATION MASTER PLAN, April, 2007.

参考文献/Reference：本案例引自《洛杉矶市洛杉矶河复兴总体规划文本》/LOS ANGELES RIVER REVITALIZATION MASTER PLAN, April, 2007.

推荐网站/Website：http://lariver.org/index.htm ;http://folar.org/ ;http://cityplanning.lacity.org/

第二章 湖泊景观生态规划设计

CHAPTER II Lake Landscape Ecological Planning & Design

第一节 圣地亚哥潟湖规划 Section Ⅰ The Plans for San Dieguito Lagoon
第二节 圣埃利荷潟湖规划 Section Ⅱ San Elijo Lagoon Planning

第一节 圣地亚哥潟湖规划

Section I The Plans for San Dieguito Lagoon

一、 项目综述

1. 项目简介

圣地亚哥潟湖是加利福尼亚州剩余不多的沿海湿地之一。与其他几个潟湖相比，它的水域面积最大，曾经是整个县最大的潟湖。水质监测表表明，与距离圣地亚哥潟湖不远的圣埃利荷潟湖相比，水质较差。鉴于当时环境保护者与城市发展者争论的困境，应圣地亚哥县环境发展署要求，莱尔领导设计实验室成员，一起研究圣埃利荷潟湖以及其他沿海潟湖规划问题，为其制定土地规划政策及指导方针。

20世纪末，南加州爱迪生电力公司与圣地亚哥河谷公园联合管理局合作实施圣地亚哥潟湖河谷恢复项目。南加州爱迪生电力公司是圣欧诺佛瑞核电厂的主要所有者与经营者。由该电力公司提出的最终恢复规划主要关注湿地恢复工作，作为加州圣地亚哥圣欧诺佛瑞核电厂污染导致鱼类减少的补偿性缓解措施，进行在圣地亚哥潟湖之内至少150公顷的沿海湿地的创新性恢复或实质性恢复工作。圣地亚哥河谷公园联合管理局在《加州环境质量法》，与美国渔业及野生动物局的管理下，担当州领导重任，在《国家环境政策法》监督下，担当联邦政府的领导机构。圣地亚哥河谷公园联合管理局为了河谷地区未来发展，将南加州爱迪生电力公司的湿地恢复项目纳入开放空间公园整

I. Overview

1. Introduction

San Dieguito Lagoon is one of the few remaining coastal wetlands in California. The San Dieguito Lagoon, which has the largest watershed of San Diego coastal lagoons, was once the largest lagoon in the County. As evidenced by measurements of water quality, the ecological condition of San Dieguito Lagoon is slightly worse than that of San Elijo, which lies just a few miles away. With the issue in a stalemate, the Environmental Development Agency of San Dieguito asked for the Laboratory for experimental Design to study the problems at hand and recommend land planning policies and guidelines for San Elijo and several other coastal lagoons where similar conflicts were emerging.

Southern California Edison (SCE), representing SONGS' owners, is working in partnership with the San Dieguito River Park Authority (JPA) to carry out the restoration project. Southern California Edison Company (SCE) is the majority owner and operator of the San Onofre Nuclear Generating Station(SONGS). The FRP focuses primarily on the wetlands restoration effort proposed by SCE to fulfill the Permit conditions, which is the creation or substantial restoration of at least 150 acres of Southern California coastal wetlands within SDL as compensatory mitigation for fish losses caused by SONGS. The San Dieguito River Park Joint Powers

Authority (JPA) took the role of state lead agency under the California Environmental Quality Act (CEQA) and the U.S. Fish and Wildlife Service (USFWS) took the role of federal lead agency under the National

体项目之中（简称公园项目）。

圣地亚哥河谷公园概念性规划（1994，2002）重点规划区域范围从沃坎山东侧的沙漠一直延展到德尔马海岸，全程达55英里（约88.5公里）。河谷公园重点规划区包括14个不同资源特征的景观单元（图2–1–1）。景观单元A（图2–1–2）就是德尔马沿海潟湖，其中一部分概念性规划为恢复圣地亚哥潟湖以及附属的湿地生态系统。

恢复过程将包括扩大既有潮汐盆地范围；创建沿海湿地多样化生境；恢复密切相关的既有潮汐湿地，从而创建出具有开放水资源、湿地、草地、山地等多种生境的景观单元，同时，该单元也兼有功能性、生态性、水文性。

应《海岸开发许可条例》的要求，南加州爱迪生电力公司在1997年9月递交了《初期恢复计划》，不久，加州海岸委员会便通过了该项计划。按照审批步骤，依据加州品质环境品质法与国家环境政策法，整个项目进入到环境审查阶段。这项工作所需提交的环

Environmental Policy Act (NEPA). The JPA incorporated the SCE wetland restoration project into their overall Open Space Park Project (Park Project) for the San Dieguito River Valley area.

The San Dieguito River Park Focused Planning Area extends for 55 miles from the desert just east of Volcan Mountain to the ocean at Del Mar. The San Dieguito River Park FPA contains a sequence of landscapes that have 14distinctly different characteristics (Fig 2-1-1). Landscape Unit A(Fig. 2-1-2) is Del Mar Coastal Lagoon. This Concept Plan endorses the proposal to restore the San Dieguito Lagoon and its associated wetlands ecosystem. Restoration would involve the enlargement of the existing tidal basin, creation of a variety of coastal wetland habitats, and the restoration of associated upland habitat in order to create a functional, ecological, and hydrological unit that will provide for tidal flushing, open water, wetlands, and grassland and other upland habitat.

As required by the Permit, SCE submitted a Preliminary Restoration Plan in September 1997 which was later approved by the CCC in November 1997. Following the approval process, the wetland restoration project entered the environmental review process, pursuant to the California Environmental Quality Act (CEQA) and the National Environmental

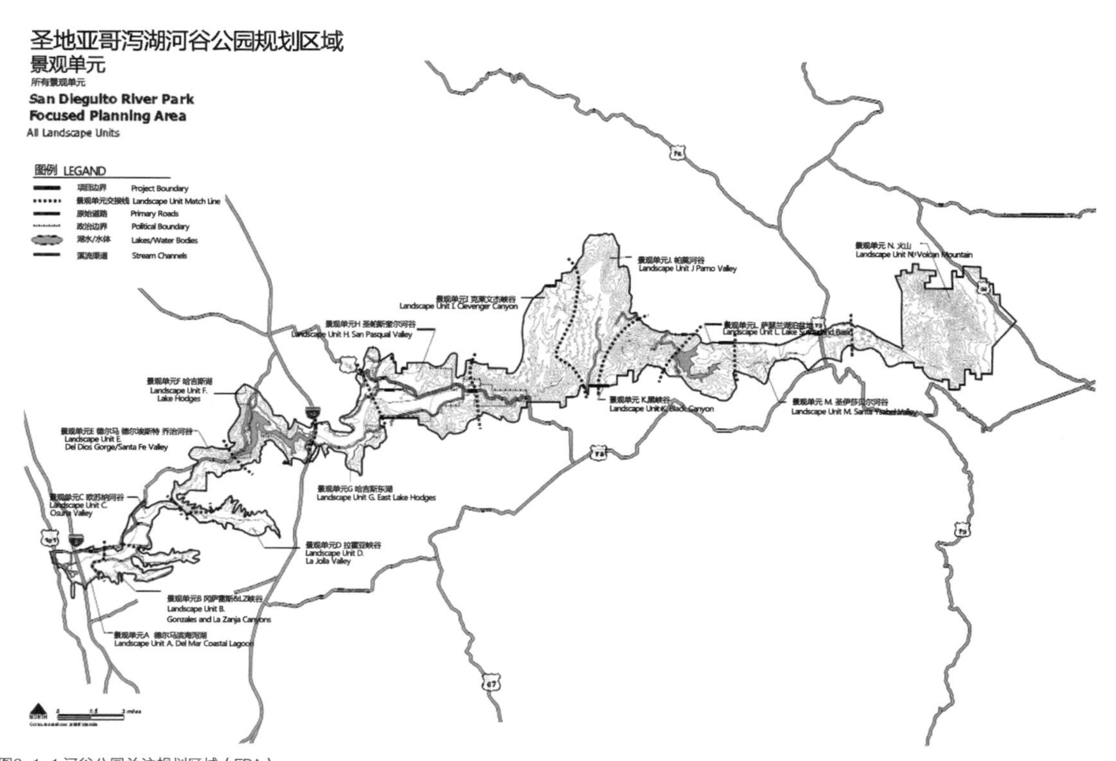

图2–1–1 河谷公园关注规划区域（FPA）

Fig.2–1–1 The San Dieguito River Park Focused Planning Area

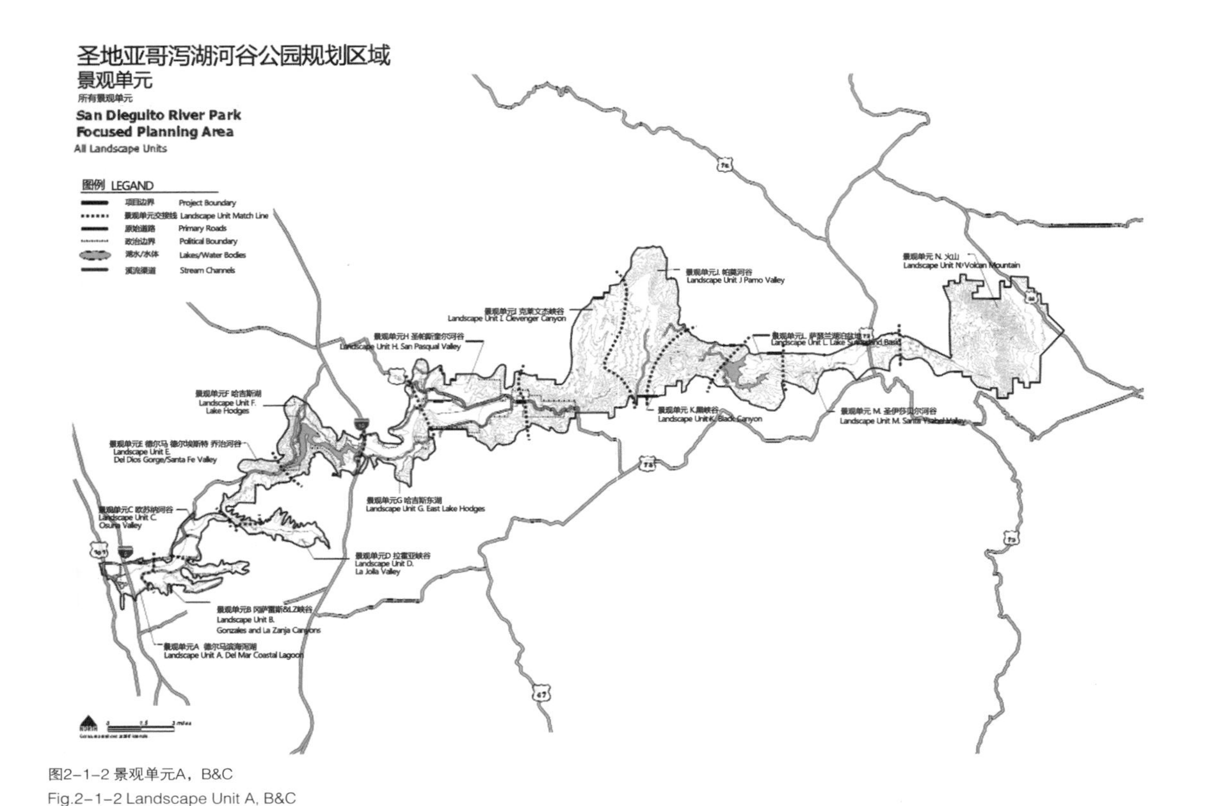

图2-1-2 景观单元A，B&C
Fig.2-1-2 Landscape Unit A, B&C

境影响报告由地方、州与联邦局（包括美国渔业及野生动物局与圣地亚哥河谷公园联合管理局以及美国陆军工程兵团）共同完成。经过包括技术专家在内的公开评议，2000年9月联合管理局授予该项目最终环境影响报告证明书。此环境影响报告为圣地亚哥湿地恢复（恢复项目）做准备。

2. 区域环境

最早描绘潟湖的地图可追溯至1887年，它描绘了数英里的潮汐甬道，沼泽、泥滩、从潟湖与内陆的入口处一直穿过现在的第5号州际公路位置（加州沿海管理局圣地亚哥县沿海湿地，1989年）。现今，潟湖缩减至I~5区西部，然而退化的沿海湿地延伸至其东部。城市开发已经悄然无声地蔓延到海岸，该地区历史性地经历了圣地亚哥河谷区域公园开放空间的重点规划区范围内最密集强度的城市发展压力。5号洲际公路，101公路，杰米·杜兰特林荫大道以及铁路都穿越了圣地亚哥潟湖。1935年

Policy Act (NEPA). This entailed a submittal of an Environmental Impact Report (EIR), completed by local, state and federal agencies including the US Fish & Wildlife Service, the San Dieguito River Park (JPA) and US Army Corps of Engineers. The JPA certified the Final Environmental Impact Report (EIR) for the project in September 2000, after considering extensive public comments, including recommendations of many technical experts. A joint environmental impact report/environmental impact statement (EIR/S) was prepared for the entire San Dieguito Wetlands Restoration (Restoration Project) component of the Park Project.

2. Environment

The earliest maps of the lagoon date from 1887 and depict several miles of tidal channels, marsh, and mudflat extending from the lagoon mouth well inland past the present location of Interstate 5 (The Coastal Wetlands of San Diego County, California State Coastal Conservancy, 1989). Presently, the lagoon is confined to the area west of I-5. However, degraded coastal wetlands continue to exist to the east. Urban development has crept much closer to its banks ; This landscape has historically experienced the

图2-1-3 圣地亚哥潟湖
Fig.2-1-3 San Dieguito Lagoon

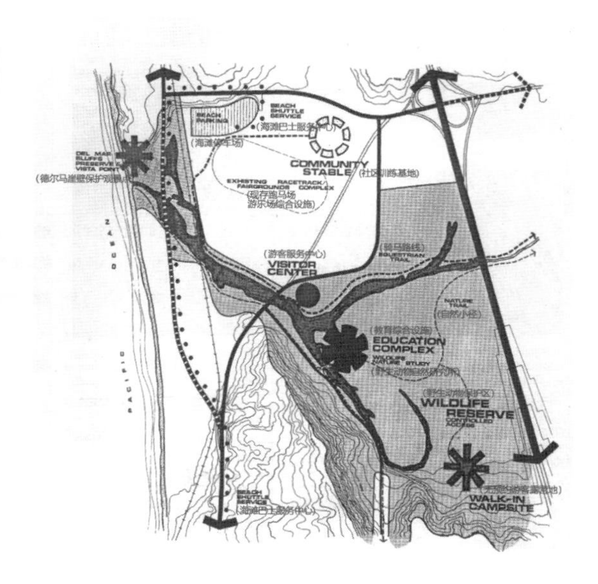

图2-1-4 主题方案 I
Fig.2-1-4 Alternative I — Diversified preserve

在圣地亚哥潟湖水域西北方向还建造了德尔马游乐场。商业迅速蔓延，沿着卢瓦尔河谷往西，到达I-5区东部，住宅开发范围集中在规划区域的北部及南部边缘。在整个景观单元的东北部最终还建造了马术运动运动设施（图2-1-3）。

二、圣地亚哥潟湖规划与设计（1985）

1. 三种不同主题的方案

本项目在规划过程中考虑了七种不同主题方案，本小标题仅介绍说明其中三种主题方案，每一种可能会附带图表，其总结概括了可预期的环境影响总体定性评价。并且以公众参与的形式向关心此项目的当地居民提供多种可能性选择及可预期影响。经过较长时间讨论，每个人都填写调查问卷来表达自己倾向的选择。所有结果立即被统计成表，并且宣布于众，便于日后更深层次的讨论。最后，每一个人填写第二次调查问卷，再一次表达自己的选择（图2-1-4～图2-1-6，表2-1-1～表2-1-3）。

2. 潟湖四大发展阶段

问卷调查结果强烈地影响了之前的优先选择，因为确切了解潟湖的动态变化不现实，发展行动可能产生意外的影响，发展阶段也需要时间进行尝试与修正，所以四大发展阶段——改善、恢复、再生

most intense development pressure of all of the landscapes within the San Dieguito River Valley Regional Open Space Park's Focused Planning Area (FPA). Interstate 5, Highway 101, Jimmy Durante Boulevard, and the railroad all cross the San Dieguito Lagoon, and in 1935, the Del Mar Fairgrounds was constructed in the northwestern portion of the San Dieguito Lagoon watershed. Commercial development occurs along Via de la Valle to the west and east of I-5, and residential development occurs within and surrounding the northern and southern edges of the planning area. Finally, equestrian facilities have been constructed at the northeastern end of this landscape unit(Fig. 2-1-3).

II. Planning and design(1985)

1. Three Different Thematic Alternatives

Seven quite different thematic alternatives were seriously considerate, three of them illustrated here. Each alternative is accompanied by a chart summarizing predictions were presented to concerned residents of the area in a public workshop. After a long discussion period, everyone attending filled out a questionnaire to express his preferences. These were tabulated immediately and the results announced to stimulate further discussion. Finally, everyone filled out a second form, again stating preferences (Fig. 2-1-4~ Fig. 2-1-6,Tab. 2-1-1~ Tab.2-1-3).

2. Stages of Realization

The results strongly influenced the preferred alternative which is shown here as a series of

可能性选择 Alternative	发展行动 Developmental actions	自然过程影响 Natural processes impacted	影响 Effects 增加 Increased	影响 Effects 减少 Decreased	人文过程影响 Human processes impacted	影响 Effects 增加 Increased	影响 Effects 减少 Decreased
多样化保护区 Diversified preserve	潟湖有利环境 Insignificant impact	野生动物 Wildlife	■		海滩通道Beachaccess		■
		植物生命 Plant life	■				
		潮汐冲刷作用 Tidal flushing	■		保持 Maintenance	■	
		海滩沙 Beach sand	■				
		补给 Replenishment	■				
		海水浸蚀 Salt water intrusion		■			
	野生动物保护区 Wildlife preserve	野生动物 Wildlife	■		可达性 Accessibility		■
		植物生命 Plant life	■				
	海滩停车场 Beach-parking	微小影响 Insignificant impact			海滩通道Beach access	■	
					交通阻塞 Traffic congestion		■
	海滩人行道 Beach boardwalk	峭壁侵蚀 Cliff erosion		■	海滩通道 Beach access	■	
	草皮道路 主要连接处 Turf road major connector	微小影响 Insignificant impact			市区阻塞 Downtown congestion	■	
	游客服务中心 Visitor center	微小影响 Insignificant impact			教育 Education	■	
					休闲 Recreation	■	
	教育中心 Education center	微小影响 Insignificant impact			教育 Education	■	
	海滩巴士 Beach shuttle	能源 Energy	■		可达性 Accessibility	■	
		大气污染 Atmospheric pollution		■	噪音		■
	自然小径 Nature trail	微小影响 Insignificant impact			教育 Education	■	
					可达性	■	
					休闲 Recreation		
	无预约游客露营地 Walk-in camp	微小影响 Insignificant impact			休闲 Recreation	■	
					可达性 Accessibility	■	
					保持 Maintenance	■	
	社区训练基地 Community stables	大地污染 Ground pollution	■		休闲 Recreation	■	
					税收 Revenue	■	

表2-1-1 主题方案 I

Tab.2-1-1 Alternatives ——Diversified preserve

可能性	发展行动 Developmental actions	自然过程影响 Natural processes impacted	影响 Effects 增加 Increased	影响 Effects 减少 Decreased	人文过程影响 Human processes impacted	影响 Effects 增加 Increased	影响 Effects 减少 Decreased
区域公园 Regional park	潟湖有利环境 Insignificant impact	野生动物 Wildlife	■		海滩通道 Beachaccess		■
		植物生命 Plant life	■		保持 Maintenance	■	
		潮汐冲刷作用 Tidal flushing	■				
		海滩沙 Beach sand 补给 Replenishment	■				
	区域公园 Regional park	水资源利用 Water usage	■		保持 Maintenance	■	
		本土植被 Native vegetation	■		休闲 Recreation	■	
		野生动物 Wildlife	■		可达性 Accessibility	■	
	野生动物保护区 Wildlife preserve	野生动物 Wildlife	■		可达性 Accessibility		■
		植物生命 Plant life	■				
	污水处理厂 Sewage treatment	有机质腐蚀 Organic matter decay	■		水资源循环利用 Water recycling	■	
					保持 Maintenance	■	
	无预约游客露营地 Walk-in camp	微小影响 Insignificant impact			休闲 Recreation	■	
					可达性 Accessibility	■	
					保持 Maintenance	■	
	野餐公园	野生动物 Wildlife		■	休闲 Recreation	■	
		耗水量 Water usage	■		可达性 Accessibility	■	
					保持 Maintenance	■	
	池塘 Ponds	野生动物 Wildlife	■		休闲 Recreation	■	
					保持 Maintenance	■	
	公共洗手间 Restrooms	微小影响 Insignificant impact			保持 Maintenance	■	
	滨海步道 Beach boardwalk	悬崖侵蚀 Cliff erosion		■	可达性 Accessibility	■	
		沙丘保护 Sand dune protection	■				
	博览园 Museum park	微小影响 Insignificant impact			休闲 Recreation	■	
					保持 Maintenance	■	
	公众训练基地 Public stables	大地污染 Ground pollution	■		休闲 Recreation	■	
					税收 Revenue	■	
	游客服务中心 Visitor center	微小影响 Insignificant impact			休闲 Recreation	■	
					教育 Education	■	
	草皮路环形线路 Turf road loop	微小影响 Insignificant impact			交通阻塞 Traffic congestion		■
					可达性 Accessibility	■	

	马术测试 Equestrian train	微小影响Insignificant impact			休闲 Recreation	■	
	徒步路线 Hiking trial	微小影响Insignificant impact			可达性 Accessibility	■	
	自行车路线 Bicycle trial	微小影响Insignificant impact			休闲 Recreation	■	
	远景点观察 Vista points	悬崖侵蚀 Cliff erosion		■	休闲 Recreation	■	
	海滩停车场 Beach parking	微小影响Insignificant impact			海滩通道 Beach access 交通阻塞 Traffic congestion	■ ■	 ■
	自然小径 Nature trail	微小影响Insignificant impact			教育 Education 可达性 Accessibility	■ ■	

表2-2-2 主题方案II ——区域公园

Tab.2-1-2 Alternatives II ——Regional Park

可能性选择 Alternatives	发展行动 Developmental actions	自然过程影响 Natural processes impacted	影响 Effects		人文过程影响 Human processes impacted	影响 Effects	
			增加 Increased	减少 Decreased		增加 Increased	减少 Decreased
多样化保护区 Diversified preserve	潟湖有利环境 Lagoon opening	野生动物 Wildlife	■		海滩通道 Beachaccess		■
		植物生命 Plant life	■				
		潮汐冲刷作用 Tidal flushing	■				
		海滩沙 Beach sand	■				
		补给 Replenishment	■				
		海水浸蚀 Salt water intrusion		■			
	市政厅/城镇群发展 Townhouse/ cluster development	水径流 Water runoff	■		可达性 Accessibility	■	
		能源消耗 Energy consumption	■		公共服务 Public services	■	
		渗透 Percolation		■	噪声 Noise	■	
		野生动物 Wildlife		■	交通阻塞 Trafficcongestion	■	
		乡土植物 Native vegetation		■			
		大气污染 Atmospheric pollution	■				
		沉降 Sedimentation	■				
	拓展潟湖 Expanded Lagoon	盐水浸蚀 Salt water intrusion energy		■	保持 Maintenance	■	
		水生动物 Aquatic fauna	■				
		水生植物 Aquatic flora	■				
		陆生动物 Terresrial fauna	■				
		陆生植物 Terresrial flora	■				

	污水处理植物 Sewage treatment plant	有机质腐蚀 Organic matter decay			水循环 Water recycling 保持 Maintenance		
	拓宽跑马场 / 游乐场利用 Expanded race track /fairgrounds use	大气污染 Atmospheric pollution 能源利用 Energy use			噪声 Noise 休闲 Recreation 保持 Maintenance 生产力 Productivity 可达性 Accessibility 劳动力资源 Labor resources		
	商业区 Commercial areas	洪泛平原危险区 Flood plain hazard			生产力 Productivity 劳动力资源 Labor resources		
	跑马场 Planted race track 停车场 parking	渗透 Percolation 径流 Runoff			视觉特征 Visual character 保持 Maintenance		
	高速公路种植 Freeway planting	植被 Vegetation 野生动物 Wildlife 水表径流 Water runoff 沉降 Sedimentation 渗透 Percolation			视觉特征 Visual character 噪声 Noise		
	潟湖区 Lagoon area 野餐区	野生动物 Wildlife 水资源利用 Water use			休闲 Recreation 可达性 Accessibility 保持 Maintenance		
	跑马场与道路缓冲区 Turf road connector	微小影响 Insignificant impact			市中心阻塞 Downtown congestion		
	断崖公园 Bluff' s Park	悬崖侵蚀 Cliff erosion			休闲 Recreation 可达性 Accessibility		
	街心花园 Pocket park	微小影响 Insignificant impact			休闲 Recreation 保持 Maintenance		
	野生动物岛屿 Wildlife Islands	微小影响 Insignificant impact			微小影响 Insignificant impact		
	划船 Boating	野生动物 Wildlife			休闲 Recreation 可达性 Accessibility		
	徒步游线 Hiking trail	微小影响 Insignificant impact			可达性 Accessibility 休闲 Recreation		
	自行车游线 Bicycle trail	微小影响 Insignificant impact			休闲 Recreation		

表2-1-3 替代方案 Ⅲ——居住综合设施

Table 2-1-3 Alternative Ⅲ ——Residential complex

阶段一 PHASE I
改善阶段 IMPROVEMENT

阶段二 PHASE II
恢复阶段 RESTORATION

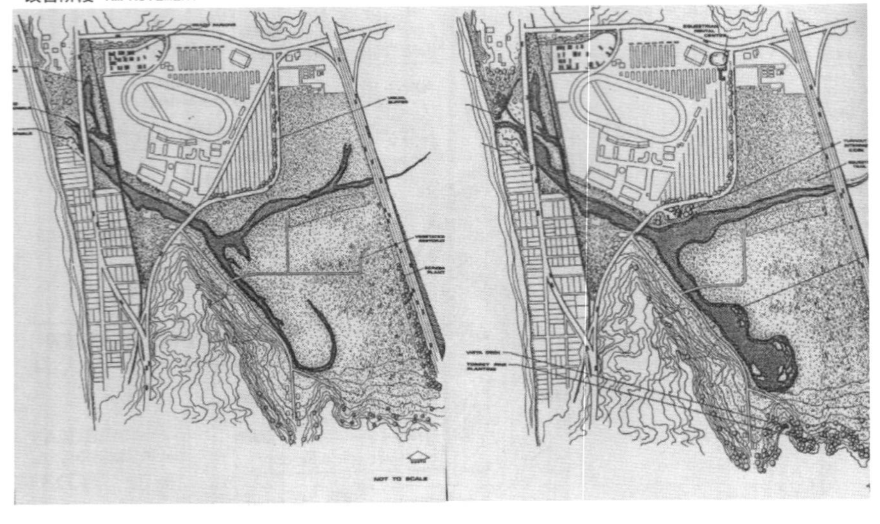

阶段三 PHASE III
再生阶段 REVITALIZATION

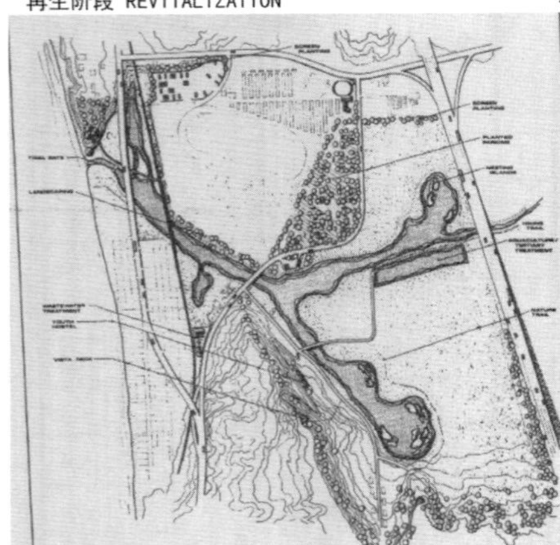

阶段四 PHASE IV
精炼阶段 REFINEMENT

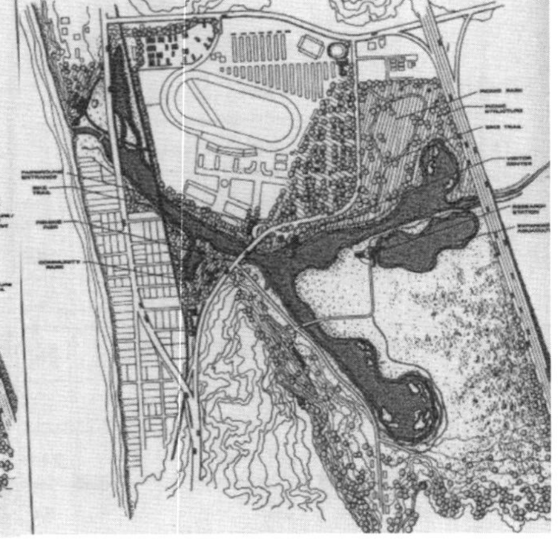

图2-1-7 规划实施阶段
Fig.2-1-7 Stages of Realization

与精炼，在明确管理目标的同时，也给予充足时间（图2-1-7）。

实现目标的每个阶段都可以描述成设计的一部分，譬如，存在一种最理想的情况是，将计划中欲实现的战略看作整体设计的要素之一。正如距离圣埃利荷潟湖仅几英里距离的小型滨海潟湖圣地亚哥这一典型案例，与之情况相似的是环境也同样被污染。两个案例存在的问题如此相似，解决问题的方法却不尽相同，涉及的政治、经济问题与自身自然环境一样，极为复杂。实现设计理想既漫长又艰

developmental stages. Since the dynamics of the lagoon are not entirely understood, developmental actions may well have unexpected effects. The staging of development allows time for experimentation and correction. The four phases: improvement, restoration, revitalization, and refinement——give definite management goals without demanding too much at once(Fig. 2-1-7).

The steps for achieving such a goal are best described as part of the design. Ideally, then, a realization strategy should be considered as an integral design component. Such was the case in the design for San Dieguito Lagoon, San Dieguito is a small coastal lagoon just a few miles south of San

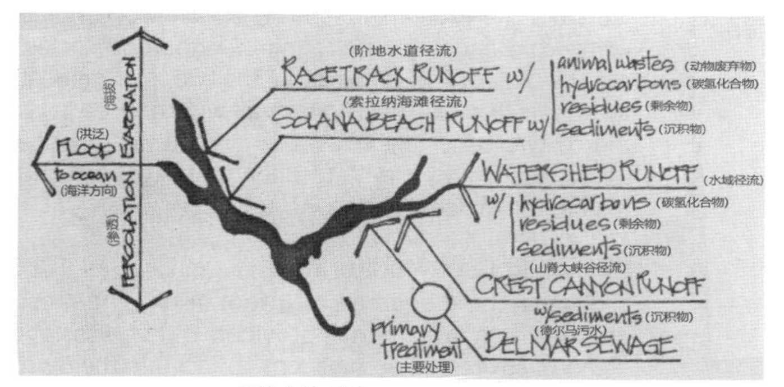

现状水流系统 EXISTING WATER FLOW

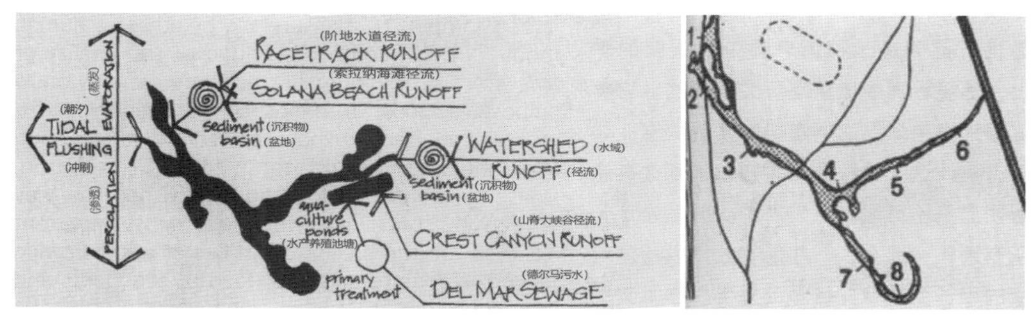

规划水流系统 PROPOSED WATER FLOW

图2-1-8 水流系统&监测站位置
Fig.2-1-8 Water Flow System & Stations position

巨。整个案例研究按照顺序分成四个步骤，其间的逻辑关系简明易懂。工作开展之前应获得相应土地并且克服障碍，以免意外发生。一旦完成第一步，最基本的潟湖再生过程就可以开始了，包括最重要的潮汐冲刷。这一步骤可以称之为“恢复”。在第三步中，可以建立一些野生动物栖息地和娱乐设施，创建与水产业相关的污水处理再利用系统实验站。最后一步拓展娱乐游憩活动范围，完成大量植被栽植项目，基于最初的实验结论，增加水产业活动范围。

此时此刻，潟湖到达自身最佳状态，在某种程度上，可以通过一定程度的管理控制实现自身可持续循环。正如图2–1–8所示，它大致上说明了输入、输出、内部流三者之间是如何有效发挥作用。城市规划部门、积极市民组成的潟湖保护委员会一起实施管理控制工作，管理依据以下三种信息来源进行决策：①利益相关的公众意愿，②众多科学家对潟湖实验所得出的研究数据，③潟湖正在进行的监测过程数据。在综合这些数据的基础之上，两管理团队将做出决策，最终由另外两个机构——城市维护部门、州渔业与娱乐管理部共同执行决策。

Elijo, and like San Elijo, it is badly degraded. The issues are quite similar in both cases, as are the proposals for resolving them, and the political and economic environments are as complex as the physical ones. Achieving the design ideal will not be quick or easy. The case study plots it in a sequence of four steps, the logic of which is fairly simple. Before anything else can happen, certain lands have to be acquired and legal obstacles removed. Once this first step is accomplished, reactivation of the basic lagoon processes, including , the most importantly, some degree of tidal flushing, can begin. This step is called "restoration". In the third step, wildlife habitats and recreational facilities are established, and experiments with waste water reclamation systems in relation to aquaculture are initiated. Finally, the last step expands the range of recreational activities, completes the extensive planting program, and based on the results of the initial experiments, increases the scope of the aquaculture activity.

At this point, the lagoon attains its climax state and is capable of sustaining itself with some degree of management control. In very general terms, how it will function_ the inputs, outputs, and internal flows(Fig. 2-1-8). Management control will be exerted by the City Planning Department and the Lagoon Preservation Committee, a particularly active citizens' group. It will receive three kinds of information to guide its decisions: goals from a larger interested public, research data from a variety of scientists who

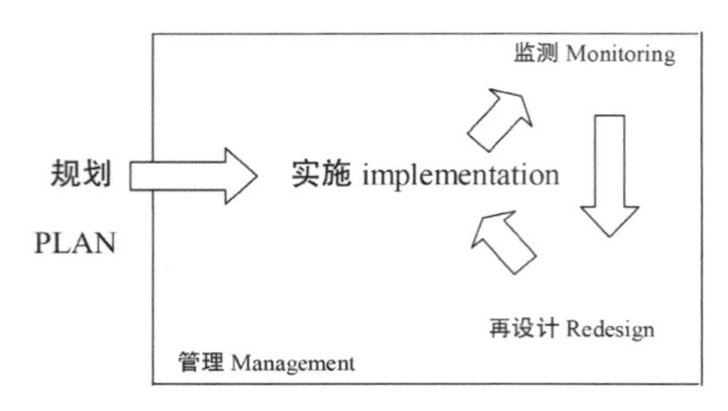

图2-1-9 监测的主要角色

F ig.2-1-9 The Critical Role of Monitoring

3. 监测的主要角色

迄今为止，监测是三种信息源中最重要的手段。事实上，环境管理中监测是至关重要的，它类似于人体健康检查中体温的测量与血常规调查。如果没有了解有机体这一系统的相关知识，那么有意义的决策很明显是不可能制定出来的。既然设计师不能监测所有因子，那么，一些能有效提供检测对象是否健康的因子变量就变得至关重要。在本案例中，水质和水量是有效的监测变量，因为这些变量可以有效地控制其他要素。然而，最重要的是，要确切了解水源水流的确切数值，以及他们的沉积物和营养物的负载情况。尽管其他监测变量无疑是很有帮助的，但是单这一信息就可以充分做出基本的管理决策。例如，周期性地深度调查，沼泽、植被生长范围（通过航拍图），鸟类数量变化等，都可以帮助设计师做出正确判断。原始数据库中信息源过多、超载问题也同等重要，它限制了监测者挑选出能确切反应该对象状况的正确指标。

监测结果显示较差的水质情况，尤其是潟湖死水地区。图2-1-9第一图表显示现状水流系统，包括水源与污染源。图2-1-9第二图表显示重新设计之后水流系统，以及增加淡水流入与降低污染物进入的各种设施。

事实上，莱尔将监测看成是原始数据库信息的拓展，尤其是原始数据信息是理想的管理兼设计手段，相对于前期阶段来说，后期阶段的反馈是每一个设计过程的基本部分。由于管理常常包含了周期性再设计，周期性的反馈就成为管理与设计之间的关键和永久的链条。

莱尔所提到的设计与管理之间的连锁、反馈、

conduct the experiments in the lagoon process, and data from ongoing monitoring of lagoon processes. On the basis of this information, the two management groups will make decisions that will then be pass on to two other agencies, the city' s maintenance department and the State Department of Fish and Game, for implementation.

3. The Role of Monitoring

Of the three sources of information, the most important by far is that provided by monitoring. Periodic structured observations, in fact, are the most fundamental activity in environmental management, the equivalent of taking temperature readings or blood samples to measure the health of a human body. Without some knowledge of the vitality of the system, meaningful decisions are obviously impossible. Since we cannot monitor everything, it is equally important to know the variables that furnish clear indications of health. In case of the lagoon, these are water quantity and quality, because they effectively control everything else. Most important to know are the precise quantities of water flowing from the sources and their load of sediments and nutrients. This information alone would probably be adequate for basic management decisions, although monitoring of other variables would undoubtedly be useful. For example, periodic depth soundings, the extent of marsh, vegetation (which could be easily determined from aerial photographs), and bird counts might contribute to decisions. The problems of information waste and overload that apply to the original information base are relevant here as well, but it is important to limit the factors monitored to those that truly significant.

Tests show poor water quality, especially in the backwaters of the lagoon. The first diagram in Fig. 2-1-9 shows the existing water flow system, including sources of water and the pollutants contributed by each. The second diagram in Fig. 2-1-9 shows the redesigned flow system and the devices proposed to increase the inflow of fresh water and reduce the inflow of pollutants.

In fact, Lyle can view monitoring as an extension of the original information base, especially since the latter ideally acts as a management as well as a design tool. As noted earlier, feedback from later to earlier stages is a basic part of every design process, Since management usually involves periodic redesign, the feedback loop can be the critical, permanent linkage between management.

The interlocking, feeding-back, ongoing relationship between design and management suggested here is quite different from the usual view of both activities until very recent times, that is ,that they are quite separate enterprises. Oddly enough, the

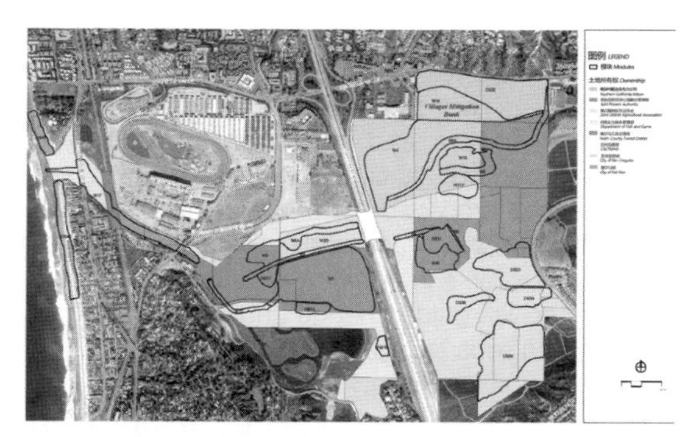

图2-1-10 圣地亚哥潟湖恢复项目土地所有权

Fig.2-1-10 Property Ownership

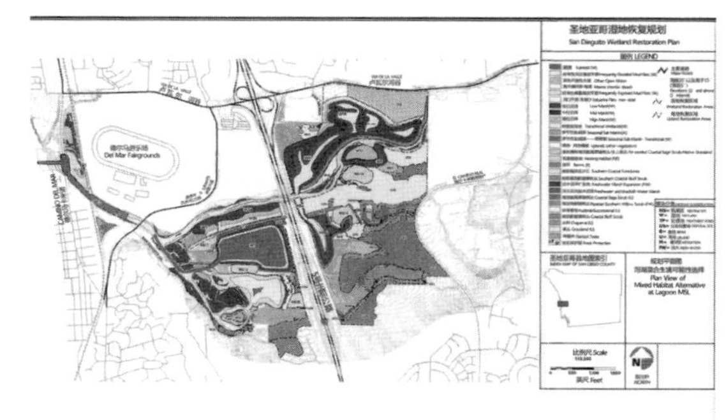

图2-1-11 圣地亚哥潟湖恢复项目总平面图

Fig.2-1-11 Site Plan

图2-1-12 圣地亚哥湿地恢复项目-南加州爱迪生电力公司要素与联合管理局要素

Fig.2-1-12 San Dieguito Wetlands Restoration Project-SCE Components and JPA Components

图2-1-13 缓冲区边界

Fig.2-1-13 Buffer Zone Boundaries

持续的联系，与我们当今所指的设计与管理的联系是截然不同的概念，当今所指的设计与管理是完全分离的两部分。说来也怪，在文献中涉及管理、设计这两大部分，传统上是相互分离，互不重视的。一方面，不考虑管理，设计会与现实相分离，而另一方面不考虑设计，管理就无法谈及未来；从长远有效性来看，设计不得不为管理提供基础研究工作，也注定要制约于变化的情况。

这是主要概念性的分水岭之一，区别于以往设计师的重要方面：充分意识到变化与不确定性。在所有错综复杂分析方法的基础上，不存在任何可以告诉设计师什么是确定的可行方法。开发一种新的模型将会逐渐意识到它本身所具有的不完善性，所有微妙的因素中伴随的是不可应对性。对未来的预期总伴随着各种各样的可能性，设计便是如此。

设计中充满不确定性，一部分原因是设计师缺

literature on each subject habitually ignores the other. Without management taken into account, however, design is divorced from reality, and without design, management can have no vision of a possible future. To be effective over the long run, design has to provide the spadework for management and recognize that the vision it presents is subjected to change.

This is one of the major conceptual divides that separates us from the designers of the past: the awareness of change and uncertainty. For all the intricacy of our analytical methods, we know that none of them tells us anything for certain. To develop a model is to become aware of its crudity, of the many subtle factors with which it cannot cope. Our predictions are always couched in terms of probability, and so are our designs.

This uncertainty is partly due to lack of sufficient knowledge and partly to the limitations of the human intellect, but it is also largely due to the way natural processes work.

Chance is usually an important factor in natural events as well. Even if we could perfect our

乏经验或相关知识，另一部分原因是人类才智的限制，因为很大程度上是自然过程在发挥巨大作用。

自然过程中一个重要的因子就是可能性。即使我们充分理解了自然过程，我们的设计预期还是受种种可能性的限制。

莱尔认为设计过程在某些方面是与自然过程相一致的，更确切地说，是与管理过程相一致。景观管理的伟大力量和希望隐藏于自然发展过程的重复之中，同时也存在于与设计不可避免的整合之中。因此人文生态系统的设计师必须指引变化方向，创造变化，而不是单一地与变化相抗衡。

三、圣地亚哥潟湖最终恢复规划（2005）

1. 项目简介

2003至2005年，该项目一直处于深化设计与施工阶段，至2005年11月，最终完成了恢复规划的设计施工过程。最终恢复规划和联合管理局的公园总体规划，都符合加州海岸委员会的相关许可规定。

恢复规划是环境影响报告提出的混合生境方案的反映。这项原始恢复规划包括由以及圣地亚哥河谷公园总体规划共同策划的一系列恢复行动。正如图2–1–10所表示的大量高地恢复行动（标示字母“U”）。除此之外，它还包括一些进行额外恢复工作的地区，这些地区往往都是在恢复过程中，削弱湿地至高地转换过程中所产生的一些影响（代表字母为“M”）。

2. 项目要素详细描述

项目要素主要包括14种（图2–1–12）：潮汐湿地栖息地、湿地、高地过渡区栖息地、高地栖息地、筑巢地、土方量平衡与分级、潮汐入口处土方量的开挖、垃圾堆置场、护坡、边坡保护、冲刷控制、其他基础设施、缓冲区（图2–1–13）、建设方法、乡村缓冲河岸带。而14条项目要素详细内容繁多，因案例篇幅限制，在此仅介绍两处，供读者参考。

1）高地栖息地

为了给乡土植物提供更适宜的土壤。从一些可以重新栽植树木的高地垃圾处理地中获取表层土，然而，

understanding of natural processes, therefore, our predictions would still be limited by factors of chance.

Lyle has seen that design process work like natural ones in some ways, and even more obviously, so does management. It is in the replication of the developmental process of nature that the power and the great promise of landscape management lie, as well as its inevitable integration with design. In doing so, designer will guide change and shape it, rather than standing squarely against it.

Ⅲ. San Dieguito Lagoon Final Restoration Plan (2005)

1. Introduction

The detailed design and engineering phase of the project took place between 2003 and 2005; a Final Restoration Plan concluded this process in November 2005. This Final Restoration Plan is consistent with the JPA's Park Master Plan and the CCC's permit requirements.

The restoration plan is a reflection of the Mixed Habitat Alternative addressed in the EIR/EIS. This original restoration plan encompassed a number of restoration activities which were planned by the JPA in association with its development of its San Dieguito River Park Master Plan. As a result, Fig. 2-1-10 shows a number of upland restoration activities (designated by the prefix "U").In addition, it includes areas where additional restoration could be undertaken to offset conversion of wetlands to uplands in the course of restoration. These areas are identified by prefix "M" .

2.Detailed Description of Project Components

Detailed description of project components includes 14 items(Fig. 2-1-12): Tidal Wetland habitat, Wetland/Upland Transitional Habitat,Upland Habitat,Nesting Sites,Excavation and Grading, Tidal Inlet Excavation, Disposal Sites, Berms, Slope Protection, Erosion Control, Other Infrastructure Considerations, Buffers(Fig. 2-1-13), Construction Methods, Villages Mitigation Bank. Because of various contents of the detailed description for 14 project components, as well as limitation of length of the case, there are only two detailed descriptions in the introduction, to be a reference for readers.

1)Upland Habitat

In order to establish suitable soils for native vegetation, the project will place topsoil that was salvaged from the site in the upland disposal areas that will be re-vegetated. It is likely, however, that this topsoil will contain a large number of weed seed. A native plant hydroseed mix will be applied to these upland areas. The project may utilize a combination of the following methods to reduce the initial

这些土壤中含有大量的野草种子。设计师因地制宜，采用乡土植物混合种植的方法恢复高地栖息地。该项目利用下列方法降低高地地区野草存活比率：

首先采用灌溉野草，等其发芽之后，再将地块分区，最后拔除野草，此过程在播种目标树种之前需要重复数次；利用从未使用过的除草剂处理表层土；并且、或者处理野草，降低其竞争力；鉴于项目尺度较大，基于实施的有效性和可行性采取适当的方法。

2）筑巢地

恢复项目包括建造四块筑巢地、恢复更新现有野草丛生的地块。筑巢地与加州海岸委员会要求南加州电力公司停止停车场、扩张游乐场设施这一要求有关，此扩张要求是南加州电力公司之前应第22届地区农业协会要求所制订。与停车场扩建相关，加州海岸委员试图寻找新的鸟类筑巢区域，探索恢复地区现有敏感鸟类筑巢区的恢复工作（图2-1-14）。

五块筑巢地位置（11号、12号、13号、14号、15号筑巢地）它们将为加州小燕鸥、环颈鸻等水鸟提供了12.3公顷平坦的筑巢地，为了不被潮汐侵蚀，筑巢地在某种程度上会比周围湿地高出一些，这样就产生了一些平缓边坡、面积较筑巢基址更小的筑巢高地。规划将为边坡提供占地总面积约20.5公顷充足的缓冲距离。四个创建的筑巢地位置及面积由美国渔业和野生动物局、加州海岸委员会协商共同决定。基地选址考虑因素包括：每块场地最小容纳1.2公顷的可使用筑巢地，从场地上可欣赏开放的全景，重要地块需多建造一些障碍物等。曾有研究表明小斑块与大斑块相比较，小斑块对鸟类生存更有利。

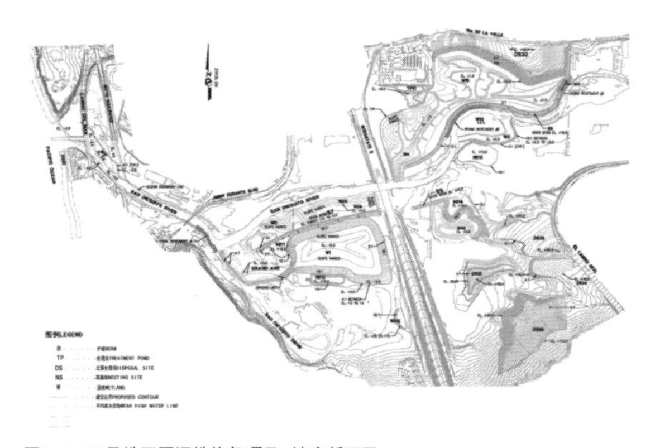

图2-1-14 圣地亚哥湿地恢复项目-坡度断面图
Fig.2-1-14 San Dieguito Wetlands Restoration Project-Grading Plan

establishment of weed seeds in these upland areas:

Utilize irrigation water to germinate weed seeds and then disc the areas to destroy weeds. This process may be repeated several times prior to hydroseeding the target species; pre-treat the topsoil with a pre-emergent herbicide; and/or implement a mowing program to reduce competition from weed species; given the large scale of the project, the method will be selected based on effectiveness and feasibility of implementation.

2)Nesting Sites

The restoration project includes the construction of four nesting sites and rehabilitation of an existing site that is now covered with weedy species. This aspect of the project was related to a request to SCE from Coastal Commission to accommodate mitigation it had previously required of the 22nd Agricultural District for wetland impacts associated with expansion of parking facilities associated with the Fairgrounds. In exchange for wetland impacts related to parking lot expansion, the CCC is seeking the creation of the new nesting areas and the rehabilitation of an existing nesting area for sensitive birds in the restoration area(Fig. 2-1-14).

The five sites will provide 12.3 acres of flat nesting area for the California least tern, western snowy plover, and other shorebirds. The nesting sites will be somewhat higher than the surrounding wetlands in order to protect the sites from tidal inundation, resulting in the creation of gentle side slopes and a nesting plateau that is smaller in acreage than the base of the nesting site. A total footprint of approximately 20.5 acres will be required to provide adequate distance for side slopes. The location and size of the four created nesting sites was determined through consultation with the USFWS and CCC. Site selection considered the ability to provide a minimum of 1.2 acres of usable nesting area per site, achieve an open panorama from the site, and establish adequate setbacks from high structures. It was determined that the creation of numerous small sites was more beneficial to nesting birds than few large sites.

The base of nesting sites will be constructed using soil excavated from other restored areas. The target heights of the nesting plateaus would vary from +10 feet, NGVD (National Geodetic Vertical Datum) at NS11 and NS12; +12 feet, NGVD at NS13; +15 feet, NGVD at NS15; and +19 feet NGVD at NS14. The quantity of the base soil needed will depend on the starting elevation for each site. Excavated soil used for the nesting site bases will be dried and compacted to 85 percent relative density. Once the bases are properly compacted, three feet of coarse white or light colored sand will be placed on top.

Sand excavated from the former naval airfield site during initial grading will be used as nesting site surface material and has been approved by the resource agencies

设计师将从其他恢复地区挖取土壤，建造筑巢地的基址。筑巢高地目标高程变化范围：11、12号地块——高程基准面10英尺（约3米）以上；13号地块——高程基准面12英尺（约3.66米）以上；15号地块——高程基准面15英尺（约4.57米）以上；14号地块——高程基准面19英尺（约5.8米）以上。表层土的土质取决于每块场地起始高程。筑巢地所需开挖的土壤必须干燥，且具有85%相对紧实的密度。基础土壤夯实之后，在表层放置3英尺（约0.9米）白色或者浅色粗沙。

原海军机场挖掘出的沙土在首次分级时就被用做筑巢地表层土，资源机构一直认可其适宜性。为了最大限度地吸引燕鸥，设计师将贝壳敲击成碎片，并与沙土混合。

中央平坦的筑巢高地，逐渐下降至沼泽平原的边坡，两者组成了筑巢地。它的分级以河流侵蚀边坡最小化为原则进行修建。建造筑巢地的首要活动是：挖掘下层基础土壤，并在上层覆盖沙土，最终将其运送至特殊地区。针对某些特殊物种来说，例如环颈鸻，其幼仔必须在没有任何障碍的情况下，才能觅食、饮水。因此，筑巢地必须处于无植被状态。规划的临水筑巢地边坡的坡度须设定在10：1，这样既满足了幼仔饮水活动需要，又避免边坡被河水冲刷。然而，其他的边坡应该以不影响开放水资源为原则，进行植被栽植。

在中心处每10英尺（约3米）间隔放置防护围栏，钢丝网围栏距离地面4英尺（约1.2米）处放置聚乙烯巢穴。钢丝网围栏须埋入地下1英尺（约0.3米），地上完成后高7英尺（约2.1米），每块基地围栏必须留有足够的宽度，以备建造未来维护设施通过。筑巢地永久性通达设施要素（如道路）的设置必须通过南加州爱迪生电力公司、圣地亚哥公园联合管理局、美国渔业和野生动物局、加州海岸委员会工作人员协商之后，做出最后决策。

四、 总结

潟湖，是海岸与长条状滨外沙坝之间的封闭或半封闭的浅水区域，常有一条或多条潮汐通道与外海相

as to its suitability. In order to optimize the attraction of terns to these sites, scattered shell fragments will be added to the sand cap.

The nesting sites will consist of a nearly flat central nesting plateau with side slopes descending to the marsh plain. Grading will be conducted in a manner that will minimize the formation of rivulets that may increase erosion of the slopes. The primary construction activity for the nesting sites will be the movement of excavated base soil and capping sand to the specified locations. For some species, such as the western snowy plover, the chicks must have unimpeded access to the waters edge for foraging so it is important for these areas to remain unvegetated. Therefore, the side slopes of the nesting sites adjoining open water areas will be graded with a 10:1 slope to allow easy access to the waters edge and also avoid erosion. However, for other side slopes, which do not border open water areas, revegetation may be proposed.

Fence posts will be placed 10 feet apart on center. Polyethylene netting will be attached on the lower 4 feet of the chain-link fence as appropriate. The chain-link fence will be buried one foot below ground level for a finished height of 7 feet above ground surface. Each site that is fenced will have an access gate large enough in width to allow construction maintenance equipment to enter. If permanent access features (e.g., roads) are required for nesting site maintenance then this issue needs to be discussed in more detail between SCE, JPA, USFWS, and CCC staff.

IV. Summary

Lagoon in English is a closed or semi-closed shallow body of water between the shore and the strip-shaped barrier beach, Often one or more tidal channel with the ocean. Barrier beach forms from the accumulation of waves around the shallow coastal waters, long stripe of gravel stone accumulation that parallel to the coast. "□"is a very special word, its pronunciation is "xi", meaning "salty immersion land", the water is not flowing or stagnant by literal interpretation.

Though San Dieguito Lagoon is very large, but its water quality is slightly worse. All that imput to the pollution of San Onofre Nuclear Generating Station (SONGS). Ecological Restoration of planning area is a tough problem. Public participation play an important role in the phase of draft design. That is the important step in government decision-making throughout the urban planning process. The results of questionnaire survey provide the basis for further work of Planners. Eventually the Lagoon's development is divided into four stages—-improvement, restoration, revitalization, and refinement. Lyle proposed the landscape regeneration is that landscape restoration and reconstruction based on Ecological aesthetics,

通。而滨外沙坝是波浪在沿岸浅水区堆积而成的，是与海岸大体平行的长条形砂砾石堆积体。“潟”是个非常用字，读音同“细”，意为“咸水浸渍的土地”，从字面可见水是不流动或流动不畅的。

圣地亚哥潟湖面积很大，但水质较差，这归咎于圣欧诺佛瑞核电厂的污染，是规划区的生态修复最棘手的问题。方案阶段，公众参与发挥了重要作用。在美国，公众参与是政府决策的重要步骤，并贯穿于城市规划的全过程。问卷调查的结果为规划师的进一步工作提供依据，最终将潟湖的发展分为四大阶段——改善、恢复、再生与精炼。项目中，莱尔提出的景观再生是以生态美学、景观形态学、景观生态学、环境伦理学及建筑学为基础的景观恢复与重建研究。“再生”一词描述的过程，恢复，更新或振兴自己的能源和原材料的来源，创造可持续的系统，整合社会的需要，与大自然的完整性。同时，通过对景观各个因子的运行状况及相互作用的检测，提高景观系统自我修复能力，保持人文及自然景观的多样性，也最终保障了整个景观系统的健康发展。

Landscape morphology, Landscape ecology, Environmental ethics, Architecture. The term "regenerative" describes processes that restore, renew or revitalize their own sources of energy and materials, creating sustainable systems that integrate the needs of society with the integrity of nature. Meanwhile, it can improve the repair ability of landscape system, keep the diversity of the human and nature landscape, eventually guarantee the healthy development of the whole landscape system, through detecting the operation status and interaction of the various landscape factors.

思考题

1. 简要叙述莱尔所提出的三个主题方案。
2. 监测的主要角色在潟湖规划中如何体现的?
3. 潟湖进行恢复规划经历哪些阶段?
4. 最终恢复规划中14个项目要素有哪些? 简要叙述一例。
5. 莱尔的核心思想在圣地亚哥潟湖规划中如何体现的?

Questions

1. What is three different thematic alternatives that proposed by Lyle ?
2. How does the role of monitoring reflect in the plans for San Dieguito Lagoon?
3. What do the phases go through in the plans for San Dieguito Lagoon?
4. What are the items in the Detailed description of project components of the San Dieguito Lagoon Final Restoration Plan? Summarize one briefly
5. How does the core idea proposed by Lyle operate in the plans for San Dieguito Lagoon?

注释/Note

图片来源/Image：本案例图片均引自Design for human ecosystems：landscape, land use, and natural resources;http://www.sce.com/PowerandEnvironment/PowerGeneration/MarineMitigation/SanDieguitoLagoonRestoration;htm?from=wetlands; san _ dieguito _ final _ restoration _ plan

参考文献/Reference：本案例引自Design for human ecosystems：landscape, land use, and natural resources;http://www.sce.com/PowerandEnvironment/PowerGeneration/MarineMitigation/SanDieguitoLagoonRestoration;.htm?from=wetlands; san _ dieguito _ final _ restoration _ plan

第二节 圣埃利荷澙湖规划

Section II San Elijo Lagoon Planning

一、项目综述

圣埃利荷澙湖位于圣地亚哥的北部20英里（约32公里）的地方，夹在太平洋和南加州沿海平原之间，是一个形态狭长，充斥着沼泽、滩涂和浅水沟渠的湖泊。太平洋边缘一处充满鸟类的滩涂与盐沼以及穿越期间的管渠网络也属于圣埃利荷澙湖的一部分。尽管现在受到周边城市化的影响，澙湖的生态作用已经严重退化，但只要有能够恢复潮汐作用的过程，它还是能够作为一个充满活力的、可持续的并且是富有成效的生态系统来影响周边的环境。实现澙湖的全部潜力是能够建立起一个复杂的、多元化用途的综合体的，在这个综合体里的每种功能都能够对其他功能起到支持作用。这些用途包括生态保护、生物生产、娱乐和城市化（图2–2–1）。

二、区域环境介绍

在10 000年前或者更早的时候，城市文明还未到来之前，每天两次的潮涨潮落混合了澙湖的淡水，这些淡水都是在冬天雨季的时候从山麓间流下来的。圣埃利荷澙湖及周边区域还是一个丰富多产的草滩湿地群落，包括软体动物，各种类型及尺寸的鱼类，大量的水禽和水鸟，以及一些小型的哺乳

I. Overview

San Elijo Lagoon is a narrow tangle of marshes, mudflats, and shallow channels that push out of the Pacific Ocean into the rolling coastal plain Of Southern California, some twenty miles north of San Diego. San Elijo Lagoon includes a network of channels winding through a tranquil, bird-filled stretch of mudflat and salt marsh on the edge of the Pacific Ocean. Although badly degraded by surrounding urbanization, it can be revitalized to work as a sustainable, productive ecosystem once tidal flushing is restored. Realizing its full potential means establishing a complex combination of uses, each located and functioning in such a way as to support the others. These uses include preservation, biotic production, recreation and urbanization (Fig. 2-2-1).

II. Regional Form Natural Pattern

For 10,000 years or so before urban civilization arrived, the tides rolled in and out twice each day mixing with the fresh water that flowed down from the foothills during the winter rainy season. A prolific community of marshgrasses, molluscs' fish of various types and sizes, an array of shorebirds and waterfowl, and a few small mammals thrived in and around the moving waters. For several hundred years, a tribe of Indians also lived on the edge of the lagoon and shared in the bounty, especially the easily harvested shellfish, which helped assure them a leisurely existence. As a result sand built up in the channel, shutting off the flow of ocean water for part of the year and isolating the waters of the lagoon. Salinity

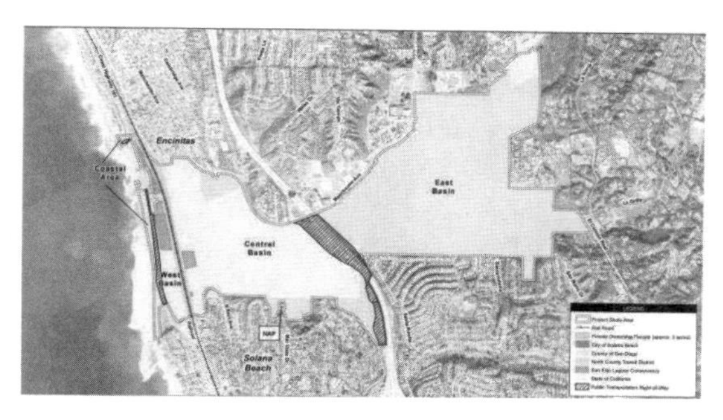

图2-2-1 圣埃利荷潟湖区位图
Fig.2-2-1 Location Map of San Elijo Lagoon

动物在这片流动的水域及其周围繁衍生息。定居于此的印第安部落数百年来与自然和谐相处，享有这些丰富的物产，特别是容易捕捞的贝类，从而得以保障他们优渥的生活。沙石在水渠中逐渐积累的结果就是阻碍了每年一部分的海水流入潟湖，隔离了海洋与潟湖水域的联系。潟湖的盐浓度逐渐升高，流入潟湖的沉淀物也开始在此处越积越多。几年后在铁路边上修建的沿海高速公路使得问题更加复杂，它进一步阻碍水的流动。所有这些问题开始有所改变是始于1884年修建在潟湖入海口处的圣达菲沿海铁路线。这条铁路架设的高架桥虽然能够保证潟湖的湖水流入海洋，但是入海口处的海浪运动不能像以前在自然条件下那样自由地与淡水进行混合了。随后几年里，人口急剧增加到需要建立一个污水处理计划来应对日常污水排放的程度。由于污水中的营养物质的含量增加，一些藻类以疯狂的速度生长并在潟湖表面繁殖与腐烂，形成了浅绿色的斑点。因为候鸟会随着太阳沿着太平洋进行迁徙，因此如果在潟湖里有鱼的话（尽管已经没有了），便会有相当数量的候鸟在此处停留休息、漫步、在泥滩上嬉戏与觅食。

三、 圣埃利荷潟湖规划与设计

1. 总体规划与适宜度模型分析

潟湖的未来已经成为一个热门的政治问题：环境保护主义者的兴趣在于鸟类的命运，因此他们主张将潟湖完整地保护下来。而开发商们则是立场坚定的强调私人财产权、税基和住房需求，更不用说这其中存在的利益了。我们很清楚潟湖的问题不能

levers rose in the trapped water, and the materials washing into the lagoon began to collect there. The problems were compounded when a coastal highway was built alongside the railroad a few years later, further restricting the movement of water. All this begun to change in 1884 when the Santa Fe Railroad line was built along the coast over the lagoon inlet. Though the trestle allowed water to flow beneath it, the inlet could no longer migrate freely in response to the movement of the ocean waves as it had under natural conditions. Within just a few years the population built up to such a level that a sewage treatment plan was needed. These became geometric splotches of pale green as algae fed by the nutrients in the sewage grew at far too fast a pace and decayed on their surfaces. Though there were few, if any, fish in the water, a good number of migrating birds still stopped to rest and stalk and flutter about on the mudflats, searching for worms as they followed the sun along the Pacific flyway.

III. San Elijo Lagoon Planning and Design

1. Master Planning and Suitability Models

The fate of the lagoon became a hot political issue. Environmentalists, whose main interest was the fate of the bird population, wanted the lagoon left alone. Developers, arguing private property rights and tax base and housing needs not to mention profits, stood firm. It became clear to us that they could not be sealed simply by pronouncing in favor of preservation or development. In the light of an analysis of the processes involved, political as well as ecological, the proposals of both sides were seriously flawed. Somewhat less expected are the consequences of leaving file lagoon alone. Increasing urbanization along file lagoon's upper drainage basin will bring about an inevitable increase in the siltation rate. Silts could be deposited as much as 100 times faster than they would if the lagoon remained in its pre-urbanization state. It is the common fate of a lagoon, whether people settle around it or not, to be filled with silt and become first a marsh and finally dry land. If the marina plan were to be carried out, concrete bulkheads would have to be built along the edges of the channels to hold the land in place and provide water deep enough for large boats. The result would be no tidal flats at the water's edge and therefore no marsh grasses. If there are no recurrent marsh grasses to die and decay, there will be no detritus for the shellfish and other small organisms to eat. The energy subsidy that tidal action provides by spreading the detritus around the lagoon to these stationary creatures will thus have gone for naught because there will then be nothing for the small fish and birds to eat, and no small fish for larger fish and birds to eat. Ocean fish

只是通过宣布支持哪种观点来得到解决。鉴于一个复杂的过程分析，政治与生态这两种观点都存在严重的缺陷。前者虽然将潟湖单独保护了下来，但是由于周边区域已经城市化，大量的悬浮物将会沉积于湖中。淤泥会以百倍于处在自然状态下的淤积速度堆满潟湖。无论人类是否定居于此，潟湖将会积满淤泥，成为沼泽，最终变成一片旱地，这是潟湖正常的命运。如果码头住区计划得到实施，防水设施都必须沿河道的边缘建造，并为大型船只提供足够的水深。而这样会导致水边的草滩湿地消失。一旦没有了草滩湿地上正常的生命演替，就不会有以贝类和其他小型生物为食的动物出现。对于这些静止不动的生物来说，潮汐作用所带来的散布在潟湖周围的食物将化为乌有，届时也不会有小鱼与鸟类的食物，进而也没有小鱼供捕捞与鸟类食用。海洋鱼类将不再进入潟湖产卵，潟湖最终将成为一潭死水，形成恶性循环。致命的是，无论是码头的住房建筑规划还是放手不管的做法都有一个显见的缺陷：它们各自均只关注一个单一的目的。这种狭隘的关注导致了在一个复杂的系统中过分考虑某一个方面的问题（图2-2-2～图2-2-4）。

因而在规划潟湖的过程中，我们将场地置于一个相互作用的网络之中，它将会按照一定的模式不断发展，至少其中一部分过程是可预见的。规划将潟湖及其流域分为七个不同的土地利用类型。这些都遵循着常规的模式进行，但实际的情况却有很大的不同。此外，这七个区域与传统的区域利用模式有很大的不同。在这里，我们并不希望利用功能的统一化，而是鼓励最大限度的多样性，只要是合理的利用，都可以打破传统观点适当引入，这与健康并富有成效的运作过程是一致的（图2-2-5）。

适宜性模型是一些通常由计算机获得的地图，这些地图表明了某些预期的利用方式的物理适宜性、土地适应性。这些模型使用网格地图，每一个网格代表一个单位的土地面积，约111.1英尺见方（约33.8平方米）。而反过来，这些决定土地的变量或特征，将形成一系列模型，如地理特征分布图包含了坡度变化及植被类型等信息，而且这些变量将被系统的列出。通过计算机搜索数据库， 适宜程

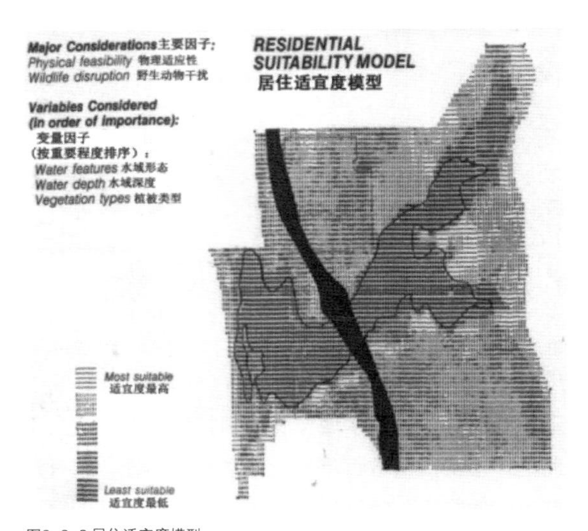

图2-2-2 居住适宜度模型

Fig.2-2-2 Residential Suitability model

will no longer swim into the lagoon to spawn, and it will eventually become a lifeless body of water. The causal chain of unhealthy effects goes on and on. The fatal, all too common flaw in both the marina housing plan and the hands-off approach is that each focuses on a single purpose. Such a narrow focus leads to an overriding concern with just one aspect of a complex system (Fig. 2-2-2~Fig. 2-2-4).

Planning divides the lagoon and its watershed into seven distinct categories of land use. These generally follow the patterns of the suitability models, but the actual configurations are quite different. Moreover, the seven uses bear no resemblance to the traditional zoning categories, because the purpose of the zoning is quite different from that of traditional zoning. Here, we are not trying to promote uniformity of use, but to encourage the greatest diversity that is consistent with the healthy and productive functioning of lagoon processes (Fig. 2-2-5).

Suitability models are maps--in this case, generated by computer--that define the physical suitability-- the fitness of the land--for certain prospective uses. These models use grid-cell mapping, each cell representing a land area 111.1 feet square, and are based on a set of listed concerns. These, in turn, determine the land variables, or characteristics that will shape the model, including geographically distributed characteristics like variations in slope and vegetation type: these variables are also listed. The computer searches through the data files, rates each grid cell for its combination of desirable attributes, and prints a map showing the aggregated results in terms of levels of suitability. In these maps, the lighter the shade of the cell, the higher the suitability rating (Fig. 2-2-6).

2. Detailed Planning

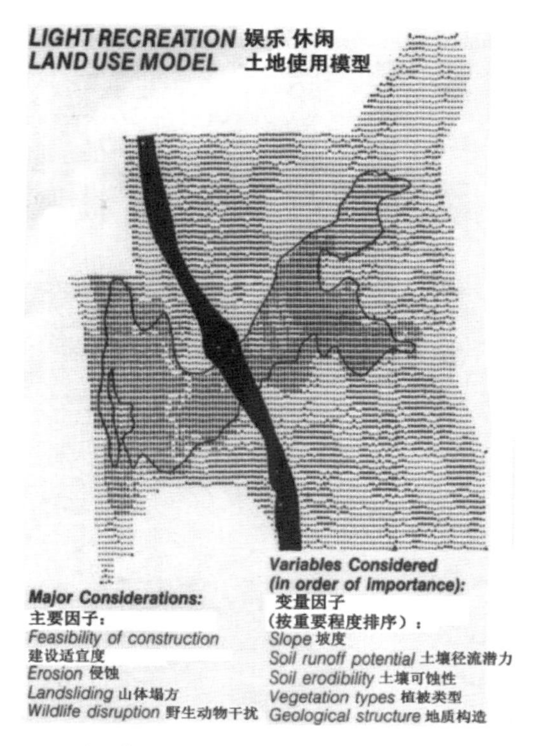

图2-2-3 娱乐休闲土地使用模型

Fig.2-2-3 Light Recreation Land Use model

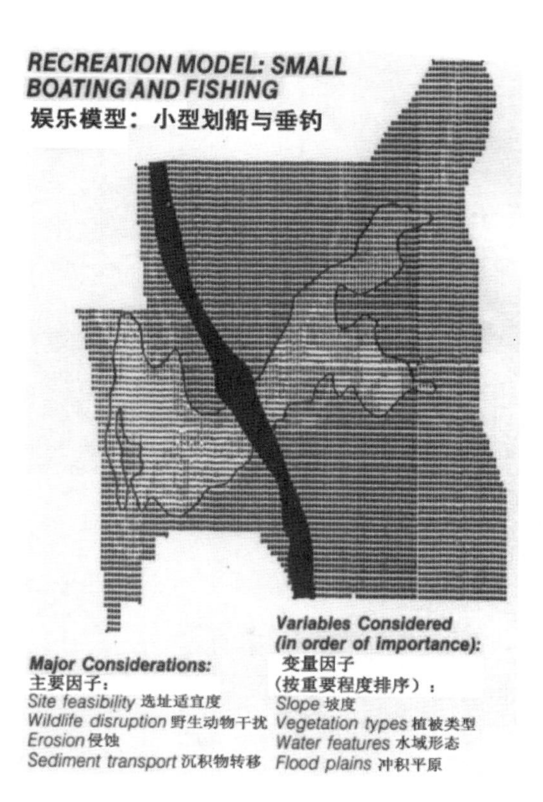

图2-2-4 娱乐模型小型划船与垂钓

Fig.2-2-4 Recreation model：Small Boating and Fishing

图2-2-5 影响预测矩阵

Fig.2-2-5 Impact Prediction Matrix

The wetland themselves are divided into three distinct zones, following the divisions already established by the railroad and the freeway. The inner zone east of the freeway--already the richest, most diverse habitat and the birds' favorite feeding and nesting place--becomes a wildlife preserve. The zone between the freeway and the railroad--where the water is deeper and wildlife far less abundant--will be devoted to biotic production and research. Areas of both natural, or protective, and productive landscape are thus important parts of the plan. The productive area is largely a man-made landscape, but one that is biologically very much alive. Some dredging and shoreline alteration may be needed here to develop the best environment for raising fish and molluscs. The old sewage treatment plant on the lagoon's edge will be refurbished as a biological treatment facility to provide an inflow of fresh water. The westerly zone--between the railroad and the ocean, already heavily altered--will become an intensive recreation area, with some commercial development. Buildings will rest on piers over the waters of the lagoon so that marine processes can go on undisturbed beneath. Surrounding the wetlands is a protective buffer zone with a hiking trail, screened from the more sensitive wildlife habitats, and outside that, on the flatter lands, will be several passive recreation areas. Although

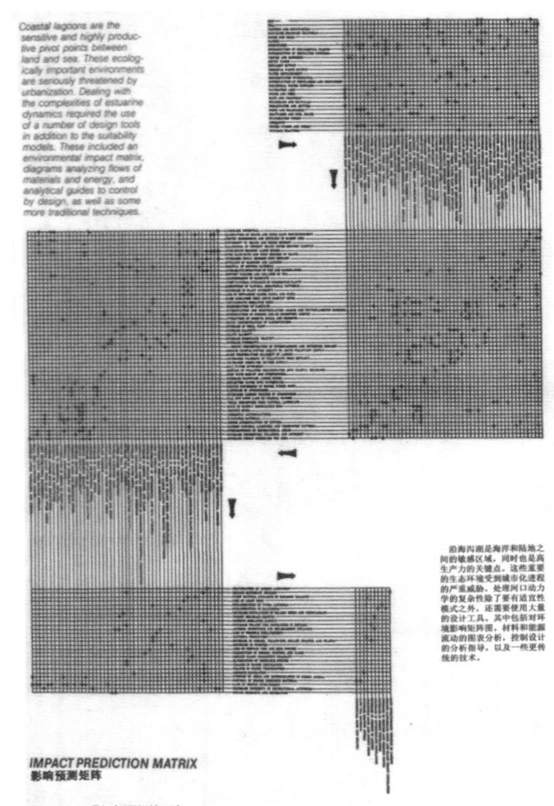

图2-2-5 影响预测矩阵

Fig.2-2-5 Impact Prediction Matrix

图2-2-6 连锁效应

Fig.2-2-6 The Chain of Effects

度为每一个网格单元的理想属性组合，然后将其汇编成一张地图，标明的其适宜程度的综合结果。在地图上，颜色越浅，适宜程度越高（图2-2-6）。

2. 详细规划

周边的铁路及高速公路将湿地分为三个区域，高速公路以东的区域是最为富饶的，包含多种类型的栖息地，也是鸟类最喜爱光顾和筑巢的地方，因此将其规划成为野生动物的保护区。而在高速公路和铁路之间的区域水位较深，野生动物的种类也不太丰富，因此这里将用于生物产品生产和研究。自然风景区、保护区和生产性景观都是该规划的重要组成部分。生产区主要为人造景观，但这个区域非常富有生命活力。一些区域的疏浚及局部海岸线的改动将会改善区域环境，从而提高鱼类和软体动物的产量。潟湖岸边旧式的污水处理将作为生态环境治理的治理对象，用以提供更加干净的水源。湿地区域的西部，即铁路和海洋之间的地方，实际上已

urban development will be encouraged on the slopes overlooking the lagoon, it will be subject to design controls that limit grading, maintain natural drainage courses and levels of runoff, and require planting for erosion. With such controls, development of these slopes will be entirely compatible with a productive lagoon.

Residential development on the overlooking slopes will do no harm if erosion is controlled and nutrient-laden runoff water is prevented from reaching the lagoon. Ideally, houses should be grouped in clusters on sites where runoff can be diverted, slowed, and allowed to percolate only very gradually into the ground water supply, thereby helping form a barrier against the progressive landward movement of underground salt water, which has been a problem in recent years. Natural drainage courses would best be left open and unimpeded so that fresh rainwater can flow into the lagoon unadulterated by contact with urban dregs.

Considering the lagoon as an ecosystem and examining its flows of materials and energy and its relationships with the surrounding urban complex, we become aware of a variety of contributions that the lagoon might make if carefully managed. Every planning and design of the area to the site is trying

经发生了很大的改变，因而该区域将成为一个密集的游憩区，并逐步发展一些商业。将建筑物建于防洪堤上，能够实现船舶的进出过程不受干扰。周边的湿地为中心区域提供了保护，起到了缓冲区的作用，为更敏感的野生动物提供了更好的保护。在外围平坦的开阔地上将建立观众休息区。虽然城市的发展会保留俯瞰潟湖的山坡地，但也要符合设计控制，限制对其改动，依然要保持自然的排水方式和径流水平，并且要求种植耐侵蚀的植物。有了这样的控制，这些山坡地的发展才完全与湖泊的生产性质相协调。

如果能够有效地控制侵蚀，防治富营养化的径流流入潟湖，那么在山坡上发展住宅项目并不一定会造成破坏。理想的情况下，房屋组团布置，这样可以释放径流，减慢速度，使它能够逐渐下渗并补充地表水，从而有助于形成一个对地下咸水逐步向陆地运动的屏障，而地下咸水向陆地运动一直是近年来的一个问题。自然排水管道最好保持开放和通畅，这样新鲜的雨水能不接触城市废渣而纯净的流进潟湖。

考虑到潟湖是一个生态系统，并通过研究其物质和能量的流动、与周围的城市综合体的关系，我们意识到如果通过精心的管理，潟湖会为我们做出巨大的贡献。场地的每一区域的规划设计力图将生态系统的生态价值、社会价值、经济价值最大化。首先，河口生态系统是一个丰富的蛋白质工厂。如果潮汐边缘保留一定的湿地，并能够健康成长，潮汐将分解潟湖的草，有利于鱼类和贝类的繁殖。事实上，潮汐沼泽可能是大自然中最有效的食品生产系统，比最好的麦田每英亩多产出约7倍多的生物量。人类可以提高潮汐的生产力，日本的一些河口已成功提高了潮汐生产力。

由于调查问卷的受访者认为，一个健康的潟湖一定要有足够的休闲功能。如果平静的水面能够在潮水的作用下保持一定的清洁，这里的水面可用于开展划船和游泳活动，而沿岸平坦的土地上可以提供一些固定设施，用来捕鱼、散步、骑自行车、野餐和四处闲逛等活动。

甚至适当密集排列的商业娱乐设施也是能够兼容的，它能很好地隔离野生动物的栖息地。一部分商店可以架空在水面上，或以驳船的形式漂浮在水面上，

to maximize the ecological value, economic value and social value of ecosystems. First of all, the estuarine ecosystem is a bounteous protein factory. If the marsh grasses are left to grow around the tidal edges, and the tides are allowed to do their work in spreading the feast of decomposing grasses about the lagoon, then enormous quantities of fish and shellfish will grow there. In fact, a tidal marsh is probably the most efficient food-producing system in nature, about seven times more productive in tons of biomass per acre that the best wheat field. This productivity can be increased severalfold, as has been accomplished in some Japanese estuaries.

As the respondents to the questionnaire perceived, a healthy lagoon has obvious recreational uses as well. Calm waters, if kept reasonably clean by tidal flushing, can be used for small nonpower boating and swimming, while the flatter land along the shore provides a fine setting for fishing, strolling, bicycling, picnicking, and lounging about.

Commercial recreation, even of a fairly intensive sort, is compatible provided that it is well separated from wildlife habitats. A cluster of shops can be built on stilts over the water, or on barges floating in it, to allow estuarine life to go on underneath. Marine businesses will fit nicely here, as will seafood restaurants, which might feature the products of the lagoon on their menus (Fig. 2-2-7).

3. Special Planning — Water Problem

At San Elijo, as in most unintentionally designed man-made ecosystems, the material flows have long been in a state of perpetual dysfunction, not for lack of materials, but because they are directed to the wrong places. During the long period when primarily treated sewage effluent was dumped into the lagoon, the enormous concentration of nutrients from the sewage brought about rapid growth of algae, which used enormous quantities of oxygen from the water, thus denying it to fish and molluscs and depleting their populations. When the algae died at a faster rate than the waters could absorb them or the tides move them out, they decayed on tile surface, causing unsightly masses of green scum and unpleasant odors. The flows of water and nutrients through the lagoon environment have long been intimately related to the treatment and discharge of sewage .These diagrams compare the ecological character of the flows under four different sets of conditions, beginning with the natural state. The solution eventually implemented for the "water pollution problem" was file four mile-long ocean outfall (Fig. 2-2-8).

The alternative that we propose would redirect the flows to reuse both water and nutrients through biological sewage treatment Thus, by feeding the primarily treated effluent into a series of ponds in

让河口中的生物从下面经过。与航海相关的商业活动很适合在这里开展，比如海鲜餐馆，他们菜单上的食品可能是以潟湖的产品为特色的（图2-2-7）。

3. 专项规划——水资源规划

在圣埃利荷潟湖，由于大多数都是未经专门设计的人工生态系统，物质长期处在一种功能失调的状态，但这并非由于物质的缺乏，而是因为缺少适宜的规划。在很长的时间里，只经过初级处理后的污水直接倾入潟湖，污水中的养分大量集聚从而导致藻类的快速增长，这一过程消耗了水中大量的氧气，也就使鱼类和软体动物的数量大量减少。当藻类的代谢速度使水体无法承受时，藻类的生长就会进一步加快，进而出现大量的绿色泡沫和难闻的气味。在潟湖的环境中，水分和养分的流动一直与污水的排放和处理密切相关，这些图比较了在四个不同的条件下水分和养分流动的生态特征，首先从自然状态开始。而规划最终解决"水体污染问题"的方案是一条四英里（约6.4公里）长的海洋排污渠（图2-2-8）。

新的污水处理方案将通过生物污水处理系统重新设计水分和养分的再利用，将经过初级处理的污水排入一系列种植水风信子等水生植物的池塘。其中水生

which water hyacinths and other aquatic plants will take up the nutrients, the water will eventually reach a level of purity that will permit its use for irrigating recreational areas and its eventual return to the lagoon. The hyacinths can be harvested for cattle feed and thus eventually be returned to the system as well. Such a pattern of water and nutrient flow is more like that of a natural ecosystem, more efficient and economical. The outfall, incidentally, would still be needed for overflows and emergencies

Such a system will not operate itself. Management will have to take the place of the self-regulating mechanisms of a natural estuarine system, which means that a high level of ongoing, creative management of file sort mentioned earlier will be needed (Fig. 2-2-9).

4. Planning Control Management

Such a system, however, will continue to work well only if it is managed--man-aged--well. Once the system begins to take shape, ongoing management is to be instituted as one of its essential components. Only management can control the feedback loops needed to augment those that have evolved as internally functioning mechanisms in natural systems. Certain kinds of control are needed to prevent foreign and potentially damaging materials like fertilizers, pesticides, oil residues, or phosphates from entering the lagoon. Human activities can be regulated in such a way as to prevent their interfering with sensitive lagoon processes or populations. Critical indicators of

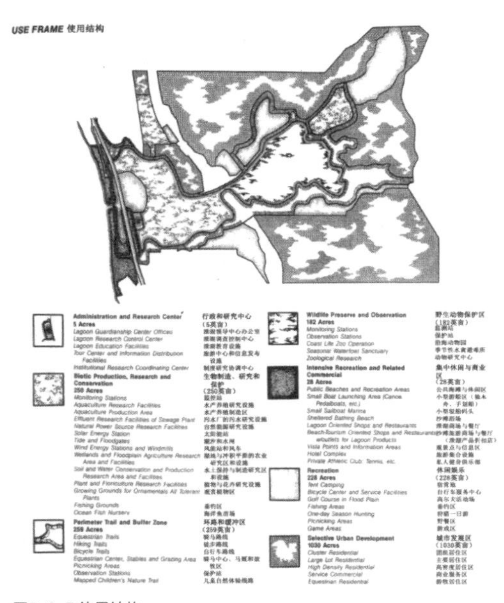

图2-2-7 使用结构

Fig.2-2-7 Use Frame

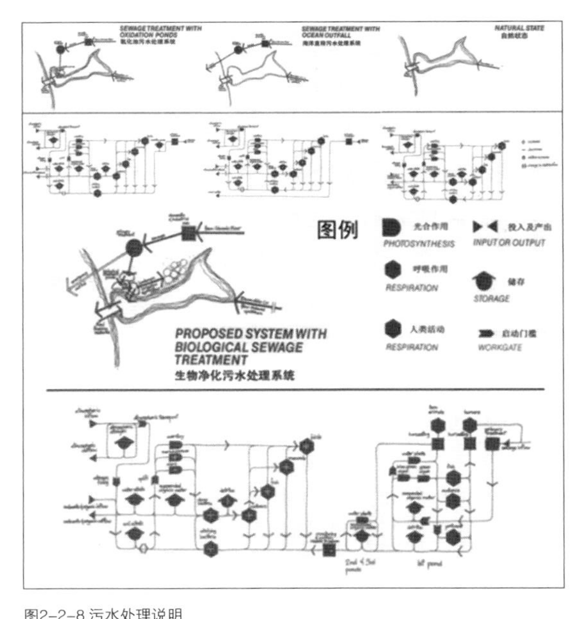

图2-2-8 污水处理说明

Fig.2-2-8 Sewage treatment illustration

植物会吸收污水中的营养素，处理后水质将达到一定的水平，能够用来灌溉游乐区，并最终流回到潟湖。使用后的风信子可以作为牛的饲料，也最终返回到生态系统之中。这样水分和养分循环系统更像是一个有效和经济的自然生态系统。另外，在污水处理池饱和及紧急情况时仍然需要排污管。

这样的系统当然无法完全自行运行，还需要管理者在自然的河口系统采取一个自动运行机制，这就意味着潟湖的保护需要一个持续的、创造性的高水平管理体系（图2-2-9）。

4. 规划控制管理

对于这样一个系统，只有具有丰富经验的管理行为才能够保证系统的良性运转。一旦系统开始初具规模，可持续的良性管理必须是系统重要的组成部分。只有管理可以控制反馈信息，而这些反馈信息能够增强自然系统的内部运作机能。为了防止一些外来的或潜在的有害物质，比如化肥、农药、石油残余物、磷酸盐等进入潟湖，规划区域需要一些特别的控制。人类可以用这样的方式来调节活动，以防止干扰潟湖敏感的自然系统和群落。环境质量的关键指标，特别是潟湖的水质，需要进行监测，以保持系统的稳定性。当出现不平衡或者恶化的迹象时，可以采取一些行动来干预。然而无论最初的设想是多么的小心谨慎，一旦出现上述的情况，潟湖都将回到像现在一样糟糕的状况甚至更糟。

人类设计和管理生态系统的行动代表了城市和自然过程的共生。粮食生产，野生动物栖息地，休闲娱乐，住宅，节约资源，水和养分循环，还有视觉美化都融入了一个相互依存的网络当中。这种整体的复合系统与圣埃利荷潟湖这种人类未触及过的河口生态系统完全不同，它的形式更加多样化，其活动也更多变。它的稳定性依赖于人类的能源和智慧，但在一定程度上，反之亦然。如果一切正常，如果我们的模型是正确的，我们的设计也按照预想的运转，如果我们的管理既富有想象力又是合理的，那么，人类和自然过程将融合为一个不分彼此的有机整体，这才是对人文生态系统的最好诠释。

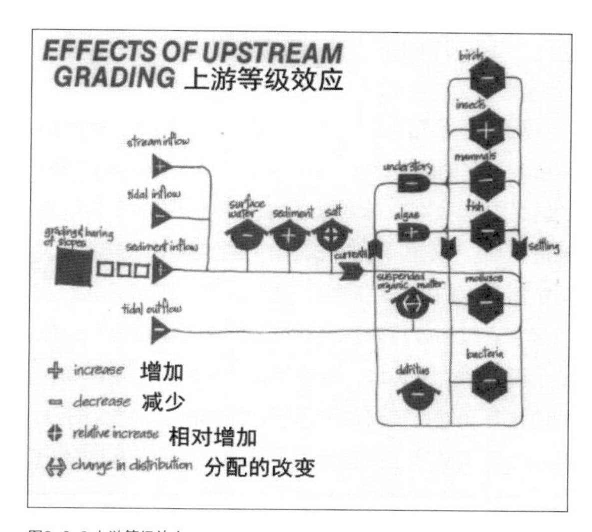

图2-2-9 上游等级效应

Fig.2-2-9 Effects of upstream grading

environmental quality, especially water quality, in the lagoon need to be monitored to maintain the stability of the system. When an imbalance appears, or if there is evidence of deterioration or conflict somewhere, some corrective action can be taken. In the absence of such a program, however carefully the initial design may be conceived, the lagoon will eventually return to its present sorry state or worse.

This intentionally designed and managed ecosystem represents a symbiosis of urban and natural processes. Food production, wildlife habitats, recreation, dwelling, resource conservation, water and nutrient recycling, and visual amenity are joined in a network of interdependence. The composition as a whole is very different from the estuarine ecosystem that would still exist at San Elijo had man never arrived on the scene, being more varied in its forms and more intense in its activities. Although it is dependent on human energy and ingenuity for its stability, the reverse is also, to some degree, true. If all goes well, if our models are correct and our design works as it should, and if the management is both imaginative and sound, then human and natural processes will merge indistinguishably into an organic whole, a human ecosystem in the best sense of the term. That, hard as it may be to achieve, is the ideal.

IV. Case evaluation

San Elijo Lagoon is a biodiversity hot spot, one of few remaining wetlands in Southern California. The 915 acres reserve hosts more than 700 species of plants and animals, many rare and endangered. Seven miles of trails for hiking, bird watching, and wildlife photography offer people a natural environment for solace and inspiration.

四、 案例评价

圣埃利荷潟湖是一个生物多样性的热点，是南加州所剩不多的湿地之一。915英亩（约370公顷）的保护区之内包括了700余种动植物，并且有许多都是珍稀濒危物种。7英里（约11.3公里）的游线中，游客可以观赏鸟类，拍摄野生动物，这个纯粹的自然环境给人们带来安慰和鼓舞。圣埃利荷潟湖修复工程修复和保护了潟湖的生态环境，保护了南加州的乡土动植物，同时恢复和维持了河口海水潮汐的冲刷。湿地对于人类的生活质量是至关重要的。特别是在人口稠密的地区，这些原生栖息地提供了一片宁静的土地是人类能够与自然亲密接触。只有通过娱乐体验，我们才能对保护自然生态有一个更加深刻的了解（表2–2–1）。

五、圣埃利荷潟湖目前保护情况（2012）

圣埃利荷潟湖保护组织修复了圣埃利荷潟湖生态保护区及其流域的生态资源和相关的生态系统，不仅为了当前的利益，还有未来几代人的利益。这一机构参与教学有保护价值的群落，以促进社会的生态意识和环保责任；进行科学研究，推进生态恢复规划和保护科学；设计和实施生态管理计划；管理和收购土地及地役权；以及与社会组织和政府机构合作（图2–2–10）。

1. 土地利用管理

由于植物的生长和动物的迁徙都不受边界的限制，首要的任务就是水陆交界处土地的保护，野生动物栖息地的保护和重要生态系统的保护。保护组织目前拥有超过60英亩（约24.3公顷）的栖息地，并对其拥有绝对所有权，还拥有45英亩（约18.2公顷）以上的土地保护权，这其中包括私人和公共的土地。

2. 外来植物入侵控制

圣埃利荷潟湖保护组织与圣地亚哥公园在圣埃利荷潟湖生态自然保护区内共同进行外来入侵植物的控制工作。工作人员持续监控入侵植物的出现，协调处理并清除外来入侵植物，同时使用本地的乡

The San Elijo Lagoon Restoration Project (SELRP) will restore and maintain San Elijo Lagoon Ecological Reserve to perpetuate native plants and animal's characteristic of Southern California, as well as to restore and maintain estuarine and brackish tidal flushing. Wetlands are essential to our quality of life. Especially in densely-populated areas, these native habitats provide peaceful places to connect with nature. It is only through experience, and recreation, that we gain a better appreciation for conserving natural history and ecology (Table. 2-2-1).

V. Current Protection Situation (2012)

San Elijo Lagoon Conservancy protects and restores the resources of San Elijo Lagoon Ecological Reserve, its watershed, and related ecosystems for the benefit of current and future generations. This institution engage and educate the community about the value of the reserve in order to promote ecological literacy and environmental responsibility; conduct scientific research to advance restoration planning and conservation science; design and implement ecological management plans; manage and acquire land and easements collaborate with the community, organizations, and government agencies (Fig. 2-2-10).

1. Land Use Management

Because plants and animals are unrestricted by borders of the reserve, a key Conservancy priority is to protect lands in the watershed; managing habitats for wildlife and to protect important ecosystem services. The Conservancy currently owns more than 60 acres of habitat through fee simple ownership, and holds more than 45 acres of conservation easements over private and public property.

2. Invasive Plant Control

Conservancy staff members work in conjunction with County of San Diego Parks and Recreation to control invasive plants in San Elijo Lagoon Ecological Reserve. Staffs continually monitor invasive plant occurrences, coordinate treatment and removal of invasive plants, and conduct re-vegetation efforts using native plants from local sources. Since 2004, San Elijo Lagoon Conservancy has administered a regional invasive species control program on behalf of Carlsbad Watershed Network (CWN). The CWN is a consortium of nine non-governmental organizations, seven cities in north-coastal San Diego County, County of San Diego, and more than 10 other governmental agencies and academia working to restore Carlsbad Hydrologic Unit (CHU).

These plants do extensive damage to private property and natural resources. They contribute to flood damage, increase risks associated with fire,

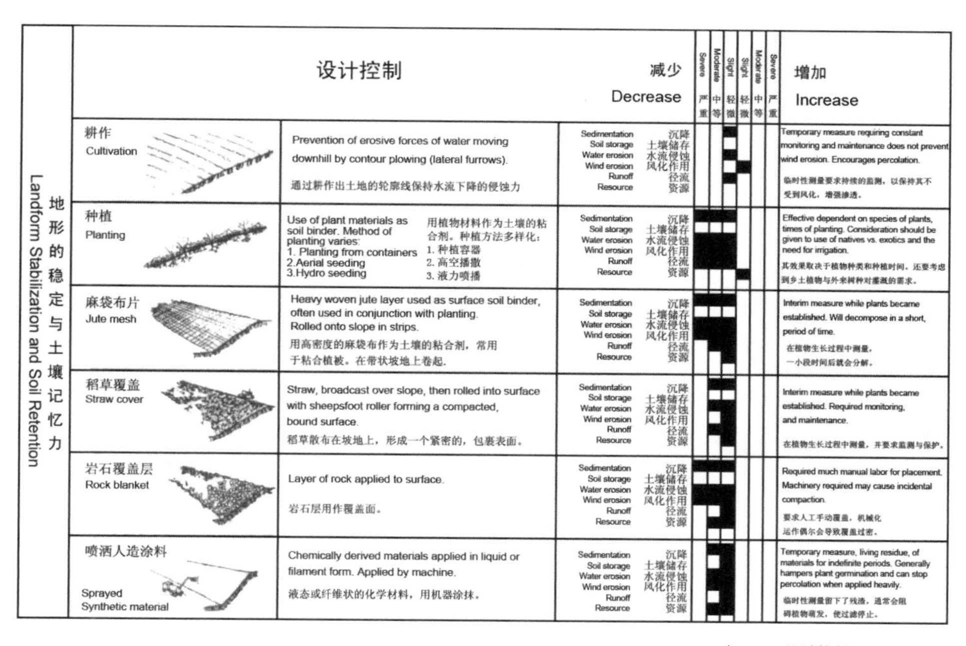

地形的稳定与土壤记忆力 Landform Stabilization and Soil Retention	设计控制		减少 Decrease	Severe 严重	Moderate 中等	Slight 轻微	Slight 轻微	Moderate 中等	Severe 严重	增加 Increase
	耕作 Cultivation	Prevention of erosive forces of water moving downhill by contour plowing (lateral furrows). 通过耕作出土地的轮廓线保持水流下降的侵蚀力	Sedimentation 沉降 Soil storage 土壤储存 Water erosion 水流侵蚀 Wind erosion 风化作用 Runoff 径流 Resource 资源							Temporary measure requiring constant monitoring and maintenance does not prevent wind erosion. Encourages percolation. 临时性测量要求持续的监测，以保持其不受到风化，增强渗透。
	种植 Planting	Use of plant materials as soil binder. Method of planting varies: 1. Planting from containers 2.Aerial seeding 3.Hydro seeding 用植物材料作为土壤的粘合剂。种植方法多样化：1. 种植容器 2. 高空播撒 3. 液力喷播	Sedimentation 沉降 Soil storage 土壤储存 Water erosion 水流侵蚀 Wind erosion 风化作用 Runoff 径流 Resource 资源							Effective dependent on species of plants, times of planting. Consideration should be given to use of natives vs. exotics and the need for irrigation. 其效果取决于植物种类和种植时间。还要考虑到乡土植物与外来树种对灌溉的需求。
	麻袋布片 Jute mesh	Heavy woven jute layer used as surface soil binder, often used in conjunction with planting. Rolled onto slope in strips. 用高密度的麻袋布作为土壤的粘合剂，常用于粘合植被。在带状坡地上卷起。	Sedimentation 沉降 Soil storage 土壤储存 Water erosion 水流侵蚀 Wind erosion 风化作用 Runoff 径流 Resource 资源							Interim measure while plants became established. Will decompose in a short, period of time. 在植物生长过程中测量，一小段时间后就会分解。
	稻草覆盖 Straw cover	Straw, broadcast over slope, then rolled into surface with sheepsfoot roller forming a compacted, bound surface. 稻草散布在坡地上，形成一个紧密的，包裹表面。	Sedimentation 沉降 Soil storage 土壤储存 Water erosion 水流侵蚀 Wind erosion 风化作用 Runoff 径流 Resource 资源							Interim measure while plants became established. Required monitoring, and maintenance. 在植物生长过程中测量，并要求监测与保护。
	岩石覆盖层 Rock blanket	Layer of rock applied to surface. 岩石层用作覆盖面。	Sedimentation 沉降 Soil storage 土壤储存 Water erosion 水流侵蚀 Wind erosion 风化作用 Runoff 径流 Resource 资源							Required much manual labor for placement. Machinery required may cause incidental compaction. 要求人工手动覆盖，机械化运作偶尔会导致覆盖过密。
	喷洒人造涂料 Sprayed Synthetic material	Chemically derived materials applied in liquid or filament form. Applied by machine. 液态或纤维状的化学材料，用机器涂抹。	Sedimentation 沉降 Soil storage 土壤储存 Water erosion 水流侵蚀 Wind erosion 风化作用 Runoff 径流 Resource 资源							Temporary measure, living residue, of materials for indefinite periods. Generally hampers plant germination and can stop percolation when applied heavily. 临时性测量留下了残渣，通常会阻碍植物萌发，使过滤停止。

表2-2-1 设计控制

Table.2-2-1 Control by Design

土植物进行植被的恢复。自2004年以来，圣埃利荷潟湖保护组织对这一区域内入侵物种的控制管理项目，属于卡尔斯巴德流域网络。这是一个联合组织，致力于恢复卡尔斯巴德流域网络，它包括了9个非政府组织，7个圣地亚哥县北部沿海城市，圣地亚哥县，以及十多个政府机构和致力于修复卡尔斯巴德水文单位的学术组织。

这些外来入侵植物严重破坏了个人财产和自然资源。它们会诱发洪水，增加用火风险，并侵蚀自然栖息地，还会引入老鼠和蛇。这些植物繁殖迅速，经常侵犯周边植物，难以控制。使用本地灌木和乔木重新覆盖可以降低被再次入侵的风险，并增加本地物种的栖息地。在河岸或溪边的栖息地，原生植被可以使进入潟湖的水质得到提升（图2–2–11）。

3. 潮汐循环

位于潟湖南部海滩营地的一个小型入海口，是潟湖流入太平洋的唯一通道。保持这一入海口的开放是保持圣埃利荷潟湖生态系统正常运行的关键。保护组织努力保持入海口的开放也就同时提升了栖息地的质量，特别是相对于规划之前入海口长期被封闭的状况。这些生态环境的改善，也促进了公众

and degrade native habitat. They also host rats and snakes. These plants spread aggressively, often onto neighboring properties, making them difficult to control. Revegetating the area with native shrubs and trees reduces the risk of re-infestation, and improves habitat for native species. In riparian, or streamside habitat, native vegetation ultimately improves the quality of water entering the lagoons of the CHU, including San Elijo Lagoon (Fig. 2-2-11).

3. Tidal Circulation

The small opening, just south of San Elijo State Beach Campground, is the lagoon's only access to the Pacific Ocean. Keeping the inlet open to the ocean is critical to maintaining the health of San Elijo Lagoon. The Conservancy's efforts to keep the inlet open to tidal flushing have substantially improved habitat quality relative to the stagnant conditions that previously developed when the inlet was closed for prolonged periods. These ecological improvements have also fostered significant public enjoyment of the lagoon and beaches. Keeping the inlet open improves trail access to all regions of the reserve (Fig. 2-2-12).

4. Scientific Monitoring

Conservancy scientists and trained volunteers are in the field: monitoring the conditions of multiple plant communities, wildlife populations, water quality, and hydrologic changes in the reserve. Our current monitoring activities include physical, chemical, and biological studies:

Physical Conditions—Conservancy staff maintain

图2-2-10 定期清理
Fig.2-2-10 Regular cleaning

图2-2-11 湖上飞禽
Fig.2-2-11 Birds on the lagoon

图2-2-12 垃圾回收处理
Fig.2-2-12 Litter collection and disposal

来潟湖和海滩休闲娱乐。同时也提高了所有保护区的可达性（图2-2-12）。

4. 科学监控

保护组织的科学家和经过培训的志愿者在保护区内监测多种植物群落状况、野生动物种群数量、水质、水文变化等。目前的监测活动，包括物理、化学和生物学研究：

物理——组织工作人员通过保持埃斯孔迪多河和潟湖的河水位标，来确定水质和由于上游流域及周边发展活动所引起的水文变化。

化学——组织收集数据记录器中的连续数据，来研究水质参数，如溶氧量、盐度、温度、浑浊度和显示潮汐波动的深度。

生物——进行每月的志愿者鸟类调查，以确定保护区内的鸟类物种丰富度和数量，同样定期进行的还有鱼类和无脊椎动物的调查。科研人员进行调查以确定保护方案，同时还结合外来入侵物种控制的植被恢复项目。

六、 总结

圣埃利荷潟湖位于太平洋沿岸，与圣地亚哥潟湖距离不远，水质优于后者，但污染问题已不容忽视。城市化进程为这片曾经生态良好的湿地带来威胁，铁路的修建阻断了潟湖与海洋的水流循环，人口的骤升加剧了环境污染，这几乎是世界各个角落

stream gauges on Escondido Creek and in San Elijo Lagoon to determine water quality and hydrologic changes due to activities in the upper watershed and surrounding development.

Chemical Conditions—We also collect continuous data from deployed data loggers to study water quality parameters like dissolved oxygen, salinity, temperature, turbidity, and depth which show tidal fluctuation.

Biological Conditions—Our dedicated birders conduct monthly volunteer bird surveys to determine bird species richness and abundance in the reserve. Fish and invertebrate surveys are also conducted regularly to determine species richness and abundance. Staff scientists conduct surveys to determine treatment and re-vegetation success in conjunction with the Invasive Species Control Program.

VI. Summary

The San Elijo Lagoon locates at the edge of the Pacific Ocean, not far from the San Elijo Lagoon, and its water quality is better than the later, but now the pollution in San Elijo Lagoon is quite serious. Urbanization has threatened this wet land which once had fine zoology. The railway construction has stopped the flow recycling with the ocean, and the population boom has intensified environment pollution, which is similar to zoology crisis in almost every corner of the world. Currently there are many flaws in the urban landscape system, and these damaged landscape spots are specific reflections of landscape pollution. However, the reasons for this landscape pollution are not natural but anthropic factors.

In the planning, designers have adopted many

都在上演的生态危机。目前城市景观体系中存在许多缺陷，而这种受损的景观斑块正是景观污染的具体体现之一。并且，造成景观污染的原因往往不是自然因素，而是人为因素。

规划中，采取多项措施来解决人为因素造成的污染，例如，建设一条4英里（约6.4公里）长的海洋排污渠，解决之前污水排入潟湖的污染问题。另外，修建生物污水处理系统，通过动植物的生态循环过程净化水质，保证水质达到要求后再排放，这与我国的人工湿地污水处理系统非常相似。规划特别提出要控制外来植物入侵，外来植物入侵的危害主要体现在影响当地生态系统的结构和功能、生物多样性、改变物种进化方向、土壤的理化性质、气候和人类的生产与生活等方面。而原生植被的保护有利于保护绿地空间的完整性、连续性和原生性生境特征，有利于营造一个鸟类栖息、觅食的乐园。

steps to solve these pollutions caused by human. For example, they build a 4 miles long pollution discharge channel to the ocean to unload the sewage that was once discharged into the San Elijo Lagoon. They put up biological sewage disposal system to purify the water through flora and fauna biological recycling. Only after the purification which ensures that the discharged water meets the requirement can the sewage be unloaded, which is similar to the artificial wet land sewage disposal system in China. The planning highlights that introduced plants must be controlled because they will affect the structure and function of local biological system, biodiversity, evolution direction, edaphic physicochemical property, climate and aspects of human life and production. The protection for protogenesis plants can preserve the integrity and continuity of green land space and original habitat features, and it is helpful to build a bird paradise for inhabitation and foraging.

思考题

1. 通过其他途径了解潮汐作用对于圣埃利荷潟湖的影响，以及圣埃利荷潟湖的生态保护、生物生产、娱乐和城市化各个用途之间是如何相互支持与相互影响的。

2. 适宜度模型的建立是否能够解决政治和生态这两种观点各自的问题？为什么？

3. 从多个方面来论证海洋排污渠对于解决“水土污染问题”的有效性。

4. 从圣埃利荷潟湖的案例分析人为参与的规划控制管理对于自然生态能够起到什么样的积极影响。

5. 科学监控是否能够有效的反映圣埃利荷潟湖的状态？是否能够敦促人们对于紧急情况作出相应的应对措施？是否能够有效防止圣埃利荷潟湖重蹈被污染的覆辙？

Questions

1. Try to find more information about the tidal effects on the San Elijo Lagoon. Try to understand the relationship among the usages such as ecological protection, biotic production, entertainment and urbanization by learning how these usages support and affect each other.

2. Can the building of a suitable model solve the problems in a political view as well as an ecological view? Why?

3. Demonstrate with different examples the effectiveness of ocean outfall in solving the problem of water and land pollution.

4. Discuss the positive effects of human planning control management towards nature ecology using San Elijo Lagoon as an example.

5. Can scientific monitoring effectively show the condition of San Elijo Lagoon? Is it able to be an alarm to warn people to take an activite response towards an emergncy and prevent the lake from being poluted again?

注释/Note

图片来源/Image：本案例图片均引自Design for Human Ecosystems: Landscape, Land Use, and Natural Resources

参考文献/Reference：本案例引自Design for Human Ecosystems: Landscape,Land Use,and Natural Resources

第三章 绿道网络生态规划设计

Chapter III Greenways Network Planning & Design

第一节 佛罗里达绿道网络 Section Ⅰ Florida Statewide Greenways Plan
第二节 迈阿密河流绿道概念规划 Section Ⅱ The Concept Plan of Miami River Greenway

第一节 佛罗里达绿道网络

Section I Florida Statewide Greenways Plan

一、 项目概况

从1995年开始，美国佛罗里达大学已经开始和该州的环境保护部门共同从事发展佛罗里达绿道网络方面的工作。佛罗里达大学主要负责发展决策模型的建立，并以此来把握最佳的连接点，以保证绿色生态网络可以覆盖整个佛罗里达州。人们使用地理信息系统来分析所有土地利用情况以及重要地区的已得数据，这些区域包括当地物种栖息地、重要的自然群落、湿地、不可达地区、泛洪区以及重要的水生生态区。这些数据是在全国绿道网络建设过程中积累而来的，该网络将所有重要的大型生态和自然资源以及联系这些资源必要的生态景观涵盖在一个广泛的功能性网络中。整个过程是由三个州立绿道委员会彼此协作并互相监督完成的。在整个模型的建立过程中，得到了来自佛罗里达绿道委员会、佛罗里达绿道协调委员会、州立机构、区域机构、联邦机构、科学家、高校人事部、保育团体、规划师以及公众等超过20个部门的技术支持。当模型完成后，其作为绿道网络实施计划的一部分在1999年全州范围的公众会议上通过了彻底的审查。结果表明，在全国范围内有大约50%的州适宜根据此模型建立绿道网络。为了强调保护力度，佛罗里达大学应邀在开发和实施过程中评估绿道网络各个

I. Introduction

Since 1995, The University of Florida has been working with the Florida Department of Environmental Protection to assist in the development of the Florida Statewide Greenways Plan. The University of Florida was asked to develop a decision support model to help identify the best opportunities to protect ecological connectivity statewide. Geographic information systems (GIS) software was used to analyze all of the best available data on land use and significant ecological areas including important habitats for native species, important natural communities, wetlands, roadless areas, floodplains, and important aquatic ecosystems. All of this information was then integrated in a process that identified a statewide Ecological Greenways Network containing all of the largest areas of ecological and natural resource significance and the landscape linkages necessary to link these areas together in one functional statewide network. The process was collaborative and overseen by three separate state-appointed greenways councils. During the development of the model, technical input was obtained from the Florida Greenways Commission, Florida Greenways Coordinating Council, state, regional, and federal agencies, scientists, university personnel, conservation groups, planners and the general public in over 20 sessions. When the modeling was completed, the results were thoroughly reviewed in public meetings statewide as part of the development of the Greenways Implementation Plan completed in 1999. The results indicated that approximately 50 percent of the state is potentially suitable for inclusion within a statewide ecological

功能的相对重要性。

greenways system. In order to focus protection efforts, the University of Florida was asked to develop and apply a process to assess the relative significance of features within the Ecological Network.

二、 规划设计过程

1. 生态绿道优先等级制定过程

制定生态绿道的优先次序分为两个步骤。首先，环境保护部、佛罗里达鱼类和野生动物保护委员会、佛罗里达自然地域管理局、水域管理部以及其他机构和团体的工作人员举行了两次会议来讨论优先次序选择的标准和数据。基于这些会议上讨论的结果，佛罗里达大学制作了一个地理信息系统模型，该模型在原始的生态绿岛模型基础上进行了修改和完善，并通过这个模型对保护全州范围内的连接度制定了高、中、低三个优先等级。

下一步是将相互分离的区域按照常规的设置标准按高、中等级的优先次序进一步精确。尽管早期的优先次序也可以用来计算这样的结果，但是佛罗里达绿道计划的实施过程以及佛罗里达长期计划中的潜在保护地区优先等级的确定需要更加精确的优先等级（佛罗里达自然地域管理局）。更加精确地布置潜在景观连接点和廊道计划的优先次序将参照以下的标准：

维持或恢复分布广泛的物种种群数量的潜在重要性（例如，佛罗里达黑熊和佛罗里达美洲豹）；维持州际连接保护网络（从南佛罗里达起的狭长地带）的重要性；提供额外的维持全州范围连接度机会的其他重要景观连接点，特别是那些从属更高优先等级连接点的；能够保护水资源，提供连续的功能性栖息地并能够维持与其他州的连接度的重要水系廊道。人们依据这些标准将生态网络的优先等级分成了6类（图3-1-1）。

2. 关键连接点的识别

1998年的佛罗里达绿道网络计划执行报告中明确表明，在确定全州范围的生态绿道网络保护的优先次序之后，就是确定识别关键的连接点。将关键连接点作为明确的计划区域看待，对于保护佛罗里达州的生态绿道网络有着重要的意义。佛罗里达绿道与游径委员会根据保护或优先级别的变化批准确定关键的连

II. Planning and design process

1. Ecological Greenways Prioritization Process

The ecological greenways were prioritized in a two-step process. First, two meetings with staff from the Department of Environmental Protection, Florida Fish and Wildlife Conservation Commission, Florida Natural Areas Inventory, the Water Management Districts, and other agencies and groups were conducted to discuss criteria and data for selecting priorities. Based on these meetings, the University of Florida developed a GIS model that refined and modified the original ecological greenways model process to identify features within the results that were high, moderate, or lower priorities for protecting statewide connectivity.

The next step involved separating areas identified as high and moderate priorities into even more refined classes of priority using a general set of criteria. Though the original prioritization was used to support this effort, more refined priorities were needed to serve as a better planning tool both for the Florida Greenways Program implementation process and to support the prioritization of potential conservation areas for the Florida Forever Program (Florida Natural Areas Inventory 2001). The following criteria were used to place potential landscape linkage and corridor projects into more refined priority classes:

Potential importance for maintaining or restoring populations of wide-ranging species (e.g., Florida black bear and Florida panther); Importance for maintaining a statewide, connected reserve network from south Florida through the panhandle; Other important landscape linkages that provide additional opportunities to maintain statewide connectivity especially in support of higher priority linkages; Importance as a riparian corridor to protect water resources, provide functional habitat gradients, and to possibly provide connectivity to areas within other states. The application of these criteria resulted in the separation of the Ecological Network into 6 priority classes(Fig. 3-1-1).

2. Identification of Critical Linkages

The Florida Greenways Program implementation report (1998) included the identification of critical linkages as the next step following prioritization in the process of protecting an ecological greenways network across the state. Critical linkages serve as more defined project areas that are most important for protecting the Florida Ecological Greenways Network. Such critical linkages are to be approved by the Florida Greenways and Trails Council on an iterative basis as linkages are protected or

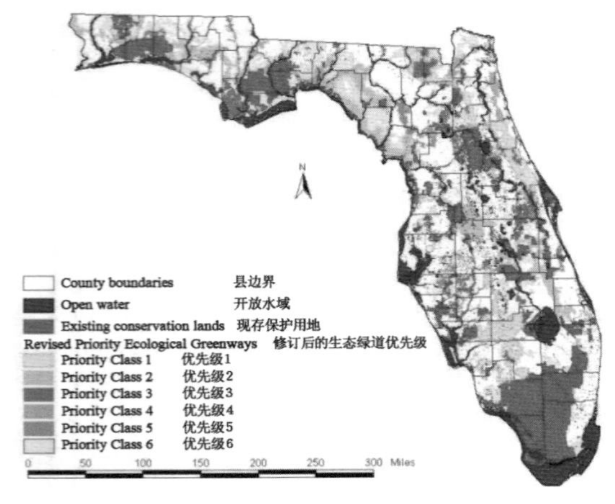

图3-1-1 优先级的分级

Fig.3-1-1 Separation of priority classes

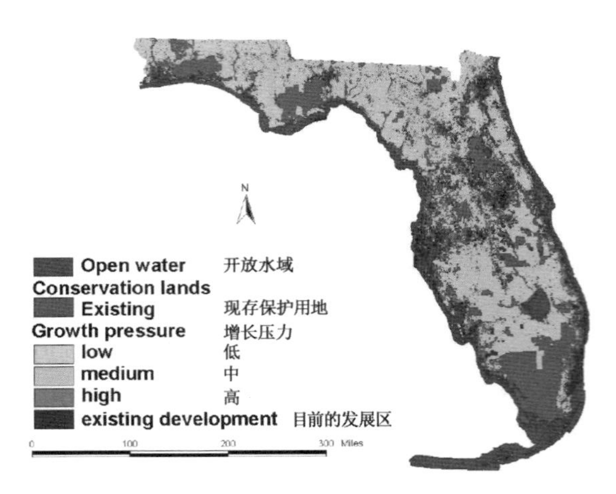

图3-1-2 增长压力分级

Fig.3-1-2 Separation of growth pressure classes

接点。人们用两个主要的数据集来界定这些关键连接点的关联。为了确定哪些连接点对于保护佛罗里达生态绿道网络是重要的，那么同时基于生态标准与保护发展压力威胁水平的优先等级就是必要的。上文描述的将佛罗里达生态绿道网络分成6个优先等级的过程就是基于生态优先等级的（图3-1-1）。贾森·泰森吉通过如下模型研究了发展压力。

1）发展压力模型

佛罗里达大学的地理规划中心已经开发出这种能够显示整个佛罗里达州的增长潜力的模型。这项研究的基础工作是贾森·泰森吉在城市与区域规划系的硕士学位项目。建立这个模型的目的是为了确定哪些区域最有可能从非城市用地向城市用地转换，从而可以告知民众包括农业用地和保护节约土地的决策。该模型的原型最早出现在林业部门关于《乡村和家庭土地保护法》的报告中。

从该分析可以得出增长潜力地图，在这张图上人们可以看到最近由非城市用地转变成居住或商业用地所具有的潜力。该模型有四个组成部分：区位增长潜力、历史增长、现有闲置住宅用地以及未来预计增长。

区位的影响要从两部分来分析：城市设施和城市枢纽。城市设施的分析说明增长潜力受到区位的影响。诸如靠近海岸或内陆水体的道路以及现有居住用地都能够说明区位可以影响潜力的增长。区域

priorities change over time. Two primary data sets were used to delineate the first iteration of critical linkages. To define linkages that are most critical to the protection of the Florida Ecological Greenways Network, prioritization based on both ecological criteria and level of threat by conversion to development (development pressure) is needed. For ecological-based prioritization, the prioritization process described above that categorized the Florida Ecological Greenways Network into six priority levels was used (Fig. 3-1-1). Development pressure was modeled by Jason Teisinger in a process summarized in the following section.

1)Development Pressure Model

The University of Florida's Geoplan Center has been developing a decision support model that indicates growth potential across the state of Florida. The basis of this work is a Master's degree project in the Department of Urban and Regional Planning being completed by Jason Teisinger. Its purpose is to identify areas most likely to be converted from non-urban to urban land use in order to inform land use decisions including agricultural and conservation land protection. A prototype of this model was recently used in the Division of Forestry report for the Rural and Family Lands Protection Act.

This analysis resulted in a Growth Potential map that displays the potential for parcels currently in non-urban land uses to be converted to residential or commercial land uses. The model has four components: Growth Potential based on Location, Historic Growth, Existing Vacant Residential and Projected Future Growth.

The Location Influence component is comprised of two analyses: Amenities and Urban Hub Influence. The Amenities analysis illustrates the effect of locational drivers on growth potential. A locational driver is an amenity that drives growth such as roads, proximity to

驱动能够促进道路、近海或内陆水域以及现存居民区的发展。城市设施约占居民区发展的10%，由此为中心向外辐射，依照与区域驱动的距离，划分出10个等级，越靠近中心等级越高。对每个城市设施进行同样的划分过程并把结果整合起来，便能够得到城市设施的分析结果。城市枢纽的影响分析使用大都市规划组织对枢纽的定义和相关人口作为影响的衡量标准。城市枢纽影响分析和城市设施影响分析共同组成区位影响分析。

历史增长潜力是通过居住单元百分比的变化和1992年至1999年之间每区每镇居住单元的直接变化得到的。这是根据公共土地调查系统数据（该系统将州打散成乡镇和几平方英里的区域）和税收部门的表格得到的。

现有闲置住宅用地是由1999年每区域每镇总的闲置住宅用地分析得到的。区域分为10个等级。

未来预计增长是通过1990年人口普查得到的。该分析测算了1990年至2020年之间的预计人口密度。

每四个部分为一组加权合并之后，再除去现有的湖泊、湿地以及保护用地得到的就是最终增长潜力分析。增长潜力地图上标示出了10个等级，1类地区表示了由非城市用地向城市用地转化的最低潜力，10类地区则表示了由非城市用地向城市用地转化的最高潜力。为了识别关键连接点，通过统计学的优化过程而把这10个等级混入到高、中、低三个增长潜力级的做法叫做自然断裂（图3–1–2）。

2）生态绿道优先等级与增长压力模型的整合

通过矩阵的使用可以将生态绿道的优先等级与增长压力模型联系起来。这些矩阵中的每一栏表示绿道优先等级与增长压力的所有可能的组合。在进行优先级与增长压力的整合时，倾向于更高优先等级与拥有更高增长压力的区域应该被赋予更高的优先等级（表3–1–1）。这样做的理由是保护工作的重点应该首先放在含有最高优先等级的资源上，因为这些资源在不久的将来极有可能会消失殆尽。

在关键连接点配对过程中最后使用的矩阵包含了生态绿道优先等级的6个优先级和3个层次的增长

the coast or inland water bodies and existing residential land uses. Areas were ranked based on distance from locational drivers. Ten bands of area radiating out from the amenity capturing 10% increments of residential development were delineated. These radiating bands were ranked 1~10 with the bands closer to the amenity having higher ranks. This was done for each amenity and results were combined to produce the Amenities analysis. The Urban Hub Influence analysis used Metropolitan Planning Organization boundaries to define hubs and the associated population as a measure of influence. The Urban Hub Influence analysis and Amenities analysis were combined to produce the Location Influence component.

The Historic Growth Potential component was derived through an analysis of the percent change in residential units and the direct change in residential units between 1992 and 1999 per section per county. This was done using the Public Land Survey System dataset that breaks the state up into townships, ranges and square mile sections and the Department of Revenue tax data tables.

The Existing Vacant Residential component was derived by an analysis of the total vacant residential units per section per county for 1999. Sections were ranked 1~10.

The Projected Future Growth component utilized the 1990 census growth projections. The analysis measured the projected change in density between 1990 and 2020.

Each of the four final data sets were weighted and combined. Lakes, wetlands, and existing conservation lands were removed resulting in the Final Growth Potential Analysis. The growth potential map is ranked with values of 1~10 with the value of 1 representing areas with lowest potential for conversion to urban land uses and the value of 10 representing areas with the greatest potential for conversion to urban land uses. For identifying critical linkages, the values of 1~10 were lumped into three categories of high, medium, and low growth potential using a statistical optimization procedure called natural breaks(Fig. 3-1-2).

2)Combination of Ecological Greenways Priorities and Growth Pressure Model

The Ecological Greenways priorities and the growth pressure model results were combined using a matrix. The matrix contains boxes that represent all possible combinations of greenway priorities and growth pressure. When combined, the tendency should be to give higher priority to areas that are part of high priority greenways and have high growth pressure (Tab. 3-1-1). The rationale is that the focus of protection efforts should first be on areas containing the highest priority resources that are most in danger of being lost in the near future.

The final matrix used in the critical linkage process paired all potential combinations of the six priority

Ecological-based Prioritization
生态-基础 优先级

	Low 低	Medium 中	High 高
Low 低	Low 低	Low 低	Medium 中
Medium 中	Low 低	Medium 中	High 高
High 高	Medium 中	High 高	High 高

表3-1-1 生态绿道优先等级与增长压力模型整合表

Table.3-1-1 The combination of the Ecological Greenways Priorities and Growth Pressure

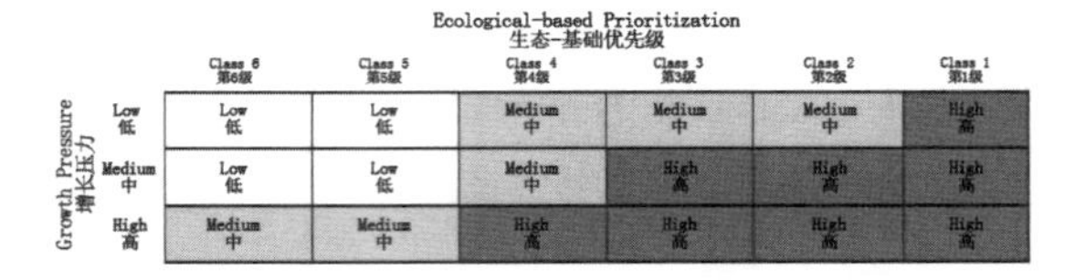

图3-1-3 生态绿道优先等级与增长压力整合后的结果

Fig.3-1-3 The result of the combination

压力所有可能的组合，这18个组合都是独一无二的。随后将高、中、低三个等级的优先级赋值予这些组合，通过这些来找到在全州范围内拥有最显著的生态绿道的地区（图3-1-3）。

3) 关键连接点备选区域的鉴别

人们使用矩阵中的值创建了一个新的地图数据层，这个数据层融合了生态绿道的优先等级和增长压力模型，并将佛罗里达生态绿道网络优先等级划分出了高、中、低三个优先等级区域（图3-1-4）。下一步，在合并后的优先级数据层的基础上来确定潜在的项目区域的边界，这些区域会包含高优先级的区域，同时也将作为现存保护用地枢纽之间的连接点。这么做的意图是希望这样的图纸能够具有相当大的包容性，从而可以确保所有的潜在连接点都能够至少包含具有高优先级的大片区域，这就表示随着未来的发展包含连接点的关键区域会分裂出来作为备选区域。经过这样一个过程后，我们得到了24个关键连接点的备选区域（图 3-1-5，图 3-1-6）。由于较高的整体发展压力，大部分的备选区域主要集中在南佛罗里达州的中北部，个别的备选区域处在大本德西至彭萨科拉地区。

4) 从候选区域中选择关键连接点

从候选区域中选取关键连接点有三个标准，这三个标准适用于所有的候选区域。佛罗里达大学的工作人员参与了最初的选择过程，并与佛罗里达州自然保护协会的工作人员制定了一份潜在关键连接点的最终目录。在筛选过程中采用以下标准：

特定的候选区域对于完成佛罗里达州的生态绿

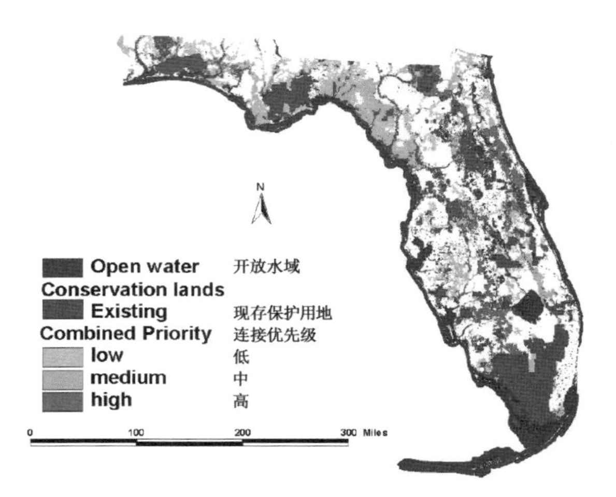

图3-1-4 新的佛罗里达生态绿道网络优先等级划分

Fig.3-1-4 The new combined prioritization of the Florida Ecological Greenways Network

levels of ecological greenway priorities and the three levels of growth pressure, which resulted in eighteen unique combinations. Then values of high, medium, or low priority were given to combinations to identify areas with the most significant ecological greenways linkages statewide(Fig. 3-1-3).

3)Identification of Candidate Areas for Critical Linkage Delineation

Using the values in the matrix, a new map data layer was created that combined the Ecological Greenways Priorities and the Growth Pressure Model results into a new combined prioritization of the Florida Ecological Greenways Network into high, medium, and low priority areas (Fig. 3-1-4). The Combined Priorities data layer was then used as the base for determining the boundaries of potential project areas that contained areas of high priority and served as linkages between major hubs of existing conservation lands. The intent was to be fairly inclusive so that all potential linkages that contained at least fairly large blocks of high priority, which often represent key areas within a linkage that could be fragmented by development in the near future, were identified as candidates. The result of this process was the delineation of twenty-four critical linkage candidate areas (Fig. 3-1-5, Fig. 3-1-6). Due primarily to higher

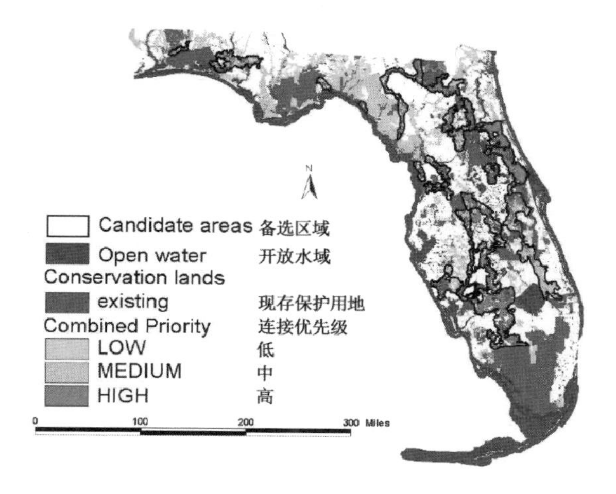

图3-1-5 关键连接点备选区

Fig.3-1-5 The critical linkage candidate areas

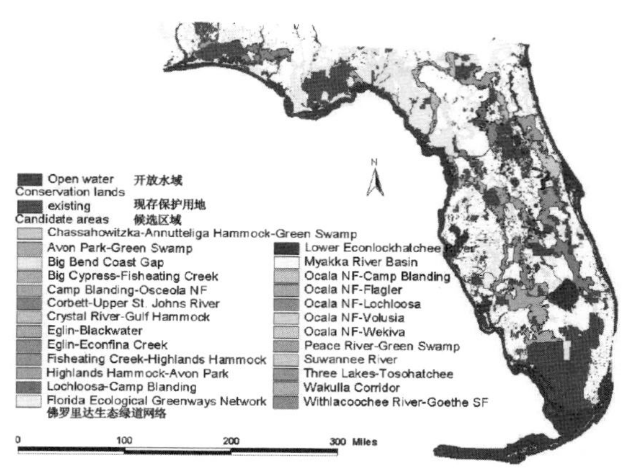

图3-1-6 24个备选关键连接点

Fig.3-1-6 The 24 Critical Linkage candidates

道网络是多么的关键，以及每个候选区域所代表的现存保护用地之间的联系是多么的重要。

候选区域存在怎样可能性会导致他们的大部分关键地区在不久的将来呈现出不协调的用途?

土地所有制的不同形式能否为保护候选区域中的可行性连接点提供一个合适的机会? 尽管我们没有全州范围内每块地区的数据，但是某个数据显示出土地拥有者的每块区域（州际范围内以平方英里为单位）的数量也可以用来进行这种评估。

在选择水域的关键连接点时要参考另外一种标准。由于水域管理区的前身是该州的保护用地，因此在每块区域内选择至少一处关键连接点用以提高对佛罗里达绿道网络内水资源保护关键连接点的保护显得尤为重要。但是基于上文提到的三个标准，与水域管理区相关的关键连接点都已入选，且并未专门为满足第四个标准而增加候选区。

24个候选区域中有10个被选定为拟议的关键连接点。关键连接点的选择并没有具体的数量目标，因此，从完全基于上述标准的24个候选区域中筛选出的10个拟议关键连接点是能够在不久的将来决定出最适合集中保护活动的联系的。佛罗里达绿道和游径协会于2002年4月提出了拟议的关键连接点和筛选过程，并同意其作为第一代佛罗里达生态绿道网络的关键连接点筛选标准（图3-1-7）。

目前大约有2.7百万英亩（约1.1万平方公里）的地区含有关键连接点，大概占现存保护用地的

overall development pressure, most candidate areas are in north-central to south Florida, but several are found from the Big Bend west to Pensacola.

4)Selection of Critical Linkages from Candidate Areas Pool

Critical linkages were selected from the pool of candidate areas using three general criteria applied to all of the candidate areas. Staff at the University of Florida went through an initial process of selection and then developed a final list of potential critical linkages working with the protection staff from the Florida Chapter of The Nature Conservancy. The criteria used in the selection process included:

How critical is a particular candidate area to completing the Florida Ecological Greenways Network and how important is the linkage between existing conservation lands represented by each candidate area?

What is the likelihood that much of or a key segment of the candidate area could be converted to incompatible uses in the near future?

Do land ownership patterns appear to provide a suitable opportunity to protect a feasible linkage within the candidate area? Although we do not have a statewide land parcel database, a data set indicating the number of landowners per section (square miles areas statewide) was used to make this assessment.

Selection of critical linkages within each Water Management District was another criterion that was considered during the selection process. Since the Water Management Districts are one of the primary agencies involved in protected conservation lands throughout the state, inclusion of at least one critical linkage in each district to promote protection of key linkages within the Florida Ecological Greenways Network consistent with water resource conservation objectives was considered vital.However, based on the three criteria listed above,

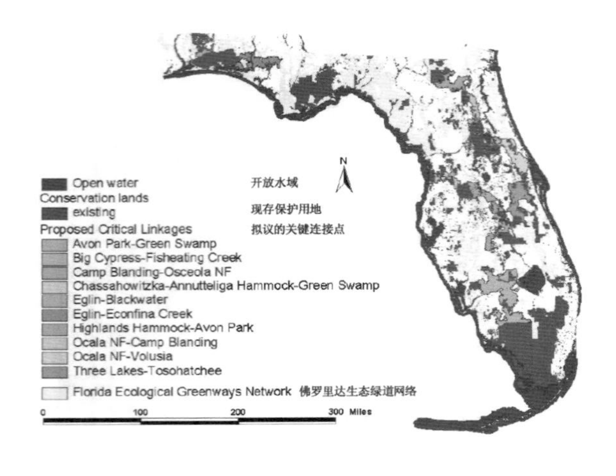

图3-1-7 第一代关键连接点

Fig.3-1-7 The first iteration of Critical Linkages

Category 类别	Acres 英亩	Percent of Critical Linkages 关键连接点所占百分比
Existing Conservation Lands 现存保护用地	465351	16.97
Proposed Conservation Lands 拟议保护用地	809257	29.51
Open Water 开放水域	46053	1.68
Other Private Lands 其他私人用地	1421726	51.84
Total 总计	2742387	10.00

表3-1-2 各类用地面积和关键节点比例

Table. 3-1-2 Land areas & percent of critical linkages

17%，拟保护用地的30%（佛罗里达永久项目，保存我们的河流计划和水域管理区研究区域），以及开放水域的2%（表3-1-2）。

3. 节点廊道设计

佛罗里达生态绿道的建立主要是为了：保护佛罗里达州当地生态系统和景观的关键要素；恢复和保持原生生态系统和生态过程的联系；促进这些生态系统和景观作为动态系统的功能性；保持这些生态系统的各个组成部分的进化潜力，以适应未来的变化。

因此这些廊道要具有一定的景观功能梯度，具体梯度表现为水域—湿地—中生高地—旱生高地。同时大部分的廊道可分为开敞地区—森林边缘—森林内部—河岸—水道—河岸—开敞地区（图3-1-8）。其中森林作为大部分物种栖息地，因此要具有一定的宽度。而河岸—水道—河岸段的设计要注意控制洪水与盐碱化的侵蚀，同

critical linkages were selected that are relevant to each Water Management District, and no candidate area was added to the list of proposed critical linkages specifically to meet this fourth criterion.

Ten of the twenty-four candidate areas were selected as proposed critical linkages. There was no numerical goal for the number of critical linkages selected and therefore a number of ten proposed critical linkages was based solely applying the criteria consistently to all twenty-four candidate areas to determine linkages most suitable for concentrating protection activities in the near future. The proposed critical linkages and the selection process were presented to the Florida Greenways and Trails Council in April, 2002 and were approved as the first iteration of Critical Linkages for the Florida Ecological Greenways Network (Fig. 3-1-7).

There are approximately 2.7 million acres within Critical Linkages with 17 percent within existing conservation lands, 30 percent in proposed conservation lands (Florida Forever projects, Save Our Rivers projects, and Water Management District study areas), and 2 percent in open water (Tab.3-1-2).

3. Corridor Design

The objectives of the establishment of Florida Ecological Greenway Network are as follows: To protect critical elements of local ecological system and landscape; To restore and maintain the link between the original ecological system and ecological process; To promote the function of these ecological systems and landscape as a dynamic system; To maintain the evolution potential of these ecological system and components to adapt to future changes.

So these corridors must reflect some landscape function gradients with the specific manifestation as water district-wetland-mid highland-dry highland. And meanwhile, most corridors can be divided as open area-forest edge-forest interior-river bank-watercourse-river bank-open area (Fig. 3-1-8). Among all these elements, forest serving as the habitat for most species thereof must hold particular breadth. The design for river bank-watercourse-river bank should focus on control over flood and salinization, prevent soil erosion and nutrition losses, and protect water quality.

1)Eglin-Blackwater River

This Critical Linkage connects Eglin Air Force Base to the Blackwater River State Forest (Fig. 3-1-9). Both conservation areas have very significant longleaf pine sandhill and flatwood communities. The linkage is approximately 53,000 acres with 24 percent in existing conservation lands and 23 percent in proposed conservation lands and a total of approximately 40,000 acres of private land. There is an important Florida black bear population found within and around Eglin

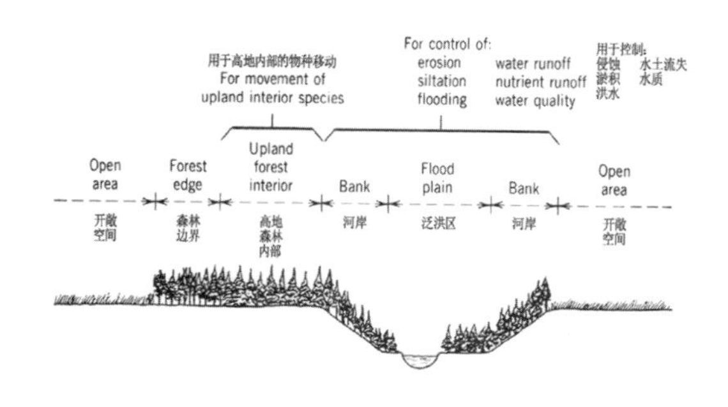

图3-1-8 大部分廊道的结构

Fig.3-1-8 The structure of most corridors

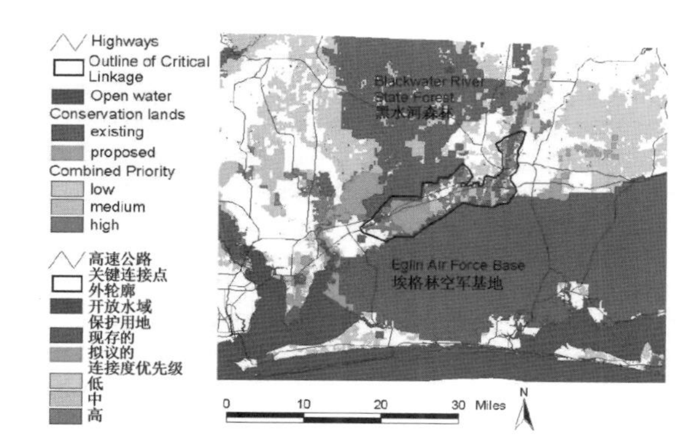

图3-1-9 埃格林-黑水河节点

Fig.3-1-9 Eglin-Blackwater river

时还要防止水土流失、营养流失并起到保护水质的功效。

1）埃格林—黑水河节点

该关键连接点连接了埃格林空军基地和黑水河森林（图3-1-9）。两处保护区均有长叶松沙地和平木林地。该关键连接点占地约5.3万英亩（约214.5平方公里），其中约24%位于已有保护区，23%位于计划保护区，私有土地约占4万英亩（约162平方公里）。两块保护区之间的连接点实行永久保护，得益于此的有埃格林空军基地及其周边的佛罗里达黑熊群落。该关键连接点内还覆盖了黄河大部分地区，且包含黄河水管理区域。埃格林黑水河曾经是二级生态绿地，并覆盖了中高生长压力地区。

2）埃格林—伊康菲纳溪流节点

该关键连接点起于埃格林空军基地，止于巴拿

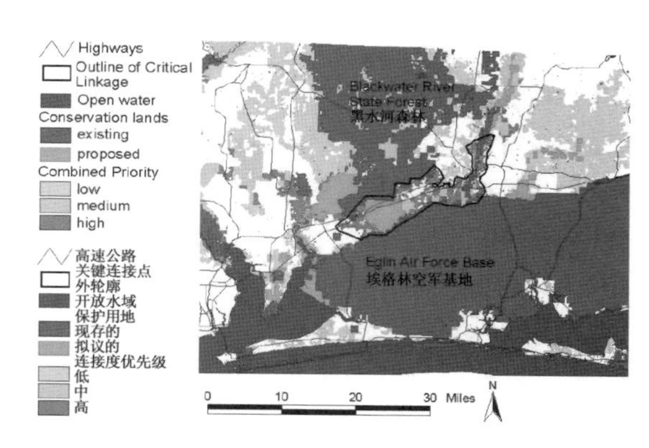

图3-1-10 埃格林-Econfina溪流节点

Fig.3-1-10 Eglin-Econfina Creek

Air Force Base that may benefit significantly from the permanent protection of the linkage between these two large conservation areas. A significant portion of the Yellow River is also within the Critical Linkages including portions of the Yellow River Water Management Area. Eglin-Blackwater River was originally part of a Priority Class 2 Ecological Greenway and contains large areas with moderate to high growth pressure .

2)Eglin-Econfina Creek

This Critical Linkage runs from Eglin Air Force Base to the headwaters of Econfina Creek north of Panama City (Fig. 3-1-10). It connects Eglin Air Force Base to the Econfina Creek Water Management Area and also includes most of the lower Choctawhatchee River and significant portions of the Choctawhatchee Water Management Area. The Sand Mountain Florida Forever project is also an essential part of this Critical Linkage. The linkage is approximately 330,000 acres with 20 percent in existing conservation lands, 6 percent in proposed conservation lands, 2 percent open water, and a total of approximately 258,000 acres of private land. The linkage represents an essential part of the ecological connection between Eglin Air Force Base and the Apalachicola National Forest. Both conservation areas harbor significant Florida black bear populations whose security will be enhanced through protection of a functional linkage and large areas of additional habitat. Eglin-Econfina Creek was originally part of a Priority Class 2 Ecological Greenway and contains large areas with moderate to high growth pressure.

3)Camp Blanding-Osceola National Forest

This Critical Linkage connects the Osceola National Forest and the Camp Blanding Military Reservation (Fig. 3-1-11). The linkage runs from Camp Blanding Military Site, the Jennings State Forest and Cecil Field Conservation Area east to the Osceola National Forest through the New River headwaters and Raiford Wildlife

马城北部的伊康菲纳河源头（图3-1-10），连接了埃格林空军基地和伊康菲纳河水域管理区域，并涵盖了查克托哈奇河下游绝大部分区域以及查克托哈奇河水域管理区域最重要的部分。佛罗里达永恒沙山项目是该关键连接点的重要组成部分。该关键连接点占地约33万英亩（约1335平方公里），其中20%位于已有保护区，6%位于计划保护区，且有2%开放水域和总计约25.8万英亩（约1044平方公里）私有土地。该关键连接点是埃格林空军基地和阿巴拉契科拉河国家森林公园之间重要的生态纽带。这两块保护区内有佛罗里达黑熊群落生活，通过对功能连接点和其他栖息地的保护，这些黑熊的安全有了保障。埃格林—伊康菲纳河曾经是二级生态绿地，并覆盖了中高生长压力地区。

3）布兰丁营—奥西奥拉国家森林节点

该连接点联系了奥西奥拉森林与布兰丁营军事用地（图3-1-11），横跨布兰丁营军事基地，詹宁斯州属森林和塞西尔·菲尔德保护地的东面与奥西奥拉国家森林，并穿过新河的源头地区与雷福德野生动画管理区。这个连接点占地约23万英亩（约930平方公里），其中现有保护土地约12%，拟议的保护用地占14%，1%的开放水域以及大约20万英亩（约809平方公里）的私人土地。这个连接点最初是第一优先级的生态绿道，因此也是该州范围内三个最重要的连接点之一。这意味着该连接点北面的一半区域都需要将奥卡拉国家森林和奥西奥拉国家森林、奥克弗诺基国家野生动物保护区联系起来以保护这些地区的复杂性。这些保护区都能够支持重要的佛罗里达黑熊种群在连接点的散布。关键连接点内的这些区域发展最大的威胁都在301号高速公路附近，确保横跨301号高速公路的保护区域不再受到进一步的侵蚀是非常必要的。在301号高速公路附近最具潜力的交汇地带主要集中在佛罗里达森林东北部的佛罗里达永久保护区内，对于完成关键连接点的建立，保护这些地区是非常必要的。

4）高地硬木林—碧湖公园节点

这个关键的连接点连接了高地硬木林州立公园与碧湖公园爆炸范围（图 3-1-12）。此连接点占地约

Management Area. The linkage is approximately 230,000 acres with 12 percent in existing conservation lands, 14 percent in proposed conservation lands, 1 percent open water, and a total of approximately 200,000 acres of private land. This linkage was originally a Priority Class 1 Ecological Greenway and is therefore one of the three most important ecological linkages in the state. It represents the northern half of the linkage needed to connect the Ocala National Forest to the Osceola National Forest/Okefenokee National Wildlife Refuge conservation complex. These conservation areas both support important Florida black bear populations that appear to be connected by dispersal through this Critical Linkage. The area most threatened by development within this Critical Linkage occurs directly around US 301, and securing a protected crossing area across US 301 before further encroachment by development occurs is essential. The best potential crossing areas around US 301 are primarily within the Northeast Florida Timberlands Florida Forever project, which is essential to protect in order to complete this Critical Linkage.

4)Highlands Hammock-Avon Park

This Critical Linkage connects Highlands Hammock State Park to the Avon Park Bombing Range (Fig. 3-1-12). The linkage is approximately 77,000 acres with 13 percent in existing conservation lands, 59 percent in proposed conservation lands, 4 percent in open water, and a total of approximately 65,000 acres of private land. The linkage represents the portion of a Priority Class 1 Ecological Greenway from Fisheating Creek to Avon Park Bombing Range that is most threatened by development. The US 27 corridor crosses the linkage and is developing rapidly to the north and south of the crossing area. The Oldtown Creek Watershed Florida Forever project is an essential part of this critical linkage, and its protection would be greatly enhanced if the boundaries of Oldtown Creek Watershed was expanded to encompass more of the Charlie Creek basin to connect to Highlands Hammock State Park and across US 27 to connect to the Lake Wales Ridge State Forest adjacent to Avon Park Bombing Range. If the area crossing US 27 is effectively protected, a wildlife underpass will need to be considered. Habitat within this linkage could help support panther re-establishment in south-central Florida in the future.

III. Summary

Greenway has been a researching focus and leading edge of Protective Biology, Landscape Biology, City Planning and Landscape Planning, which is called "A Greenway Move."The research for Florida Greenway begins in 1995. The researching group estimates a geographic information system model to ensure the high, medium and low prior levels in the biological network and then establishes the critical linkage and

7.7万英亩（约311.6平方公里），其中有13%的现存保护用地，59%的拟议保护用地，4%的开放水域以及大约6.5万英亩（约263平方公里）的私人土地。这个连接点代表了从查理河盆地到碧湖公园爆炸范围内的部分的第一优先级的生态绿道，这部分是受到城市发展威胁最大的区域。美国27号高速公共路横穿了这个连接点，同时由北向南的这个交汇地带还在迅速发展。佛罗里达常青项目旧镇流域是该关键连接点的必要组成部分，若能扩大该流域范围，覆盖更多查理河盆地地区，实现跨越27号州际公路连接HH州立公园和毗邻碧湖公园的威尔士湖岭国立森林公园，则能够更好地保护该关键连接点。如果美国27号公路的交汇地区要得到有效的保护，野生动物的地下通道就必须被纳入考虑。连接点内的栖息地能够帮助黑豹群落在以后重新在佛罗里达中南部建立起来。

三、总结

20世纪90年代以来，绿道一直是保护生物学、景观生态学、城市规划和景观设计等多个学科交叉的研究热点和前沿，这种热潮被称为绿道运动。而美国佛罗里达绿道的研究也正始于1995年，研究机构建立了一个地理信息系统模型，以确定生态网络中的高、中、低三个优先等级。然后，确定关键的连接点，将其作为重点地段设计。通过生态绿道优先等级与增长压力模型的整合，锁定全州范围内拥有最显著的生态绿道的区域，将其作为保护工作的重点。

佛罗里达州生态绿道网络在2003年7月被修改，反映了该地区自1995以来土地利用所发生的变化，并且在具有生态战略意义的地区方面纳入了最新的信息。虽然在这个过程中会对佛罗里达生态绿道网络的边界造成一些重要的变化，但这都是预计之中的，因为这些变化的区域都能够支持关键连接点的认定，而且不会产生大幅的影响。尽管现有关键连接点的边界会在一定程度上产生变化，但是目前的边界都只是为了一般开发项目区域所指定的，而且这些边界会根据具体土地所有制形式、土地利用和自然资源的信息进行修改，同时在这个过程中会得到一些新的保护建议。

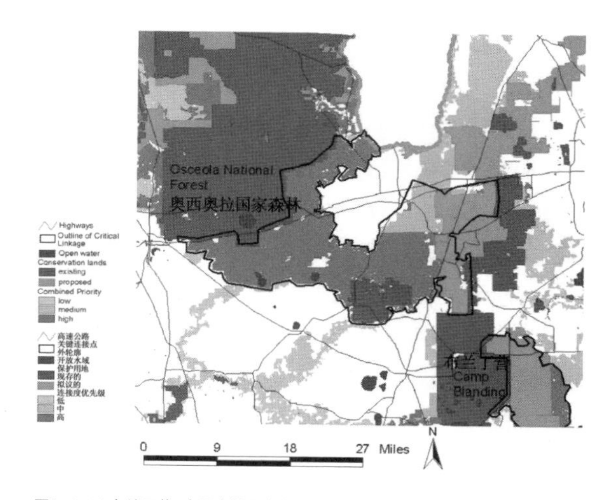

图3-1-11 布兰丁营-奥西奥拉国家森林节点

Fig.3-1-11 Camp blanding-Osceola National Forest

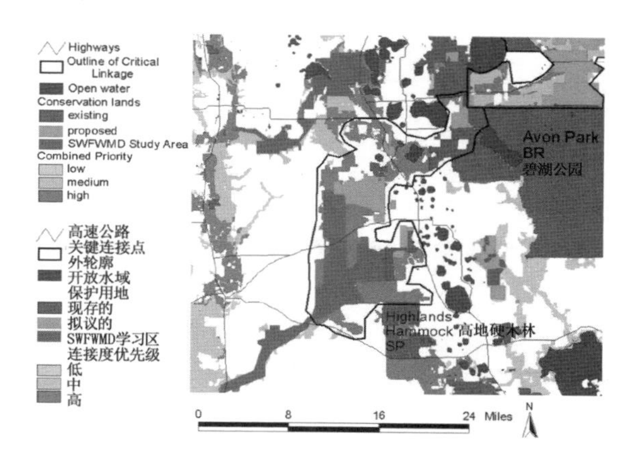

图3-1-12 高地硬木林-碧湖公园节点

Fig.3-1-12 Highlands Hammock-Avon Park

makes them the key design area. They select the most obvious biological greenway area in the state through the combination of biological greenway prior levels and increasing pressure model and then make it their focus of the protection.

The Florida Ecological Greenways Network was modified on July 2003 to reflect changes in land use that have occurred since 1995 and to incorporate new and updated information on areas of ecological significance. Although there may be some significant changes to the boundaries of the Florida Ecological Greenways Network through this process, it is expected that the areas supporting Critical Linkages identified will not be substantially affected. Though boundaries of the existing Critical Linkages could change to some extent, the current boundaries are meant only to serve as general project area boundaries that should be modified as more specific information on land ownership patterns, land uses, and

该项目的可借鉴之处是在偌大的州域范围内通过科学方法确定生态网络重点片区，同时聚焦关键连接点，通过有针对性的生态修复与织补，实现全州绿地生态系统网络化，提高网络的连接度，强化绿道的生态功能、休闲游憩功能、经济发展功能、社会文化和美学功能。

上世纪末以来，美国在绿道建设中积累了很多经验，对我国绿道建设具有重要的参考价值。目前，珠三角地区以及杭州、成都等地正全力发展绿道规划建设，建立起城市和乡村的多层级绿道网络系统，美国绿道建设对我国具有启示意义。

natural resources are obtained during the process of developing land conservation proposals.

The reference of this program is that they select key areas of the biological network by scientific means in the wide territory, and perform targeted biological restore and darning over the critical linkage to realize a biological network of the Greenland in the state, improve the connection in the network and strengthen the greenway function for biology, relaxation, economic development, social culture and aesthetics.

USA has accumulated much experience in greenway construction since 1990s, which is a good reference for the greenway construction in China. Now greenway planning and construction in the Pearl River Delta, Hang Zhou and Chong Qing is on the way. It will establish a multiple level greenway network system in both city and countryside. The US greenway construction can be a kind of inspiration for that in China.

思考题

1. 关键连接点对于佛罗里达州的生态绿道网络来说是最为重要的区域，试论述关键连接点对于生态绿道网络为何是最为重要的。

2. 在生态绿道网络建设与保护的过程是非常持久的，请根据自己的了解来为政府与非政府组织制定一个长远的建设与保护战略计划。

3. 进一步的了解发展压力模型之后，任意选取一区域，并汇出该区域的增长潜力图。

4. 尝试为文中介绍的四个节点绘出廊道结构图。

5. 根据推荐网站继续了解其他节点廊道，并给出介绍与分析。

Questions

1. Critical Linkages is the most important area in Florida Statewide Greenways. Try to discuss the reason

2. The system construction and conservation of ecology greenways are long lasting work. Try to work out a long-term construction and conservation strategy for the government and NGO according to your own understanding.

3. Further study the Development Pressure Model. Choose an area in the model to plot its Growth Potential map.

4. Try to draw the Corridor chart for the four linkages introduced in Chinese.

5. Study, introduce and discuss other corridors on the recommended websites.

注释/Note

图片来源/Image：文中图片均来自University of Florida,GeoPlan Center,Identification of Critical Linkages Within the Florida Ecological Greenways Network

参考文献/Reference：University of Florida,GeoPlan Center,Identification of Critical Linkages Within the Florida Ecological Greenways Network ;Julius Gy. Fábos, International greenway planning: an introduction,Landscape and Urban Planning 68 (2004) 143–146 ;Randall Arendt,Linked landscapes Creating greenway corridors through conservation subdivision design strategies in the northeastern and central United States,Natural Lands Trust, Media,PA 19063, USA

推荐网站/Website：http://www.dep.state.fl.us/gwt/

第二节 迈阿密河流绿道概念规划

Section II The Concept Plan of Miami River Greenway

一、项目概述

迈阿密河（图3-2-1）是迈阿密市宝贵的自然资源，其位于佛罗里达州的迈阿密市中心。在过去的20年里，数个规划的编制已经确定了河流廊道为迈阿密河的最佳利用方式。这个行动计划报告提供了具体建议和实施策略，将加速整个河流廊道的空间环境的提升。由于这些行动，河流廊道将变得更容易让居民和游客使用，海洋工业及运输活动将继续发展和繁荣，土地价值将不断提高，新的休闲设施的设置将会把这条河变成一个景观目的地，迈阿密自然与文化遗产将会得到保护并造福后代。

虽然迈阿密河是一条较短的河流，但是其历史悠久。迈阿密河经过几千年的进化从潮汐河道变成了一条淡水溪流，将水从埃弗格莱兹沼泽地输送入比斯坎湾。这条河是佛罗里达东南部最古老的自然地标。从1909年到1933年间，这条河流被延长和拓宽。在1909年迈阿密运河被建设用于埃弗格莱兹沼泽地排水工程时，造成了著名的迈阿密急流断流。残余急流可以在位于西北部的南滨水车行道附近的天堂点城市公园里看到。当开始挖掘迈阿密运河时，河水水位大幅度下降，这导致沼泽地污泥流进曾经清澈的河道。

到1940年以后，人们日益关注迈阿密河的环境

I. Overview

The Miami River(Fig.3-2-1) is a valuable natural resource situated in heart of Miami,Florida.For the past two decades,numerous plans have been produced to determine optimal uses for the river corridor.This Action Plan report provides specific recommendations and implementation strategies that will hasten physical improvements throughout the river corridor. As a result of these actions, the river corridor will become more accessible to residents and visitors,marine industrial shipping activity will continue to thrive and prosper,land values will steadily improve,new recreational amenities will make the river a destination landscape,and an important element of Miami's natural and cultural heritage will be protected and enhanced for future generations to enjoy.

Although the Miami River is a short river,its history is long.The Miami River evolved over thousands of years from a tidal channel into a freshwater stream that carried water from the Everglades to Biscayne Bay. The river is the oldest natural landmark in southeast Florida.From 1909 to 1933 the river was lengthened and widened.The famous Miami rapids ceased in 1909 when the Miami Canal was built as part of the Everglades drainage project.Remnants of the rapids can be seen at the City of Miami's Paradise Point Park at NW South River Drive.When digging for the Miami Canal began,the water table dropped dramatically and Everglades muck slid into the once clear waters of the river.

Concern over environmental degradation,water pollution,bridge openings and the generally unkept appearance of the Miami River has been voiced

图3-2-1 迈阿密河鸟瞰
Fig.3-2-1 Bird's-eye view of Miami River

退化、地下水污染、桥梁的缺失和一般景观外观等问题。20世纪70年代早期的研究是由商会、迈阿密市政府、戴德县污染控制部、美国地质研究所、中南部防洪管理部门、美国陆军工程兵团、美国国务卿理查德·斯通的办公室和戴德县规划顾问委员会共同负责的迈阿密河河道及周边环境恶化情况的调查。这最终促成了强制性的河流清理。1976—1986年的迈阿密社区综合规划认为迈阿密河及其周边是一个特殊的区域，同时阐明，迈阿密河是一条"流动"的河流和一处重要的资源。

迈阿密河目前虽然没有得到充分使用，但是提供了许多工作机会。当我们进入21世纪，迈阿密河开始意识到其作为重要的社区资源的潜质需要通过绿道规划来发掘。

二、 场地介绍

迈阿密河绿道规划项目的研究区域包括迈阿密河最东面的5.5公里的河段，从西北36号道路和西北40大道附近的塞里尼提大坝一直延伸到到位于比斯坎湾的河口。研究区域包括了从河流两侧向外延伸最远达到2500英尺的区域。在康福特运河、瓦格纳溪流和太米阿米运河处，其两侧的用地向外延伸最远达到500英尺（152.4米）（图3-2-2）。

迈阿密河并不是一个同质的景观，实际上，这条河流是一个由功能性的土地美学特质联系起来的多样化景观。根据绿道行动方案，迈阿密河按照土地利用、特征及功能的不同划分成了三个部分。

since the 1940s.In the early 1970s studies were undertaken by the Chamber of Commerce,the City of Miami,the Dade County Pollution Control Department,the US Geological Survey,the Central and Southern Flood Control District,the US Army Corps of Engineers,Secretary of State Richard Stone's Office and the Dade County Planning Advisory Board on the deplorable conditions along and within the River.As a result there were code enforcement sweeps to clean up the river. The 1976-1986 Miami Comprehensive Neighborhood Plan recognized the River as a special district and stated that the Miami River is a working river and a major resource.

It is presently underutilized and offers many redevelopment opportunities.As we enter the 21st Century,the Miami River is beginning to realize its potential as a major community resource through this Greenway Plan.

II. Introduction of the Site

The Miami River Greenway project study area consists of the easternmost 5.5 miles of the Miami River,from the Salinity Dam located at approximately NW 36 Street and NW 40 Avenue,to the mouth of the River at Biscayne Bay.The Study Area consists of those lands on both sides of the River,extending out approximately 2,500 feet from the river at the longest point on either side.The land on both sides of the Comfort Canal,Wagner Creek and the Tamiami Canal extends out approximately 500 feet at the longest point on either side (Fig.3-2-2).

The Miami River is not a homogeneous landscape.In reality,the river is a series of different landscapes that are linked by its functional land aesthetic qualities.For the purposes of this Greenway Action Plan,the Miami River has been divided into three distinctive sections that reflect the different land uses,characteristics and functions found along the river.

"Lower River"-Biscayne Bay to the 5th Street Bridge. This section of the river is best characterized by dense urban development comprised of high-rise residential land ,commercial,single family and multi-family ho using,retail,hotel,office,waterfront restaurants,and institutional land uses."Middle River"-5th Street Bridge to the NW 22nd Avenue Bridge.Land use is primarily characterized by single family and multi-family housing that abuts the river and extends throughout adjacent neighborhoods."Upper River"-22nd Avenue Bridge to Palmer Lake.Land use is primarily characterized by the industrial complex (Fig.3-2-3).

1. Land Ownership

For purposes of the Inventory,the Study Area focused only on those parcels of land directly adjacent to the River and its tributaries.In addition,parcels along

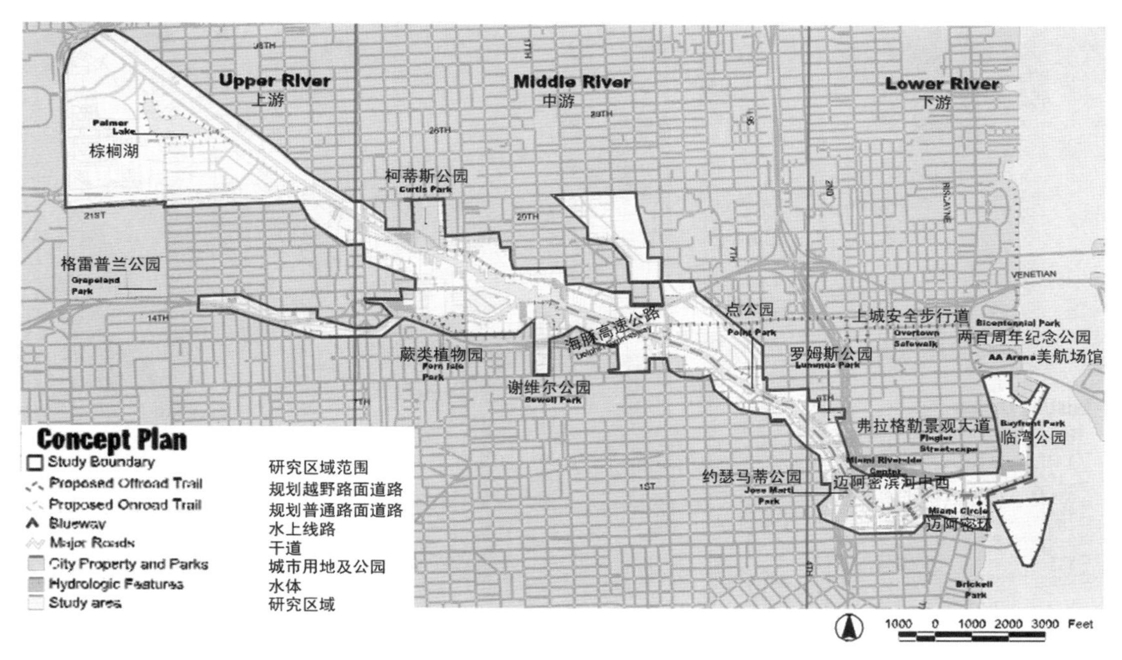

图3-2-2 迈阿密绿道概念规划

Fig.3-2-2 The concept Plan of Miami River Greenway

下游——从比斯坎湾到第五大道的大桥。这一部分以高层住宅、商业、独栋住宅和多层住宅、零售业、酒店、办公楼、滨水餐馆及各种机构的高密度开发为特征。中游——从第五大道的大桥道西北22大道的大桥。土地使用以紧靠河流的独栋住宅和多层住宅以及从河流向外延伸的住宅为主。上游——从西北22大道的大桥岛棕榈湖。土地使用主要以工业综合设施为主（图3-2-3）。

1．土地权属

从土地存储的角度出发，只注重研究那些直接临近河流及支流的用地。此外，沿着河流南北两岸与河流基本平行的道路也是研究的重点。研究区域内大约有854块土地，其中215块为公家所有，639块土地是个人财产。

2. 土地及水资源利用现状

位于迈阿密城市中心的研究区域内的河段主要是一条城市化的“动态”河流，河流廊道周边土地利用方式多种多样。迈阿密河在迈阿密市区的布里

North River Drive and South River Drive,roadways which parallel the River for much of its length,were also part of the focus area.There are approximately 854 parcels of land within the Study Area,215 of which are in public ownership and 639 are privately owned.

2. Existing Land and Water Uses

The Study Area,located in the heart of the City of Miami,is predominantly an urban working river with a wide variety of different land uses along the corridor. The Miami River enters Biscayne Bay in the Brickell area of downtown Miami. Land uses in this area consist of high-rise office, condominium, and apartment buildings.

3. Environmental

Vegetation:Many of the native plants that were originally found along the Miami River have been destroyed.However,some native and exotic vegetation does still exist in the parks and a few other undeveloped parcels found in the Study Area.

Geology: The dominant geological feature of the Study is Northern Biscayne Bay, which is 2 to 6 miles in width and about 24 miles in length.A shallow bedrock basin of Pleistocene limestone underlays the Bay.An eastern ridge,made of Key Largo limestone and the politic Miami limestone forms the Atlantic Coastal Ridge,defining the western bay shore and separating Biscayne Bay from the Everglades.

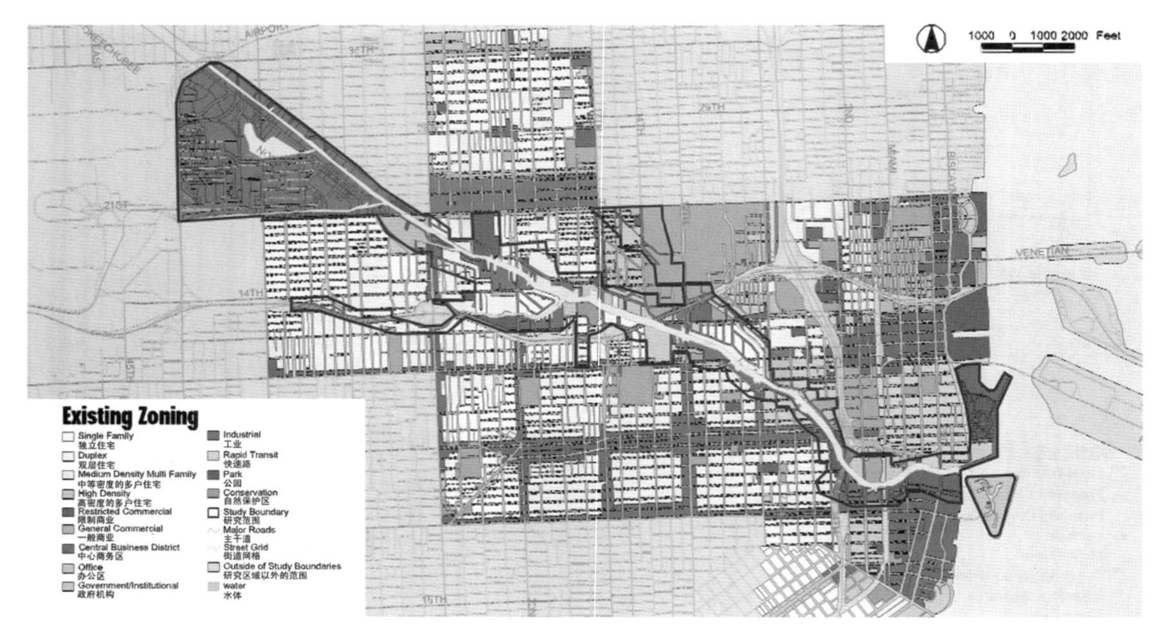

图3-2-3 土地区划现状

Fig.3-2-3 Existing Zoning

克尔地区进入比斯坎湾。这一地区的土地利用主要由高层办公楼和公寓建筑组成。

3. 环境情况

植被：许多之前沿着迈阿密河可以搜寻到的本土植被现在已经遭到破坏。尽管如此，在研究区域内的公园及其他一些未开发地块中仍然能看到一些本土植被与外来植物共生的情况。

地质：研究区域内最明显的地质特征就是北部的比斯坎湾，它的宽度为2到6公里，长度为大约24公里。其底部是一个更新世石灰石盆地。由基拉戈石灰石和迈阿密石灰石构成的东部海脊构成了大西洋沿海脊的一部分，这也界定了比斯坎湾西部的边界，将其与沼泽地分隔开来。

土壤：根据土壤调查地图，当地的土壤类型主要有：大卫细沙、洛克戴尔细沙和阿泽尔细沙。

水文：迈阿密河起源于佛罗里达埃弗格莱兹湿地，终止于位于比斯坎湾的河口，这条河的年流量为69立方千米。在研究区域内的迈阿密河有三条主要支流，从东至西依次是席博德运河、康福特运河和太米阿米运河。

地形：研究区域内除了西维尔公园以外的区域

Soils: According to the soil survey map, the following types of soils are located within the Study Area: Davie Fine Sand, Rockdale Fine Sand and Arzell Fine Sand.

Hydrology: The Miami River originates in the Florida Everglades and ends at the River's mouth, where it empties into Biscayne Bay.The watershed of the River covers a total drainage area of 69 square miles. The Miami River has three main tributaries within the Study Area.From east to west,these tributaries are the Seybold Canal, the Comfort Canal and the Tamiami Canal.

Topography:Little change in topography exists in the Study Area,which has little relief,except at Sewell Park.This is due in large part to the urban nature of the Study Area,development over the last century has significantly altered much of the original topography.

Wildlife:In spite of the urban character of the Study Area and the resulting habitat destruction, some native animal and plant species that were originally found along the Miami River are still in existence. Furthermore, the Miami River still provides habitat value for aquatic species, particularly the West Indian Manatee.In addition,certain migratory species are known to use the Study Area during migration.

4. Access and Transportation

Roadways and Proposed Roadway Improvements:The Miami River is surrounded by a network of local streets,collector streets,arterials,minor and major thoroughfares,bridges,and the Interstate Highway System.

地形起伏不大，这主要是研究区域内的自然环境造成的，在经过上个世纪的开发之后，这一区域的原始地形已经有了明显的变化。

野生动植物：尽管研究区域的开发和建设活动使得动植物生境遭到破坏，但是在沿河的一些区域仍然能够找到一些本土生物的踪迹。甚至迈阿密河还为某些物种提供了宝贵的栖息地，特别是儒艮（俗称海牛）。此外，规划区域内还有某些迁徙物种。

4. 道路交通

现有及规划道路改善情况：迈阿密河周边被密集的道路网络所包围，这些道路有地方性道路、人行道、城市干道、小型或者大型的通道、桥梁及州级公路网等。公共交通情况：当人们使用戴德县的公共交通工具时，他们有机会俯瞰迈阿密河。公共铁路交通在位于布里克尔与政府行政中心车站之间的西南第五大街处及位于第二大道与第一大道交界处横跨迈阿密河。

三、 规划主题

从利益相关者的会议和土地存储的现状出发，数个不同的主题存在于迈阿密河流廊道中。这些河流主题有助于定义项目的独特之处、文化影响力和自然条件。在全国范围内的其他河流项目中，项目主题有助于各方理解和长期支持项目设想。

1. 迈阿密河是我们的家园

河流绿道体系将会使河流与周边社区重新紧密结合，最终这条河流将成一个提供休憩、保健和舒适环境的景观，同时它也提供交通运输、教育、社会服务，并能为周边社区的居民提供工作岗位。

为使项目最终取得成功，引导市民参与规划设计的全过程是必不可少的。绿道系统将会使社区通过公园、公共空间、历史性场地及其他已知的目的地连接起来。绿色通道也将使得有可能更全面系统地连接滨河人口集中的地方。这将使人们可以步行或者骑车沿着5.5英里（约8.85公里）的河边到达目

Mass Transit:Passengers utilizing Miami-Dade County's mass transit system,Metro Rail and Metro Mover,have an opportunity to see the River from a magnificent overhead view.Metro Rail crosses the River near S.W.5th Street between the Brickell and Government Center Stations,and between 2nd Avenue on the west and S.W.1st Avenue on the east.

III. Planning Theme

From the stakeholder meetings and inventory of existing conditions, it is apparent that a number of different themes exist within the Miami River corridor.These river themes help to define the unique features,cultural influences and physical conditions of the river. For other significant river projects around the nation,river themes have helped to achieve a better understanding and longterm support for project initiatives.

1. The Miami River is Our Home

A river greenway system would serve to reconnect to the river the diverse ethnic neighborhoods adjacent to this important community asset. The river is after all a landscape that can serve the recreation,health and fitness, transportation,education,social and employment needs of the adjacent residential neighborhoods.

This is happening by involving the residents in a planning and design process that would result in successful development of a safe and accessible river greenway system.The greenway would link together neighborhoods with the significant parks,public open spaces, historic sites and other destinations found along the river.The greenway will also make possible a more comprehensive system of access to the river from adjoining population centers.This will enable people to transport themselves,by walking and cycling,to a multitude of key destinations along the 5.5-mile river (Fig.3-2-4).

2. The Miami River is a Working River

The river's navigation and commercial shipping directly generate more than 4 million tons of cargo each year with an estimated value of more than$6 billion annually.The river constitutes one of the largest employers in downtown Miami,and most importantly offers a range of jobs hard to find in other employment sectors of the downtown area. Scheduled to begin in the year 2001,will remove contaminated sediment from the river bottom and result in, more cargo volume and more jobs in the industry.The river greenway system will serve to sustain maritime shipping and recreational boating activities along the Miami River.The greenway will provide improved access to the river to view the shipping industry at work and help to raise awareness

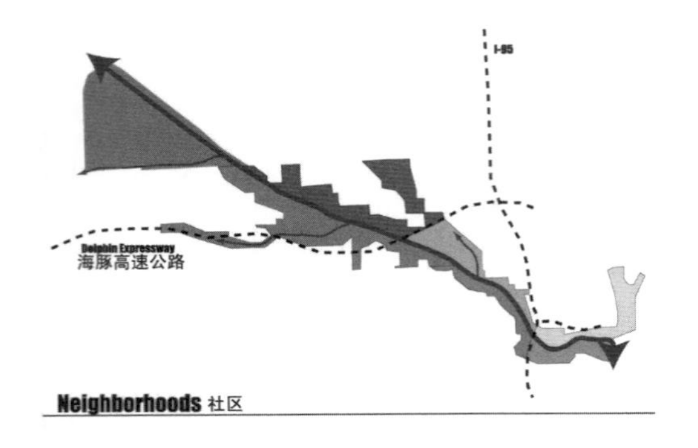

图3-2-4 概念规划描述的沿河社区
Fig.3-2-4 Conceptual diagram depicting neighborhoods along the Miami River

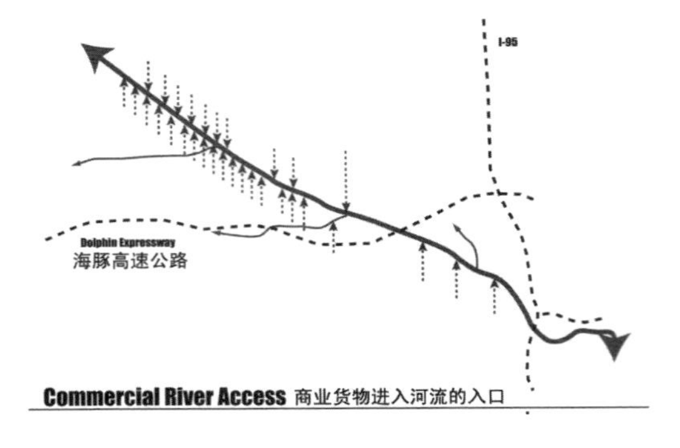

图3-2-5 概念规划描述的迈阿密河的商业运输频率
Fig.3-2-5 Conceptual diagram depicting frequency of commercial marine use along the Miami River

的地（图3-2-4）。

2. 迈阿密河是一条“流动”的河流

河流运输业每年的运输量为400万多吨，估计运输货物价值超过60亿美元。河流变成了迈阿密市区最大的雇主。最重要的是，它为在闹市区很难找到其他就业岗位的人们提供了一系列的工作机会。开始于2001年的河道清理计划通过清理河流底部的沉积污染物来增加货运量，以便提供更多的就业机会。河流廊道系统将有助于推动迈阿密河的海洋运输和水上休闲活动的开展。河流廊道将提供一种新的视角去看待航运业，并使人们意识到迈阿密河作为工业河流景观的重要性（图3-2-5）。

3. 迈阿密河是一个目的地景观

迈阿密河流绿道和沿迈阿密河的滨河步行道是使迈阿密河成为一个目的地景观的关键，此外，与市郊规划绿道、弗拉格勒景观大道、哈瓦那溪流廊道及布里克尔运河廊道的连接也是项目成功的关键。为给游客及市民的日常活动提供一个友善和创造性的环境，对公共及私人空间的改造投入是不可或缺的。这一切都以使社区居民认识到河流绿色基础设施建设投资的重要性。最终的河流廊道建设需要考虑到入口设施、自行车道及步行道路的设计，使在市区与河流周边区域的人流能够进入到河流绿道中。借鉴国内其他成功的滨河景观建设活动，迈阿密一定要清楚地认识到目的地景观是

of the importance of the river as a working river landscape (Fig.3-2-5).

3. The Miami River is a Destination Landscape

The proposed Miami River Greenway and the long standing Miami Riverwalk are the key elements to creating a successful destination landscape along the river. Additionally,connections to the river are critical as well,such as the proposed Overtown Greenway,the Flagler Streetscape improvements,and linkages to East Little Havana and the Brickell Corridor.New investment in public and private spaces is needed to create a more friendly and conducive atmosphere for tourism and daily activity.This can begin to happen as the community recognizes the value of making investment in the"green infrastructure"of the river. Finally,access to the river needs to be improved with gateways,bicycle and pedestrian paths that flow from the downtown and adjacent residential neighborhoods to the river.As with other successful river landscapes throughout the nation,Miami must understand that a destination landscape is a blend of scenic attraction and robust commerce (Fig.3-2-6).

4. The Miami River is an Important Environmental Resource at Risk

The word"Miami"means sweet water,and for as long as humans have occupied the landscape,the Miami River has been an important source of water,animal life and plant life.Even though it is polluted today and supports a large industrial marine complex,the river ecosystem remains an important functional element of the riverine landscape.Important species of plants and animals still inhabit the river,with the manatee being a symbol of an ecosystem at risk.The Biscayne Bay is a direct recipient of the fresh water flowing from the Miami River.The Bay is a valuable resource for residents of South Florida, supplying recreation,tourism and economic opportunities. Less

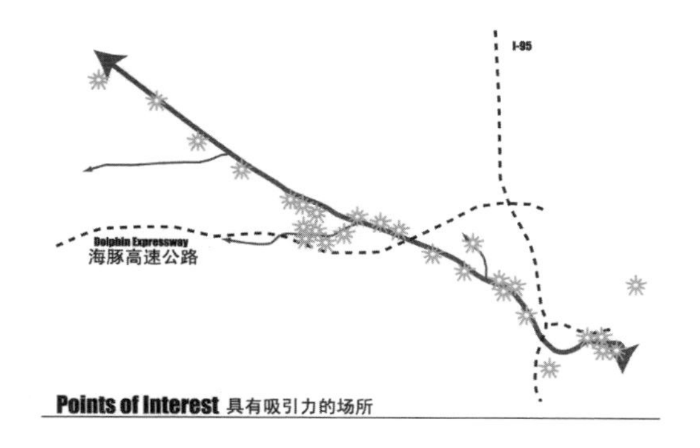

图3-2-6 概念规划描述的沿着迈阿密河的有吸引力的场所
Fig.3-2-6 Conceptual diagram depicting points of interest found along the Miami River

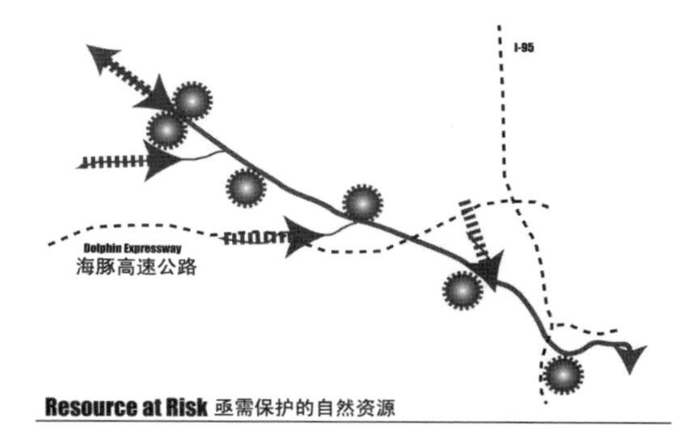

图3-2-7 概念规划描述的沿着迈阿密河的环境资源的位置
Fig.3-2-7 Conceptual diagram depicting locations of environmental resources along the Miami River

一个融合优美风景与繁荣商业于一体的集合（图3-2-6）。

4. 迈阿密河是一处亟须保护的自然资源

迈阿密的原意为甘甜的水，当人们在这里定居之初，这里就是一个水资源和动植物资源丰富的地方。尽管当前已被严重污染，但它今天依然拥有支持大型航运工业的综合设施，河流生态系统仍然是一项重要的河流景观功能因素。重要的植物和动物仍然居住在这里，儒艮就是一个濒危生态系统的象征。比斯坎湾承接着来自迈阿密河的淡水。对南佛罗里达的居民而言，河流是一种宝贵的资源，因为它提供了娱乐、旅游和产生经济收入机会。不到15%的南佛罗里达州的雨水汇入比斯坎湾，作为主要饮用水源供应着更广大的地区。其余的降水或蒸发掉，或排入运河和大海。从1990年到2010年，公共供水需求在南佛罗里达的预计将增加69%。现在是时候去评估迈阿密水资源和实施行动计划来进一步保护和恢复这些资源了。绿色廊道能为保护这个至关重要的资源提供新的视角。这可以通过整体河道疏浚和岸线稳定计划实现（图3-2-7）。

5. 迈阿密河是一个经济源

国际贸易、旅游业是大迈阿密经济最重要的组成部分。迈阿密是全世界公认的旅游胜地，旅游业及相关产业雇佣了250 000多名当地居民，游客在一

than 15% of South Florida's rainwater reaches the Biscayne Aquifer,the primary drinking water supply for the metro region.The remainder is evaporated by the sun or drained by canals to the sea. Between 1990 and 2010,the demand for public water supply in South Florida is expected to increase by 69%.Now is the time to take stock of important water resources like the Miami River and implement programs of action to further protect and restore these resources. The greenway can instill a new stewardship ethic for this critically important resource.This can be realized through a comprehensive river dredging and shoreline stabilization program (Fig.3-2-7).

5. The Miami River is an Economic Resource

After international trade,travel and tourism is the most important industry to the Greater Miami economy.Miami is recognized around the world as a leading tourist destination,employing over 250,000 residents and generating$10.9 billion in overnight visitor expenditures.The Miami River landscape has not realized the full benefits of the tourism industry.

While Miami is the second greatest international financial center in the United States,it is also the 4th poorest city in the nation.Historically,property values adjacent to the Miami River have not kept pace with waterfront property values at oceanfront and bayfront locations around Miami-Dade County.This is clearly reflected in the economic values currently found in adjacent residential neighborhoods.However,the Miami River is quickly becoming a sought after address for residential, commercial and retail development. Property values in the downtown section of the river are already beginning to increase dramatically. In fact,the future Florida Marlins baseball stadium on the will spur economic revitalization to the lower section of the river,and will in turn begin to transform the river into a destination landscape.While the"working river"continues to be the most significant economic

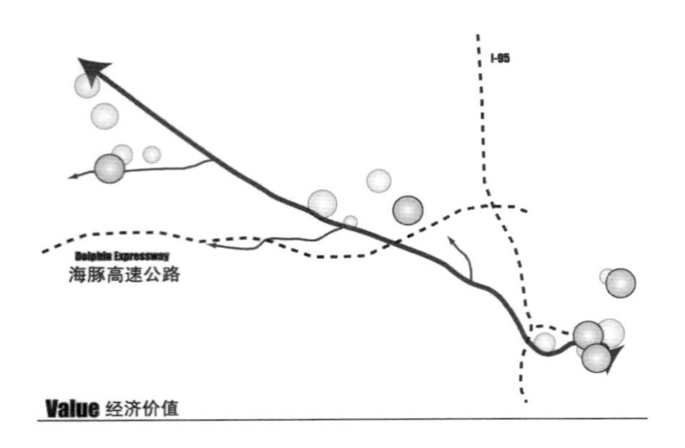

图3-2-8 概念规划描述的沿着迈阿密河的经济开发价值的分布
Fig.3-2-8 Conceptual diagram depicting distribution of economic value along the Miami River.

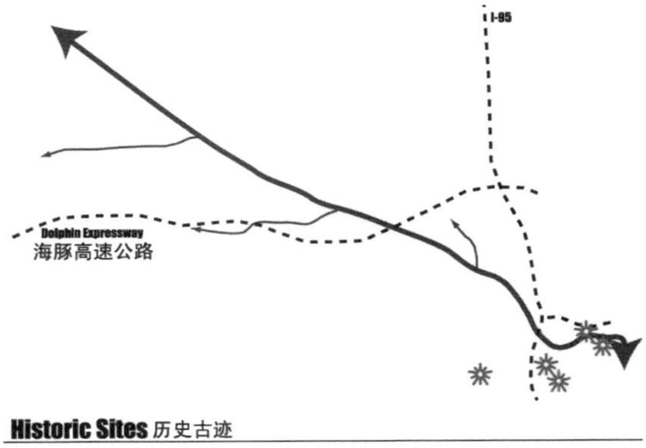

I-95
Dolphin Expressway
海豚高速公路
Historic Sites 历史古迹

图3-2-9 概念规划描述的沿着迈阿密河的历史遗迹的分布
Fig.3-2-9 Conceptual diagram depicting distribution of historic sites along the Miami River.

夜之间能够支出109亿美元。迈阿密河流景观没有认识到旅游业的发展的全部机遇。

当迈阿密成为美国第二大的国际金融中心时，它也是全美排名第四的贫穷城市。从历史上看，在戴德县，毗邻迈阿密河滨水区的土地价格未能跟上海滨和靠近比斯坎湾的土地价格的增长。这很明显地反映在当前邻近的住宅社区的经济价值上。然而，迈阿密河流廊道为住宅、商业和零售业的快速发展提供了场地。在市中心毗邻河流的部分土地的价值已经开始大幅上升。事实上，未来的佛罗里达旗鱼棒球体育场将刺激河流下游区域的经济振兴，并开始反过来将河流转变为景观目的地。而“流动”之河仍然是河流景观、旅游业的最重要的经济要素，更高的土地价值将会使这条河流成为当地经济的重要组成部分。并可以用一种不影响海洋产业生计的方式运作。河流绿道系统将进一步强化经济活动扩张。绿色廊道将会使这条河的景观游赏变得更加便利和有吸引力，同时这也会推动并增加毗邻河堤的土地价格的提升（图3-2-8）。

6. 迈阿密河是我们的遗产

迈阿密河口作为人类定居及活动的场地已经有2 000多年的历史了，当前，位于迈阿密河口的2.2英亩（约8903平方米）的考古遗址就是早期本土美

element of the river landscape,tourism and higher property values will enrich this economic base to make the river a major economic force in the local economy. This can occur in a way that does not replace or threaten the livelihood of the marine industry.The river greenway system will further enhance this expansion of economic activity.The greenway will make the river landscape accessible,attractive,and connected,which in turn will continue to increase the value of property adjacent to the riverbanks (Fig.3-2-8).

6. The Miami River is Part of Our Heritage

The mouth of the Miami River has been a focal point for human settlement and activity for more than 2,000 years.Currently,the 2.2 acre site at the mouth of the river,known as the"Miami Circle,"is all that remains of a native American civilization archaeological record.Efforts are currently underway to properly interpret this unique landscape and the artifacts that have been found on the site.This is but one of many interesting historical sites along the river.There are many important sites and feature landscapes along the river that should be identified and interpreted.Local residents and visitors would gain a greater appreciation for the significance of the river landscape through these endeavors.The Greenway can become a land use that supports the interpretation of this heritage,providing pedestrian and water-based access to sites along the river corridor (Fig.3-2-9).

IV. Planning Goals

The Miami River is a valuable natural historic and cultural resource .We envision the Miami River as an ecological system that serves to enrich the lives of residents throughout the Miami metropolitan area.The

国文明存在的记录。我们需要为正确理解这一独特的景观和在这个地点上发现史前遗迹作出努力。这只是沿着这条河的许多有趣的历史古迹之一。有很多重要的地点和具备独特特征的沿河景观应被识别和理解。通过这些努力，当地居民和游客将理解这些河流景观的重要意义。绿色通廊道能诠释这些遗产的意义，为行人提供亲水路径，并使他们沿着河流廊道去游览这些场地（图3-2-9）。

四、 规划目标及措施

迈阿密河是一个宝贵的自然、历史和文化资源。我们将其视为一个服务于大迈阿密都市区的生态系统。通过增加公众进入迈阿密河流廊道的机会，保持其作为“流动河流”的活力，恢复河道水质，将其打造成服务于游人及市民的旅游目的地，推动毗邻土地的合理利用，并构建动植物生境景观和促进社区的多文化的交流。最终，周边社区的环境及经济发展将会得到改善。

目标1：改善到达河流的通道，具体包括清除妨碍人们进入和接近河流景观的阻碍系统；提高河道的视觉识别性和河流的公众意识；改善游艇码头；利用公共土地构建一个绿道步行网络；通过防卫环境设计促进整个河流绿道的安全使用；连接现有的公园、历史名胜和主要自然景观节点；鼓励土地所有者和土地开发商将私人土地开发与绿道建设相结合；制定一个有效的可替代的由自行车和步行设施构成的交通体系，内相邻的绿色通道，使人们可以方便地到达河流两岸的目的地（图3-2-10，图3-2-11）；建设北戴德绿道总体规划中的迈阿密河游步道；同私有土地所有者合作，以共同改善公众进入河流廊道的途径。

目标2：维持迈阿密河作为“动态河流”的发展。具体包括支持船舶运输业与海洋产业的发展；与当地、州和联邦官员一起完成迈阿密河的全面疏浚项目。协调好绿道建设与未来船舶运输业与海洋产业发展之间的关系；继续清除河中的遗弃船只；改善导航设施、严格管理河中船只的航行活动；改善河流附近的道路和桥梁，鼓励可持续和协调发展，同时减少车辆和船舶交通之间的冲突；增加邻

Miami River Greenway will help to improve the future economic well-being of our community by increasing public access to the waterway, sustaining the"working river"maritime shipping industry,restoring water quality in the river channel,serving as an attractive destination for local residents and visitors,encouraging appropriate adjacent land use,fostering an ethic of stewardship for plants and animals native to the river landscape,and celebrating the multi-cultural ethnicity of our community.

Goal 1: Improve Access to the River.Remove the system of fences that has served to restrict access to and along the river landscape.Improve the visual identity and public awareness of the river. Provide for improved access to marinas.Construct a comprehensive system of greenway trails on publicly owned properties.Promote safety and security throughout the river greenway system through the adoption of CPTED principles.Link existing parks,historic and significant natural sites throughout the river landscape.Encourage landowners and land developers to link private facility development to the greenway. Develop an efficient alternative transportation system of bicycle and pedestrian facilities within and adjacent to the greenway that will enable people to connect to destinations on both sides of the river (Fig.3-2-10, Fig.3-2-11). Implement the Miami River Trail element of the North Dade Greenways Master Plan.Work with private landowners to voluntarily acquire parcels of land that will provide for improved public access to the river.

Goal 2: Sustain the"Working River"Industries of the Miami River.Support the growth of the shipping and marine industries.Work with local,state and Federal officials to complete a comprehensive dredging program for the Miami River.Coordinate greenway development with future expansion of the shipping and marine industry.Continue the removal of abandoned and derelict vessels from the river.Improve navigational and safety code enforcement on the river. Improve roads and bridges in the vicinity of the river to encourage sustainable and compatible development and minimize conflict between vehicular and shipping traffic.Increase adjacent residential,commercial and industrial property values and increase the tax base through the development of the greenway system. Contribute to the economic well-being of the community by providing employment opportunities resulting from development of the greenway. Define,quantify and tout the economic benefits of the Miami River Greenway.Clean up toxic areas and hazardous waste found within the river corridor. Improve law enforcement activities throughout the Miami River corridor.

Goal 3: Restore Water Quality throughout the River Ecosystem.Work with Federal,State and Local

图3-2-10 自行车道的设计
Fig.3-2-10 Bicycle Lanes Design

图3-2-11 道路景观提升
Fig.3-2-11 Streetscape Improvement

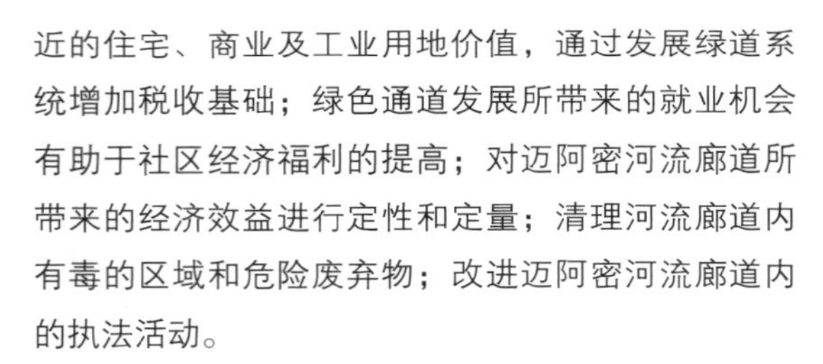

近的住宅、商业及工业用地价值，通过发展绿道系统增加税收基础；绿色通道发展所带来的就业机会有助于社区经济福利的提高；对迈阿密河流廊道所带来的经济效益进行定性和定量；清理河流廊道内有毒的区域和危险废弃物；改进迈阿密河流廊道内的执法活动。

目标3：恢复整个河流生态系统的水质。具体包括与联邦、州、当地政府和地区企业一起，通过综合系统的措施减少点源和面源污染以改善水质；根据联邦和州的法律、法规和程序，清理经确认的污染场所；在可能的情况下提供运河堤植被缓冲带，以净化坡面雨水径流和污染（例如土壤生物工程技术）；在可能的情况下通过利用植物、当地的石灰石岩石和其他水工程技术稳定河堤；通过严格的管理措施减缓地表径流和污染物对主要支流的污染；加强对上游水质的保护，以保证水体的源头；教育当地居民改善水质重要性，并将其作为管理措施之一。

目标4：将其打造成为服务于迈阿密大都市区的目的地景观。具体包括把这条河打造成为服务当地居民和游客的目的地景观；完成河流绿道项目规划区

governments and area businesses to improve water quality through a comprehensive program aimed at reducing.point source and non-point source pollution.Target identified polluted sites for clean up in accordance with Federal and State laws,regulations and programs. Where possible provide riverbank vegetative buffers,an element of cleaning up overland stormwater runoff and pollution. Where possible stabilize the banks of the river through the use of plants, native limestone rock and other proven water engineering techniques(i.e.soil bioengineering) Utilize best management practices to slow runoff and pollutant loading on tributary streams that feed the main river stem. Develop local upstream water quality programs for the feeder streams of the river. Educate local residents as to the importance of best management practices to improve water quality.

Goal 4:Serve as a Destination Landscape for Metro Miami. Make the river an attractive destination for local residents and tourists. Complete the existing Riverwalk program in applicable sections of the river greenway. Diversify the land use and destinations found along the river to include more restaurants,retail shops and other river related business uses. Implement a blueway(i.e.water transportation program)for the Miami River that would foster water-based linkages to key destinations. Program river greenway events for each month of the year. Incorporate art into the river greenway landscape.

域内已有步行路网规划项目的实施；丰富土地利用方式并增加沿河景观节点的数量以使更多的餐馆，零售商店和其他河流相关业务与廊道发展紧密联系；建设水上航道（例如内河运输），加强景观节点之间的水路连接；规划每月一次的河流绿道活动项目；将艺术的元素应用到河流廊道的景观设计中。

目标5：鼓励土地的兼容性利用。促进土地利用的多样性，增加景观节点的数量；振兴邻里公园等关键目的地景观，以鼓励更多的使用；考虑由迈阿密河流委员会、迈阿密市政府和戴德县共同管理的河流规划重叠区的利用；发展社区规模零售、商业、旅游业；实施由城市设计开发委员会制定的迈阿密市区设计准则；通过步行及自行车道路将迈阿密河流廊道延伸至迈阿密市区、布里克尔廊道及相邻住宅区；与迈阿密的财政界一起筹集该项目的启动和运行资金，这将确保有序和兼容的开发当前未被充分利用的土地；鼓励土地利用的区划管理，保护自然资源。

目标6：为迈阿密河培养一个管理伦理的概念。促进居民长期参与和实施河流廊道系统；促进继续保护濒危的儒艮和在河流廊道发现的濒危动物和植物物种；恢复和保护河岸沿线环境以适当为植物和动物提供栖息地；通过河岸带的植被恢复和在适当的地方种植本土植物以提高生物多样性；鼓励公私合作来保护河流资源；鼓励基于课题，试验，监测项目和其他目的的教育活动；与当地的学校一起提供露天的生物学、生态和地质类课堂。

目标7：促进相邻街区文化的多样性。创造支持文化庆典的公共空间，并邀请当地团体的参与这些空间的设计；连接包括公园、休憩用地和历史遗迹在内的河流景观资源；创造具备文化认同和社区吸引力的廊道入口；通过一个社区警务程序和可防卫环境设计保护绿道使用者的健康、福利和安全；尊重周边土地拥有者的隐私；在沿着廊道并毗邻居民居住和工作的区域开发更多的娱乐设施。

五、概念规划

1. 中心和辐射轴

从规划框架结构、河流的主题及规划设计的想

Goal 5:Encourage a Compatible Land Use Vision for the River.Promote a diversity of land uses and destination landscapes that support increased activity along the river.Revitalize key destination landscapes such as neighborhood parks to encourage more use. Consider the use of an overlay planning district for the river,jointly prepared by the Miami River Commission,the City of Miami and Miami-Dade County.Develop neighborhood scale retail,commercial and tourist destinations.Implement the Downtown Miami Design Guidelines as developed by the Urban Design Committee. Develop strong bicycle and pedestrian connections to Downtown Miami,the Brickell Avenue corridor and adjacent residential neighborhoods.Work with financial community of Miami to establish financial incentives for the Miami River Greenway that will ensure the orderly and compatible redevelopment of vacant or underutilized properties along the river. Encourage land use and zoning regulations that protect natural resources.

Goal 6: Foster an Ethic of Stewardship for the Miami River.Promote the long-term involvement of local residents in the implementation of the Miami River Greenway system.Promote the continued protection of the endangered manatee and other endangered species of animals and plants found within the river corridor. Restore and enhance the riparian environment along the riverbanks to support plant and animal habitat where appropriate.Encourage biodiversity through revegetation of riparian zones and reintroduction of native plant species where appropriate. Encourage public/private partnerships for stewardship of river resources. Encourage education based projects,experiments,monitoring programs and other activities.Work with local school system to provide outdoor classroom settings for biology,zoology and geology classes.

Goal 7: Celebrate the Multi-Cultural Diversity of Adjacent Neighborhoods. Create public spaces that will support cultural celebration and invite local groups to help program the space for such celebrations and festivals. Link together the resources of the river landscape,including parks,open spaces and historic sites. Create gateways between local neighborhoods and the greenway that recognize cultural make-up and influences of that neighborhood. Protect the public health, welfare and safety of greenway users through a community policing program and through the implementation of CPTED (Crime Prevention Through Environment Design). Respect the privacy of adjacent landowners. Develop additional recreation facilities along greenway lands close to where residents live and work.

V . Concept Planning

1. Hubs and Spokes

From the framework,river themes,vision,goals and objectives emerges a concept for future

象力出发，规划目标和理念将未来的河流廊道发展作为核心概念。这个概念也在一定程度上受到佛罗里达州绿道规划中“中心和辐射”概念的影响。在佛罗里达州的州绿道系统中，中心的概念由生态功能显著的土地，公园和目的地自然景观组成。辐射轴线的概念由诸如小溪、河流、交通走廊等线性景观组成。对于迈阿密河流绿道而言，其中心为现有及规划的公共区域公园和邻里公园，辐射轴线则是滨河步行道、滨水道路、自行车道、步行设施和水上航道（图3-2-12，图3-2-13）。

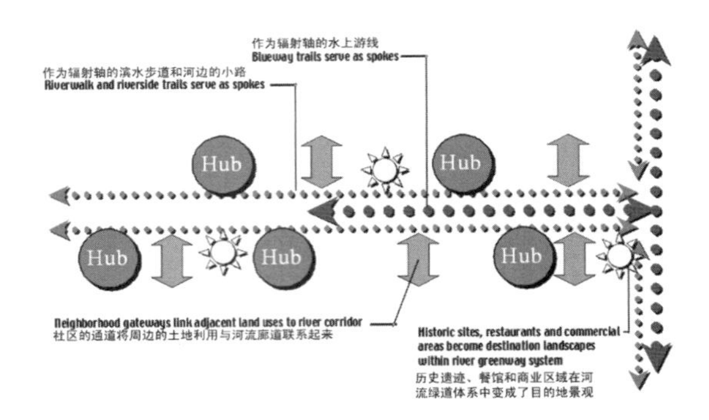

图3-2-12 中心与辐射轴
Fig.3-2-12 Hub and Spokes

2. 一体化的新都市

基于利益相关者的参与和公共投入，一体化的新都市的概念被应用于绿道行动计划中，这是社区共同的愿望。具体而言，当河流廊道景观持续发展、土地利用发生变化时，公园作为结构节点，其未来的发展会辐射周边。在这一理念下，未来的发展将使孩子们可以在整个社区内，安全的从自家门前到达旅游目的地。未来复合利用土地将具有支持包括公共场所、商业、办公和居住等功能。这些利用方式将通过一个步行网络和河流廊道网络连接起来。此外，它还可以舒缓交通压力。建筑物的前面将会是有着人行道和树木的街道。建筑有助于构建室外空间，这些空间将会被作为整体道路骨架和街区的一部分。泊车位将会在视线上被建筑遮掩。在社区，建筑会提供不同的用途，并呈现出不同的规

river greenway development.The concept is also based in part on the State of Florida Greenway concept of"hubs and spokes."Under Florida's statewide greenway system,hubs can be comprised of ecologically significant lands,public parks and destination landscapes.Spokes are comprised of linear landscapes such as streams and rivers,utility corridors and transportation corridors.For the Miami River Greenway,the hubs are represented by existing and proposed public regional and neighborhood parks,the spokes would be the Riverwalk, riverside trails,on-road bicycle and pedestrian facilities and the blueway elements of the river (Fig.3-2-12,Fig.3-2-13).

2. Integration of New Urbanism

Further,based on stakeholder meetings and public input,it is the desire of the community that concepts of new urbanism be represented in this Greenway Action Plan.Specifically,as the river corridor landscape continues to evolve and land use changes occur,it is assumed that the parks would also serve as nodes from which future development would emanate.Under

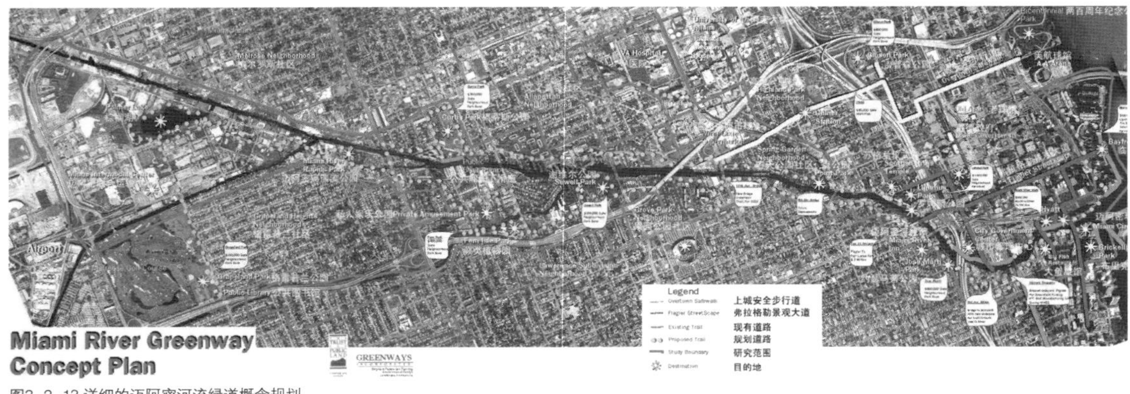

图3-2-13 详细的迈阿密河流绿道概念规划
Fig.3-2-13 Detailed Miami River Greenway Concept Plan

模和外观。建筑外观会反映迈阿密的风土文化。公园将成为主要的户外公共空间，足以服务于音乐会会场，露天市场和河流节庆活动场地。

六、 详细设计

1. 迈阿密滨水游览步道

滨水游步道的完善是迈阿密河下游河段绿道项目中最关键的要素。通过对滨水游步道的完善，从比斯坎湾到拉莫斯公园和约瑟马蒂公园之间的公共空间将会变得连续起来。预想的滨水游步道将会以两种景观形式存在：

（1）与沿河宽阔的路径、铺装或步行道路公享空间；

（2）作为一种单独的路径和人行道铺装沿河布置。

在某些位置，由于滨水商业用地的存在，滨水游步道的设施变得不太可能实现，在这些位置，滨

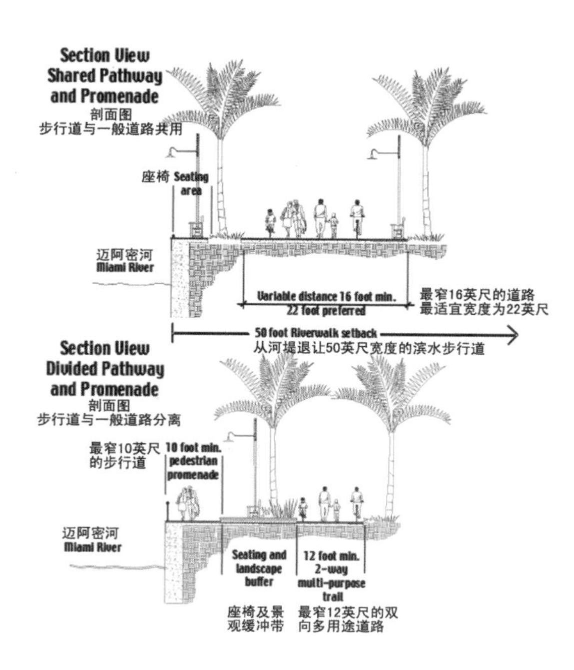

图3-2-14 滨水步行道

Fig.3-2-14 Riverwalk Trails

the concept of new urbanism future development would be safe enough to support children walking from their front door to local destinations throughout the community.Future land use development would support integrated mixture of uses including public places,shops,offices,and places to live.These uses would be linked together by a network of walkable streets and the riverwalk greenway system. Additionally,streets would serve to diffuse traffic patterns.Buildings would front along streets lined with sidewalks and trees.Buildings would help to create the effect of outdoor rooms and would be arranged as part of an overall pattern of streets and blocks.On-site parking would be discretely hidden from view behind buildings.Buildings in the neighborhoods would be constructed for different uses,scale,and architectural expressions.Architectural expression of the buildings would reflect the climate and traditions of central Miami.The parks would serve as the principal public outdoor space,and be large enough to accommodate concerts,open air markets and river festivals.

VI. Detailed Design

1. Miami Riverwalk

Completion of the Miami Riverwalk is one of the most critical elements of the greenway program for the lower river.Through completion of the riverwalk,continuous public access will be provided from Biscayne Bay to Lummus Park and Jose Marti Park.The riverwalk is envisioned to continue as one of two developed landscapes:

(1)As a broad shared pathway and promenade or boardwalk along the river's edge;

(2)As a divided pathway and pedestrian promenade along the river's edge.

In certain locations a riverside location is not possible due to existing water dependent businesses. At these locations, the riverwalk would be linked through both sidewalks and bicycle facilities around these landscapes and back to riverwalk development. The riverwalk is developed within a protected setback zone of 50 feet that is measured from the top of the river bank.No privately owned buildings are allowed to be constructed within this 50 feets setback area (Fig.3-2-14).

2. Miami River Trail Blueway

Many local residents expressed an interest in seeing a water trail established in the lower section of the Miami River to link activity/destination centers on Biscayne Bay to destinations along the Miami River.The photo right has been digitally enhanced to suggest how such a blueway could be established.Critical to its success,the blueway would be located outside the Miami River navigational

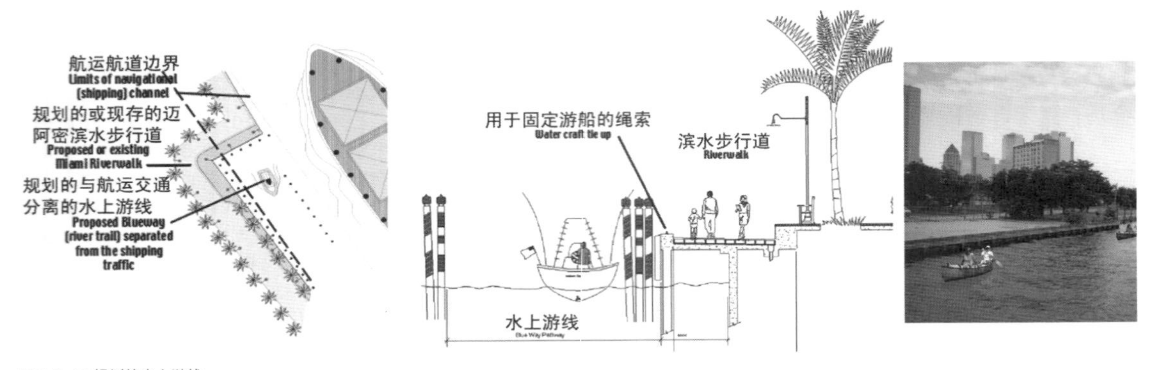

图3-2-15 规划的水上游线
Fig.3-2-15 Proposed Blueway Trails

水游步道将会采用与这些景观周边的人行道和自行车道相连接的方式设置。滨水游览步道需要保证拥有50英尺（15.24米）的宽度，这一宽度的测量需要从河堤的最高点开始，私人所属的建筑不允许在这50英尺的范围内进行建设（图3-2-14）。

2. 迈阿密河水上游憩路径

许多当地居民期望能在河流下游有一条可以连接比斯坎湾周边的活动场地与沿河活动场地的水道。右面的图片经过数码图像处理后展现了如何实现这样一条水上游憩路径。实施的关键在于这条水上游憩路径将会被设置在迈阿密河航运通道的外围，这样就不会影响到河流日常的航运交通。水上游憩路径将会通过在水中设置标志物的方式进行限定（图3-2-15）。

3. 与弗莱格勒街、城市中心商业区的连接

迈阿密河流绿道中一个最重要的连接存在于弗莱格勒街和城市中心商业区，连接河流北面位于城市中心区域的位置。未来滨水游步道的延伸以及与现存环湾道路体系的融合将会为迈阿密河流廊道的视觉及实体空间连接提供绝好的机遇。在海特旅馆，现有的人行道出于安全原因已经被阻断，这一道路可以通过再次开放作为沿着布里克尔大道进入迈阿密中心城区的连接。沿着南迈阿密大道和西北第二大道的人行道将会为弗莱格勒街与河流廊道间提供连接。随着未来旗鱼棒球场的建设，很有可能在河流与城市中心区域之间形成一条依托地铁系统的连接路径，也可能存在一条从河流廊道沿着弗莱格勒大街延伸的直接连接城市

图3-2-16 绿道与迈阿密市中心的连接
Fig.3-2-16 Greenway Connections to Downtown Miami

channel so as not to interfere with daily shipping traffic.The blueway would be designated with markers placed in the water (Fig.3-2-15).

3. Flagler Street/Downtown Bussiness District Linkage

One of the most important links to the Miami River Greenway corridor will be connections to the Flagler Street corridor and the Downtown Business District. Links can occur at several locations north of the river into the downtown area. Future extensions of the Riverwalk trail to the existing Baywalk trail system will provide the best opportunity for visual and physical connections to the Miami River corridor.At the Hyatt Hotel,an existing walkway connection has been fenced off for security purposes,which could be opened to establish an important link along the west side of Brickell Avenue into downtown Miami.Sidewalks along South Miami Avenue and NW 2nd Avenue would provide links to the Flagler Street corridor.With the future development of the Marlins Baseball stadium,it may also be possible to have a link from the river to the downtown under the Metro-Mover system.A direct link is also possible from the river corridor along Flagler Street (Fig.3-2-16).

中心区与廊道的路径（图3–2–16）。

4. 路上自行车及步行设施

由于缺少滨水的公共土地和大量滨水商业的存在，发展诸如路上设施之类的道路组成要素，以多元化的构建迈阿密河流绿道交通是十分必要的。为了实现这一目标，建议实施两个不同类型的自行车和人行道设施：一种是自行车车道和人行道设施，另一种是自行车游憩路线和人行道设施。自行车道将通过道路上的条纹图案的铺装进行界定，5英尺（约1.5米）宽的游步道将被分离出来以便游人沿车行道双向使用。自行车游憩路线将只通过标识牌进行界定，而不是通过相关的铺装标记（图3–2–17）。

5. 罗姆斯公园

迈阿密市已经出台了大量的文件用于支持罗姆斯公园周边滨水步行道的改造。规划和剖面图展示了迈阿密河上宽阔的步行道，其在西北大街南段的三个进入罗姆斯公园的路径上为为坐椅和行道树提供了足够的空间。在沿河处保留出了为将来发展商业和零售业预留的土地（图3–2–18）。

七、 总结

迈阿密河流经迈阿密市中心，是一处珍贵的自然资源，被确定为河流绿道而予以保护。绿道是源于欧美发达国家的先进理念，引用查理斯・莱托在其经典著作《美国的绿道》中的解释，“绿道就是沿着诸

4. On-road bicycle/pedestrian facilities

It will be necessary to develop much of the Miami River Greenway trail elements as on-road facilities due to the lack of publicly owned riverfront properties and the amount of water dependent businesses that reside along the river channel.To accomplish this,two different type of bicycle and pedestrian facilities are recommended:a bicycle lane and pedestrian facility;and a bicycle route and pedestrian facility.The bicycle lane will be designated by on-road striping of the road pavement.Five foot travel lanes would be designated for travel in both directions along the roadway.The bicycle route facility will be designated by signage only and will not have associated pavement markings (Fig.3-2-17).

5. Lummus Park

The City of Miami has developed extensive construction documents for the installation of riverwalk improvements at Lummus Park.Plan and section drawings show a boardwalk extending over the Miami River,with ample space for seating and shade trees provided at three principal points of entry off NW North River Drive.Land is reserved along the waters edge for future commercial and retail business development (Fig.3-2-18).

VII. Summary

Miami river is a precious natural resources flowing through downtown Miami,and now it is protected was identified in the form of river green way.Green way is an advanced concept from the developed western countries, quoted the explanation of Charles. Leto (Charles Little) in his classic "American green way,"the explanation, "green way is a kind of linear open space constructed along the river,valleys, ridges line and other natural corridor,or along the abandoned railway line used for recreation activities, the ditches and scenic roads, it includes the natural landscape lines and artificial

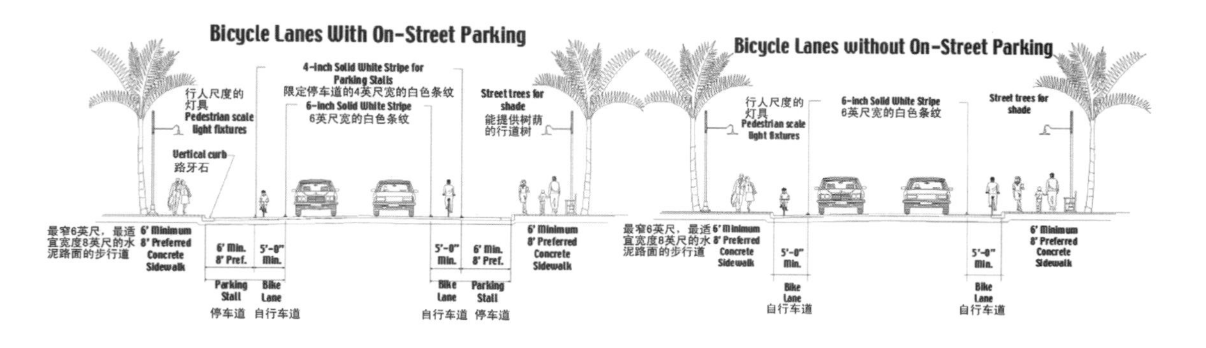

图3–2–17 自行车道详细设计

Fig.3–2–17 Bicycle facility details

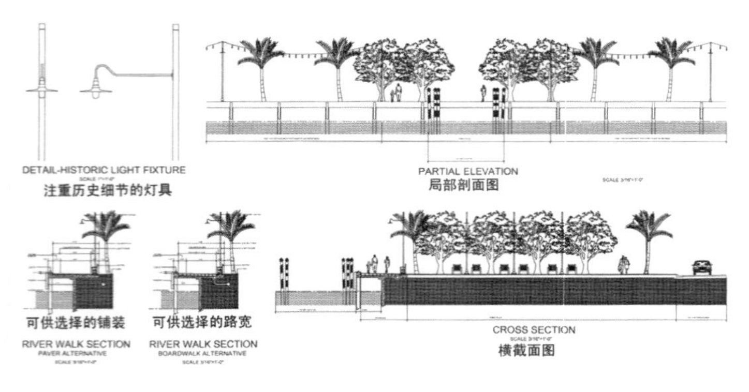

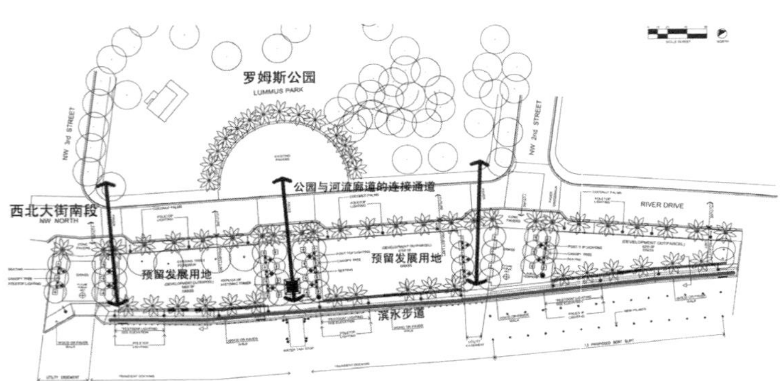

图3-2-18 罗姆斯公园总体规划
Fig.3-2-18 Lummus Park Master Plan

如河滨、溪谷、山脊线等自然走廊，或是沿着诸如用作游憩活动的废弃铁路线、沟渠、风景道路等人工走廊所建立的线型开敞空间，包括所有可供行人和骑车者进入的自然景观线路和人工景观线路”。迈阿密河流绿道即是这种连接了公园、工业用地、居住区、商业区等用地的开敞空间纽带。

河流绿道作为线性的、不同于两侧基质的狭长景观单元，其结构特征对一个景观的生态过程具有重要影响。规划中，上游、中游、下游三个部分依据不同特征被赋予不同功能，而数个规划主题，涵盖了社会、经济、环境效益发展等多重目标，包括：提高河流的公众可达性，支持水运和海洋产业的发展，恢复水质，打造休闲、游憩、餐饮等为一体的综合性目的地景观，倡导土地兼容性利用，提高土地利用率，保护文化多样性，加强后期管理，引导公众参与，实现迈阿密河绿道的可持续发展等。

迈阿密河流绿道建设能有效解决生态斑块破碎、休闲空间串联度低、绿色经济发展缓慢等环境

landscape line which are available for pedestrians and cyclists". Miami river green way is the very linear open space connecting the park, the industrial land, the residential, and the commercial together.

Rivers green way which is a long and narrow landscape units and has different matrix from both sides of it, its structure characteristics has important influence on landscape ecological process. Planning endow the upper,middle and downstream with different function according to different characteristics , and several planning themes cover multiple development goals of the social, economic, and environmental benefits including improving public accessibility of the river,supporting the development of water transportation and Marine industry, and restoring the water quality,making a comprehensive destination landscape with leisure and recreation, catering services,advocating land compatibility use,improving the land utilization, protecting the diversity of culture, strengthening the later-period management, guiding the public participation, realizing sustainable development of Miami river green way, etc.

The construction of miami river green way can effectively resolve the ecological patches fragmentation,low connection degrees of leisure

问题，对迈阿密市城市品位提升，社会经济发展，生物多样性保护，提高居民生活质量有着重要的意义和作用。

space and the slow development of green economy,at the same time,it will make a notable impact for promoting taste of the city,the development of social economy,the protection biological diversity and improving the quality of life.

思考题

1. 结合对迈阿密河流绿道项目的理解，谈一下对绿道内涵的认识。

2. 当前国内外存在大量绿道建设项目，这些项目在层级及类型上存在一定的差异，以广东省为例，现有广东省绿道体系规划，而作为我国各类规划试点的重要城市，深圳市拥有较完善的市域绿道体系规划，这两者之间也存在很好的关联性。试从尺度的角度的出发，结合迈阿密河流绿道的建设，分析不同尺度条件下绿道建设的侧重点及其之间的衔接关系。

3. 绿道规划与一般性的城市规划项目相比较而言，更加贴切居民的日常生活，但作为一种非法定规划，试讨论一下设计者需要考虑如何充分调动公众的充分参与，以在绿道建设中尽可能的融入公众的意愿。

4. 从绿道的起源及发展来看，生态修复与生态保护是时代赋予它的新的内涵，在具体的项目中，如何将这一内涵落到实处，而这一新的内涵又如何与其他内容进行协调。

5. 绿道不单单是一种纯粹的带状绿色空间的营造，因此设计者需要考虑的不单是其公共效能的发挥与生态服务机能的正常运转，市场经济的内在机制要求绿道建设具备经济回报机能（直接创收或间接经济带动），结合迈阿密河流绿道的内容，试从经济回报的角度对绿道建设进行分析。

Questions

1. According to the understanding of Miami river greenway, talk about your understanding to the meaning of greenway.

2. There are many domestic and foreign greenway construction projects at present,they have some differences in the types and hierarchy, with an example of Guangdong province, the existing Guangdong greenway system has been planned, and as the important pilot city of various types of planning, Shenzhen has the perfect urban greenway system,there are also very good relevance between them.From the point of view of the scale, combine the construction of the Miami river green way and analyze the emphasis of the construction and the cohesion relationship in the different scales.

3. Compared with general urban planning projects, the greenway planning focuses more on the daily life of the residents, but as a non-statutory planning,please consider how to fully arouse the public's full participation in the construction of greenway and integrate the public's will.

4. From the perspective of the origin and development of greenway,ecological restoration and protection is endowed new connotation by times.Please consider how will the connotation be implemented and how to coordinate the connotation with other content in specific projects.

5. Greenway is not only a kind of pure green space,so what designers need to consider is not only the normal operation of public performance and ecological service function. The internal mechanism of market economy request the greenway construction can creat profits(directly creat income or indirect economic driver).Please refer to the content of Miami river greenway to analyze the greenway construction from the point of view of economic returns.

注释/Note

图片来源/Image：图片均来源于《迈阿密河流绿道概念规划文本》/*Miami River Greenway Action Plan*

参考文献/Reference：案例来源于www.miamirivercommission.org

第四章　绿色基础设施规划设计

CHAPTER IV GREEN INFRASTRUCTURE PLANNING & DESIGN

第一节 马里兰绿色基础设施评估 Section Ⅰ Maryland's Green Infrastructure Assessment
第二节 哈罗新城绿色基础设施规划 Section Ⅱ The Green Infrastructure Plan (GIP) for the Harlow Area
第三节 环城翡翠项链：亚特兰大新公共领域 Section Ⅲ The Beltline Emerald Necklace: Atlanta' s New Public Realm

第一节 马里兰绿色基础设施评估

Section I Maryland's Green Infrastructure Assessment

一、介绍

1. 绿色基础设施概念

在韦伯新世界字典中将基础设施定义为："结构和内在基础，特别是支撑社区持续下去和发展的基础设备和设施"。当听到基础设施一词时，大多人会想到道路、排水管、公用线路或其他灰色基础设施；或者是医院、学校、监狱和其他社会基础设施。现在，许多人和机构都在讨论另外一种对"社区持续下去和发展"非常重要的基础设施，即绿色基础设施。

绿色基础设施是一个新名词，但不是一个新理念，它扎根于150年前的规划和保护措施中。绿色基础设施起源于两个重要的概念：

（1）造福人类而连接公园和其他绿色空间；

（2）有益于生物多样性和避免生境破碎而预留和连接的自然区域。

绿色基础设施系统帮助保护和恢复自然运作的生态系统并为未来的开发提供一个框架。为了达到这个目的，它们提供了一个生态的、社会的和经济功能和效益的多样性：丰富生境和生物多样性，维护自然景观过程；净化空气和水；增加娱乐机会；改善健康，也加强了自然与场所感的联系。

2. 马里兰绿色基础设施评估运动掀起的背景

I. Overview

1. Green Infrastructure

21Webster's New World Dictionary defines infrastructure as "the substructure or underlying foundation, especially the basic installations and facilities on which the continuance and growth of a community depends". When they hear the term infrastructure, most people think of roads, sewers, utility lines, and other gray infrastructure; or hospitals, schools, prisons, and other social infrastructure. Today, many people and organizations are talking about another type of infrastructure that is critical to the "continuance and growth of a community": green infrastructure.

Green infrastructure is a new term, but it's not new idea. It has roots in planning and conservation efforts that started 150 years ago. Green infrastructure has its origin in two important concepts:

(1) linking parks and other green spaces for the benefit of people;

(2) preserving and linking natural areas to benefit biodiversity and counter habitat fragmentation.

Green infrastructure systems help protect and restore naturally functioning ecosystems and provide a framework for future development. In doing so, they provide a diversity of ecological, social, and economic functions and benefits: enriched habitat and biodiversity; maintenance of natural landscape processes; cleaner air and water; increased recreational opportunities; improved health; and better connection to nature and sense of place.

2. Background of Maryland's GIA

Before colonization by Europeans, Maryland was

在被欧洲殖民化之前，马里兰95%覆盖着林地，5%是沼泽地。到1993年的时候林地和湿地都减少了一半。现在被开发的土地面积增长的速度更加快，每年有15 000英亩（约60.7平方公里）的土地被开发（图4–1–1，图4–1–2）。如果目前的趋势继续下去，到2030年将会有额外的800 000英亩（约3237.5平方公里）的土地被开发，而且大部分是农村地区。至少有180种植物和35种动物已经在马里兰灭绝。另外有310种植物和165种动物是稀有的，和受到威胁或濒临灭绝的。

幸运的是，马里兰有一个历史最久的土地保护项目。开放空间项目成立于1969年，到2003年的时候，这个项目基金已经购买了面积共250 000英亩（约1011.7平方公里）的国家公园、野生动物栖息地、自然区，以及36 000英亩（约145.7平方公里）的地方公园。

乡村遗产项目，是马里兰"聪慧增长活动"的主要项目之一，它成立于1997年。这个项目鼓励地方政府和私人土地信托去认识乡村遗产区域并且为完成现有的土地保护活动或者开始一个新保护活动提供资金。

政府在1991年任命马里兰绿道委员会为负责规划和协调绿道实施的主要国家级机构。绿色基础设施评估开始于1990年代中期，它主要研究一些非常关注绿道项目的团体的看法。自然资源部的生态系统委员会建议成立一个国家级综合自然资源管理计划，其中一部分就是绿色基础设施评估。

95% forested, the other 5% being tidal marsh. By 1993, both forest and wetlands had decreased by half. The area converted to development is increasing even more rapidly now, 15,000 acres of land are developed in Maryland each year (Fig.4-1-1, Fig.4-1-2). By 2030, if current trends continue, 800,000 additional acres of land will be developed, much of it in rural areas. At least 180 plant and 35 animal species have been extirpated from Maryland. Another 310 plant and 165 animal species are rare, threatened, or endangered.

Fortunately, Maryland has one of the oldest land protection programs in the country. Program Open Space was created in 1969, and by 2003, program funding had purchased about 250,000 acres of state parks, wildlife habitat, and natural areas, and 36,000 acres of local parks.

The Rural Legacy Program, a keystone of Maryland's "Smart Growth Initiatives," was established in 1997. The Program encourages local governments and private land trusts to identify Rural Legacy Areas and to competitively apply for funds to complement existing land preservation efforts or to develop new ones.

The governor appointed the Maryland Greenways Commission (the Commission) in 1991 as the primary statewide entity responsible for planning and coordination of greenways implementation. The GIA (Green Infrastructure Assessment) was begun in the mid-1990s to address the perception among some groups that the greenways program was largely recreation-focused. DNR (Department of Natural Resources)'s Ecosystem Council recommended the development of a statewide Integrated Natural Resources Management Plan, part of which evolved into the green infrastructure assessment.

3. The Overview of GIA

The GIA could identify and prioritize Maryland's green infrastructure, it was based on principles of landscape ecology and conservation biology, and

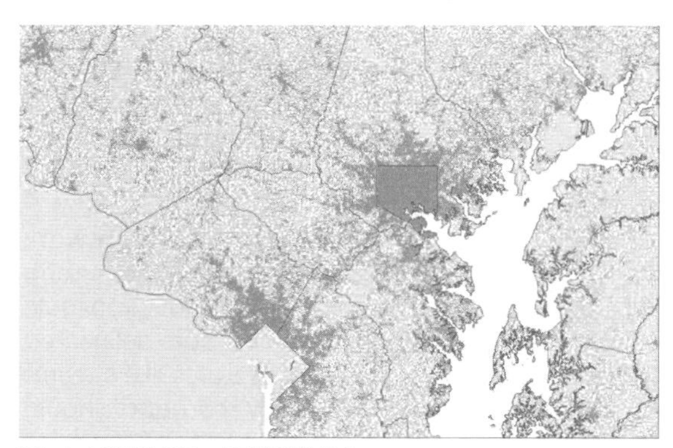

图4–1–1 马里兰1900—1960年开发模式
Fig.4-1-1 Development patterns 1961—1997 in Maryland

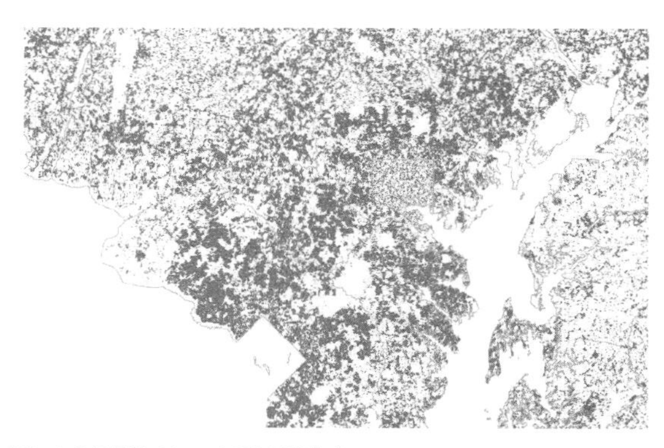

图4–1–2 马里兰1961—1997年开发模式
Fig.4-1-2 Development patterns 1900—1960 in Maryland

3. 绿色基础设施评估概述

绿色基础设施评估可以识别和排序马里兰绿色基础设施，它以景观生态和生物保护学原则为基础，并提供一个可以评估马里兰土地保护和恢复活动的一致性方法。这个评估识别一个与生态价值最高的未开发区域相连接的生态中心和廊道网络。它也为要进行保护的土地的排序和优选提供方法。网络中心和廊道被一系列的生态参数评估，然后在它们的地理区域进行排序。绿色基础设施评估也可以帮助评价特殊的地方区域，个体网格是由我们使用的卫星图形的结果来决定的像素点。绿色基础设施的网络中心和廊道的保护状况、管理地位和开发风险也都被调查过。一个网络中心或者廊道的开发风险可以与其生态分值联系起来帮助优先保护活动。

绿色基础设施评估的研究结果就是得到一个具有生态价值的网络中心和廊道相互连接 的网络的景象，这个连接网络如果得到保护将会帮助保护马里兰所有生物所依赖的自然生态系统功能。

二、相关方法介绍

1. 网络的识别

1）构件的识别

绿色基础设施保护的言下之意就是连接大的、连续的生态价值显著的自然区域（网络中心）与创造自然地连接网络的廊道连结在一起（图4-1-3）。这些连结可以弥补破碎化导致的功能损失。

绿色基础设施评估识别两种重要的资源地——网络中心和廊道。在绿色基础设施网络中的中心是指马里兰剩余的最重要的大生态斑块。这些斑块包括以下的一点或及几点内容：存在敏感的植物或者动物种类的地区；包含大块的连续的内部林地（至少有连续的250英亩[约101公顷]，加上300英尺[约91.4米]的过渡区）或者是包含有至少250英亩的原始湿地；有溪流或者是河流，或是已经被公共（主要是自然资源部或者联邦政府）和私人组织（如自然保护协会或者马里兰鸟类学会）保护的区域。

绿色基础设施网络中的廊道具有线性特征，至

provides a consistent approach to evaluating land conservation and restoration efforts in Maryland. The assessment identified a network of hubs and corridors that contained the most ecologically critical remaining undeveloped lands. The GIA also provides an approach for ranking or prioritizing land protection efforts. Hubs and corridors were assessed for a variety of ecological parameters, and then ranked within their physiographic region. The GIA can also help evaluate specific local areas, individual "grid cells" were pixels determine by the resolution of the satellite imagery we used. Green infrastructure hubs and corridors were also examined for their level of protection, management status, and risk of development. A hub or corridor's risk of development can be combined with its ecological score to help prioritize conservation efforts.

Work on the GIA resulted in the vision for an interconnected network of ecologically valuable hubs and corridors that if protected would help preserve the natural ecosystem functions on which all life in Maryland depends.

II. Related Methods Introduction

1. Identification of the Network

1)Identification of the Components

The concept underlying green infrastructure protection is to link large, contiguous blocks of ecologically significant natural areas (hubs) with natural corridors that create an interconnecting network of natural lands across the landscape(Fig.4-1-3). Such connection can help to offset the functional losses caused by fragmentation.

The GIA identified two types of important resource lands - "hubs" and "corridors." Hubs in the green infrastructure network represent the most important large ecological patches remaining in Maryland. Hubs contain one or more of the following:

Areas containing sensitive plant or animal species;Large blocks of contiguous interior forest (at least 250 contiguous acres, plus a 300 foot transition zone)or wetland complexes with at least 250 acres of unmodified wetlands;Streams or rivers, or conservation areas already protected by public (primarily DNR or the federal government) and private organizations like the Nature Conservancy or Maryland Ornithological Society.

Corridors in the green infrastructure network are linear features, at least 1,100 feet wide, linking hubs together to allow animal and plant propagule movement between hubs. The corridors delineated in many cases follow prominent features like streams or ridges. To function effectively, corridors should be wide enough to provide interior conditions for habitat specialists, as well as protecting the hydrology

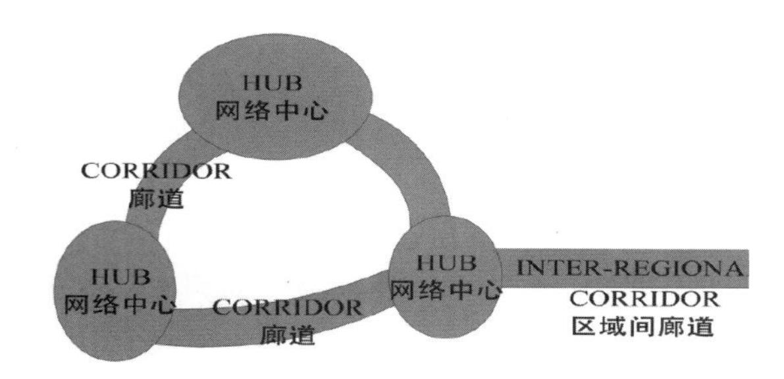

图4-1-3 网络中心——廊道概念图
Fig.4-1-3 Hub-corridor conceptual diagram

少1 100英尺（约335.3米）宽，将网络中心联系在一起供动植物繁殖、迁徙。廊道在很多情况中根据主要的特征如溪流或者山脊被描述。为了更加实用，廊道应该足够宽，为生境专家的研究提供内部环境，同时也保护与它们相接地溪流和湿地水文状况和水质量。廊道的识别和描述根据大量的数据，包括土地覆盖、土地利用、湿地、道路、溪流和鱼汛。连接被分成三种不同的类型：陆地的、湿地的和水生的。

2）马里兰绿色基础设施网络的构成

在全州境内，马里兰绿色基础设施在自然区域（林地、湿地和裸岩地、沙地、粘土）内由1 777 475英亩（约7 193平方公里）的网络中心和252 997英亩（约1 023.8平方公里）的廊道组成，总共2 030 471英亩（约8 217平方公里）（图4-1-4）。这些计算中不包括开敞水面。另外改变过的开敞地区（农业、草地、矿场及清理过的土地）在潜在的绿色基础设施网路中共有375 546英亩（约1 520平方公里）。这些“间断处”是指那些可以被恢复成自然覆盖类型的土地。

被开发的地区（25 240英亩[约102平方公里]）也不被计算在内，因为这些地区往往很难去恢复。

马里兰绿色基础设施占马里兰土地总面积的33%（若包含间断区域，所占面积为39%）；63%的马里兰林地，包括90%的州级内部森林；87%的马里兰天然湿地，保括99%的国家特殊关注的湿地；91%的马里兰内部森林溪流；99.7%的马里兰自然遗产地；89%的马里兰陡坡（坡度25%以

and water quality of streams and wetlands contained within them. Corridor identification and delineation were based on many sets of data, including land cover/land use, wetlands, roads, streams, slope, floodplains, Maryland Biological Stream Survey (MBSS) aquatic resource data, and fish blockages. Linkages were tailored to three different ecotypes: terrestrial, wetland, and aquatic.

2)The Constitution of Maryland Green Infrastructure

Within state boundaries, Maryland's green infrastructure is comprised of 1,777,475 acres of hubs and 252,997 acres of corridors in natural land cover (forest, wetland, and bare rock/sand/clay); totaling 2,030,471 acres(Fig.4-1-4). Open water was excluded from these calculations. In addition, altered open areas (agriculture, lawns, quarries, and cleared lands) comprise 375,546 acres in the potential green infrastructure land network. These "gaps" represent areas that could potentially be restored to a natural cover type. Developed areas (25,240 ac) were excluded from these calculations: they are usually difficult to restore.

Maryland's green infrastructure contains 33% of Maryland's total land area (39% when gaps are included); 63% of Maryland's forest land, including 90% of the State's interior forest; 87% of Maryland's remaining unmodified wetlands, including 99% of the Wetlands of Special State Concern; 91% of Maryland's streams within interior forests; 99.7% of Maryland's Natural Heritage Areas; 89% of Maryland's steep slopes (25%); 44% of Maryland's highly erodible soil and so on.

In general, the green infrastructure model is relatively efficient at capturing most of the state's biodiversity and natural resources. However, it missed some areas, such as isolated natural heritage elements, some streams and their riparian buffers, some steep slopes, and some wetlands.

2. Ecological Ranking of Hubs and Corridors

1)Ranking of Hubs

The researchers calculated a wide variety of statistics for each hub. Twenty-seven parameters (Table. 4-1-1) were selected and given an importance weighting according to feedback from biologists and natural resource managers; literature reviews; minimization of redundancy, area dependence, and spatial overlap; balancing different ecotypes; data reliability; and examination of output from different combinations. None of these parameters were highly correlated (>80%).

Hubs were ranked within their physiographic region from best to worst for each parameter in Table 4-1-1. These rankings were calibrated by converting

上）；44%的马里兰高度易侵蚀性土壤等。

总之，绿色基础设施模型在锁定全州生物多样性和自然资源地时是比较有用的。然而，它遗漏了一些区域，如隔离的自然遗产区、一些溪流及其缓冲区、陡坡、湿地。

2. 网络中心和廊道的生态等级

1) 网络中心等级

研究者为每个网络中心计算大量数据。总共有27个参数（表4–1–1）被选中并被给予一个重要加权，这个加权从以下方面考虑得到：生物学家和自然资源管理者的反馈；文献综述研究；冗余、地方独立性、空间重叠状况的最小化；不同生态类型的衡量；数据的可靠性及不同组合的结果检查。参数之间不具有高度的相关性（>80%）。

网络中心在它们的区域中按照最好到最坏的顺序在表4–1–1中列出。我们将这些等级转换成百分数而进行校准（百分数=等级×100/最大等级）。为了获得一个综合的生态学排序，分别给27个生态参数百分数再乘以一个重要加权，并且将每个网络中心相加。这个重要的加权具有这些参数的实用性和数据可靠性功能（图4–1–5）。

2) 廊道片段的等级

廊道也被评估并且在地质范围内根据一系列生态参数进行排序。节点也被看做是廊道中的一部分。用来排序廊道的生态参数中有许多是和网络中心使用的参数相似，但是参数加权更加强调廊道连接的是什么以及连接的效益是怎样的。廊道状况也是非常重要的。存在断裂、道路穿梭（特别是主道路）或者宽度不足，会让野生动物及种子的传播更加困难。廊道的长度也是一方面因素，如果廊道质量等级一样的话，短一点的连接更适合。

每个廊道等级用22个参数来计算（表4–1–2）。参数之间不具有高度的相关性（>80%）。廊道片段在它们的地质区域和尺度等级内根据每个参数从最好到最坏进行排序。为了得到一个综合的生态等级，我们将这些等级转换成百分数，并且将这些参数的百分数与一个重要的加权相乘（图4–1–5）。

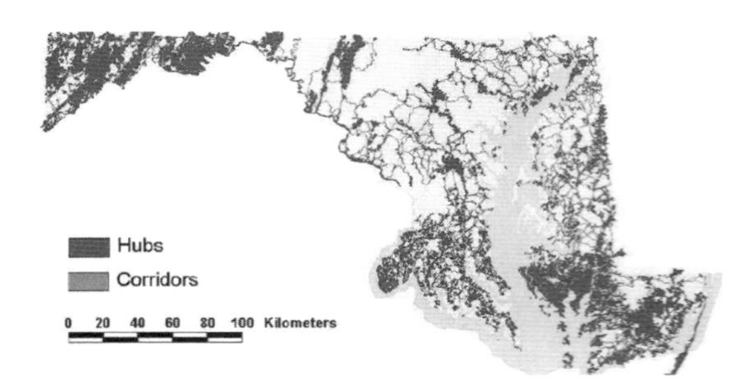

图4–1–4 马里兰绿色基础设施网络
Fig.4-1-4 Maryland's green infrastructure network

to percentiles (percentile = rank × 100 / maximum rank). To derive a composite ecological ranking, the percentiles for the 27 ecological parameters were multiplied by an importance weighting, and added together for each hub. The importance weightings were a function of the parameter's utility and data reliability (Fig.4-1-5).

2) Ranking of Corridor Segments

Corridors were also evaluated and ranked within their physiographic region for a variety of ecological parameters. Nodes were considered part of their corridors. Many of the ecological parameters used to rank corridor segments were similar to those used to rank hubs. However, the parameter weights emphasized more what the corridors linked, and how effective that linkage was. Corridor condition was also important. Corridors with breaks, road crossings (especially if they were major roads), or insufficient width were considered more difficult for wildlife and seeds to traverse. Corridor length was also a factor, shorter connections were preferable if quality was otherwise equal.

There were 22 parameters calculated for each corridor segment (Tab.4-1-2). None of these parameters were highly correlated (>80%). Corridor segments were then ranked from best to worst for each parameter, within their physiographic region and size class. To derive a composite ecological ranking, we converted these rankings to percentiles, and multiplied the percentile for each parameter by an importance weighting (Fig.4-1-5).

3. Ranking Threats to Maintaining Green Infrastructure

1)Protection-Development Restrictions

There are many factors that together influence the risk that portions of the green infrastructure will be lost, primarily through their development into urban type uses. And the potential for changes in the management status of lands protected by public or quasi-public ownership that might affect green

Parameter 参数	Weight 加权
Heritage and MBSS (Maryland Biological Stream Survey) element occurrence 遗产及马里兰生物流调查要素的出现	12
Area of Delmarva fox squirrel habitat 德玛瓦半岛黑松鼠栖息地面积	3
Fraction in mature and natural vegetation communities 成熟及天然植被群内的部分区域	6
Area of Natural Heritage Areas 自然遗产地	6
Mean fish IBI (Index of Biotic Integrity) score 鱼类生物完整性平均值	1
Mean benthic invertebrate IBI score 海底无脊椎动物生物完整性平均值	1
Presence of brook trout 河鳟的出现	2
Anadromous fish index 溯河鱼类指数	1
Proportion of interior natural area in hub 网络中心内陆自然地的比例	6
Area of upland interior forest 山地内陆森林地面积	3
Area of wetland interior forest 湿地内陆森林地面积	3
Area of other unmodified wetlands 其他天然湿地的面积	2
Length of streams within interior forest 内陆森林的溪流长度	4
Number of stream sources and junctions 溪流源和交叉处的数量	1
Number of GAP(Gap Analysis Project) vegetation types 间断分析项目中的植被类型数量	3
Topographic relief (standard deviation of elevation) 地形起伏(标准海拔偏差)	1
Number of wetland types 湿地类型数量	2
Number of soil types 土壤类型数量	1
Number of physiographic regions in hub 网络中心内部地质区域数量	1
Area of highly erodible soils 高度易腐蚀性土壤面积	2
Remoteness from major roads 与主要道路的偏远度	2
Area of proximity zone outside hub 网络中心外邻近区域面积	2
Nearest neighboring hub distance 与最近的网络中心的距离	2
Patch shape 斑块形状	1
Surrounding buffer suitability 周边缓冲区适宜度	1
Interior forest within 10 km of hub periphery 网络中心周边10km内的内陆森林	1
Marsh within 10 km of hub periphery 网络中心周边10km内的沼泽	1

表4-1-1每个网络中心在其地质区域内的生态价值排序使用的参数和加权
Table. 4-1-1 Parameters and weights used to rank overall ecological significance of each hub within its physiographic region.

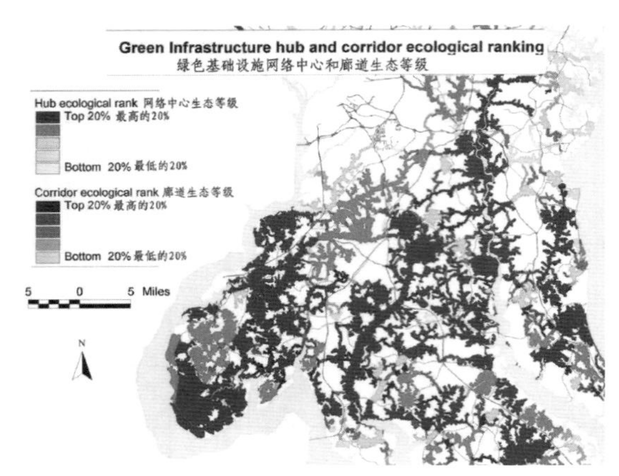

图4-1-5 马里兰南部部分网络中心和廊道综合生态等级
Fig.4-1-5 Hub and corridor composite ecological ranking, for a portion of southern Maryland

3. 维护绿色基础设施的风险等级

1) 保护——开发的限制因素

很多因素共同存在，影响着部分绿色基础设施消失的风险，主要通过它们开发成的城市使用类型而表现出来。而由公共或者私营事业保护的土地管理状态的潜在变化，可能会影响绿色基础设

infrastructure function.

Level of protection-development restrictions through ownership rights in land: Various restrictions on development are imposed through two primary mechanisms: public ownership rights of some sort and public regulation of private action.Level of Protection-development restriction through regulation: A number of state and local regulations and state incentives were identified as helping protect natural areas from development, to one degree or another.

To derive a composite score, restrictions on development, both through ownership rights and through regulation, were given a score between 0 and 1 for each mechanism, with 0 being the most restrictive (no development permitted), and 1 being the least restrictive (no restrictions on development) (Fig.4-1-6).

2) Current Management Status

In order to assess the potential for green infrastructure that is protected by ownership mechanisms to retain its ecological function, the researchers examined the management status of

Maryland's green infrastructure, according to the version published in Maryland Greenways Commission (2000), the GAP management criteria from Scott et al (1993), and protection data layers developed by Maryland DNR and others. Ownership, easement, and regulatory mechanisms vary in their level of environmental protection, allowing different activities

Parameter 参数	Weight 加权	% of score 百分数
Does corridor link hubs in top ecological tier? 廊道是否连接高生态等级的网络中心	8	14.5%
Top ecological ranking of hubs connected by corridor 生态等级高的网络中心通过廊道连接	4	7.3%
Mean upland impedance 山地阻碍平均值	4	7.3%
Mean wetland impedance 湿地阻碍平均值	4	7.3%
Mean aquatic impedance 水域阻碍平均值	4	7.3%
Total area 总面积	1	1.8%
Number of corridor breaks 廊道断裂数	4	7.3%
Road crossings, weighted by road type 道路穿过，根据道路类型加权	8	14.5%
Percent of gap area 间断区域的比例	2	3.6%
Sum of rare species scores 稀有物种的总分值	2	3.6%
Area of Delmarva fox squirrel habitat 德玛瓦半岛黑松鼠栖息地面积	1	1.8%

Fraction in mature and natural vegetation communities 成熟及天然植被群内的部分区域	2	3.6%
Fish IBI 鱼类生物完整性	1	1.8%
Benthic invertebrate IBI 海底无脊椎动物生物完整性	1	1.8%
Presence of brook trout 河鳟的出现	1	1.8%
Area of upland interior forest 山地内陆森林面积	1	1.8%
Area of wetland interior forest 湿地内陆森林面积	1	1.8%
Area of other unmodified wetlands 其他天然湿地面积	1	1.8%
Length of streams within interior forest 内陆森林的溪流长度	1	1.8%
Area of highly erodible soils 高度可腐蚀性土壤面积	1	1.8%
Mean distance to the nearest primary or secondary road 与最近主道路或次道路的距离	1	1.8%
Surrounding buffer suitability (within 300' of hub) 周边缓冲区的适宜性（网络中心300英尺以内）	2	3.6%

表4-1-2 每个廊道片段在其地质区域内生态价值排序使用的参数和加权
Table.4-1-2 Parameters and weights used to rank overall ecological significance of each corridor segment within its physiographic region

Data layer数据层	Restriction score 限制分值
Public and privately owned conservation lands (as of 2000) 公共和私人所有的保护土地（截止2000年）	0.0
Other public ownership (as of 2000) 其他公共所有权	0.5
Conservation easements (as of 2000) 保护地役权	0.0
Agricultural easements (as of 2000) 农业地役权	0.0
Open water 开敞水域	0.0
Wetlands 湿地	0.2
Steep slopes (25%) 陡坡（25%）	0.3
100 year floodplains 历时100年的食物链	0.8
SSPRA（Sensitive Species Project Review Areas） 敏感物种检阅区	0.9
Critical Area resource conservation areas 重要资源保护区	0.8
Zoned by county for conservation (as of 1994) 由县划成保护区的地方	0.7-0.9
Zoned by county for agriculture (as of 1994) 由县划成农业区的地方（截至1994年）	0.7-0.9
Agricultural districts (as of 2000) 农业区（截至2000年）	0.7

表4-1-3 开发限制及其相对限制强度（分值在0至1之间）
Table.4-1-3 Restrictions on development, and their relative strength (a score between 0 and 1)

Parameter 参数	Importance 重要性	Weight 加权
Mean level of protection from development 避免开发的平均保护水平	High 高	5
Percent of hub in inside designated Priority Funding Areas 在指定的优先资助区域内的网络中心百分比	High 高	3
Percent of hub with existing or planned sewer service 现存有或者规划有下水道的网络中心百分比	High 高	3
Population growth or loss 1990-2000 1990-2000年增长或下降人口	Medium 中	2
Number of parcel centroids in the hub, divided by hub area 在网络中心内部的地块中心数，被中心分割的面积	Low 低	1
Commuting time to urban centers 到市中心的上下班时间	Low 低	1
Land demand from proximity to Washington DC and Baltimore 邻近华盛顿特区和巴尔的摩的土地需求	Medium 中	2
Mean market land value 土地市场的平均价值	Medium 中	2
Mean distance to nearest major road 离最近主道路的平均距离	Medium 中	2
Area of waterfront property 滨水地产的面积	Medium 中	2
Mean proximity to preserved open space 与保护开敞空间的平均邻近值	Medium 中	2

表4-1-4 每个网络中心在其地质区域内综合开发风险排序使用的参数及加权
Table. 4-1-4 Parameters and weights used to rank overall development risk of each hub within its physiographic region

施的功能。

开发保护的限制水平通过土地所有权实现：不同的开发限制通过两种主要的途径，即一些分类中的公共所有权和私人行为的公共法规；开发保护的限制水平通过法规实现：很多国家和地方法规及国家激励政策在某种程度上被认为是帮助保护自然区域免于被开发。

为了得到一个综合值，开发限制都通过土地所有权和法规来实现，并为每种途径赋予0到1之间的分值，0是指限制性最强（不允许开发），1是指限制性最低（不限制开发）（图4–1–6）。

2) 目前管理状态

为了评估通过所有权途径来保持绿色基础设施生态功能的潜力，研究者调查了马里兰绿色基础设施的管理状况，根据马里兰绿道委员会2000年公布的版本，GAP管理标准来源于斯考特等人及马里兰自然资源部及其他部门的保护数据层。所有权、地役权及调节机制在环境保护水平各不相同，允许不

or management.

3)Projected Forest and Wetland Loss 1997-2020

2020 land use was projected by Maryland Dept. of Planning at the 12-digit watershed (sub watershed) level. Comparing 2020 land use with MDP 1997 land use. MDP predicted no net loss of wetlands (an optimistic projection), so the researchers used the projected percent loss of forest by 12-digit watershed. Changes <5 ac in the entire watershed were rounded to 0%.

We converted percent forest loss to a grid, and summarized by hub. For consistency, we used MDP 1997 land use to determine the relative proportion of upland forest within each hub.

Projected hub forest loss by 2020 was estimated by multiplying the percent projected upland forest loss by the percent upland forest not protected by public ownership, easements, or steep slopes (>25%) (Fig.4-1-7).

4) Overall Ranking of Hubs by Relative Risk of Development

Hubs from highest to lowest for each parameter were ranked in Table 4-1-4 within their physiographic region, and converted these rankings to percentiles. The researchers multiplied these percentiles by the parameter importance value, and added these together to give an overall ranking.

同的活动和管理。

3）1997—2020年林地及湿地的预计损失

2020年的土地利用由马里兰规划部门通过12位的流域标准预测出来。对比2020年的土地利用和马里兰规划部1997年的土地利用状况。马里兰规划部预计没有湿地的净亏损（一种积极的推测），所以研究者使用了12位流域预测的林地损失百分比，整个流域中变化小于5英亩（约2公顷）的地方将其损失率忽略不计。

可以将部分损失的林地转换成网格，通过网络中心进行概括。为了保持一致性，使用1997年MDP土地利用数据来确定每个网络中心高山林地的相对比例。预计2020年损失的网络中心林地量是通过预计部分损失的高山林地量乘以没有得到公有制及地役权保护或者在陡坡（>25%）上的林地量估算得到的（图4–1–7）。

4）网络中心相对开发风险综合评级

网络中心的开发风险等级在其地质区域内根据每个参数从最高到最低在表4–1–4中排序出来，并且将这些等级转换成百分数。研究者将这些百分数与参数的重要值相乘，并将这些值相加得到一个综合评级。

5）依据开发的相对风险对廊道分段分级

廊道片段的排序使用的开发风险参数与网络中心使用的一样。廊道趋于比网络中心穿过更多的私人土地，而且可能有更大的开发风险。如果廊道丢失的部分是非常重要的，破坏整个连接效益的时候体现得更加明显。一种可能性的解决方法就是在关注廊道的保护时更加关注开发搁置地、生态友好型设计或者地役权而不是关注土地获得。后者可能更加适用于网络中心的保护（图4–1–8）。

6）单个网格单元风险评估

这些单元格是面积为0.314英亩（约1207.7平方米）的方形区域，马里兰的每个单元格都根据它的地方特性及景观环境来赋予相应的生态分值。首

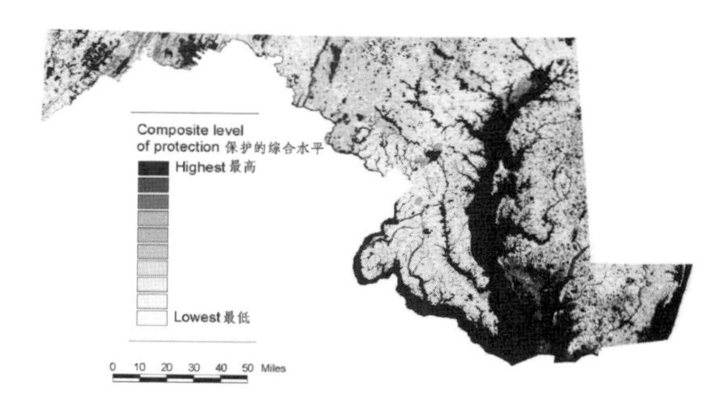

图4–1–6 马里兰开发限制及强度
Fig.4-1-6 Restrictions on development in Maryland, and their relative strength

5) Ranking Corridor Segments by Relative Risk of Development

Corridor segments were ranked using the same development risk parameters as hubs. Corridors tend to cross more private land parcels than hubs, and may be at greater risk of development. This is especially true because loss of part of the corridor, if the break is significant, destroys the effectiveness of the entire linkage. One potential solution is to focus corridor protection more on development set-asides, eco-friendly design, or easements, rather than on land acquisition. The latter may be more effective and appropriate for hubs (Fig.4-1-8).

6) Assessment of Individual Grid Cell Risk

The cells were squares corresponding to an area of 0.314 acre. Each cell in Maryland was given an ecological score based on both its local significance and its landscape context. First, the researchers identified land that was un-developable (conservation land or open water). Public land not managed for conservation was considered publically developable. They used the same development risk parameters to assess each grid cell that used to assess hubs and corridors, only at a finer scale. They reclassified these risk parameters to an equal-area percentile distribution on developable land. Zero was considered the lowest probability of development for each parameter; 100 the highest.

Then the researchers added the calibrated parameters, multiplied by the importance weights. They examined three different models, using different combinations of parameters and weights. Output from models 1 and 2 was very similar. Model 3's output was slightly different, placing greater emphasis on areas with existing or planned water and sewer (Fig.4-1-9).

4. Conservation Efforts

1) Defining Focus Areas

The researchers combined the ecological and risk

图4-1-7 预计2020年至少会损失10%的网路中心和廊道被开发
Fig.4-1-7 Hubs and corridors projected to lose at least 10% of their cover to development by 2020

先，研究者识别出未开发的土地（保护地或者开敞水域）。没有计划保护的公共土地被认为是可公共开发的。他们使用与评估网络中心和廊道相同的开发风险参数来评估每个网格细胞，只是在一个更小的尺度上。他们将风险参数重新分类成分布在可开发土地上等面积的百分位。0是指参数开发可能性最低，100是最高。

然后这些研究者将校准后的参数相加，与重要加权相乘。他们检验使用参数和加权组合的不同方式而得到的三种模型。模型1与模型2的结果非常相似，模型3的结果有点不同，它更加强调现有或者规划有水域和下水管的地区（图4-1-9）。

4. 保护措施

1) 确定焦点区域

研究者们将网络中心、廊道的生态和开发风险等级联系起来，帮助聚焦保护措施。有关这些焦点区域描述的其中一个版本为：在高生态等级（在地质区域中综合生态等级为前三名）的网络中心，不管是在全州还是在它们的地质区域内，未受保护的部分的开发风险值在较高的50%范围内；连接高生态等级网络中心的廊道，开发风险值在较高的50%范围内；与现存被保护的绿色基础设施土地线连接的廊道，开发风险值在较高的50%范围内（图4-1-10）。

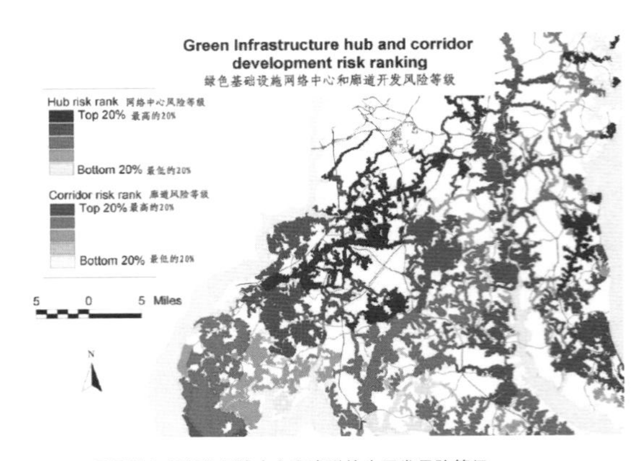

图4-1-8 马里兰南部部分网络中心和廊道综合开发风险等级
Fig.4-1-8 Hub and corridor composite development risk ranking, for a portion of southern Maryland

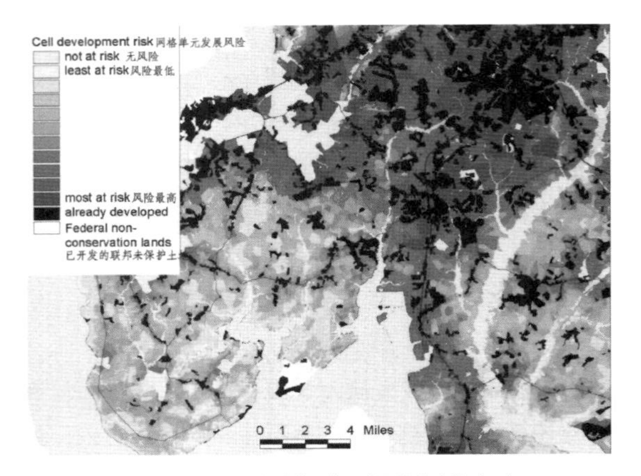

图4-1-9 马里兰南部查尔斯县部分单元格开发风险值（模型三）
Fig.4-1-9 Cell development risk (model 3) scores for part of Charles County in southern Maryland

ranks of hubs and corridors to help focus protection efforts. One version of these focuses areas was:hubs in the top ecological tier (composite ecological rank in the top third of hubs in the physiographic region), and with their unprotected portion in the top 50% development risk either statewide or within their physiographic region;corridors linking hubs in the top ecological tier, and in the top 50% development risk; and corridors linking existing protected green infrastructure land, and in the top 50% development risk (Fig.4-1-10).

2)Maryland's GreenPrint program

In 2001, the Maryland General Assembly passed legislation (House Bill 1379) establishing the GreenPrint program. GreenPrint is a targeted program that attempts to preserve the most ecologically valuable natural lands in Maryland, its green infrastructure hubs and corridors, by purchasing land from willing sellers. These purchases can be either fee simple (ownership transferred to the state or a county) or conservation easements (original owner keeps the

2）马里兰绿图计划

在2001年，马里兰人民大会通过建立绿图计划的法案（众议院法案1379）。绿图是通过从志愿买家那里购买土地来保护马里兰最有生态价值的自然区域、绿色基础设施网络中心及廊道为目标的项目。这些购买可以仅是费用上的（将所有权转变成州的或者县的）也可以是保持地役权形式的（原所有者保留地产但是将开发的权利卖掉）。地役权必须是永久性的，并且地产应该保持它们的生态价值。

在第一个财政年度——2002年中（2001年7月1日~2002年6月30日），马里兰立法机构分配2 250万美元来获得绿色基础设施土地及地役权。自然资源部有权使用75%的资金，马里兰农业地保护基金会有另外的25%的使用权。地方政府需要去批准土地收购和地役权。

3）衍生项目

德尔马瓦半岛廊道保护：马里兰绿色基础设施评估方法被用于德尔马瓦半岛与特拉华州和弗吉尼亚州的合作项目中（图4-1-11）。议员韦恩·吉尔格雷斯特介绍了相关理念，他提出一个工作地、农场与林地相互联系的网络并且管理的与自然风景很和谐。这个廊道保护项目的目标有：支持工作地，保护乡村风光及德尔马瓦半岛的生物多样性。项目组决定使用现有的绿色基础设施评估方法学，他们了解现有的资料，用电脑分析，想出一个超出半岛政治边界的网络中心与廊道网络的模型。协办单位正在研究给美国农业部门提交的一个计划，这个计划是关于怎样让每个州都能执行议员吉尔格雷的提议。

安妮阿伦德尔县绿道计划：安妮阿伦德尔县是全州范围内第一个在绿色基础设施概念及全州绿色基础设施评估结果的基础上进行绿道规划的。他们将全州GIA的大部分程序改编以适应他们自己的县级评估（图4-1-12）。例如，县城将全州GIA中使用的最小网络中心面积由250英亩（约101.2公顷）改成50英亩（约20.2公顷）。这个县的绿道总体规划在2003年早期从美国规划联盟中的马里兰分会那获

property, but sells the rights to develop it). Easements must be perpetual, and properties should retain their ecological values.

For fiscal year 2002 (July 1, 2001 - June 30, 2002), the Maryland legislature allocated $22,500,000 for green infrastructure land acquisitions and easements. The Department of Natural Resources (DNR) was given authority to spend 75% of the funds, and the Maryland Agricultural Land Preservation Foundation (MALPF), the other 25%. Local governments are required to approve acquisitions or easements.

3) Spin-off Programs

Delmarva Peninsula Conservation Corridor: Maryland's GIA approach was used on the Delmarva Peninsula in cooperation with Delaware and Virginia(Fig.4-1-11). Congressman Wayne Gilchrest introduced the concept; he envisioned a network of working lands--farms and forests--linked and managed in harmony with the natural landscape. The conservation corridor project has the goals of sustaining working lands and protecting the rural character and biodiversity of the Delmarva Peninsula. The program team decided to utilize the existing GIA methodology. They determined what data existed, ran the computer analysis, and came up with a hub and corridor network model that stretches across the political boundaries of the peninsula. The cooperators are working on a plan to be submitted to the U.S. Department of Agriculture on how each state will meet Congressman Gilchrest's vision.

Anne Arundel County Greenways Plan:Anne Arundel County was the first in the state to base its greenways plan on the concept of green infrastructure and the results of the statewide green infrastructure assessment. They adapted many of the procedures developed in the statewide GIA to their county-level assessment(Fig.4-1-12). For example, the county

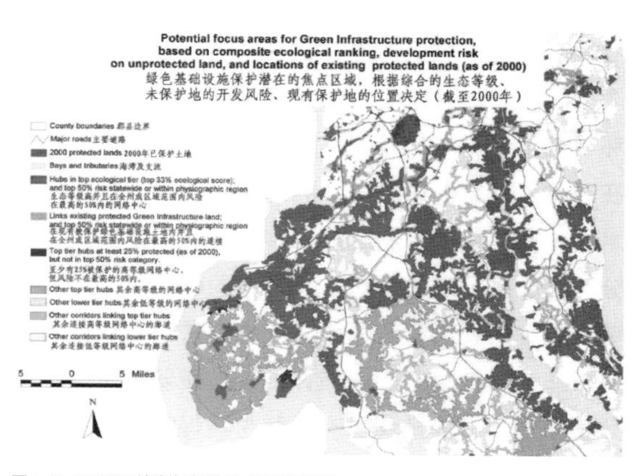

图4-1-10 马里兰南部部分焦点区域样图
Fig.4-1-10 Sample version of focus areas for a portion of southern Maryland

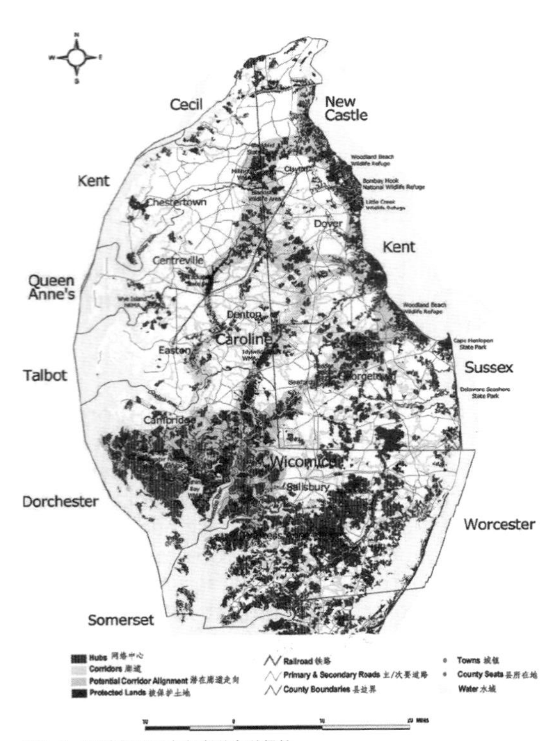

图4-1-11 德尔马瓦半岛廊道廊道保护
Fig.4-1-11 Delmarva Peninsula Conservation Corridor

图4-1-12 安妮阿伦德尔县的绿道规划——网络中心和廊道网络
Fig.4-1-12 Anne Arundel County Greenways Plan — Hub and corridor network

得了一个奖项，也因政府在2002年的创新活动而获得了管理者精明增长奖项。

三、 总结

绿色基础设施是指一个相互联系的绿色空间网络，与我国绿地系统类似，其组成要素包括绿道、湿地、公园、风景区等，这些不同自然要素组成一个相互联系、有机统一的网络系统。城市化进程破坏了马里兰原本优越的自然资源，而且这种威胁还在进一步加剧。研究中，利用绿色基础设施评估技术识别出网络中心和廊道两种重要的元素，进而划分等级，进行风险评估，采取不同的保护和发展策略，这种严谨调研和科学分析的方法具有说服力和可操作性。绿色基础设施评估技术是研究中的重要分析方法，它以景观生态学和生物保护学原则为基础，提供一个评估方法，将众多片段化的、无序的生态中心和廊道网络按照一定标准排列有序，分出

set the minimum threshold size for hubs at 50 acres instead of the 250 acres used in the statewide GIA. The county won an award in early 2003 from the Maryland chapter of the American Planning Association for their greenways master plan. The county also won the Governor's Smart Growth award for government innovation in 2002.

III. Summary

Green infrastructure is an interconnected network of green space, similar as China's green space system ,its elements including green way, wetland, the park, the scenic area, etc.These different natural elements form a network system which is connected with each other and organic unity .Urbanization process destroys the favorable natural resources in Maryland , and this threat also intensifies in further.In the study, using GIA technology to identify two important elements,the hub and corridor ,then to rank and assessment the risk.,take different protection and development strategy.This rigorous research and scientific analysis method is persuasive and operational.GIA is the important analysis method in the study,it based on the principles of landscape ecology and biological protection,provide a assessment methods, set

主次，是本研究结论形成的重要前提。绿色基础设施评估技术在应用上具有一定科学性和合理性，这种分析方法在我国同类规划中并未普遍应用，仅有少量学术论文提及。

简言之，马里兰绿色基础设施由各种开敞空间和自然区域组成，该系统的良好运转可为野生动物迁徙和生态过程提供起点和终点。相应的规划措施对降低暴雨和洪水危害，改善水系质量，节约城市管理成本都有显著作用。

many of the fragmented, disordered ecological hub and corridor network orderly according to certain standards,and prioritize them,it was important prerequisite for forming conclusions of the study.GIA has certain scientific and reasonable in application ,this analysis method is not commonly used in similar planning in our country, only a few academic paper have mentioned.

In brief,Maryland green infrastructure is constituted by all sorts of open space and natural areas ,the system can provide the beginning and the end for wild animal migration or ecological process if it was in good operation.The corresponding planning measures have significant effect in reduce heavy rains threat and flood hazards, improve the quality of water system and save the city management cost.

思考题

1. 绿色基础设施规划在不同空间尺度上（宏观尺度、中观尺度和微观尺度）确立的战略重点和规划目标相应的依据应包括哪些方面?
2. 绿色基础设施的保护与发展利用在综合评估方面怎样更加体系化?
3. 绿色基础设施规划与相关利益主体之间的博弈，政府应扮演怎样的角色?
4. 绿色基础设施与我国绿地系统在概念、规划方法上有哪些异同?
5. 绿色基础设施评估在动态监测上还需哪些技术支撑?

Questions

1. In green infrastructure planning,what aspects should be included in the corresponding basis of establishment of strategic emphasis and planning objective in different spatial scales (macro scale, medium scale and micro scale)?
2. How does the comprehensive evaluation of protection and development of green infrastructure more systemized?
3. what's the role of government should play in the game between the green infrastructure planning and stakeholders ?
4. what are the similarities and differences between green infrastructure and Chinese green space system in the concept and planning methods?
5. what technical support still be needed in dynamic monitoring of green infrastructure assessment ?

注释/Note

参考文献/Reference：[1]Ted Weber Maryland's Green Infrastructure Assessment—A Comprehensive Strategy for Land Conservation and Restoration2003 ；[2]The Conservation Fund2004,Maryland' s Green Infrastructure Assessment and GreenPrint Program ；[3]Mark A. Benedict,Edward T. McMahon,Green Infrastructure: Smart Conservation for the 21st Century,2001；[4]付喜娥.吴人韦，绿色基础设施评价（GIA）方法介述一以美国马里兰州为例，中国园林，2009.25（9）：41~45

推荐网站/Website：http://www.dnr.state.md.us/greenways/gi/gi.html; http://www.dnr.state.md.us/greenways/gi/gidoc/gidoc.html; http://www.greenprint.maryland.gov/; http://www.greeninfrastructure.net/.

第二节 哈罗新城绿色基础设施规划

Section II The Green Infrastructure Plan (GIP) for the Harlow Area

一、项目综述

哈罗位于英国埃塞克郡西部，“大伦敦规划”中12.5公里绿环东北部的边缘，是伦敦的卫星城。哈罗地区的绿色基础设施规划是由克里斯·布兰福德协会独立研究，并在国家，区域和地方组织的代表组成的督导组的密切监督下工作。规划为绿地的保持、养护及扩张提供指导，其中包括公园、花园、树林和自然保护区或不对公众开放的区域；另外还包括一些线性空间，如越野道路、公路、河流、小溪、绿篱、野生生物通道及与公众相联系的开放空间。绿色基础设施强调的是地球上各类土地资源之间的联系与互通，由此人类构建出一个对保护地球生物多样性最有利的网络框架。

绿色基础设施规划的概念是基于一种战略性方法，用来确保在早期的施规划中，将环境资源的自然、文化价值与土地开发、增长管理和建成的基础设整合在一起。这项方法使得土地管理更加积极主动，并在各个规划层面上更好地与增长和发展的相结合。绿色基础设施规划是在城市增长区域发展可持续发展社区和创造高质量生活的一个重要的手段。

I. Overview

Harrow is located in western Essex County of UK, "Greater London Plan ", the northeast edge of 12.5 km of Green Circle, which is the satellite town of London. The Green Infrastructure Plan (GIP) for the Harlow Area is an independent study by Chris Blandford

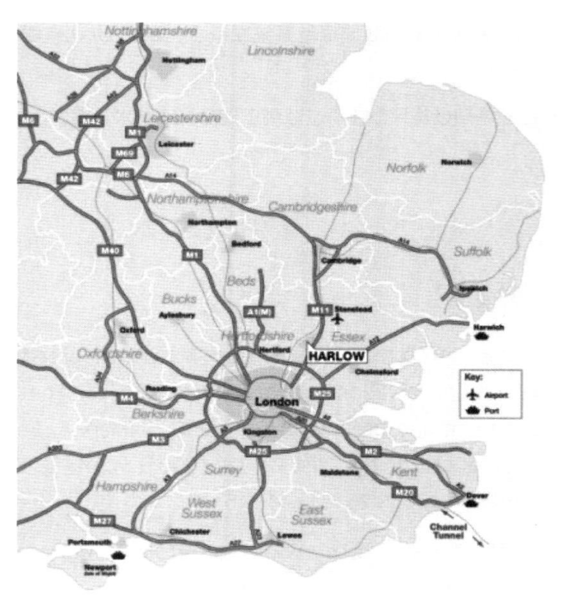

图4-2-1 哈罗新城区位图
Fig.4-2-1 Location map of Harlow Area

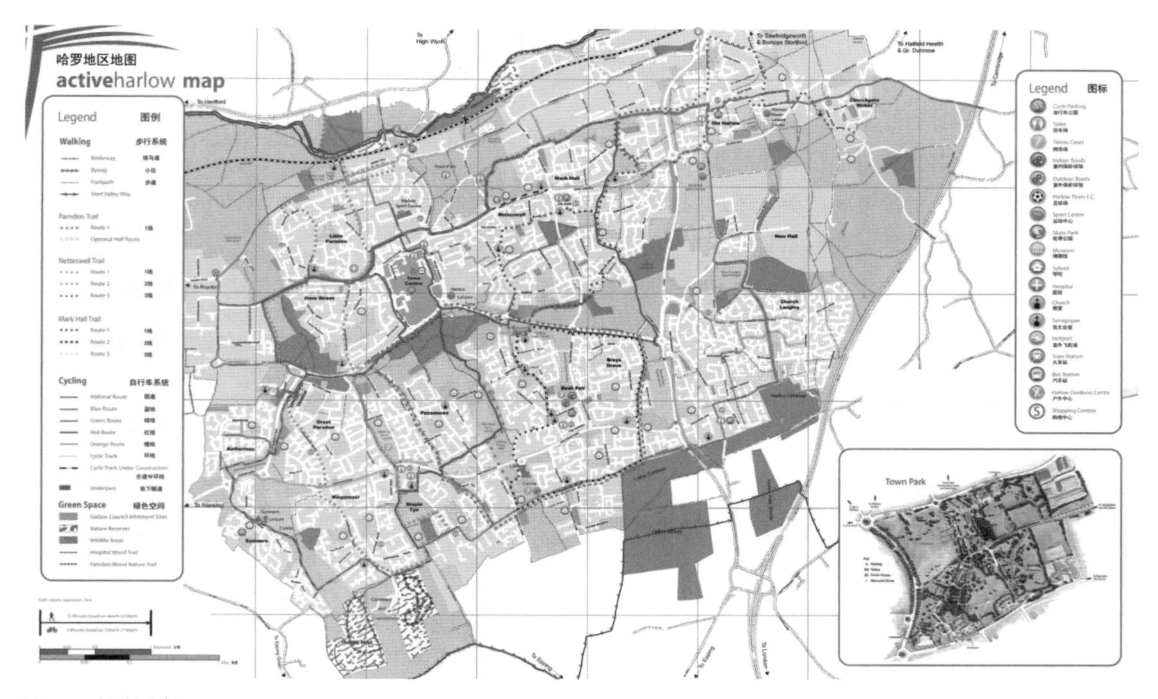

图4-2-2 哈罗新城总平面图
Fig.4-2-2 Site-plan of Harlow Area

二、 绿色基础设施现状资源

1. 物理资源和自然生态系统

哈罗地区的物理资源和自然生态系统对其区域的独特性、土地利用、生物多样性、市容景观产生了影响。哈罗地区内的主要物理资源是李河和斯托特河流及其流域。河流廊道、洪泛区和支流是重要的绿色基础设施构成部分，因为它们为当地提供了重要的社会、经济和环境效益，成为野生动物的重要栖息地。绿色基础设施规划提供一个重要的机会，即创造了一个有创新性的协调方法最大限度地利用这些物理资源和自然生态系统，也同时克服气候变化的挑战。

2. 生态资源与生物多样性

哈罗地区生物多样性较高，因此生态资源丰富。重要的自然生态资源包括大量的自然重点保育区，核心栖息地的多样性（水域、河岸和陆地），以及野生动物栖息地之间，起关键作用的廊道的大

Associates, working under the close supervision of a Steering Group consisting of representatives from national, regional and local organizations. The GIP provides guidance on how the green areas should be protected, enhanced and where appropriate extended. Green spaces can include areas such as parks, gardens, woods and nature reserves with or without public access; linkages include linear features such as off-road paths, highways, rivers, streams or hedgerows, which can provide corridors for wildlife and connect people to open spaces. Green infrastructure is the link and connectivity between the various types of land resources for which humans construct a most favorable network framework to protect of the planet's biodiversity.

The concept of green infrastructure planning is based on a strategic approach to ensuring that environmental assets of natural and cultural value are integrated with land development, growth management and built infrastructure planning at the earliest stage. This approach enables land management to be more proactive, less reactive, and better integrated with efforts to manage growth and development at all spatial planning levels. Green infrastructure planning is therefore a key mechanism for delivering sustainable communities and quality of life benefits within growth areas.

量的线性空间，尤其是珍稀保护物种。一个关键的规划项目是在斯托特河谷中建立一个景观尺度的多功能湿地系统，这将对生物多样性行动计划的实施起到重要的作用。其他资源的机遇主要为一般性的恢复，生态资源功能的加强和改善，通过保持该地区核心栖息地（包括林地、灌木丛和池塘）的连通性与活力。绿色基础设施规划的关键是，通过敏感性设计，新城发展的定位和布局，农业活动，减轻洪水和蓄水，休闲娱乐设施的详细规划与使用来完成对重点自然保护地区的保护。

3. 自然、城市及滨河景观特征

景观及视觉体验的多样性对自然风景、城镇景观及河流景观具有重要作用。哈罗地区的主要景观资源包括由地貌景观构成的景观结构框架，特别是河谷、山脊、高原、开放的农田和林地；农村居民点中的居住个体和他们分散的景观格局；历史景观区；以及体现地方特色和地域感特征的标志性建筑及其他景观要素。在哈罗地区的绿色基础设施网络的发展中，加强地方特色和独特景观的主要方法是通过积极的土地管理恢复、加强和改善那些与绿色基础设施相关的景观特色的生态价值，历史景观和其他文化价值。另外，确保吉伯德的“景观导向”设计方法能够持续应用于周边新开发的城镇，并在绿色基础设施规划有所反映。新的发展需要与一个强大的景观框架相协调，以确保其城市形态和建筑设计的形式与周围乡村的特征相和谐，并且有所呼应。

4. 考古、历史和文化资源

考古、历史和文化资源的一个主要特征是通过展示过去的遗迹来体现地域性。该地区的主要资源包括历史悠久的公园和花园、工厂、中世纪的聚落遗址、史前遗址、二十世纪的军事用地和机场、历史建筑及寺院遗址。其他幸存的历史景观包括古老的林地、公共用地、历史通道、草甸和历史地区。这些资产大多为私人所有，而不向一般公众开放。然而，绿色基础设施规划提供了一次机遇，使历史环境资源在生活中的潜力得到最大化的发挥，并借此大力提高了哈洛地区考古、历史和文化资源的进

II. Existing Green Infrastructure Assets and Opportunities

1. Physical Resources and Natural Systems

The physical resources and natural systems within the Harlow Area influence local and regional distinctiveness, land use, biodiversity, and landscape and townscape character. The main physical resource within the Harlow Area is the Rivers' Lee and Stort and their catchments. The river corridors, floodplains and tributary streams are considered to be critical green infrastructure components, as they supply key social, economic and environmental benefits for local communities and provide important habitats for wildlife. The GIP provides a significant opportunity to create a coordinated and creative approach for maximising the utilisation of the opportunities provided by these physical resources and natural systems, while overcoming the challenges of climate change.

2. Ecological Assets and Biodiversity

There is a wide range of ecological assets that contribute to the biodiversity of the Harlow Area. The Key Ecological Assets include the relatively large number of sites protected for their nature conservation importance, the diversity of Key Habitats (aquatic, riparian and terrestrial), and the numerous linear landscape features acting as Key Links providing dispersal corridors for wildlife between these sites and habitats, including for Rare and Protected Species. There is a major opportunity to create a fully functioning wetland system on a landscape scale within the Stort Valley, which would make a significant contribution to the achievement of important Biodiversity Action Plan targets. Other key opportunities include the general restoration, enhancement and improvement of features of ecological value such as woodlands, hedgerow field boundaries and ponds throughout the area to maintain the connectivity and viability of key ecological habitats. The protection of sites of nature conservation importance through the sensitive design, layout and location of new urban development, agricultural activities, flood alleviation and water storage, and the careful planning of recreational uses and facilities, is key to the GIP.

3. Landscape, Townscape and Riverscape Character

The diversity of landscapes and visual experiences are a key contribution to the landscape, townscape and riverscape character of the Harlow Area. The Key Landscape Assets of the Harlow Area include the major physiographic and landscape features providing the structural framework for the landscape generally,

一步宣传。除了一些著名的遗迹，从广义的时代角度来看，大量的景观具有考古发掘的潜力，同时还可以进一步将这一类资源作为教育资源。

5. 绿色网络和娱乐设施

哈罗地区具备广泛的游憩机会及设施，为当地社区和游客提供乡村中各种休闲活动的享受，既可以是正式娱乐活动，也可非正式。这包括大范围覆盖并联通的公共步行网络、李河谷地区公园、哈特菲尔德森林郊野公园和埃平森林等核心资源。另外，为了解决北哈罗社区和索布里吉沃斯社区之间缺乏绿色公共空间的问题，哈罗地区建设了一个新的带状公园——斯托特滨海公园。此外还有进一步提高哈罗及周边地区现有网络的连通性和质量，包括萨斯特兰提案中在绿色基础设施的建设中扩展哈罗地区的国家循环网络。

三、 新城绿色基础设施总体规划

1. 创建与联通野生动物栖息地

哈罗地区的物种丰富多样，包括多种自然栖息地。有些为成熟的林地，如潘顿林地、洛兹林地和嘉里希尔林地，分散遍布于哈罗地区。位于该地区南部的埃平森林的北部，包括不同的林地、草地、荒原和湿地生境。在一个现代农业普遍对生物多样性造成一定冲击的地区，田野、树篱和农田边界为野生动物提供了宝贵的栖息地和避难所，也同时为生物多样性的保护提供了保障。这些生态价值较高的栖息地防止了现代农业劳作的入侵。在哈罗地区的栖息地内有一些珍稀保护物种，如大冠蝾螈（欧洲保护物种），爬行动物（常见的蜥蜴、蛇蜥和草蛇）和繁殖的鸟类。现有的栖息地还需要加强其连通性，以形成镶嵌于整个哈罗地区的生物多样性网络，这需要通过鉴别和保护野生动物的栖息地之间的“野生动物走廊”来完成。

2. 乡间绿道建设

乡间绿道由道路和公共道路（哈罗主城区外围通往风景区的道路）组成，并有几条旅游路线穿过

and for Harlow in particular, (the river valleys, ridges, plateau, open farmland and woodland blocks); the individual identity of rural settlements and their dispersed pattern within the landscape; areas of historic landscape; and the range of key views, landmarks and other landscape elements and features that contribute positively to local distinctiveness and sense of place. The main opportunities for strengthening the local character and distinctiveness of landscapes through the development of the Green Infrastructure Network within the Harlow Area primarily relate to the restoration, enhancement and improvement of landscape features of ecological value, historic designed landscapes and other landscape of cultural value through positive land management. There is also an important opportunity to ensure that Gibberd's 'landscape-led' approach to the planning and development of Harlow is continued in the design of new development in and around the town, and reflect in the GIP. New development needs to incorporate a strong landscape framework to ensure that its urban form and building design is shaped by, and responds to, the character of the surrounding countryside.

4. Archaeological, Historical and Cultural Assets

There are a wide range of sites and features of archaeological, historical and cultural value within the Harlow Area that contribute to sense of place through a link with the past. Key assets include historic parks and gardens, mills, medieval settlement sites, prehistoric sites, 20th century military sites and airfields, historic buildings and townscapes and monastic sites. Other surviving historic landscape features include ancient woodlands, common land, historic routeways, water meadows and historic field systems. Many of these assets are on private land and not accessible to the general public. However, there is a major opportunity through the GIP to develop a strategic approach to maximizing the potential quality of life benefits of the historic environment through enhanced physical and intellectual access to, and improved presentation of, key archaeological, historical and cultural assets in the Harlow Area. Beyond the known sites and features, much of the wider landscape has potential for the discovery of archaeology from a range of periods that can add further value to the GIP as an educational resource.

5. Access Networks and Recreational Facilities

The Harlow Area has a wide range of opportunities and facilities to enable local communities and visitors to gain access to and enjoy the countryside, for both formal and informal recreational activities. This includes an extensive and generally well-connected public footpath network,

区域内（例如走谷地幽谷，哈坎娄路，斯托特河谷路，埃塞克斯路和森林路）。在某些情况下，由于边界、沟渠和树篱的缺失，乡间道路的特征正在消失。不同风格标牌形成混杂的景观，使得视觉景观上十分混乱，并造成游步道与旅游路线使用者的混乱。可以通过改善游步道和道路网络的连接性，为行人提供完整的景观道路系统。乡间道路为野生动物提供多样化的廊道，建立较宽的树篱也可以为种植野生花卉提供一个理想的场所。

3. 河道治理

区域内主要河流（李河和斯托特河）和较小的支流（包括品希溪流，费德勒溪流，亨德逊溪流和

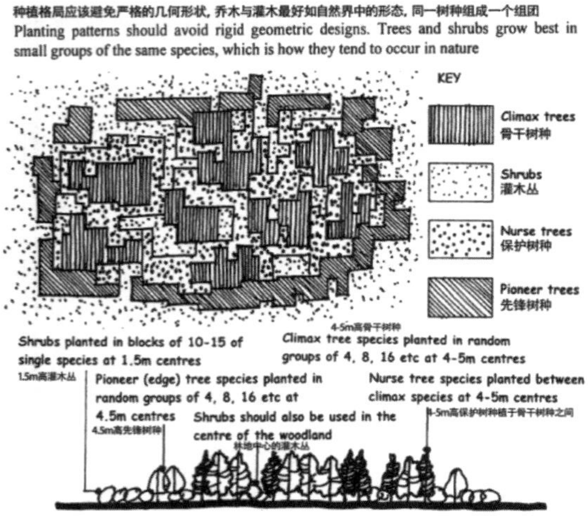

图4-2-3 野生动物栖息地种植格局
Fig.4-2-3 Planting pattens of wildlife habitat

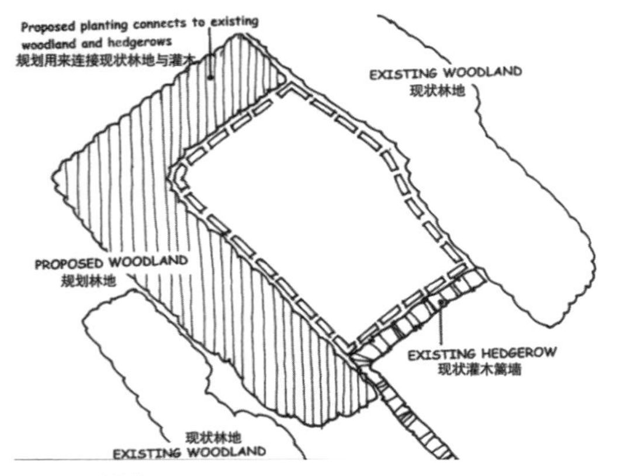

图4-2-4 种植格局
Fig.4-2-4 Planting pattens

and key resources such as the Lee Valley Regional Park, Hatfield Forest Country Park and Epping Forest. There is a key opportunity to address an identified shortfall in public green space provision for the communities of North Harlow and Sawbridgeworth through the creation of a new linear park – the StortRiverpark. There are also opportunities to further enhance the connectivity and quality of the existing access etwork in and around Harlow, including incorporation of the Sustrans proposals for extension of the national cycle network in the Harlow Area into the GIP.

III. Green Infrastructure Master Planning for the Harlow Area

1. Wildlife Habitat Enhancement, Creation and Links

The biodiversity of the Harlow area is rich and varied, comprising a spectrum of habitats. Several areas of mature woodland, such as Parndon Wood, Lord's Wood and Galleyhill Wood, are scattered throughout the Harlow Area. The northern extent of Epping Forest (comprising a diverse mixture of woodland, grassland, heathland and marshland habitats) is located within the south of the area. Fields, hedges and agricultural boundaries offer a source of biodiversity and provide a valuable habitat and wildlife refuge within a landscape where modern agricultural practices have generally reduced biodiversity across the area. These ecologically-rich pockets of habitats require protection from encroachment of modern farming practices. Habitats within the Harlow Area contain several rare and protected species such as great crested newt (a European protected species), reptiles (common lizard, slowworm and grass snake) and breeding birds. There is a need for connection and enhancement of existing habitats to form an inter-connected mosaic of biodiversity throughout the Harlow Area by identifying and protecting 'wildlife corridors' between wildlife sites.

2. Countryside Access Routes

Countryside Access Routes consist of a combination of roads and public rights of way which provide access to landscapes outside the major urban areas within the Harlow Area. Several recreational routes run through the area (e.g. Lea Valley Walk, Harcamlow Way, Stort Valley Way, Essex Way and Forest Way). In some cases, the character of rural lanes is being eroded through the loss of verges, hedge banks, ditches and hedgerows. A variety of different styles of signage can provide visual clutter and confusion to users of footpaths and recreational routes. There is potential for greening and improving the connectivity of this network of footpaths and roads, to provide comprehensive public access to the landscape. Rural roads can provide diverse wildlife corridors, with wide

大哈林布里溪流）形成了哈罗地区的主要天然湿地，包括湿草地，芦苇塘，开阔水面和沼泽。溪流对于水鼠、白爪小龙虾和底栖无脊椎动物来说是重要的栖息地。一些规模较小的支流能很好地融入周围的景观。然而，一些区域的河岸处理过于粗糙，未得到足够的重视，河口和植被管理不善。提高农业生产、娱乐活动和林地种植，这些都可能导致由地表排水和化肥使用所引起的湿地栖息地丧失。因此，集约农业、城市发展及工业所引起的污染物排放和泄漏，对现有的水系构成了巨大的威胁。

4. 道路交通的绿化

大多数居民和游客对区域内道路的认知源于几

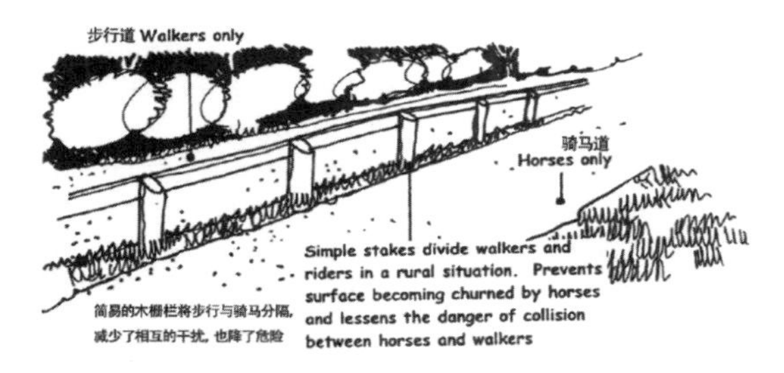

图4-2-5 道路结构
Fig.4-2-5 Structure of road

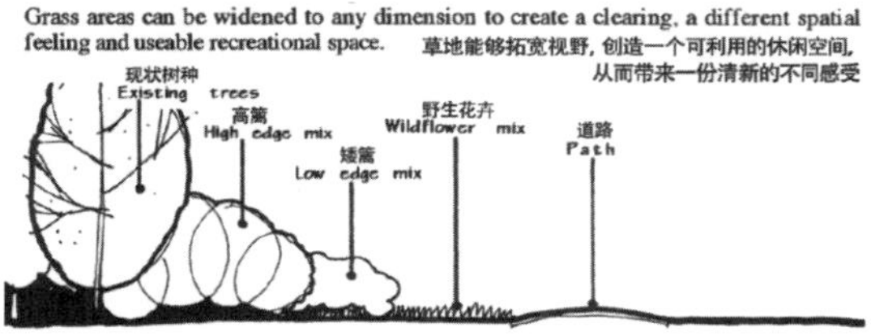

图4-2-6 道路剖面
Fig.4-2-6 Section of road

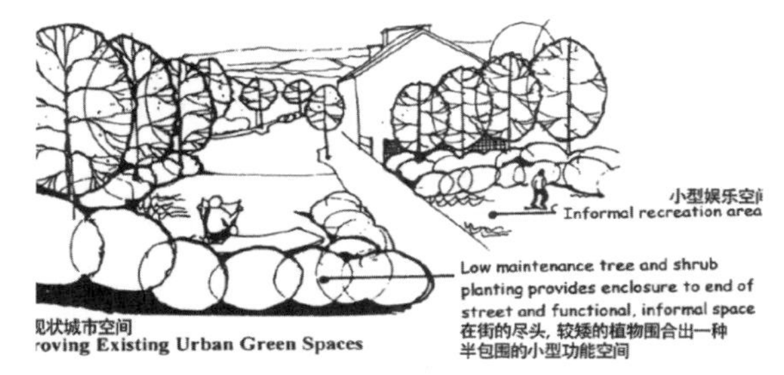

图4-2-7 城市空间改善
Fig.4-2-7 Urban green spaces improving

verges providing an ideal habitat for wildflowers.

3. Riverways

The main rivers (Lee and Stort) and smaller tributaries (including Pincey Brook, Fiddlers Brook, Hunsdon Brook & Great Halling bury Brook) within the Harlow area retain apredominance of natural wetland habitats, including wet grassland, reedbeds, open water, water margin and swamp. Rivers within the area are important for water voles, white-clawed crayfish and benthic invertebrates. Several of the smaller tributaries are well integrated into the surrounding landscape. In places, however, the edges of the riverways are harsh and neglected, with vegetation that is poorly managed and a lack of managed access to the river corridor. Pressures to increase agricultural production; recreational pursuits and the planting of woodland could lead to wetland habitat loss through land drainage, cultivation and nitrification from fertilizers. Run off, discharge and spillage of pollutants from intensive agriculture, urban areas and industry pose a significant threat to the existing river systems.

4. Greening of Road Transport Corridors

The majority of visitors' and residents' perception of the area stems from the principal road corridors such as the A414 (connecting A10 near Ware to North Weald Bassett- via Harlow), A1184 (connecting Harlow to Bishops Stortford) and several B - roads. Some road corridors are well integrated into the landscape, however several are exposed, open and poorly integrated, resulting in wider visual impacts. There is a need to develop stronger links between road corridors and the surrounding landscape and also enhance transport links to project a quality image.

5. Open Space and Recreation Facilities

Open space within the Harlow area consists of a combination of two main categories:

(1) Rural recreation areas such as Hatfield Country Park and Lee Valley Regional Park, community woodlands and smaller-scale recreational areas;

(2) Urban fringe and internal green spaces such as playing fields, allotment gardens, children's playgrounds, town and pocket parks.

There is reasonable provision of urban fringe and internal urban green spaces, particularly within Harlow urban area. Levels of accessibility, state of dereliction and quality of spaces, however, vary throughout Harlow Area. There is a need to improve open space and recreational facilities within the urban fringe and internal urban areas to form useful, attractive and functional spaces popular for recreational and amenity use by locals and visitors. There is also a need to improve accessibility to existing open spaces and recreational facilities to create an interconnected

条主干道，如A414，A1184和几条B编号公路。一些公路与景观相得益彰，然而另外几条却十分简陋，视觉景观不佳。因此有必要加强道路和周围景观之间的联系，提升交通系统的景观形象。

5. 开放空间和娱乐设施

哈罗地区的开放空间分为两大类：

（1）田园休闲区——哈特菲尔德国家公园和李河谷地区公园，社区林地和规模较小的休闲区；

（2）城市边缘和内部的绿地，如运动场、小型公共花园、儿童游乐场、街心公园。

在哈罗市区，特别是在城市边缘和内部的绿地分布数量是合理的，然而，可达性、废弃程度和空间质量却参差不齐。因此，有必要改善开放空间及康乐设施，使其形成具有吸引力，受到当地居民和游客欢迎的休闲和景观能空间。另外，还要提高现有的开放空间及康乐设施的可达性，创建城乡绿色空间网络。

6. 新住宅开发

通常情况下，哈罗的新住宅开发计划选址在条件恶劣、几乎没有植被覆盖的城市边缘地带。这种开发计划使城市边缘的视觉景观恶化，并影响了周边地区景观。因此新住宅开发计划，特别是在城市边缘地区，有必要综合考虑周围的景观的融合。如果有可能，应该将景观特色和重要的野生动物栖息地融入新的发展计划。

network of urban and rural green spaces.

6. New Housing Development

It is often the case that new housing developments within the Harlow Area exhibit harsh urban edges, with little vegetation or screening along boundaries of the development sites. Such developments affect the visual character of settlement edges and expand suburban character into surrounding landscape character areas. There is a need to ensure that new housing developments, particularly on the edges of major urban areas, are well screened and integrated into the surrounding landscape. Wherever possible, it is necessary to integrate significant landscape features and key wildlife habitats into new developments.

7. Industrial and Commercial Development

There are several industrial and commercial areas within the larger urban areas. These developments often exhibit harsh edges, with little vegetation screening, determining that they are highly visible from surrounding roads, footpaths and also from landscape at the edge of settlements. Often, an amalgamation of different signage creates a confusing and untidy image. Buildings are often out of context with the scale, pattern and colors of the surrounding buildings and landscape features. There is a general need to enhance existing industrial and commercial sites within the Harlow Area to integrate them more with the surrounding landscape and townscape and potentially make them more attractive to investors and customers alike. There is also a need to ensure that new industrial and commercial development is designed to cause minimum visual intrusion and is in keeping, wherever possible, with local landscape character. Brownfield sites are also particularly important for urban wildlife habitats.

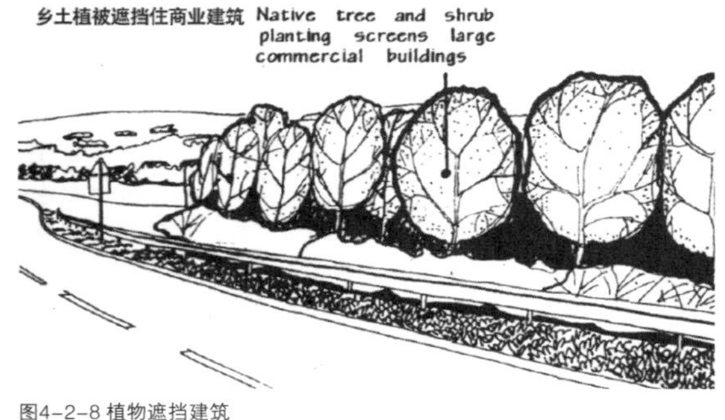

图4-2-8 植物遮挡建筑
Fig.4-2-8 Green screens buildings

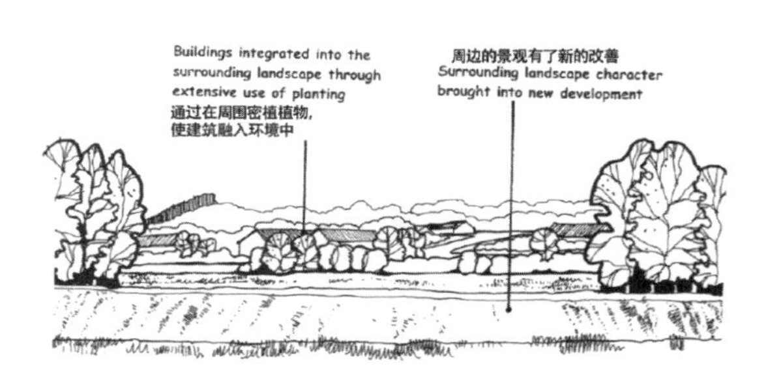

图4-2-9 植物改善景观
Fig.4-2-9 Plantsimprovelandscapes

7. 工业与商业发展

有一些工业和商业领域在更大范围的市区内。这些项目往往选址在条件恶劣没有绿化的城市边缘地带，这就使他们在周边道路和地区中十分醒目，通常各种标识牌混合在一起形成一种杂乱无章的形象。建筑物的尺度、布局和颜色与周围的背景十分不合。因此，目前将现有的工业和商业项目融合到周围的环境中十分重要，这样也可以吸引更多的投资者和客户。同时，未来建设的工业和商业应该尽量减少其对周围环境的视觉冲击，尽可能保持当地的景观特色，同时棕地也是重要的城市野生动物栖息地。

8. 城乡边界地带发展

在哈罗地区的几个城乡边界地区的景观与周围的景观往往不合，其特点是：无论是视觉景观上还是物质上都没有缓冲地区，没有真正的过渡连接；另外，短期的发展并没有充分考虑到地形地貌、植被和周围的景观。因此有必要整合城乡边界地区与周围景观的发展，通过提供强大的景观结构，根据该地区的景观特色，建立积极和谐连贯的大都市边缘视觉景观。

四、专项规划

1. 物质资源和自然生态系统规划

发挥湿地系统的潜力，保持地下水位，通过存储冬季期间过剩的水来补充夏天水的不足；保护河流作为重要的线性空间的完整性，以保证野生动物的迁移，扩散和繁殖；节约不可再生资源，充分发挥可再生能源在建设中的利用。

8. Development Edge Treatment

Several urban edges within the Harlow Area are harsh and visually intrusive within views from surrounding landscape character areas. The juxtaposition between development and countryside is often characterised by: an abrupt edge; no real connection, either visually or physically between development and countryside; adhoc development with little respect for existing landform, vegetation and the surrounding landscape. There is a need to integrate the edges of development with the surrounding landscape through provision of a strong landscape structure, in accordance with the landscape character of the area, helping to create a positive and coherent image upon arrival at the edge of a major urban area.

IV. Special Planning

1. Physical Resources and Natural Systems Principles

Take account of the potential of wetland systems to recharge the water table by storing water during periods of excess (winter) and for periods of deficit (summer);Protect the integrity of river corridors as important linear corridors for the migration, dispersal and genetic exchange of wildlife; Conserve non-renewable resources and promote the use of renewable sources of construction material and energy.

2. Ecology and Biodiversity Principles

Protect and enhance all existing Key Ecological Assets (statutory and non-statutory designated sites of nature conservation importance), key habitats and species as key components of the Green Infrastructure Network; Contribute to relevant BAP habitat and species targets in the Harlow Area; Be informed by ecological surveys and BAP priorities to guide the design and implementation of green infrastructure improvements and development schemes.

3. Landscape, Townscape and Riverscape Character Principles

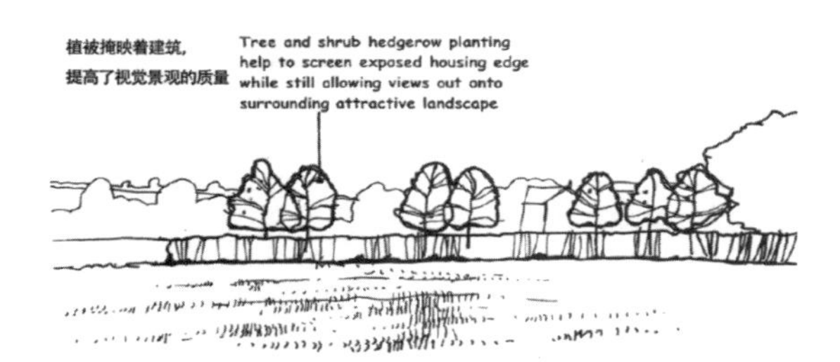

图4-2-10 植物改善景观
Fig.4-2-10Plantsimprovelandscapes

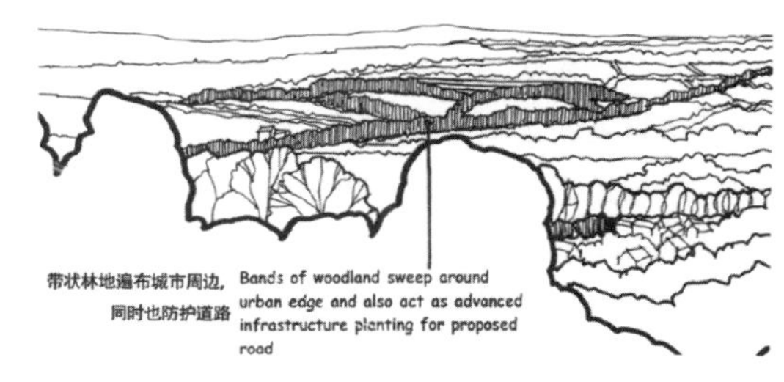

图4-2-11 道路防护林带
Fig.4-2-11 Road protection forests

2. 生态资源与生物多样性规划

保护和加强全部现有的生态资源，将核心的保护物种及栖息地作为绿色基础设施网络的重要组成部分；在哈罗地区保护BAP的栖息地和物种；使用生态调查引导绿色基础设施的改善和发展计划的设计和实施。

3. 自然、城市及滨河景观规划

反映吉伯德的“景观导向”设计方法在哈罗及周围地区新发展中的应用，设计将一个强大的景观框架融入建筑设计之中，使其于周围的环境相融合；细节的地方特色和视觉景观能够反映出独特的自然景观、城市及滨河景观的特点，并创造新的，体现地方特色和地方感的标志性建筑；尊重当地独特的风格，在建设中促进适当的软硬材料的使用。

4. 考古、历史和文化资源规划

加强保护与展示文化遗产，优先将文化遗产景观纳入绿色基础设施网络；通过调查来评估历史景观、考古和文物建筑的保护、设计和建议。

5. 绿色网络和娱乐设施规划

在绿色基础设施网络中，为行人、骑自行车者、骑马者及河滨区游客提供具有吸引力且方便安全的路线；新线路的提出应该基于调查和可行性研究，以评估新线路如何更好地与现有的道路网络相结合；改善绿色基础设施网络内的所有主要的道路及交通枢纽的环境质量——包括绿道，公园道，铁路和地区出入口。

五、 总结

1944年，由阿伯克龙比主持制订了《大伦敦规划》，该规划受到“田园城市”理论的深刻影响，其主导思想是把伦敦区域分为内外四层，以分散人口、工业和就业。规划在离伦敦市中心约35英里（约56.3公里）处呈放射线建成8座新城，哈罗新城便是其中一座。英国在二战后的几十年中，共规划建设了三十多座新城镇，而哈罗新城被誉为第一代

Reflect Gibberd's 'landscape-led' approach in the design of new development in and around Harlow by incorporating a strong landscape framework into new developments to ensure that urban form and building design is shaped by, and responds to, the character of the surrounding countryside; Be informed by a detailed local characterisation and visual assessment to reflect distinctive landscape, townscape and riverscape characteristics, and to inform the creation of new viewpoints, landmarks and places of distinction; Promote the use of hard and soft building and landscaping materials that respect locally distinctive styles where appropriate.

4. Archaeological, Historical and Cultural Heritage Principles

Protect, enhance, promote and interpret heritage assets, giving priority to the 'heritage landscapes' identified in this GIP as key components of the Green Infrastructure Network;Be informed by historic landscape, archaeological and built heritage surveys to evaluate conservation priorities, designs and proposals.

5. Access and Recreation Principles

Promote attractive, distinctive, accessible and safe routes for pedestrians, cyclists, horse-riders and river users within the Green Infrastructure Network; Ensure that proposals for new routes are informed by a survey and feasibility study to evaluate how proposed paths integrate with the existing path network and connect the destinations and gateways within the wider Green Infrastructure Network; Improve the environmental quality of all key strategic access links and gateways within the Green Infrastructure Network – including Greenways, Parkways and Railways and at the key gateways to Harlow.

V. Summary

The Greater London Plan of 1944, drawn up by Patrick Abercrombie. Deeply influenced by Garden Cities, its dominant idea is that London area are divided into four layers inside and outside, aiming to disperse population, industry and employment. The Harlow Area was one of the eight new cities which were built about 35 miles from the center of London in the radial pattern. England planned and built more than 30 new towns in the decades after World War II. The Harlow Area was honored as the representative of the first generation new towns. Owing to the famous application of "Neighbourhood unit" in Harlow Area, it still has enlightenment for residential district planning in China, however the green infrastructure planning is highlights in Harlow Area planning as well.

The green infrastructure planning of the Harlow

新城的代表。哈罗新城被我们熟知多源于其“邻里规划理论”的应用，该理论对我国居住区规划至今仍具有启示意义，而本文中介绍的绿色基础设施规划也同样是哈罗新城规划的亮点。

哈罗新城是伦敦的卫星城，其绿色基础设施规划保留和利用了原有的地形和植被条件，采用了与地形相结合的自然曲线，造就了一种绿地与城市交织的宜人环境。绿色基础设施重视保护乡村间野生生物栖息地、公共开放空间和绿色廊道的相互联系，力求构建多功能的绿色基础设施网络。

哈罗新城绿色基础设施采用多项措施来确保其实施的效果，包括：加强并连接野生生物栖息地、乡村游线设计、河道设计、道路交通廊道绿化、开敞空间和游憩设施、住宅开发、工业和商业开发、可持续的城市排水系统和开发区边缘处理等。事实证明，其完善的绿色基础设施和良好的生态环境很好地解决了由于工业化带来的伦敦人口激增问题。

Area--the satellite towns in London, is based on the original terrain and vegetation conditions. Through the combination of natural curve with terrain, then the pleasant environment that grassland is interweaved with city are created. The green infrastructure planning put a high value on the relationships among the conservation of rural wildlife habitats, public open space and Greenway, in order to build multifunctional green infrastructure network.

The green infrastructure of Harlow Area carries out several measures to guarantee the effect, including: enhancement and link of wildlife habitats, design of countryside access routes, riverways design, road corridors greening, open space and recreation facilities, new housing development, industrial and commercial development, sustainable urban drainage system, and edge area of development zone, etc. It has been proved that, the population boom resulting from industrialization in London are solved very well through perfect green infrastructure and a favorable ecological environment.

思考题

1. 哈罗新城绿色基础设施规划主要受到了哪些理念与思潮的影响?
2. 哈罗新城绿色基础设施规划的主要目的是什么?
3. 哈罗新城绿色基础设施规划涉及了哪些方面的内容?
4. 哈罗新城绿色基础设施规划对目前中国的那一类规划具有借鉴意义?
5. 哈罗新城绿色基础设施规划对当代规划有哪些借鉴意义?

Questions

1. What impacts HL is influenced by?
2. What is the main objective of HL?
3. What aspects does the HL operate from?
4. Which variety plan does the HL could Influence in China?
5. What are the use for reference of HL to contemporary plan?

注释/Note

图片来源/Image：本案例图片均引自http://www.harlow.gov.uk/

参考文献/Reference：本案例均引自http://www.harlow.gov.uk/

推荐网站/Website：http://www.harlow.gov.uk/

第三节 环城翡翠项链：亚特兰大新公共领域

Section III The Beltline Emerald Necklace: Atlanta's New Public Realm

一、 项目综述

1. 项目简介

“环城翡翠项链”（图4–3–1）是21世纪的亚特兰大将要发展和繁荣的公共领域框架。通过将拟建及现有的道路相结合，这条长达23英里（约37公里）的环城带将会带给该城每一位居民不同于美国其他任何地方的漫步、跑步、轮滑以及自行车骑行的机会。紧密联结着46个毗邻社区以及新建的3个亚特兰大都市捷运站，通过长达20英里（约32.2公里）的环城轨道交通系统将衔接亚特兰大每一个主要目的地，包括组成这条面积达2 544英亩（约10.3平方公里）的翡翠项链的13个节点公园（4个扩建公园、4个新建公园、5个多用途公园）。

总之，环城翡翠项链将在613英亩（约2.5平方公里）的现状基础上沿环城带新增1 401英亩（约5.7平方公里）公共用地。在这2 014英亩（约8.15平方公里）的公共用地之外，本规划还拟开发另外530英亩（约2.15平方公里）的多用途公共或私有用地，其中的三处将新建亚特兰大都市捷运站。

I . Overview

1. Introduction

The Beltline Emerald Necklace (Fig.4-3-1) is a public realm framework around which 21st Century Atlanta will grow and prosper. By tying together proposed and existing trails, the 23-mile-long Beltline Trail will provide every resident of the city with strolling, jogging, rollerblading, and cycling opportunities unequaled anywhere in the country. By tying together 46 neighborhoods with each other and with three new MARTA stations, the 20-mile-long Beltline Transit system will provide access to every major destination in Atlanta, including the thirteen jewels(4 Expanded Park Jewels, 4 New Park Jewels, 5 Mixed-Use Jewels)that make up the 2,544-acre Emerald Necklace.

Altogether, the Beltline Emerald Necklace will add 1,401 acres of new parkland to 613 acres of parkland currently along the Beltline. In addition to these 2,014 acres of parkland, the plan proposes an additional 530 acres for mixed-use, public/private developments, three of which will grow around new MARTA stations.

The Beltline Emerald Necklace provides Atlanta with an opportunity which far exceeds that of any major American city: to create a city-wide system of parks and transit, to create stronger, more attractive communities, and to actively shape a new and

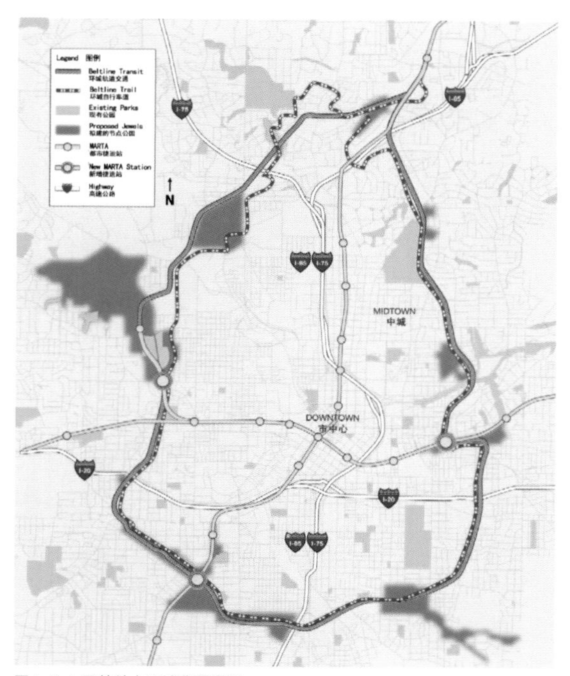

图4-3-1 亚特兰大环城翡翠项链
Fig.4-3-1 The Beltline emerald necklace of Atlanta

环城翡翠项链将带给亚特兰大人远超其他任何美国大城市的机遇：创建一个遍及全市的公园和交通系统，营建更为强大和更具吸引力的社区，并积极地塑造一个崭新的良好的公共领域框架，它将积极地影响后世居民的生活质量。

2. 规划目标

环城绿带将不局限于一系列的旅游目的地，整个系统将会把当前被地形（高差）和地理（不相干的道路网）割裂到环城廊带两侧的社区连接起来。来自各地的人将因各种目的而汇聚到环城带上，同时参与到各种活动中去。通过创建一个能使城市得以发展的新公共领域，将持久又广泛地刺激私有市场。环城带将会成为一个强化邻里社区关系和改善居民生活质量的催化剂。最重要的是，这一变化将沿着整条环城带而发生——而不仅仅局限于某一特定社区或城镇。总之，环城带将会把亚特兰大从一个由高速公路架构起来的城市带向一个由壮观的公共领域支撑起来的城市（图4-3-2）。

规划目标如下：通过对铁路遗产的利用来转变

improved public realm framework that will positively impact residents' quality of life for generations to come.

2. Goals

The Beltline will be more than a series of destinations. The entire system will connect communities that are currently divided both topographically (with different grades) and geographically (with disparate road grids) on either side of the corridor. People will converge on the Beltline from different places for different purposes, simultaneously engaging in different activities. It will stimulate a sustained and widespread private market reaction by creating a new public realm around which the city can grow. The Beltline will become a catalyst for strengthening neighboring communities and improving the daily life of their residents. Most importantly, this change will occur all along the Beltline—not merely in one particular neighborhood or section of town. In short, the Beltline will reorient Atlanta from a city framed by highways to a city framed by a magnificent public realm (Fig.4-3-2).

The goals including: transforming Atlanta via visionary use of our rail legacy; improving quality of life for all residents; connecting neighborhoods with parks, trails, transit and transportation; Ensuring growth across livable neighborhoods; engaging the community in shaping Atlanta's future.

II . Context

1. Transportation

1)Regional Transportation

Atlanta is currently a city of highways. 76% of the city's working population commutes to work by car, whereas only 15% utilizes the MARTA system. The regional use of mass transit is even less, and the Hartsfield-Jackson Atlanta International Airport, the busiest airport in the nation, is not easily accessible via mass transit for the large portion of Atlanta's population that is not within walking distance of a MARTA rail station (Fig.4-3-3).

2)Rail Transportation

In Atlanta, rail traffic peaked immediately after World War II, when over 1,000 trains per day were scheduled to or through the Atlanta rail hub. In the post-war period, industries sought locations where they could operate single-floor production lines and obtain easy highway access. Consequently, both rail lines and the adjacent facilities they serve have been abandoned. This change has enabled creation of the Beltline and has opened the opportunity to develop

亚特兰大；改善所有居民的生活质量；通过公园、自行车道、轨道交通和运输交通连接各社区；确保宜居社区的正常发展；让社区参与塑造亚特兰大的未来。

图4–3–2 环城带鸟瞰模型
Fig.4-3-2 Aerial model of the beltline

二、 项目背景

1. 交通

1) 区域交通

亚特兰大目前是一个高速公路之城。该市76%的工作人口驾驶汽车上班，而仅15%的人使用都市轨道交通系统。公共交通的区域性使用甚至更少，大部分亚特兰大人难以通过公共交通到达美国最繁忙的亚特兰大哈茨菲尔德－杰克逊国际机场，因为亚特兰大捷运站不处于他们的步行范围内（图4–3–3）。

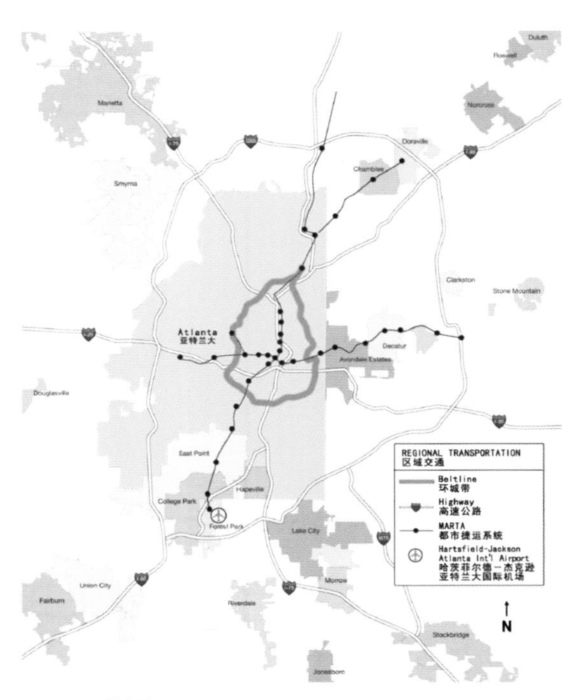

图4–3–3 区域交通
Fig.4-3-3 Regional transportation

2) 铁路交通

二战之后，亚特兰大的铁路交通迅速发展到高峰，当时每天有超过1000列次火车抵达或经过亚特兰大铁路枢纽。战后期间，工业公司开始寻找能够运作地面生产线并拥有便捷的高速公路接入口的基地位置。因此他们遗弃了铁路及其附近的设施。这一转变使得环城带的形成成为可能，并带来了开发沿线一些废弃地的机遇。

3) 自行车道

自行车道基金会有助于亚特兰大自行车道网络的建设。但是，如今亚特兰大市核心区内的自行车道仅限于街道自行车道和斯通山–亚特兰大绿道自行车道，该自行车道始于亚特兰大西部，穿越市中心，经过自由公园道，最后进入迪卡布县。除这条自行车道以外，自行车道基金会的主要成就在亚特兰大市中心以外的郊区已经实现。这些现有的自行车道目前仍破碎间断，但随着环城带的加入，它们将很快成为一个广阔又连续的系统（图4–3–4）。

2. 环城带状况

环城带自身是对隶属于不同单位的铁路通行权

some of the abandoned properties along the route.

3)Bike Paths

The PATH Foundation has been instrumental in building Atlanta's network of bike paths. Today, however, PATH's trails within Atlanta's inner core are limited to on-street bike lanes and the popular Stone Mountain/Atlanta Greenway Trail, which runs from Atlanta's Westside, through Downtown, and out Freedom Parkway into DeKalb County. With the exception of this trail, PATH's major successes have occurred in suburbs outside of Atlanta's core. These

的一个整合，这些单位对其所拥有的土地和轨道进行了不同程度的维护，包括仍然使用的货运线路、闲置的铁路以及用作非法垃圾场的废弃土地（图4–3–5）。

与规划图所显示的相反，环城带并非连续不断的——实际上，铁路线有五处并不相连。其中三个间断点很容易从地图上辨别出来。剩下的两个点看起来似乎是连续的，但一条都市轨道交通线从其中一点穿过，另一点则被一座35英尺（约10.7米）宽的垂直桥梁所隔断（图4–3–6）。

目前，环城带廊道中已宣布或在建的开发工程有接近35个。这些工程将会为亚特兰大房屋储量新增超过6 600个单元——大大增加了该市的人口——并增加了超过110万平方英尺（约10.2公顷）的零售面积和20万平方英尺（约1.86公顷）的办公空间。无论环城带工程能否成为现实，这一地区的发展都将继续，但环城带的加入则将提升发展的质量，提升这些开发区居民的生活质量，以及提升整个亚特兰大市居民的生活质量。

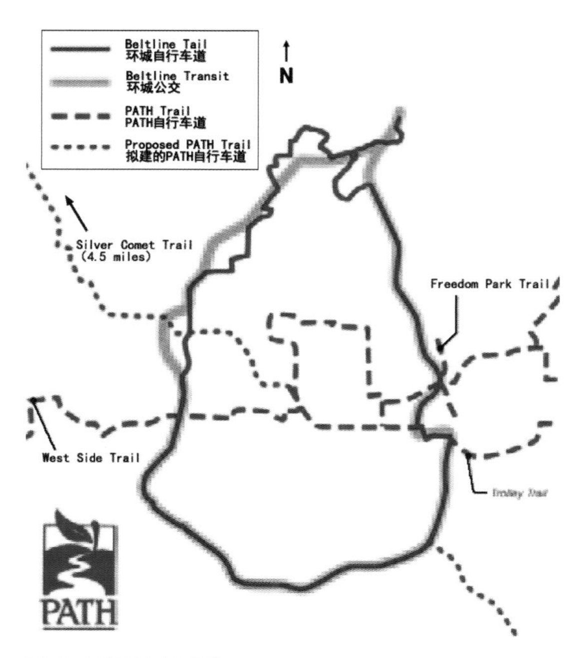

图4–3–4 亚特兰大自行车道
Fig.4–3–4 Bike paths of Atlanta

existing trails are fragments of what will soon become, with the addition of the Beltline, an expansive, continuous system (Fig.4-3-4).

2. Beltline Conditions

The Beltline itself is a compilation of rail rights-of-way that are owned by different parties who have maintained their property and their tracks to varying degrees, from active freight lines to inactive tracks to abandoned property that serves as an illegal garbage dump (Fig.4-3-5).

Contrary to what maps may display, the Beltline is not continuous—indeed, at five points the rail lines simply do not connect. Three discontinuities are easily discernable on a map. The remaining two may look continuous, but MARTA now runs a line through one, and the other is separated by a 35-foot, perpendicular bridge (Fig.4-3-6).

Currently, approximately 35 development projects have been announced or are under construction in the Beltline corridor. These projects will add over 6,600 units to Atlanta's housing stock—further increasing the city's population—and add more than 1.1 million square feet of retail and 200,000 square feet of office space. Development will continue in these areas regardless of whether the Beltline becomes a reality, but the addition of the Beltline will increase the quality of the development, the quality of life for residents in these developments, and the quality of life for residents in Atlanta as a whole.

3. Population and Community

1)Population

The population of metropolitan Atlanta doubled between 1950 and 1980 and has nearly doubled since. Although the city continues to expand, its suburbs are approaching a saturation point; residents are starting to move back into the city to avoid the long commute and traffic delays en route to jobs in Downtown and Midtown Atlanta, the city's most dense business districts. Now that residents are moving back into the City of Atlanta, a new demographic trend is emerging. The new residents moving into city limits are more likely than previous city residents to be childless and seeking apartments, condos, or other high-density housing. The Beltline corridor is positioned to meet this market demand. While census tracts that run along the Beltline corridor (roughly a half-mile in each direction) declined in population until 1990, this corridor has since increased in population at a faster rate than the rest of the city.

2)Community Planning

The Beltline engages 46 of Atlanta's historic neighborhoods, 15 of the city's 24 Neighborhood Planning Units (NPU's) and 11 of the 12 city council

3. 人口及社区状况

1）人口

亚特兰大市的人口在1950年到1980年间翻了一番，后来几乎又翻了一倍。虽然该市一直在扩张，其郊区人口却正接近饱和；人们开始搬回市区以缩短前往市中心和环中心区（该市最密集的商业区）的上班路线和避免交通阻塞。如今，人口开始回迁亚特兰大市区，一个新的人口流动趋势开始出现。与以前搬回市区的居民相比，新搬回市区的居民很可能是无子女并欲寻求公寓或其他高密度住宅的人，环城带将满足这一市场需求。由于沿环城带的人口普查区（两侧大概各0.5英里[约805米]）的人口在1990年之前一直减少，这条廊道上的人口后来以比该市其他地区更快的速度增长。

districts within a quarter mile of either side (Fig.4-3-7). The historic freight railroads that make up the Beltline generally predate the adjacent neighborhoods. As the city expanded, the railroads became barriers between communities of different socio-economic backgrounds. The Beltline provides a unique opportunity to knit these communities together. Trading an unsightly and dangerous industrial rail line for a well-maintained park, these neighborhoods will reorient themselves toward a shared common space, greatly strengthening community relationships.

4. Schools and Parks

1)Schools

There are 14 schools (7 elementary, 3 middle, and 4 high schools) within a half-mile of the Beltline. The Beltline will create transportation opportunities and additional open space for the 8,537 students, 925 teachers, and scores of community groups who share

图4–3–5 环城带状况
Fig.4-3-5 Beltline conditions1-An actively used freight line along the Beltline

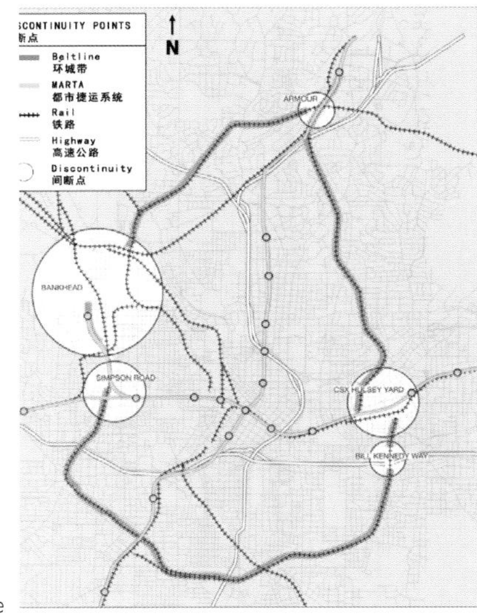

图4–3–6 环城带间断点
Fig.4-3-6 Discontinuity points of the Beltline

2）社区规划

环城带关系到46个亚特兰大历史社区、24个社区规划单元中的15个单元以及12个环城带两侧0.25英里以内的议会区中的11个地区（图4-3-7）。组成环城带的历史货运铁路普遍建于其毗邻社区之前。随着城市的扩张，这些铁路成为了不同社会经济背景社区之间的障碍。环城带带来了一个联系这些社区的绝佳机会，将一条难看又危险的工业铁路转变为一个维护良好的公园，这些社区将重新转变为一个公共空间，从而极大地增进社区关系。

4. 学校及公园

1）学校

距环城带0.5英里（约805米）以内有14所学校（7所小学、3所初中和4所高中）。环城带将为这些学校中的8 537位学生、925位教师以及无数每天共享这些设施的社区团体创造交通条件并增加开放空间（图4-3-8）。

2）公园和游憩设施

亚特兰大市拥有345座公园和180个运动场（3 235英亩），其中的82座公园和32个运动场（988英亩）位于距环城带0.5英里的范围内，11座公园和6个运动场（613英亩）直接与环城带毗连（图4-3-9）。

these facilities on a daily basis (Fig. 4-3-8).

2)Parks & Recreation Facilities

The City of Atlanta has 345 parks and 180 playgrounds (3,235 acres), of which 82 parks and 32 playgrounds (988 acres) are within a half mile of the Beltline and 11 parks and 6 playgrounds (613 acres) are directly adjacent to the Beltline (Fig.4-3-9).

III . Planning and designing

1. Creating a Continuous Beltline

1)A continuous Beltline is Essential

An uninterrupted, Beltline Trail and Transit loop will provide an easy orientation for anyone traveling to one of the Beltline jewels, or connecting with the MARTA rail system. It will also change neighborhoods on both sides of the right-of-way. For the first time, people living on one side of the Beltline will be able to arrange a ballgame, ride the Beltline, or walk to dinner at friends' homes on the other side. The Beltline will bring together residents of long-separated neighborhoods to share activities that, in many instances, have been prohibitively inaccessible. At the same time, residents of neighborhoods with relatively few recreation opportunities will gain access to parks and facilities throughout the city by using the Beltline.

A continuous Beltline provides the opportunity to unite its mass transit with a completely expanded and connected park system. A connected park system that unites a city's great parks is something most cities can only dream about. When combined with the transit and development opportunities, the

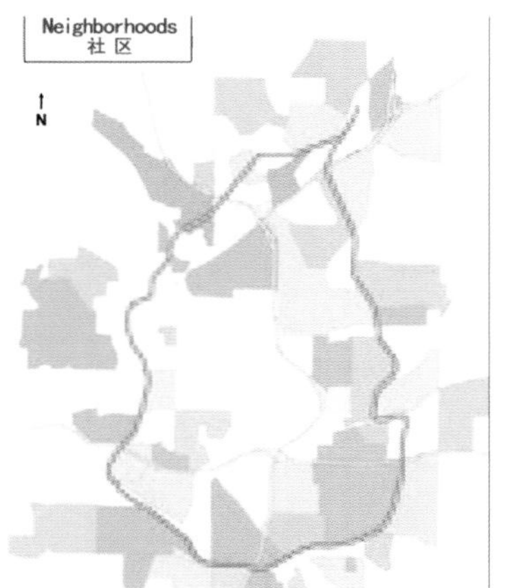

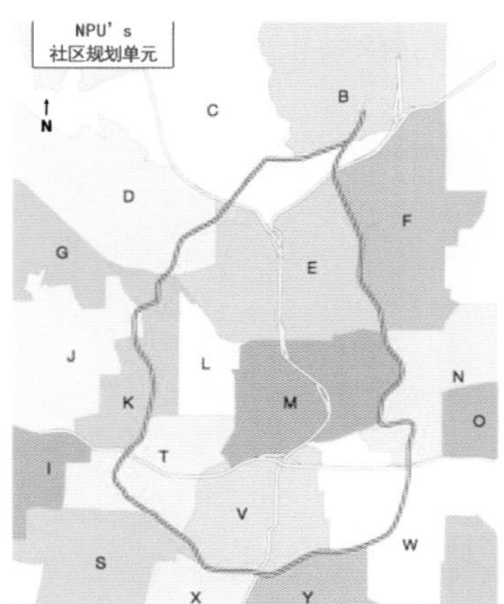

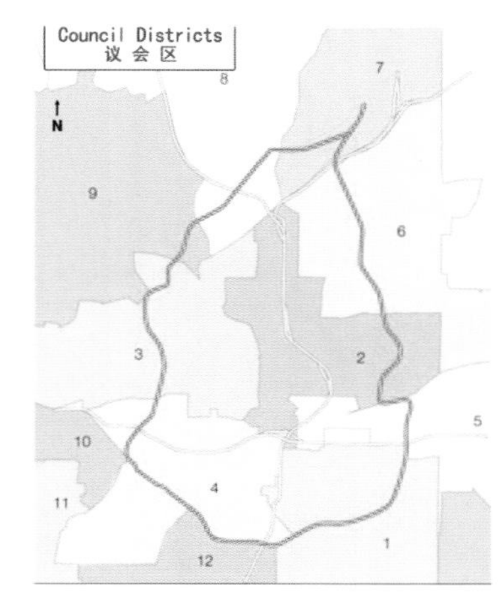

图4-3-7 沿环城带分布的社区、社区规划单元和议会区
Fig.4-3-7 Neighborhoods, Neighborhood Planning Units, Council districts along the Beltline

4-3-9）。

三、 规划设计介绍

1. 创造一条连续的环城带

1) 一条连续环城带的必要性

一条不间断的环城自行车道和轨道交通环线，将给任何一位想去某个节点公园游玩或去往捷运系统的人一个简明的导向，也将改变道路通行权两侧的社区。环城带两侧的居民将首次能够组织一场球赛、沿环城带骑行、或步行去环城带另一侧的朋友家里吃饭。环城带会将长期分隔的邻里社区聚集在一起共享一些许多情况下难以实现的活动。同时，通过环城带，拥有相对较少游憩机会的社区居民将在全市范围内获得更多的公园和公共设施可达性。

连续的环城带为联合公共交通的机遇带来了一个扩张彻底并紧密联系的公园系统。一个联合了全市各大公园的公园系统是大多数城市梦寐以求的事

continuous Beltline provides a 21st century public realm framework around which Atlanta will grow for the next 100 years. To realize the goal, this report recommends splitting development into two phases: the Beltline Trail and the Beltline Transit(Fig.4-3-10).

2)Phase I: The Beltline Trail

There are three interrelated reasons for creating the multi-use trail first: constituency building, timing, and funding. The momentum and public support currently behind the Beltline will not last if implementation does not begin within the next year. If there is no visible progress, support will evaporate. The Beltline Transit will take at least three years before construction can begin, five years before any section opens, and as many as 10 to 15 years before the transit loop is complete. The Beltline Trail, on the other hand, can begin construction in sections within a year's time and can be completed before the first section of the Beltline Transit is actively running. This action is critical to building a stronger and more widespread constituency for the Beltline (Fig.4-3-11& Fig.4-3-12).

Although the Beltline Trail must come first, its construction must not prevent the subsequent introduction of the Beltline Transit. Most importantly,

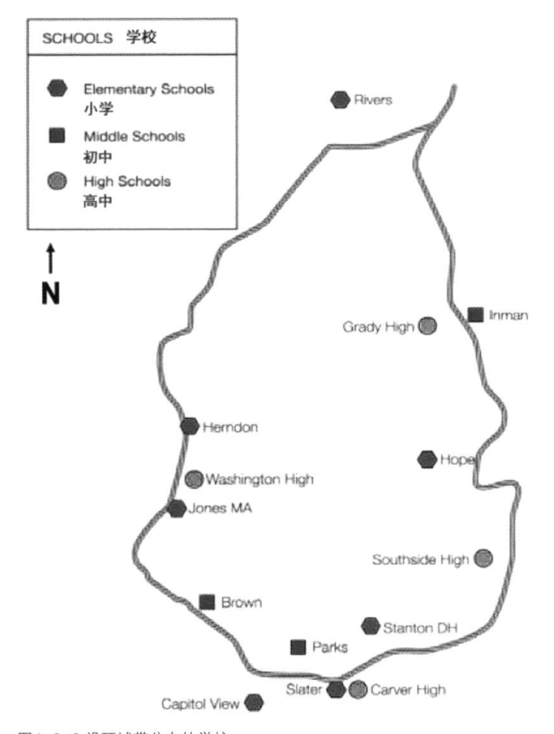

图4-3-8 沿环城带分布的学校
Fig.4-3-8 Schools along the Beltline

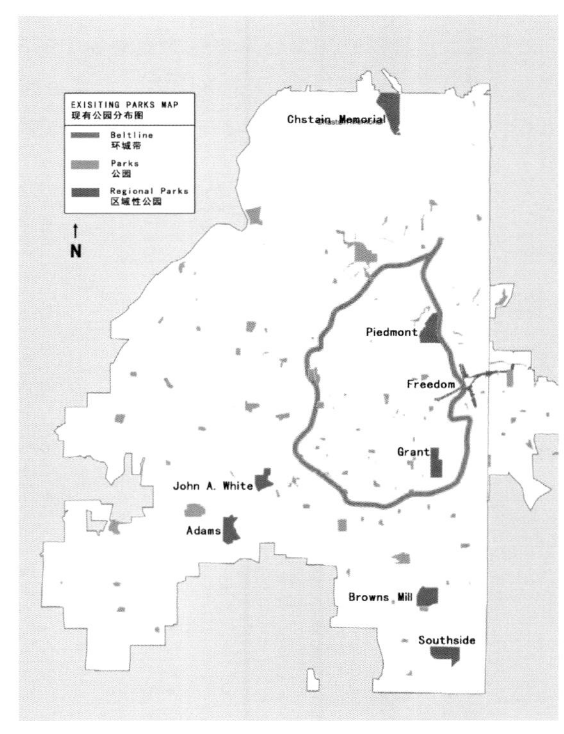

图4-3-9 亚特兰大现有公园
Fig.4-3-9 Parks of Atlanta

情。一旦与交通和开发机遇结合起来，这条连续的环城带能提供一个21世纪的公共领域框架，亚特兰大将在以后100年中围绕这一框架发展壮大。为了实现这一目标，本报告建议将工程划分为两期：环城自行车道和环城轨道交通（图4–3–10）。

2）I期：环城自行车道

之所以先建设多功能自行车道有三个相互关联的原因：支持者的争取、时间和资金。如果明年内还没有开始实施，环城带目前的势头及其背后的公众支持将不再持续。如果没有明显的进展，这些支持也会消失。环城轨道交通在开始施工建设之前至少还将花费3年时间，任意路段投入运行之前则还需5年，整个交通环线的完成则需要10到15年之多。另一方面，环城自行车道却可以在1年的时间内开始分段建设，而且可以在环城轨道交通的第一条路段投入运行之前完全竣工。这一行动对建立一个对环城带更为有力和广泛的支持来说显得至关重要（图

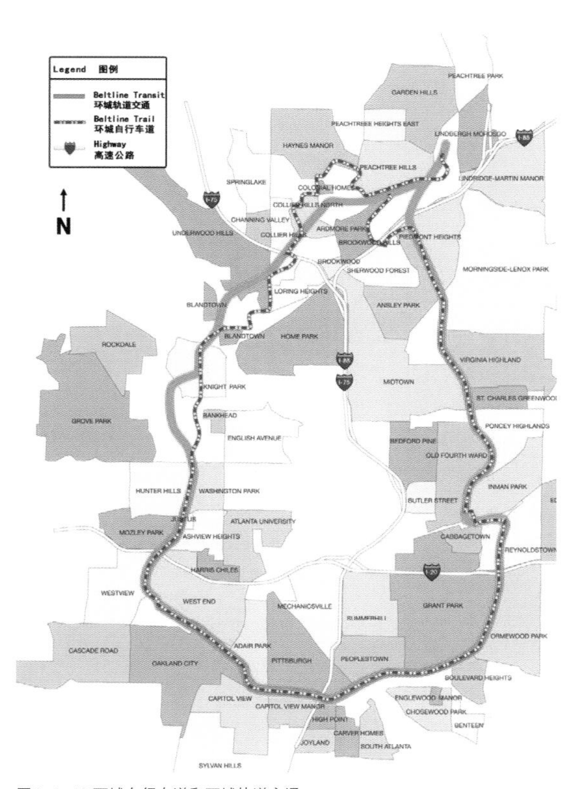

图4–3–10 环城自行车道和环城轨道交通
Fig.4-3-10 The Beltline Trail and the Beltline Transit

the land for the transit must be reserved, where feasible, for the future transit, and users must be informed that a transit line is planned for the future.

3)Phase II: The Beltline Transit

Atlanta, like virtually every American city, has been transformed by interstate highways. Everybody hoped that MARTA would play a significant role in reversing the trend toward automobile dependency. MARTA did accelerate the growth of high-density subcenters in Midtown and Buckhead, but its impact has yet to be felt in most of the city's residential neighborhoods. The Beltline will change that. A single, continuous transit loop will connect Atlanta's inner neighborhoods with one another and with MARTA. In doing so, the Beltline will provide a mass transit connection to the airport and with employment and shopping opportunities in Downtown, Midtown, Buckhead, and throughout the city (Fig.4-3-13& Fig.4-3-14).

4)New MARTA Stations

In order for the Beltline Transit to successfully transform daily life in Atlanta, it must help MARTA fulfill its potential as an effective transportation alternative for all residents. The existing MARTA stations are not sufficient. This report recommends constructing three new MARTA stations at key points along the Beltline so as to secure the Beltline Transit and MARTA as a viable transportation network. These locations are Simpson Road, Murphy Crossing and Hulsey Yard (Fig. 4-3-15).

2. Creating the Beltline Jewels

1)Opportunities to Create Jewels

The goal of adding hundreds of acres of new parkland is within the grasp of many American cities, but none so notably as Atlanta. The reason is that cities are continually changing. As one land use disappears, property becomes available for other activities. This is especially true of warehousing and manufacturing, which during the 19th and early 20th centuries depended on shipping by rail. By the 1930s, many companies were beginning to ship by truck. As a result of the creation of the interstate highway system during the second part of the 20th century, most companies shifted freight operations to truck, eliminating the advantage of a location along a railroad.

The vacant buildings that spread across the urban landscape at first appeared to be a major problem. In fact, they were an opportunity. Almost as quickly as factories, lofts, and warehouses were abandoned by their owners, inventive entrepreneurs began to experiment with adaptive reuse of these buildings. During the 1960's Ghirardelli Square in San Francisco demonstrated the possibilities of conversion to retail shopping. At the same time, artists in SoHo in

4-3-11和图4-3-12）。

虽然环城自行车道必须先行建设，但却不能妨碍随后的环城轨道交通建设。最重要的是，在可行的地方，用于环城轨道交通建设的土地必须保留，而使用者必须再规划一条轨道交通线路。

3）2期：环城轨道交通

亚特兰大几乎和每一个美国城市一样，都因州际高速公路而发生了转变。每个人都希望亚特兰大都市捷运系统能够扮演扭转汽车依赖趋势的角色。都市捷运系统确实加速了中城和巴克海特高密度的次中心区的成长，但其影响仍未被大多数居住社区感受到。环城带将改变这一事实。一条单程的连续轨道交通环线将把亚特兰大内部社区相互连接起来，并将它们与都市捷运系统相连。一旦如此，环城带将提供连接机场的公共交通，并为市中心区、中城、巴克海特及整座城市提供就业和购物机会（图4-3-13和图4-3-14）。

4）新增亚特兰大都市捷运站

为了使环城轨道交通能成功转变亚特兰大的日常生活，必须促使都市捷运系统拥有使其成为一项对所有居民来说都实用的交通选择的潜力。现有的捷运站数量不足，本报告建议在环城带的关键节点新建三个捷运站，以确保环城公共交通和都市捷运系统是一个可行的交通网络。这三处位置便是辛普森路、墨菲交叉口和哈勒西货栈（图4-3-15）。

2. 兴建环城带节点公园

1）兴建节点公园的机遇

许多美国城市都有增加数百英亩公共用地的目标，但却没一个城市有亚特兰大这般显著。原因在于，城市是持续变化的。随着一种土地利用的消失，土地又可以为其他活动所利用。这对仓储和制造来说显得尤为真切，它们在19世纪和20世纪早期之间主要就是依靠铁路运输。到20世纪30年代，许多公司开始用卡车运输。20世纪下半叶，州际高速公路系统的建设，导致大部分公司转用卡车运输货物，削弱了铁路沿线的位置优势。

New York City pioneered the conversion of lofts into apartment buildings.

Atlanta is one of the few American cities that developed around railroads, not ports. For this reason, it is uniquely situated to develop its derelict railyards and tracks into public parks or recreation destinations. The nation's rail system is now less than half its 1916size, and decreasing. More than 2,000 miles are abandoned annually. Passenger service is a fraction of what it was prior to the creation of the interstate highway system. Railroad companies everywhere have deferred maintenance and concentrated traffic on a few main lines, making available for reuse enormous amounts of land.

The thirteen park jewels proposed in this report are the result of Atlanta's extraordinary good fortune. Four are expansions of current parks, four are entirely new parks, and another five are mixed-use opportunities that will combine open space with the development of new communities. All together, the Beltline Jewels will combine with the Beltline Trail and right-of-way to create the 2,544-acre Beltline Emerald Necklace (Fig.4-3-16).

2)Utilization and Development Pattern

When viewed collectively, the Beltline Jewels represent a wide range of opportunities. They involve reuse of abandoned property at Simpson Road and Murphy Crossing; acquisition of railroad rights-of-way owned by CSX, Norfolk & Southern, and the Georgia

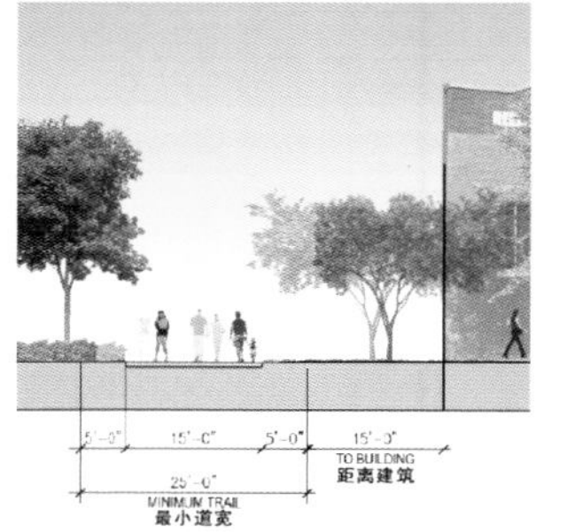

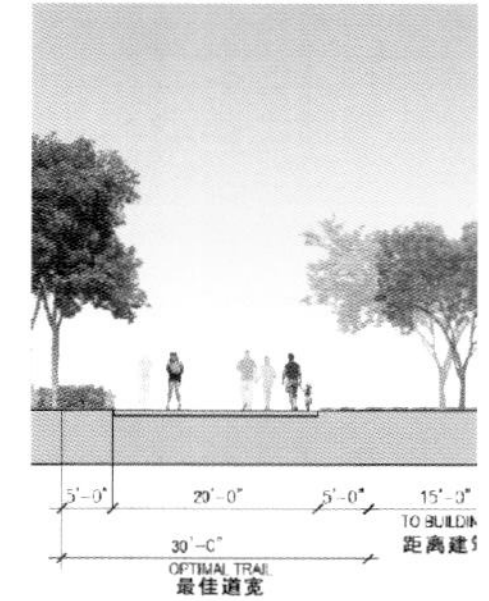

图4-3-11 与住宅区相邻的自行车道设计（最小与最佳宽度）
Fig.4-3-11 Beltline Trail (minimum &optimal width) adjacent to development

图4-3-12 与公园相邻的自行车道设计
Fig.4-3-12 Beltline trail adjacent to park

最初，散布在城市各处的空置建筑表现为一个主要问题。实际上，它们是一个机遇。几乎在工厂、高层建筑和仓库被其所有者遗弃的同时，别出心裁的企业家便开始尝试对他们的建筑进行适应性的再利用。在20世纪60年代期间，旧金山的巧克力广场证明了转变为零售店的可能性。同时，纽约艺术家索霍成为了将高层建筑转变为公寓的先锋。

亚特兰大是美国少数几个围绕铁路而非港口发展起来的城市之一。因此，将其废弃的铁路站和轨道开发为公园或游憩目的地正是绝佳的途径。美国现在的铁路系统比其在1916年的规模还小，并还在逐年减小。每年有超过2 000英里（约3 218.7公里）被废弃。由于州际高速公路系统的兴建，铁路客运服务只占到以前的一小部分。各地的铁路公司推迟铁路维护并将交通集中在少数几条主要路线上，使得大量土地可以再利用。

本报告中的13座节点公园得益于亚特兰大非同寻常的时运。其中4个是对现有公园的扩建，另外4个是完全新建的公园，而其他5个则是将开放空间与新社区结合在一起的多用途地区。总之，环城带节点公园将结合环城自行车道和道路通行权以创建这条2544英亩的环城翡翠项链（图4–3–16）。

2) 利用与开发模式

总的来看，环城带节点公园代表着广泛的机遇。其中包括对辛普森路和墨菲交叉口废弃地的再利用；对CSX、Norfolk & Southern和乔治亚州交通部所有的铁路通行权的收购；对CSX哈勒西货栈（图4–3–17）和富尔顿县采石场（图4–3–18）的适应性再利用；对将继续属于亚特兰大市自来水厂、州有河床、乔治亚电力公司输电路线和高尔夫球场的土地的联合利用；以及对阿尔德莫公园、墨菲交叉口、堑壕河、北大街和皮德蒙特公园毗邻区的政府土地的提升利用。

环城带节点公园将创造一条总面积达2544英亩的环城翡翠项链，其中包括613英亩现有开放空间与超过1400英亩的新建开放空间直接相连，以及超过500英亩的综合开发区。亚特兰大几乎一半的公共用地将直接沿环城带相连，亚特兰大将在以后100年中

Department of Transportation; adaptive reuse of the CSX Hulsey freight yard (Fig.4-3-17) and Fulton County quarry(Fig.4-3-18); joint use of properties that will continue to be part of the City of Atlanta's

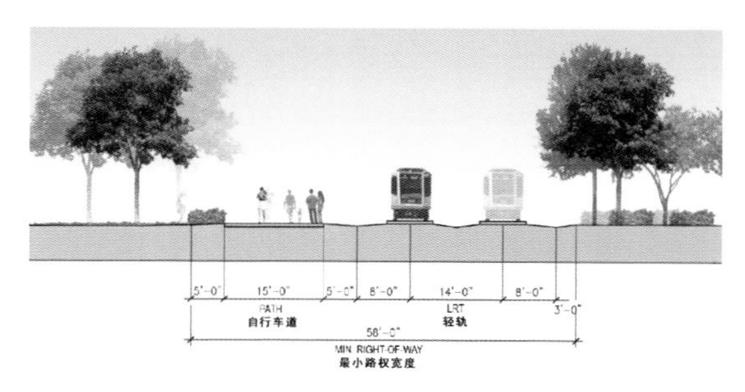

图4–3–13 自行车道与轨道交通相邻
Fig.4-3-13 Beltline Trail Adjacent to Transit

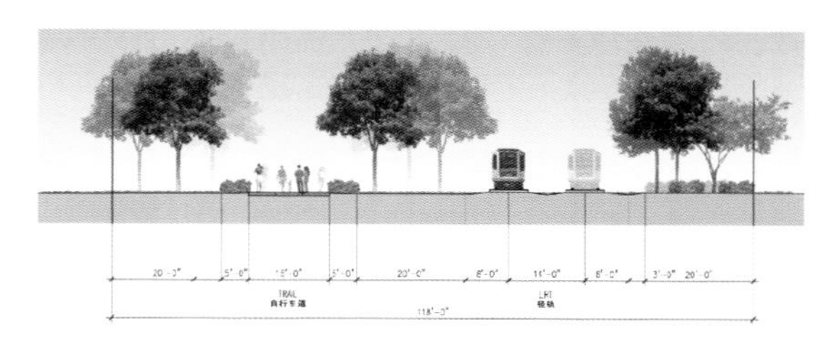

图4–3–14 最佳环城带轨道交通截面设计
Fig.4-3-14 Optimal Beltline Trail Section

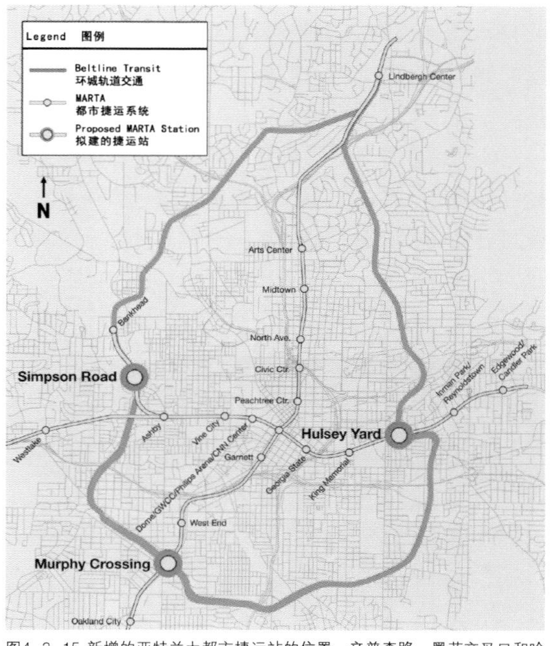

图4–3–15 新增的亚特兰大都市捷运站的位置：辛普森路、墨菲交叉口和哈勒西货栈
Fig.4-3-15 Location of the new MARTA stations Simpson Road, Murphy Crossing and HulseyYard

围绕这条连续的公共领域框架发展壮大。

3. 行动方案

大部分准备购置的土地目前都处于空置、未充分利用或可以再利用的状态。对环城自行车道所需的道路通行权的购买可通过环城公共轨道的资金来补贴。此外，环城带周围潜在的非同寻常的私有市场暗示着亚特兰大崭新的21世纪公共领域前途一片光明。

1）后续计划

对环城带来说，第一步也是至关重要的一步便是全力推进亚特兰大人的新公共领域这一愿景。这一愿景将使整个城市而不仅是其沿线的社区受益。因而环城带必须始终作为一个单项工程来考虑——一个完整的交通系统和一个完整的开放空间系统——将在以后100年中给亚特兰大市带来积极影响。

亚特兰大市的上述转变将需要以下步骤：发起对主要土地所有者和政府机构的协商洽谈；为早期土地购置、规划、施工、设计、成本预算、法律事务以及公共参与募集私人资金；启动公共规划设计程序以创造节点公园雏形；建立环城带税收分配区；建立环城公园管理局；建立环城带社区改善区。

公共用地基金会已经领头开始了环城翡翠项链的初步规划。它应当继续扮演这一角色直到税收分配区开始运作，从而推动实施进程。此外，公共用

图4-3-16 马多克斯公园扩建前和扩建后
Fig.4-3-16 Maddox Park Before and After

Waterworks, state-owned creek beds, Georgia Power Company electric lines, and golf courses; and enhanced use of government-owned property at Ardmore Park, Murphy Crossing, Intrenchment Creek, North Avenue, and adjacent to Piedmont Park.

Altogether the Beltline Jewels will create a 2,544-acre Beltline Emerald Necklace which includes 613 acres of existing open space directly connected to over 1,400 acres of new open space and over 500 acres of new mixed-use development. Nearly half of Atlanta's parkland will be directly connected along the Beltline, a continuous public realm framework around which the city will grow for the next 100 years.

3. Action Plan

Most of the land slated for acquisition is currently vacant, underutilized, or can be made available for reuse. The acquisition of the rights-of-way necessary for the Beltline Trail can be reimbursed by funding for the Beltline Transit. In addition, the extraordinary private market potential surrounding the Beltline suggests that the future is bright for Atlanta's new 21st century public realm.

1)The Next Steps

The first and most critical step for the Beltline is to promote the vision of a new public realm for Atlanta. This vision benefits the entire city, not just the neighborhoods along its path. The Beltline, therefore, must always be considered as a single project—an entire transit system and an entire open space system—that positively impacts the City of Atlanta for the next 100 years.

Transforming the City of Atlanta as outlined above will require the following steps: Initiate the necessary negotiations with major property owners and political constituents; raise private funds for early site acquisition, planning, engineering, design, cost estimating, legal work, and public participation; begin public planning and design process to create initial jewels; establish a beltline tax allocation district; establish a beltline park governance; establish a beltline community improvement district

The Trust for Public Land (TPL) has taken the lead in initial planning for the Beltline Emerald Necklace. It should continue to play this role until the Tax Allocation District (TAD) is in operation, thereby jumpstarting the implementation process. In addition, TPL should continue to promote initial Beltline Jewels in all sections of the city. Full implementation, however, necessitates an independent entity that is specifically charged with creating and managing the Beltline Emerald Necklace. This report recommends creating such an agency and a Community Improvement District to raise operating revenues.

地基金会应当继续推进全市各地区的环城带节点公园建设。但是，行动的全面落实需要一个具体负责建造和管理环城翡翠项链的独立实体。本报告提议创建这样一个代理机构和一个社区改善区来筹集经营收入。

2）早期行动

最早的土地购置应当由公用土地基金会和PATH基金会与亚特兰大市紧密合作来牵头。这种公私合作关系应开始于以下起步的初步规划、设计和建设：环城自行车道的创建；恩诺特、马多克斯和阿尔德莫公园的扩建和提升；霍兹克劳公园的兴建；哈勒西货栈和林荫大道交叉口的多用途开放空间的规划及开发土地购置应当集中在环城带和那三个（恩诺特、阿尔德莫和霍兹克劳公园）能在最短时间内完全投入公共使用的开放空间节点上。与此同时，皮德蒙特公园的扩建和市政厅东北大街公园的新建也在进行中。最后，哈勒西货栈、林荫大道交叉口和马多克斯公园的扩建规划也应开始。这将发起必要的早期行动以允许税收分配区投入运营。

3）环城公园的管理

没有一个合法实体做支撑，环城带将仅仅是一个不切实际的概念。因此，亚特兰大市应当考虑建立一个独立的非营利性机构，以创造、维护和运营环城翡翠项链为唯一目的。这个组织中应当包括主要的股东、公园和开放空间拥护者的代表、房地产公司、非

2)Early Action

The earliest property acquisitions should be spearheaded by TPL and the PATH Foundation in close coordination with the city. This public-private partnership should begin the initial planning, design, and engineering for the following initial steps: creation of the beltline trail; expansion and enhancement of Enota, Maddox, and Ardmore parks; creation of Holtzclaw park; planning for mixed-use open space and development at Hulsey yard and Boulevard crossing.

The acquisitions should focus on the Beltline Trail and the three open space jewels that can be completed for public use in the shortest possible time: Enota, Ardmore, and Holtzclaw Parks. Meanwhile, the expansion of Piedmont Park and the creation of North Avenue Park at City Hall East will already be underway. Finally, planning should begin on Maddox Park expansion, Hulsey Yard, and Boulevard Crossing. This would initiate the early action necessary to allow the TAD to hit the ground running.

3)The Beltline Park Governance

Without grounding itself to a legal entity, the Beltline will remain nothing more than a visionary idea. Therefore, Atlanta should consider creating an independent, non-profit institution for the sole purpose of creating, maintaining, and operating the Beltline Emerald Necklace. This organization should include the key stake-holders, including representatives of park and open space advocates, the real estate industry, the non-profit sector, the design community, and the neighborhoods adjoining the Beltline.

IV. Summary

Atlanta's opportunity lies in the Beltline. It is an

图4-3-17 CSX 哈勒西货栈
Fig.4-3-17 Hulsey Yard

图4-3-18 富尔顿县采石场
Fig.4-3-18 Fulton County quarry

营利性部门、设计团队以及环城带的毗邻社区。

四、总结

亚特兰大的机遇在于环城带，这在建设完备的城市中是鲜有如此机会的，因为在这些城市中要建造一座大型公园或任何公共领域的附属实体通常需要迁移无数的土地所有者。谓为奇迹的是，亚特兰大存在这样的空白：一条铁路及其权限带环绕该市内部社区而过，距市中心仅2英里（约3.2公里）。如果抓住这个机遇，亚特兰大则可以沿一条长达23英里的环城自行车道创造一条2 544英亩（约10.3平方公里）的“翡翠项链”——同时在此过程中完成其自身的转变。

毫无疑问，亚特兰大拥有一个远胜其他任何美国大城市的机遇：创造更强大、更具吸引力的社区并积极地塑造一个崭新的良好的公共领域框架，它能积极地影响后世居民的生活质量。公共土地基金会主席威尔·罗杰斯曾写道：“伟大的城市因伟大的公园而知名”。亚特兰大是一个伟大的美国城市，而环城带将成为亚特兰大应有的伟大公园。

环城绿带作为城市森林的一部分，对维持城市生态系统的平衡、改善城市人居环境有着重要的作用。我国也有众多城市重视环城绿带的建设，如上海市环城绿带、南京明城墙环城绿带、西安城墙环城绿带等，此类环城绿带多依托道路、护城河或城墙等已有环状基础设施而建，多以市中心为圆心呈环状围绕，景观共享性极高。23英里（约37公里）长的“环城翡翠项链”是亚特兰大市环城绿带响亮的称号，结合废弃的铁路线而建，绿带串联起众多居住社区、公园和重要的交通设施站点，创建了一个遍布全市的公园的交通系统，探索了城市新公共领域框架设计的理论。

这条铁路线是工业时代留下的遗产，产业的转型和城市交通格局的转变使这条曾经繁忙的铁路线成为废弃地。结合这条铁路线建设绿带，为环城自行车道的建设提供了空间，使环城带两侧的居民可以通过骑行或步行到达目的地。环城带还为环城轨道交通的建设带来机遇，转变了亚特兰大居民的日

opportunity that rarely occurs in already built-up cities in which the creation of a great park or any substantial addition to the public realm often requires displacing a myriad of property owners. Miraculously, a vacancy exists in Atlanta: a belt of railroad rights-of-way that circle through the city's inner neighborhoods, two miles from downtown. By seizing on this gift, Atlanta can, if it chooses, create an "Emerald Necklace" of 2,544 acres along a 23-mile Beltline Trail—and recreate itself in the process.

Without a doubt, Atlanta has an opportunity which far exceeds that of any major American city: to create stronger, more attractive communities and to actively shape a new and improved public realm framework that will positively affect residents' quality of life for generations to come. Will Rogers, President of the Trust for Public Land, has written: "Great cities are known for their great parks." Atlanta is a great American city, and the Beltline will become the great park Atlanta deserves.

As a part of the urban forest, the Beltline plays an important role in maintaining the balance of the urban ecosystem and improving the urban human settlement environment. There are many cities that attach great importance to the construction of the greenbelt in China, such as Shanghai greenbelt, Nanjing Ming dynasty city wall greenbelt, Xi'an city wall greenbelt, etc. Most of these greenbelts are constructed on the base of annular infrastructures such as roads, moats and city walls, and most of them take the central city as the center of the annular greenbelt, of which the landscape shared highly. The 23 miles Beltline Emerald Necklace is a resounding label of the greenbelt of the city of Atlanta; it will be built in the combination with the disused railway line, connect numerous residential communities, parks and important transport facility sites, create a transport system across the city's parks, and explore the theory of design the framework of the city's new public realm.

This railway line is the legacy of the industrial age, and the transformation of industry and the changes of urban traffic pattern make it abandoned which used to be a busy railway. Combined its construction with the railway line, the greenbelt provides space for the construction of the beltline trail, and makes it accessible for the residents on both sides of the Beltline to reach their destinations by riding or walking. It also brings opportunity for the construction of the beltline transit, changes the daily travel patterns of residents of Atlanta. At the same time, the construction of beltline jewels has been put on the agenda. These visible changes bring a new look to the urban landscape, and the more important potential role of it, is to promote the realization of vision of Atlanta's new public realm. Environmental

常出行模式。同时，环带的规划也将重要节点公园的建设提上日程。这些看得见的改变使城市景观面貌焕然一新，而其潜在的更重要的作用，是推进亚特兰大人的新公共领域这一愿景。沿整条环带发生的环境更新将蔓延至整个城市，成为亚特兰大超越美国其他城市的助推器。

renewal along the Beltline will spread to the entire city of Atlanta, and become the boosters of Atlanta to surpass the other cities of United States.

思考题

1.你认为本案例最显著的特色是什么?它对我们有何启示?

2.本规划文本篇首题“谨以此报告献给弗雷德里克•劳•奥姆斯特德”以缅怀和致敬美国景观规划设计之父奥姆斯特德；本规划取名“翡翠项链”，与奥姆斯特德规划设计的波士顿“翡翠项链”遥相呼应，试研究比较两者的异同?

3.环城翡翠项链的连续性是如何实现的，体现在哪些方面?

4.节点公园于环城带的作用和价值何在?

5.环城带被定位为“亚特兰大新公共领域”，其公共性呈现在哪些方面?

Questions

1. What do you think is the most significant features of this case? And what kind of enlightenment it brings to us?

2. The sentence “This report is dedicated to Frederick Law Olmsted” is written on the first page of the Report of this plan to recall and salute Olmsted, the father of American landscape architecture; named “The Beltline Emerald Necklace”, this plan echoes the “Emerald Necklace” of Boston planned by Olmsted, make a study and try to compare the similarities and differences of them?

3. How and where does this plan achieve the continuity of the Beltline Emerald Necklace?

4. What are the function and value of the beltline jewels to the Beltline?

5. The Beltline is planned to be the new public realm of Atlanta, so how does it present its publicity?

注释/Note

图片来源/Image：本案例及图片均引自《亚特兰大市环城翡翠项链规划文本》/The Beltline Emerald Necklace: Atlanta's New Public Realm. (Prepared for The Trust for Public Land by Alex Garvin & Associates, Inc. December 15, 2004)

推荐网站/Website：http://beltline.org/ ；http://www.atlantaga.gov/index.aspx?page=383 ；https://pathfoundation.org/

第五章　生物多样性规划设计

CHAPTER V BIO-DIVERSITY PLANNING & DESIGN

第一节　梅诺莫尼河谷项目 Section I Menomonee River Valley Project

第二节　甘尼特报业集团总部景观设计 Section II Landscape Design for Gannett

第三节　克罗斯温茨湿地 Section III Crosswinds Marsh

第四节　乌兰野生动物园 Section IV Woodland Park Zoo

第一节 梅诺莫尼河谷项目

Section I Menomonee River Valley Project

一、 项目综述

密尔沃基市的梅诺莫尼河谷是一处比较早的聚集地和商业中心，历史上该地区以肥沃的土地和丰富的野生动物资源而著称，19世纪中叶，河谷成为一处重要的商业区。20世纪河谷成为美国一处重要的重工业区。20世纪20年代，这里成为美国第三大铁路和汽车制造业基地。一个世纪的工业发展，河谷从一个环境优美生产力发达的地区变成了一个工业废弃建筑遍布其间的污染区。1998年为了实现河谷地区更新，密尔沃基市确定了包括密尔沃基市商场旧址四个优先更新区域。1999年，一个多学科专家团队为河谷的可持续发展提供了一个框架，提出了一系列的指导原则。

场地由梅诺莫尼河南北两岸组成，北部区域包括密尔沃基路铁路汽车工厂（商场旧址），南面为航线园，大量的工业构筑物和交通网络仍然存在于场地内，一条高架桥横跨场地南北，场地西部还遗留下来两个烟囱。高架桥和烟囱是表明了场地过去的工业文明和特征。针对场地现状以及城市更新的总体目标，规划人员确定了场地规划目标（图5–1–1）：通过建立一个工业园区来吸引商业，这些商业有利于强化场地与密尔沃基市劳动力市场、交通主干道及中心商业区之间的联系，保留能够为当地居民提供就业机会的工作，鼓励雇主提供新的工作岗位；制定一个可预

I. Project Introduction

Milwaukee's Menomonee River Valley served as an early gathering place and commercial center, which was prized for its variety of wildlife and agricultural fertility. It became an important commercial area in the middle of the 19th century and a heavy industrial district in the 20th century. In the 1920's, the site comprised the third largest railroad and car complex in the United States. With the industrial development for a century, the land went from being a highly productive setting with an excellent natural environment to having large expanses of contamination, decay and abandoned industrial buildings. In 1998, in order to redevelop the Valley, the City identifies four high-priority areas for redevelopment and includes the former Milwaukee Road 'Shops'. In 1999, a team including individuals from various disciplines creates a framework for sustainable development for the land and a series of guiding principles.

The site is composed of two sites on either side of the Menomonee River: the Milwaukee Road Railroad Car Manufacturing Site (the Shops) to the north, and the 'Airline Yards' site to the south. There are a lot of industrial relics and transportation structures left in the project area. A viaduct cross above the site from north to south, Just in the west of the planning area stands two brick chimneys, which together with the viaduct present unique industrial experience and character. Based on the current conditions of the project area and the goal of the city regeneration, the planners develop the plan objectives (Fig.5-1-1):Retain jobs that are accessible to our residents and encourage employers to create new jobs, by developing a business park that appeals to businesses requiring access to the Milwaukee workforce, transportation arteries and the central business

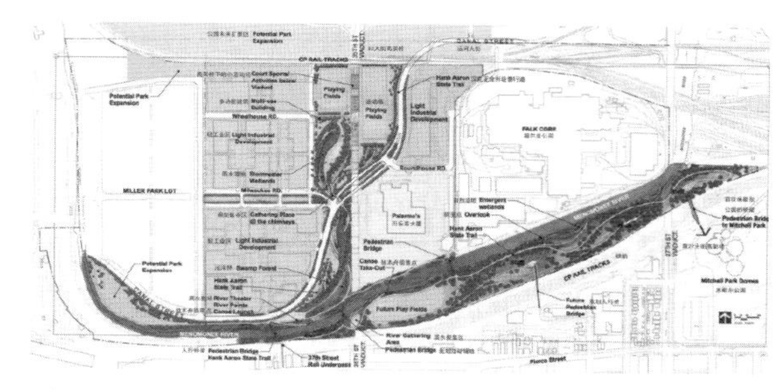

图5-1-1 总体规划
Fig.5-1-1 Master plan

见、透明、统一适用的管理机制；确保场地新开发是高质量、高效率、富有美感且与其他设施相协调；恢复规划区域内河谷原有的自然景观，增加河谷区的可达性；建立场地与周边社区的联系，为社区人员到达河岸及游憩设施提供可达的途径；将场地与城市网格及交通网络相联系。

山谷地区再开发愿景是重新焕发场地经济活力、恢复场地景观并重建场地与周边社区的联系。恢复意味着再生——注入新活力——同时再现山谷先前开发中丢失的价值和特征。环境修复、雨水管理以及本土景观恢复是恢复山谷景观的重要措施，同时这些措施能够将山谷转变为一个生产性、支持多种用途的景观。相似的、联系的重建意味着过去场地价值的恢复：通过新的城市街道和建筑肌理来重建城市南部与北部的自然联系，重建人们日常生活中同自然过程的联系，重建区域生态系统、历史与文化的联系。上述目标已经成为场地再开发规划的一部分。

二、 场地及周边背景介绍

梅诺莫尼河谷工业中心位于密尔沃基河盆地，盆地包括梅诺莫尼河，由6个水域组成，覆盖900平方英里（约2331平方公里）。它包含400英里（约644公里）左右的常流河以及超过85个的自然冰川湖或冰川池。现今，密尔沃基市区域存在工业污染，大面积不渗透地表，以及为了控制洪水而进行河流渠道化以及综合排水管道等，使得这些河流遭到污染。最近几年，改善水体质量成为政府部门的一项重要工作。梅诺莫尼河谷工业中心的上游，城市森林及公园网络成为河流的缓冲区。由于早期为了提供工业建设土地，河谷内自然河口被填埋，因而地下水已遭到污染。河流和湿地的填埋、雨水径

district;

Develop a predictable, transparent and uniformly-applied regulatory process;

Ensure all development is high-quality, efficient, aesthetically pleasing and consistent with the site's other amenities;

Revitalize and provide public access to the Valley's natural features;

Form connections with adjacent neighborhoods allowing neighbors access to the river and recreational amenities;

Connect this site to the City grid and regional transportation network.

The vision for the redevelopment of the Valley is to return it to economic prominence, regenerate its landscape, and reconnect it to the community. The notion of regeneration implies a rebirth – an infusion of new life – as well as reclamation of past values and character that were lost during the Valley's prior development. Environmental remediation, stormwater management and restoration of indigenous landscapes are intended to revitalize and transform the Valley into a productive working landscape that can support a broad range of uses. Similarly, the idea of reconnection implies a renewalof past values: from a physical reconnection of the City's north and south neighborhoods with a new urban street and building fabric, to a reconnection of people to natural processes in the context of everyday life, to a reconnection with regional ecology, history and culture. These overall objectives have been integrated into the redevelopment plan for this site.

II. Introduction of the Site and Context

Menomonee Valley Industrial Center is located in the Milwaukee River Basin, which includes the Menomonee River, consists of six watersheds, and covers 900 square miles. It contains approximately 400 miles of perennial streams and more than 85 natural glacial lakes and ponds. Today, many of the rivers in the Milwaukee area are contaminated due to industrial pollution, large areas of imperviousness, river channeling to control flooding, and combined sewers. In recent years, improvement of the water system

流、土壤侵蚀及点状分布的工业污染源等成为流域内水体质量下降、栖息地退化的祸因。商业区旧址大约有100英亩（约40.5公顷），其界限北至CP铁路南至梅诺莫尼河。大片的地上停车位隶属于西边的密尔沃基布鲁尔体育场，福克公司的机器零件制造厂位于东部。纵观场地，工业遗迹和交通网络控制着这片区域。第35街的高架桥从山谷的北面横跨到南面，高出山谷水平面60英尺（约18.3米），将场地分为东西两部分。高架桥的高度和特征公园利用方式及体验提供了一个独特的视角。高架桥的西部是过去商业区旧址的发电场遗留下的两个巨大的烟囱。整个开发区域的表面是一层8～20英尺（约2.4～6.1米）厚的混合物，由碎石、煤渣、建筑用砂和填充黏土组成，表层下面是沼泽地的有机黏土和沉积泥炭。由于场地位于一片曾被填埋的湿地上，加之曾经铁路工业场地遗留下的建筑物，项目会受到一些地理因素的影响。

由于这是一块棕地，场地污染尽管看不见，但却严格限制了场地的利用。土地表层需要加以覆盖以防止雨水渗入土壤中，同时整个区域需要大面积的滞洪区。场地不可达性及功能缺失的现状形成了一个大街区，导致米勒公园站台与山谷另一边的密尔沃基商业区、社区缺少必要的联系。

在梅诺莫尼河谷工业中心的上游段，尽管莫斯修普芳德等区域曾经作为工业用地并且导致水质的恶化，但河流的大部分区域都有森林缓冲带或者成为公园网络的一部分。梅诺莫尼河流域的大部分陆地成为城市用地（42%），其他的主要用地为草地（22%）、农业用地（17%）、森林（8%）以及湿地（7%）（《2001年密尔沃基河盆地报告》）。洪水成为密尔沃基区域的重要问题，在最近几年密尔沃基都市索拉吉区已经实施了数项防洪项目，河流截弯取直、河道通道化、这些措施仅仅是让洪水迅速排走。即使如此，梅诺莫尼河流经城市时，在其南部河岸上留下了窄窄的植被带。

三、 场地规划

项目主要实现四个环境目标：

has become a priority. Upstream of the Menomonee Valley Industrial Center, much of the river is buffered with forest or included within a network of parks. Because the natural estuary of the Menomonee Valley had been filled to provide a buildable surface for industry, groundwater there has been contaminated for years. Stream and wetland filling, runoff, overland erosion, and industrial point sources of pollution are major contributors to degraded water and habitat quality within this watershed. The Shops property is approximately 100 acres, bounded by the CP railroad tracks to the north and the Menomonee River to the south. A large surface parking lot owned by the Milwaukee Brewers Stadium District lies to the west, and the Falk Corporation's machine parts manufacturing plant is located to the east. Visually, industrial relics and transportation structures dominate the area. The 35th street viaduct spans the Valley north to south standing 60' above the Valley floor, bisecting the site into eastern and western halves. The height and character of the viaduct present unique opportunities for park uses and experiences. Just west of the viaduct are two brick chimneys that are remnants of the Shops power plant. The entire development area is underlain by the organic clay and peat deposits of the marsh, which are buried under 8 to 20 feet of mixed rubble, cinder, foundry sand and clay fill. Due to its location on a filled wetland and the existence of relic structures left from the former rail yard, the project site is affected by several geotechnical issues.

As a brownfield, the site's contamination, while invisible, provided severe usage restrictions. The subsurface would need to be capped, which would eliminate the opportunity for stormwater infiltration and necessitate large areas for stormwater detention. The site's inaccessible and functionless condition created a major block, which separated Miller Park Stadium from downtown Milwaukee, and the neighborhoods on either side of the valley.

Upstream of the Menomonee Valley Industrial Center, much of the river is buffered with forest or included within a network of parks, although there are some former industrial areas that have contributed to river quality degradation such as the Moss Superfund site. the majority of land area in the Menomonee River watershed is covered by urban uses (42 percent), while most of the remainder is grasslands (22 percent), agriculture (17 percent), forests (8 percent), and wetlands (7 percent) (The State of the Milwaukee River Basin 2001). Flooding has been great problematic in the Milwaukee area, and in recent years the Milwaukee Metropolitan Sewerage District has implemented several flood-control projects, channelizing and straightening the river to convey flood waters off the land more quickly. Even so, a slim fringe of vegetation persists along the Menomonee

首先，恢复和建设社区公园及河流廊道的自然景观，这些区域内的自然景观在栖息地改善、生物多样性提高方面发挥重要作用；其次，雨水管理的生态方法应用主要通过生物过滤及地层过滤等方法，这些方法也是传统滞洪方法的替代措施；再次，洪水控制与场地商业功能及游憩利用相结合，实现土地利用的最大化；最后，棕地现状、地表覆盖以及地下水保护需要等决定了场地的利用及结构。

项目文化方面主要围绕场地表达、娱乐休闲以及项目城市身份建立等方面展开。烟囱暗示了场地及区域的工业历史，而滨河区则唤起人们对河谷过去自然景观的记忆（图5-1-2）。视觉上直观地强调场地雨水功能则提供另一种表达方式。通过相联系的自行车道及社区运动场地，休闲活动项目实现了社区与场地的联系。密集工业用地围绕中心开放空间的构思来自于一个村庄通常被工业区包围的想法，这种形态布局也为项目城市身份提供了指导方向。

场地规划展示了两个重要的模式主题：原有的区域肌理及新规划的肌理，河流及以原有铁路线所形成的有机“丝带”。公园现有的形态通过地形塑造反映出了河流及航道的流线特征，为公园雨水管理功能创造了条件。雨水出水口周边的深水池及延展的“低洼地”距离越接近，河流越密集，而步行道所构成的网络成为“低洼地”的边界。东西向贯穿整个场地的运河街形象地反映出了这种有机肌理。工业开发区形成新的结构网，这个网络反映了周边居住区布局形态及

图5-1-2 公园鸟瞰图
Fig.5-1-2 The overlook of the park

River's south side as it runs through the city.

III. Site Planning

The program addresses four major environmental goals. First, the restoration and creation of natural landscapes within the community park space and along the river corridor was a priority for habitat and biodiversity improvement. Second, the use of ecological methods for stormwater management through biological and subsurface filtration was favored as an alternative to conventional detention methods. Third, an approach to flood control that integrates the needs of commercial function and recreational use was envisioned to maximize the use of the land. Finally, the brownfield needs associated with capping of the site and the protection of the groundwater created specific parameters for how the site could be used and structured The cultural aspects of the program revolved around interpretation, recreation, and the creation of a civic identity for the site. The chimneys presented a physical opportunity to interpret the site and region's industrial past, while the riverfront offered a connection to its natural history (Fig.5-1-2). Visual emphasis on the site's stormwater function provided another level of interpretation. A recreational program emphasis was included to provide the desired community connections to the site, through the use of linked bike paths and community play fields. The concept of dense blocks of industrial development sites around a central open space, borrowed from the idea of a village commons surrounded by individual sites, provided programmatic direction for the project's civic identity.

The plan reflects two major pattern themes: the grid of the old and new development, and the organic ribbon-like lines of the river and former rail lines. The park's primary form-giving reflects the flowing lines of the river and rails through the sculpted landform created to provide the stormwater management function of the park. Deep pools around the stormwater outfall areas and spreading basins gain intensity as they get closer to the river a, while the pedestrian paths create ribs that define the basins. Canal Street, which provides an east-west connection through the site from downtown Milwaukee to Miller Park and defines the subspaces of the park, further reflects the organic influence. The industrial development parcels are oriented to a new valley grid with a structure and density that reflects the pattern of surrounding neighborhoods and a more compact arrangement of industrial use than historically found in the area (Fig.5-1-3).

The park forms a linear open space spine that is perpendicular to the river, connecting the north and south sides of the city to the valley, through a sequence of areas that transition from active human

相对于以前更为紧凑的工业利用（图5–1–3）。

公园形成了一条与河流相垂直的开放空间轴，将城市的南北边与山谷相连。这条轴线贯穿公园的一系列连续的空间，这些空间内的活动将由北部烟囱公园积极的公众利用方式转变为南部河流草地及航线公园滨河区的消极利用方式。这些区域通过种植密度、植被表面变化以及地形来完成活动转变。

公园植物配置的目的是恢复场地在山谷中的自然生态功能，为人们提供游憩活动。在烟囱公园南部区域草地生境较少，仅有的草地以不同大小的尺度点缀于自然区，以呼应公园雨水处理功能；在北部海拔高度较高的地区则有较多的草地，包括三个运动场地，但是场地靠近河流的区域则强调生物多样性及生物过滤作用。自然区内平坦地的植物配置包括中湿性草原、湿地草原和自然混合群落。雨水出水附近的水池则保留自然生长的植物群落。这些群落将逐步将湿地草地转变为中湿性草地。

场地内的人流运动及水流流动是场地的一项重要功能。经过规划人员的详细规划，整个场地的雨水流动贯穿整个场地；而这也为场地开发提供了重要的生态功能。由于水体不能渗透到场地土壤中，这就要求水体必须慢速地在公园区内流动，经过植物的过滤作用最终排入河流中。规划人员将整个公园的雨水流动和人的移动相结合。运动场地等重要空间鼓励人流及水流的通过，形成场地形态。公园内部及周边区域的交通系统主要是面向非机动车车行及步行。此外，为了创造游憩机会，公园为社区居民增加了教育项目。

道路系统不仅贯穿公园的各个部分，而且流动的、交织的、分拆的形态模拟了河流及其支流的形态关系。这些道路并不是仅仅置于景观之中，而是决定了周边景观的形态；高出水面的道路将不同的雨水处理池分隔开。公园道路系统还与更大区域网络、山谷南北边社区相联系。

该场地曾经是国内最大的铁路工厂之一，两座80英尺（约24.4米）高的烟囱成为这段历史的遗留物。它们曾服务于铁路厂的发电站，现今成为公园绿色中心区的焦点。

use in the northern Chimney Park section, to passive uses in the southern River Lawn section and the Airline Yards riverfront area. These areas use planting density, vegetative surface variation, and landform to define this transition.

The park's vegetation is designed to bring back natural ecological function to the valley, while also providing active recreational use. There are limited lawn areas of varying scale interspersed with natural areas that mainly correspond with the areas of the site used for stormwater management. On balance, the northern, higher elevations of the park have more lawn, including the three play fields, while closer to the river there is a greater emphasis on biofiltration and a diversity of planting. The ground plane planting for the natural areas consists of a variety of seeding types including mesic prairie, wet prairie, and emergent planting mixes. The pool areas immediately surrounding the stormwater outfalls are not planted but are ringed with the emergent seed mix, which transitions to the wet prairie and then to mesic prairie.

Movement is an essential function of the site: movement both of people and of water. The carefully planned movement of stormwater through the site, as discussed earlier, provides important ecological function to the development; as the site cannot infiltrate the water, it must spread out and slowly move over the entire park area as it makes its way to the Menomonee River. This movement of water is integral to the movement of people through the park. The majority of the spaces encourage movement through their flowing, braided form and pattern. The park is primarily designed for biking and walking within and through the site. In addition to active recreational opportunities, the site is programmed to provide educational opportunities for community users.

The pathway system not only provides connectivity throughout the park, but its flowing, braided, and branched shape evokes the form of the river and its tributaries. They are not placed onto the landscape, but actually give form to the landscape, and create the ridges that separate the water treatment basins. The park path system connects to a larger regional network as well as neighborhoods to the north and south sides of the valley.

Two 80-foot-tall smokestacks are the only remnants

图5–1–3 烟囱公园
Fig.5–1–3 The Chimney Park

成为人们集会的场所。南部烟囱周边区域为硬质铺装的开敞空间，广场上安置了一些坐椅；北部烟囱则主要是草地。两个烟囱的中心地段进行了平整，用于放置艺苑35大街高架桥对山谷南北部联系造成不良的视觉影响。对于设计人员来说，公园区与高架桥平行布置也是一个戏剧性的想法；它加强了公园使用者与这个巨大基础设施的联系，同时规划利用这座高架桥界定了公园空间及结构。高架桥也为步行者及驾驶人员提供了一幅以新工业区、谷坡、米勒运动场为背景，俯瞰公园及梅诺莫尼河河的全景图。

在整个场地6个区域中，雨水排水渠道将临近的工业区雨水收集到水池中，然后排入河中。这些区域中，湿地植被提供了生物过滤作用，并且形成了湿地栖息地。每一个排水口处都有步行道，步行道的终端是特别设计的高地；借助于高地，游客可以俯瞰水池景观。这些俯瞰点都有一些与玻璃艺术品相结合的栏杆，艺术品展示了在公园中发现的各种植物及动物物种。

四、 环境恢复规划

政府已开始并将持续投入大量资金用于解决商业旧址的环境问题。修复行动计划中概述了具体的环境修复行动。一般的恢复措施包括在已经受污染的土壤上覆盖几英尺厚的土壤及长期监测地下水。在一些限制性区域恢复土地自然生产力，同时移除棘手区域内的污染土壤。政府将按照威斯康星州自然资源部的规定，对土壤和土地自然生产力进行管理；伴随场地的关闭，需要长期监测地下水（预计将持续10年）；土地移交给开发商时要保证土地是封闭的或者即将封闭；为已受污染的土壤提供放置的场地，并将其作为未来发展的一部分。

除土壤污染之外，对整个场地而言，甲烷的产生也是个潜在问题。过去10年中在山谷修建了许多新建筑物，包括被动式甲烷、土壤气体收集系统。作为场地的再开发和建设的一部分，修复行动计划要求采取措施应对甲烷产生的问题。开发商也需要将这个要求纳入建筑和设施设计当中。

of the site's long history as one of the nation's largest rail yards. They once served the rail yard power house and are now the focus of the central green area of the park, which is the site's main gathering space. The southern chimney is grounded in a paved area with several benches and the northern chimney is in the main lawn area. The chimney hearths create a frame that is eye level and planned for art installations.

The 35th Street Viaduct is a visible thread of connection across the valley. The deliberate placement of the park area in alignment with the viaduct was a dramatic move on the designer's part to engage the park user with the massive piece of infrastructure and use it to define the park spaces and organization. It also affords panoramic views for pedestrians and drivers on the viaduct down to the park and the Menomonee River, with the new industrial area, the valley slopes, and Miller Stadium as background.

In six locations throughout the site, stormwater outfall pipes release water from adjacent industrial development sites into ponds that hold the water and slowly release it toward the river. At these areas, natural wetland vegetation provides biofiltration of the stormwater and wetland habitat. Each outfall has a pedestrian path and ends in a brow that forms an overlook toward the pond, as a deliberate endpoint. The overlook has a railing that is inset with glass art panels that depict various plant and animal species found in the park's habitat areas.

IV. Environment Restoration Plan

The City has made and will continue to make significant investments to address environmental issues at the Shops site. Specific environmental activities are outlined in the Remedial Action Plan. The general remediation approach consists of placing several feet of fresh soil over the existing contaminated soil and long-term groundwater monitoring. Free product will be recovered in a very limited area and "hot spots" of soil were removed. The City will: Undertake all soil and free product management required for WDNR closure; Perform long-term groundwater monitoring (expected to last 10 years) associated with closure; Convey the property to the developer with the assurance that the property is or will be closed; and Provide an onsite placement area for impacted soil that is excavated as part of future developments.

In addition to soil contamination, methane generation is a potential issue across the entire Valley. Many new buildings constructed in the Valley during the past decade included passive methane/soil gas collection systems. Methane mitigation is required by the site-approved RAP as part of site redevelopment and building. Developers need to incorporate this component into site building and facility design.

五、 雨水处理

雨水管理区的设计为公园提供了结构框架而且创造了一系列的不同类型的景观。公园为整个街区以及运河大街提供滞洪和雨水处理系统，它是通过贯穿公园南北轴的雨水处理系统完成的。雨水处理系统依据公园地的地形而设计，提供了2年一遇、5年一遇、100年一遇的洪水储存区。处理区分为两个雨水管理区，它们横穿公园中的运河大街。较低的雨水管理区形成了沼泽森林，并在河流的边缘设计了一个排水口。当遭遇百年一遇的洪水时，整个公园除了运动区将全部被淹没。由于棕地污染，公园表面进行了必要的覆盖，因而这些区域虽然能够滞洪及处理水体，但水体却不能下渗（图5–1–4）。

雨水管理区和湿地森林水体包括三个步骤：从商场旧址及运河大街到公园雨水排水口，收集并运输雨水，大固体颗粒将会在一系列的小型水池中沉淀下来；水流缓慢流过宽阔的浅湿地草甸，然后顺利渗入路基层。路基层主要由破碎、粗糙的混凝土组成，又称之为“集水暗沟”；收集渗透后的水流，这些水体将通过沼泽森林中植被的蒸发作用排入大气中。这个系统既可以清除水体中的固体颗粒，也可以消除水中的污染物，同时可以滞留百年一遇的洪水。它避免了传统滞洪水库及箱式雨水管道系统的弊端。随着航线公园自然景观的恢复，雨水湿地、草地、沼泽森林与公园内常规的景观形成了对比。这种对比用一种深思熟虑的方式重新阐述了奥姆斯特德传统公园风格，风格的表达是通过设计过程表达及人造景观营造所形成的形态来完成

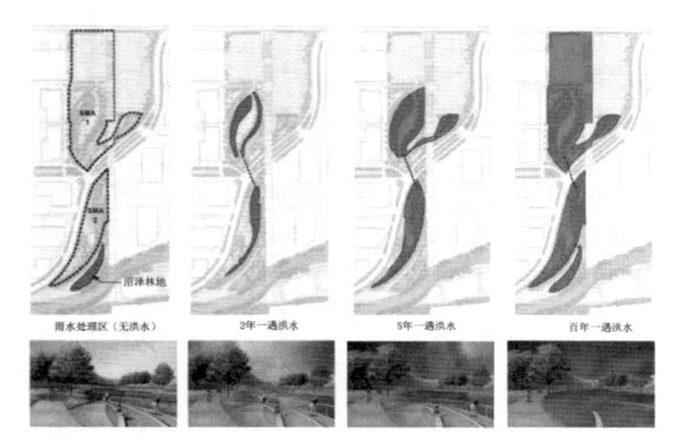

图5–1–4 雨水管理
Fig.5-1-4 Storm water management

V. Storm Water Management Plan

The design of Stormwater Management Areas (SMA) gives structure to the park and allows for a creation of a range of landscape types. The park provides flood detention and water quality treatment for the entire Shops site as well as the Canal Street extension in the form of a stormwater "treatment train" (STT) that traverses the park's north-south axis. The stormwater treatment train (STT) is designed into the landform of the park land, providing storage for 2-year, 5-year and 100-year storm events. The treatment areas are divided into two stormwater managemen areas (SMA) by the crossing of Canal Street through the park. The lower SMA leads to a “swamp forest” outlet at the river’s edge. At the 100-year storm event all of the park landscape except for one playing field would be inundated. Because the park has been capped due to brownfield contamination, these areas provide detention and water quality treatment, but do not allow for infiltration of the stormwater (Fig.5-1-4).

The SMAs and the Swamp Forest treat water in three steps: 1. Stormwater is collected and piped, from the Shops redevelopment and Canal Street, to storm outfalls located in the park. Here large particulates settle in small pools. 2. Storm flows spread out across broad shallow wetland meadows. The subgrade layer, called an‘infiltration gallery’, consists of coarse-textured crushed concrete that allows stormwater to infiltrate. 3. Infiltrated storm flows will be collected and transpired through the plant material at the Swamp Forest. The system both removes particulates and pollutants, and detains the 100-year storm, eliminating the need for traditional detention basins and a trunk storm sewer system. The stormwater wetlands, meadows, and the swamp forest, along with a natural landscape restoration across the river at the Airline Yards site, contrast with formal landscapes within the park. This contrast reinterprets Olmsted's park tradition in a deliberate manner by using forms that express processes and celebrate the "made" landscape. Through this visible process, the park will remain a dynamic place of ever changing settings with each day, each storm, and the changing seasons (Fig.5-1-5, Fig.5-1-6).

VI. Major Areas Design

1. The Chimney Park Design

The Chimney Park area will be the most active area of the park. Facilities here will include playing fields, court games, trails, parking, informal lawn areas, and picnic areas. These facilities will be interwoven with existing site features including the 35th St. Viaduct,

的。通过可加的过程，在每一天或者每一场雨水之后抑或每一个季节，公园都会发生变化，成为一个动态变化的场所（图5–1–5，图5–1–6）。

六、 重要分区设计

1. 烟囱公园设计

烟囱公园将会成为园区中最活跃的区域。这里的设施将包括运动场、网球场、步行道、停车位、自然式草坪区以及野餐区。这些设施会穿插于现有场地特征中，包括35街高架桥、商场旧址发电站的烟囱，还有第一雨水管理区的雨水处理区。位于高架桥下的网球场和烟囱附近的小型集会场地的设计都充分利用这些保留构筑物的地标特性。特别是位于再开发区入口处以及与米勒公园直接相连的步行道东端的烟囱，将会产生令人记忆深刻的识别特征，并且唤起人们对场地过去用途的记忆（图5–1–7，图5–1–8）。

2. 河流草地设计

河流草地的设计会吸引游客到梅诺莫尼河河岸处游玩。公园的这一区域包括水上剧场、集会空间、划船区、垂钓区、步道、野餐区及第二雨水管理区的雨水管理系统。水上剧场设置在以河岸草坪为基础的混凝土或石质平台上，这一区域的河岸草坪也可以供游客垂钓或划船。步道系统将包括沿河岸的汉克阿伦州际小径及跨过梅诺莫尼河的一座人行天桥。特色景观包括从运河大街到河边的一条倾斜草坡，草坡面向河流展开。河流景观是由包括重建的河岸和沼泽森林渗透区域在内的自然区域组成的，构成第二雨水管理区的草甸、湿地和池塘为本地生态的恢复发挥了一定的作用。对雨水处理领域的不断深入也为自然生态的恢

the chimneys at the former Shops powerhouse, and stormwater management areas at SMA 1. Court games beneath the viaduct and an area suitable for small gatherings at the chimneys are designed to take advantage of the landmark quality of these existing structures. The chimneys, in particular, located at the entrance to the proposed redevelopment and at the east end of a direct pedestrian connection to Miller Park, will provide a memorable identifying feature and recall the area's historic uses(Fig.5-1-7, Fig.5-1-8).

2. The River Lawn Design

The River Lawn invites visitors to the edge of the Menomonee River. This part of the park will include a rive theater and gathering space, river access for boating and fishing, trails, picnic areas, and stormwater management at SMA 2. The river theater is focused on a concrete or stone terrace at the base of the river lawn that will also provide access for fishing and boating. Trail connections will include the Hank Aaron State Trail along the river's edge and a major pedestrian bridge that crosses over the Menomonee River. Landscape features will include a sloping lawn from Canal Street to the edge of the river. The lawn open views to the river that are framed by naturalized areas including restored river banks and the swamp for infiltration area. The meadows, wetlands and ponds that make up SMA 2 contribute to the restoration of the native ecology. Maintenance access to the stormwater treatment areas will also provide limited public access to restored natural areas (Fig.5-1-9).

3. Airline Yards Design

"Airline Yards" is a 23-acre piece of land that can be seen below the 27th Street viaduct, stretching to west of the 35th Street viaduct. The narrow 2,600 foot long and 300 foot wide stretch has little development value, but, in the days of Milwaukee Road, it was ideal to store many rail cars in wait for their final destination throughout the country. Since the bankruptcy of Milwaukee Road in 1980s, this stretch has sat completely isolated and unused - an unfortunate case with 2,600 linear feet of river frontage. Through this project, this isolated

图5–1–5 第二雨水管理区出水口结构图
Fig.5-1-5 Outfall structure at SMA2

图5–1–6 沼泽森林效果图
Fig.5-1-6 View of Swamp Forest

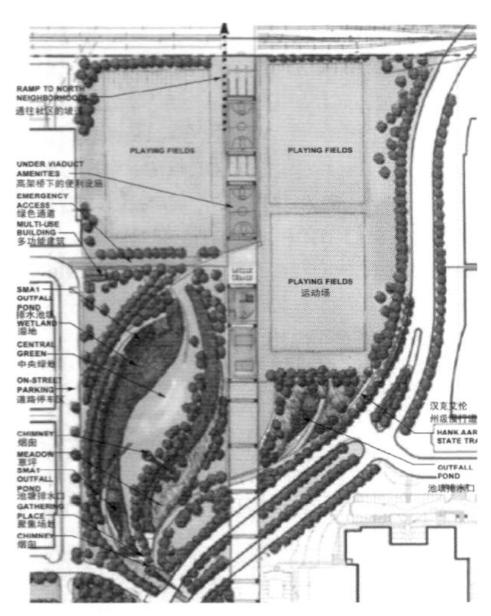

图5–1–7 烟囱公园规划图
Fig.5-1-7 The Chimney Park Plan

图5–1–8 烟囱公园效果图
Fig.5-1-8 View of the Chimney Park

复提供了有限的公众可达性（图5–1–9）。

3. 航线园设计

“航线园”面积为23英亩（约9.3公顷），位于第27街高架桥下，一直延伸到第35街高架桥的西部。这是一块长2 600英尺（约792.5米）宽300英尺（约91.4米）的狭窄地块，开发价值极小。然而在过去，密尔沃基大道却是全国许多轨道车的最终归宿处。自从20世纪80年代密尔沃基大道破产后，这一路段已经完全被孤立起来且不被利用——这一2006英尺（约611.4米）长的地段就是一个不幸的事实。借助于河谷工业区项目，这块孤立地块将向公众提供广阔的原生自然区；区域及本地道路结合步行道为游人提供亲水机会；提供室外运动场地。组成公园的三座山丘唤起了人们对场地过去圆形悬崖的记忆，这些悬崖在铁路未建造之前组成了山谷边缘。规划人员利用附近商场旧址上的碎渣重新构建悬崖，从而与南边社区联系起来，并恢复了多种多样的本土景观。汉克阿伦州际小径的一个分支将公园河边区域与山丘顶部联系起来。位于山顶的人行天桥将延伸到铁路轨道与公园南部的社区联系起来。第37街的一座历史性隧道将会被重新开放以连接南部社区从而创造娱乐和就业机会。山顶部的小露台是一个放眼远眺的好地方，视线可以跨越河流一直延伸到市中心东部。特色构筑物包括点沙坝，

area will be opened up to provide extensive native areas, regional and local trail connections with pedestrian access to the river, and expansion potential for open play fields. The park is made of three large hills, which are intended to recall the rounded bluffs that formed the edge of the Valley prior to construction of the railroad. The bluffs will be constructed using debris from the adjacent Shops site, make connections to the south side neighborhoods, and will provide variety in a series of restored native landscapes. A branch of the Hank Aaron State Trail provides a connection down to the river's edge and up to the tops of the hills. Pedestrian bridges located at the tops of the hills will extend over the railroad tracks to provide connections to the neighborhoods on the south side of the park. A historic tunnel will be reopened at 37th Street to connect the neighborhood to south to recreation and employment opportunities. Small terraces at the tops of the hills will provide a place to look out across the river and east to downtown. Built features include river point bars, dike structures at the river's edge that help restore aquatic habit and provide human access to the river. Landscape types include upland prairie and savannah on the slopes of the hills, a restored riparian forest at the river's edge, a fen at the hollow between the hills and up to two turf playing fields(Fig.5-1-10~Fig.5-1-12).

VII. Summary

In summary, the site design successfully achieves the goals as follows: 1.reconnects the valley to its place along the river with a biodiverse, repaired landscape. 2. Models a healthy, visibly connected flow of stormwater from development, including upstream off-site parcels, through a filtering open space and into the larger watershed. 3. Creates a vital and multifaceted connection between downtown Milwaukee and the city's stadium and outskirts beyond. 4. Links two sets of neighborhoods,

位于河边的堤坝结构——用于重建水生物栖息地并且便于游人亲水。景观类型包括山地草原、山坡草原、已恢复的河岸森林、山丘间的沼泽及两个草地运动场（图5-1-10～图5-1-12）。

七、总结

现状中的河谷，遍布工业废弃建筑，处处可见工业时代留下的环境污染。规划的愿景是恢复场地景观，使其重新焕发经济活力，并重建场地与周边社区的联系。规划采用栖息地改善、生物多样性提高、雨水生态管理、洪水控制、土地兼容性利用、棕地生态修复等多重措施来实现环境目标。其中雨水生态管理具有创新意义，雨水管理区的设计不仅具有滞洪和雨水处理的重要功能，更利用地形创造了丰富的景观，运用生态的方法营造了舒适的微气候。雨水生态管理确保来自梅诺莫尼河谷工业中心的雨水不会进入复合雨水处理系统，从而净化了梅诺莫尼河的水质。

棕地修复也是本案例的棘手问题，这是工业废弃地面临的共同难题，也是生态修复关键的第一

which had previously been separated by fallow and contaminated land, with recreational pathways and open space as well as economic opportunity. 5. integrates industrial, infrastructure, and open space site uses.

Nowadays industrial abandoned buildings are located along the two sides of Menomonie river. Menomonie river is heavily polluted during Industrial Age. The expecting aim of the planning is to recover the landscapes on this site, and at the same time stimulate the economy. Rebuilding the connection of the site and communities around is another goal to be reached. In this planning, habitat amelioration, biodiversity improvement, stormwater management, flood control, land compatibility utilization, and brown field ecological rehabilitation are considered to get a better environment. Among all these strategies, stromwater ecological management is especially innovative. The design for stormwater management district can not only meet the important function of flood control and stormwater treatment, but also enrich landscapes and adjust microclimate to a comfortable level. Managing stormwater ecologically can prevent the stormwater from the industrial center of Menomonie River entering the hybrid stromwater treatment system, so that the water of Menonmonie River can be treated.

Brownfiled Rehabilitation, one of important

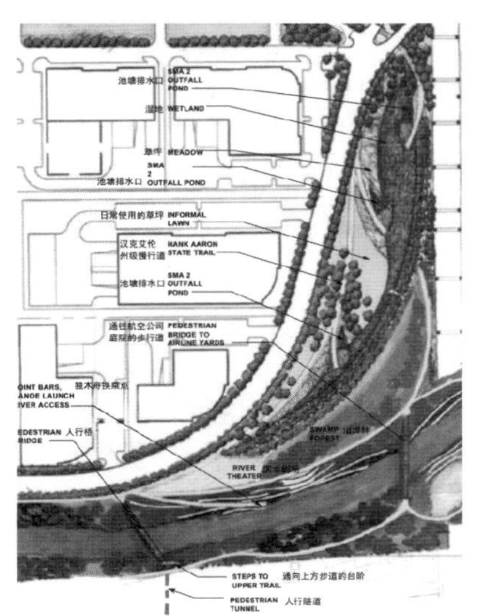

图5-1-9 河流草地规划图
Fig.5-1-9 The River Lawn Park

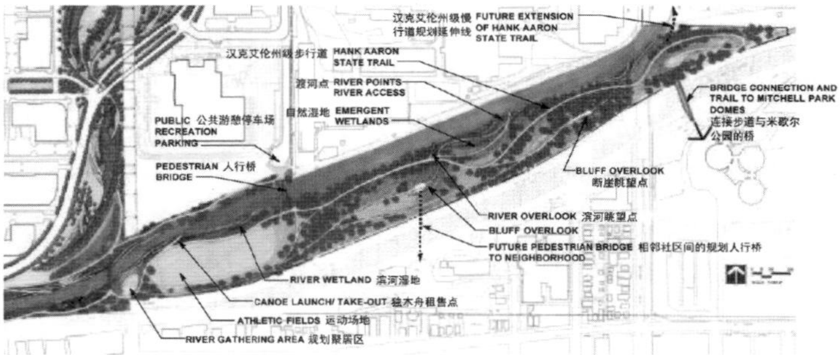

图5-1-10 航线园规划图
Fig.5-1-10 The Airline Yards Plan

图5-1-11 航线园：自然湿地剖面图
Fig.5-1-11 Airline Yards section at Emergent Wetlands

图5-1-12 航线园：河流鸟瞰点
Fig.5-1-12 Airline Yards River overlook point

步。“棕地”意译自英文单词“Brownfield”，该词最早出现于20世纪80年代美国国会通过的《环境应对、赔偿和责任综合法》中。棕地存在于城市内部会造成土地闲置、环境污染、城市景观破碎等问题。在精明增长的城市发展理念下，这些棕地亟需生态修复，恢复其场地的安全性、景观性和生态性，实现城市土地集约化利用。目前，我国也不乏棕地修复的成功案例，如中山岐江公园旧船厂遗址改造再利用、北京创意广场798工厂再利用和沈阳铁西工业区的改造等。

本项目场地设计成功达成了以下目标：通过一条已恢复的生物多样性景观将河谷与场地重新相连；建立了一个健康的、可以观看的、相互连接的雨水流，它涉及整个场地及上游的部分地区，雨水经过开放空间的过滤排入大型湿地中；创造了商业中心与城市体育场、郊区之间的多方面的重要的连接；方案为社区提供了游憩步行道、开放空间及经济就业机会，将曾经被休耕地及污染土地所隔离的社区相联系起来；将商业、基础设施及开放空间相结合。

part of industrial waste lands treatment, is also difficult issue needed to be resolved in this cases. It's the first step of ecological restoration. Chinese "brownfield" is translated from the word "brownfield". This word dates back to Comprehensive Environmental Response,Compensation,and Liability Act,CERCLA passed by the U.S.Congress in the 1980s. Brownfiled in the city can cause problems such as land waste, environment pollution, fragmentation of urban landscapes. According to the concept of smart development for cities, these brownfields urgently need to be ecological restored. The safety of these fields, landscapes in these fields, their ecology, landuse concentration should be ensured. Currently, there are successful cases of Brownfield in China, such as Qijiang Park of Zhongshang, Beijing Creative Square 798, Xitie Industrial District of Shenyang City.

思考题

1. 项目开发的愿景是什么？哪些措施能够将山谷转变为一个生产性、支持多种用途的景观？

2. 哪些要素在规划中保留下来或者再利用？哪些要素表达了场地的工业历史？

3. 场地规划展示了哪两种模式主题？它们是如何贯穿场地的？

4. 场地雨水处理过程是什么？如何利用场地现状来实现？

5. 航线园的景观设计有哪些特色？它在整个场地中有着怎样的作用？

Questions

1. What is the vision for the redevelopment of the Valley? What measures are taken to transform it into a productive working landscape which can support a broad range of uses?

2. In the redevelopment plan, what elements are reserved or recycled? What elements can express the industrial history of the site?

3. Which two pattern themes does the plan reflect? How are they through the site?

4. What is the process of storm water management? How does the storm water management plan use the current conditions?

5. What are the characteristics of the Airline Yards design? What role does it play in the whole site?

注释/Note

图片来源/Image：图5-1-1～图5-1-12来自http://www.govengr.com/

参考文献/Reference：维基百科http://en.wikipedia.org/wiki/Menomonee _ Valley

推荐网站/Website：http://www.govengr.com/ ；http://lafoundation.org/research/landscape-performance-series/case-studies/case-study/135/

第二节 甘尼特报业集团总部景观设计

Section II Landscape Design for Gannett

一、 项目综述

美国甘尼特报业集团旗下的《今日美国》报纸总部项目设计是一个最大化利用现有场地优势并改善其功能的成功案例。在商业和零售业快速发展的弗吉尼亚州泰森角，这一案例已成为生态多样性的庇护所。

甘尼特报业《今日美国》总部项目位于弗吉尼亚州麦克莱恩市泰森角270英亩（约110万平方米）的琼斯分水岭汇合处。该项目占地约25英亩（约10万平方米），其东部紧临首都环线和杜勒斯阿克塞斯大街交叉点，西接琼斯分水岭车道。景观设计师采用室内、室外空间紧密连接的设计手法，营造出一块开阔的场地，其中包括屋顶花园、露台、滨水植物区和一片保留林地（图5–2–1）。

设计团队采用了场地修复理念作为场地规划的技术策略。场地修复这一理念由克里斯托弗·亚历山大于1997年出版的书籍《模式语言》中提出，他认为在场地规划中应投入时间、成本和精力去进行生态系统的修复，并建议尽量避免影响新的地区，应在已经受破坏的地区寻找新的发展空间。

二、 场地及周边背景介绍

麦克莱恩市位于泰森角附近，横跨流经华盛顿

I. Overview

The Gannett/USA Today Headquarters is a project that maximized design efficiency by taking advantage of existing site qualities and improving their function. It is an ecologically diverse refuge in the rapidly developing business and retail center of Tysons Corner, Virginia.

Gannett/USA Today is located at the confluence of the 270-acre Jones Branch watershed in the Tyson's Corner area of McLean, Virginia. The site is a 25-acre parcel bounded on the east by the intersection of the Capital Beltway and the Dulles Access Road and to the west by Jones Branch Drive. The Landscape architect developed a site strategy that seamlessly weaves the indoor and outdoor spaces into a campus of extensive roof gardens and terraces, riparian planting and preserved woodlands(Fig.5-2-1).

The design team employed site repair as the primary site planning technique. The concept of site repair as described in Christopher Alexander's 1997 book, A Pattern Language, suggests that site planning take into

图5–2–1 甘尼特园区全景
Fig.5-2-1 The Gannett USA Today Headquarters

和美国中东部最重要的的河流波托马克河。它位于杜勒斯机场通路和1-495号首都环线的交汇点，是商业及写字楼高度发达的地区之一。同时，也是为在华盛顿工作的外交官、国会议员以及政府官员服务的高档住宅区。在最终确定选址地点之前，甘尼特报业集团为总部新址选择了不下50处备选选址地点，最终，他们选择了这块25英亩的场地——一块欠发达但土地规整、错落有致的场地。

甘尼特《今日美国》总部项目包括三种独特的未开发地貌特征：低地、草地和山坡。低地位于占地270英亩的分水岭地区的最低处，包括一个已经退化的服务于1/4泰森角的区域性雨洪调蓄池塘。草地为了附近场地发展需要，曾作为区域的垃圾填埋场。山坡上遍植挺拔的橡树林，风景宜人。在初步设计阶段，业主希望将报业集团总部的建筑及其附属园区建在山顶的最高处，使之成为标志性地标，并可以伫立在公司的建筑里远眺华盛顿。经过详细的场地分析、光线研究以及设计规划后，景观设计师与建筑设计师共同努力说服业主将大厦选址从极具生态性、绿树成荫的山顶移至曾作为垃圾填埋场的草地。这一关于选址的重大改变，标志着该项目场地开发的全方位策略转换。

甘尼特报业《今日美国》总部的场地范围如图5-2-2所示。通过区位分析，设计团队认识到了场设计过程中的挑战与潜力，将场地与区域的历史、区域文化、生态保护等联系起来。景观形式、景观功能以及景观设计方法应与这一地区综合性因素相结合，才能将场地的独特特点表现出来（图5-2-3）。

三、 场地设计形式

场地设计分为三个不同的景观空间：园区建筑旁正式的建筑景观，建筑与雨水调蓄池之间的中央空地，以及主动、被动休憩区和山丘。该区域被分成了三个处理区和功能区，在建筑和雨水调蓄池塘间呈倾斜状分布。中央空地作为整个开放空间的核心，保留了山坡上原有的植被，使之成为一处景色宜人的休闲空间。东南朝向既保证了空地朝向阳面，又可以抵御冬季的北风，屏蔽高速公路的噪音（图5-2-4）。

account the time, costs, and effort needed to restore intact ecological systems. It advises to avoid intact areas and focus new development in already damaged areas.

II. Introduction of the Site and Context

McLean lies adjacent to Tysons Corner and across the Potomac River from Washington, DC. It is part of the rapidly developing area of commercial and office uses located at a strategic intersection formed by the Dulles Airport Access Road and the 1-495 Capital Beltway. It also is home to upscale residential neighborhoods that serve diplomats, members of Congress, and government officials in Washington, D.C. The Gannett Company considered over 50sites for their new headquarters before choosing this 25-acre parcel. The site itself was a patchwork collection of undeveloped, disturbed, and regulated land.

The undeveloped site was composed of three distinct land features: lowland, meadow and hilltop. Located at the base of the 270-acre watershed, the lowland contained a degraded regional storm water management pond serving a quadrant of Tyson's Corner. The meadow had been used as a fill site for adjacent development. The hilltop stood out as a beautiful prospect covered with a stand of mature oaks. During the preliminary design phase, the

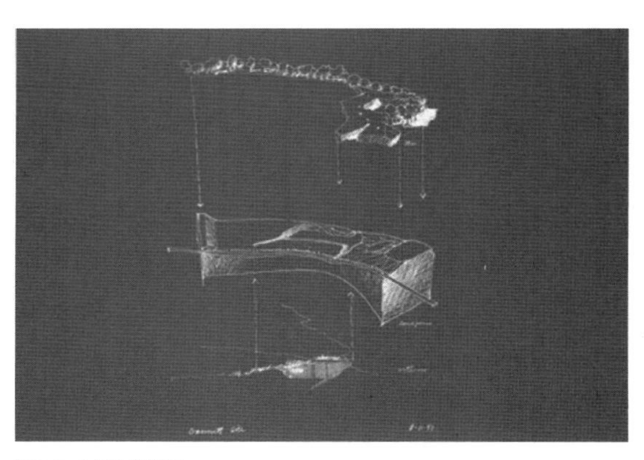

图5-2-2 场地分析图
Fig.5-2-2 The site analysis

图5-2-3 甘尼特园区场地现状
Fig.5-2-3 The site

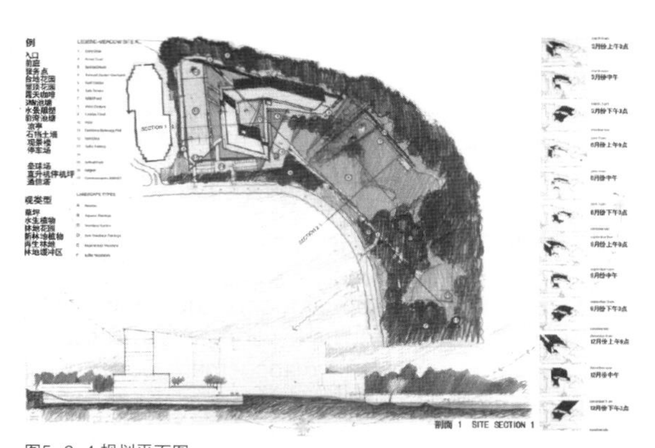

图5-2-4 规划平面图
Fig.5-2-4 The Plan of the Gannett USA Today Headquarter

空地上方有约2英亩（约0.8公顷）的屋顶花园和露台，公司员工可以从大厦中层步入室外空间。屋顶花园在减少雨水径流的同时还可以降低噪音，同时也为楼下的新闻间提供了安静的环境。

沟渠和阶梯花园的设计强调了依地形设计的理念，同时也展现了设计师精心设计的人行流线和雨水径流路径。设计灵感源于树木被砍伐后沿山坡滚下所形成的天然空间。设计师依地势在空地上构造出一层层的石阶墙，形成了阶梯状的草坪和种植区，还兼有功能性设计：包括很多沟渠、池塘和嵌槽的阶梯通道。同时，通过将河岸阶梯与区域性雨水调蓄池塘相连接来疏导雨水，紧密连接私密空间（图5-2-5）。在这里可以远望池塘和远处茂密的山林，池塘中的水通过植物丛生的水池进行氧化、净化处理以再循环使用。大厦的地基使用与石阶墙同样的石材建造，使建筑与景观在结构上遥相呼应。

四、 种植设计

图5-2-5 雨水调蓄系统
Fig.5-2-5 The stormwater basin system

client wanted to place the building and related site development at the highest elevation on the hilltop to generate a prominent corporate visual landmark and offer views to Washington, DC. After a thorough site analysis, sun studies, and programming the landscape architect convinced the client, with the support of the architect, to move the building site from the healthy wooded hilltop to the meadow/fill site. This fundamental shift in site strategy marked the beginning of a comprehensive change in the attitude toward site development.

The regional context of the Gannett/USA Today Headquarters is shown in a sequence of scales in Fig. 5-2-2.Through regional analysis, the site design process is infused with an understanding of potentials and challenges. It situate the site relative to area history, regional attitudes, and ecological health. Descriptions of landscape form, function, and philosophies of this area of Virginia explore a combination of factors to which the project design uniquely responds (Fig5-2-3).

III. Site Design Form

The site design is broken down into three distinct landscape spaces: a formal landscape adjacent to the building, the central court terrace between the building and stormwater basin, and the active/passive recreation area and knoll. Divided into three zones of treatment and function, this terrace is a sloped transition space between the building and the stormwater pond. The central commons forms the armature of the open spaces allowing the wooded hilltop to be preserved as a visual and recreational amenity. The southeast siting allows the commons to open to the sun while shielding northern winter winds and highway noise (Fig.5-2-4).

Above the commons, two acres of green roofs and garden terraces over structure provide employees with immediate access to outdoor spaces at mid-level connections. The roof gardens reduce rainwater runoff and add insulation and sound attenuation to the broad floor plates of the news rooms below.

Pedestrian and water movement through the site emphasize an engagement with the topography through a series of runnels and terraced gardens. This framework is conceptually derived from the naturally occurring pattern of fallen logs on a forested slope offering eddy spaces in the landscape. This concept is developed with angled stone walls that terrace the topography of the commons and direct rainwater through riparian step pools connecting to the forebay and regional storm water management pond. These walls form distinctive lawn and planting areas along with runnels, pools, niches and slotted stair passages linking intimate spaces that look out to the pond and

项目的植物配置方案从山顶的丛林风貌一直延伸和扩展到了地势较低的草甸风貌，并且使雨水调蓄池塘变成了一个野生动植物的庇护所和栖息地（图5-2-6）。景观设计师选择了大量野生的当地树种和灌木，并充分体现了他们的野生价值。空地上种植了大量适宜湿地和滨水区的乔灌木，如郁金香、黑紫树、蓝果树、柳栎以及弗吉尼亚木兰。植物高大的树冠提供了阴凉处，沟渠和水塘里生长着大片的蕨类植物，例如北美梭鱼草和香蒲。石阶墙上爬满了爬山虎和五叶地锦（图5-2-7）。

五、 雨水调蓄池设计

场地开发前已有一块5英亩（约2公顷）的雨水调蓄池塘，可以涵蓄泰森角270英亩地区的绝大部分的水量。景观设计师经过区域分析，对这一雨水调蓄池塘的功能、调蓄能力和现有系统净化能力进行评估，并且评估目前的功能缺失。为了使之发挥更大的作用，在设计中不宜改变池塘现有形状。

图5-2-6 植物设计
Fig.5-2-6 The planting plan

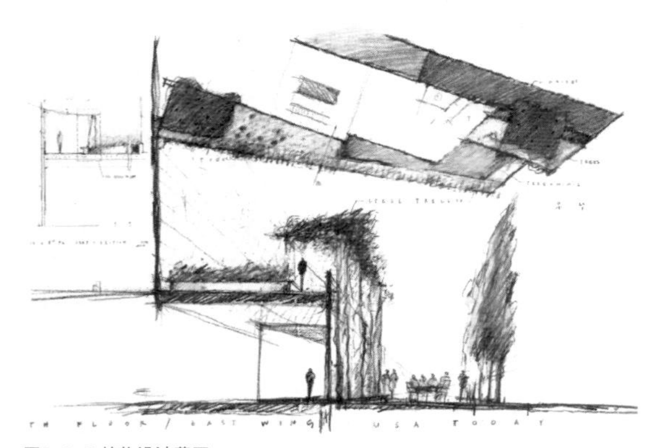

图5-2-7 植物设计草图
Fig.5-2-7 The sketch of the planting plan

the wooded hilltop beyond (Fig5-2-5). Recirculated water from the pond is oxygenated and cleanse as it drops through planted pools at wall gaps. Constructed with the same stone as the building base, these walls create a textural link between building and landscape.

IV. Planting Design

The planting plan complements and extends the wooded hilltop out into the lower meadow landscape and tums the stormwater management pond into a wildlife refuge (Fig.5-2-6). The landscape architect chose a broad range of native trees and shrubs for their wildlife value. Trees and understory plants associated with moist soils and riparian conditions, such as Tulip Poplars, Black gums, Willow Oaks, and Sweetbay Magnolias, sweep up into the commons providing shade and shelter. Drifts of ferns, pickerel weed and cattails are planted in the runnels and pools, while climbing plants such as Boston Ivy Virginia Creeper soften and green the terrace wall facedes (Fig.5-2-7).

V. Stormwater Pond Design

The predesign site contained an existing 5-acre stormwater pond that detained water from the surrounding 270-acre watershed-a large portion of Tysons Corner. Regional analysis by the landscape architects assessed the stormwater function, capacity, and health of the existing system and provided cues as to what was missing. Due to its larger function, the general pond shape could not be altered as part of the site improvement plans.

However, the landscape architects enhanced the pond by adding a forebay to improve stormwater quality. This forebay allows the introduction of a shady buffer planting that cools, slows down, and filters incoming stormwater run off. It allows water to slow and sediment to drop(Fig.5-2-8). It also raises and extends the water level, which makes the pond more visible from the building, whose views are oriented toward the pond. A stone pump house functions as a landscape feature as well as pumps water for lawn irrigation and recirculates the central court terrace runnels that aerate the system. Pond-edge bird counts by employees include green and great blue herons.

VI. Human System Design

The design provides a variety of habitable spaces that integrate the architecture with the landscape. Intensively planted roof gardens at multiple levels and outdoor terraces at ground level provide spill-out areas for gathering, outdoor dining, and working that blend seamlessly with interior spaces (Fig.5-2-9). Both the intact and renovated landscapes—from open lawn

然而在设计中，景观设计师为雨水调蓄池塘增加了一个前处理池，以改善调蓄池中的雨水水质。这个前处理池中种植植物以使雨水流入时受到缓冲，可以使流入池塘的雨水冷却、减缓速度，并且过滤流经的雨水（图5-2-8）。它可以使雨水流速变缓从而更好地沉淀沉积物。同时还能扩展水面面积，使人们能够更好地从建筑中欣赏池塘景观。石坝作为景观元素之一，还兼有灌溉草坪和循环回用水的功能。池塘边缘养了一些种类的鸟类，包括大型的苍鹭。

六、 人类活动场所设计

设计提供了一个广阔的室外空间，使建筑与景观紧密相连。屋顶花园上错落有致地密集种植植物以及在地面露台为聚集、户外烧烤以及工作等户外活动与室内空间无缝衔接（5-2-9）。不论是原有的还是装饰过的景观，从开敞的草地到郁郁葱葱的林地以及到池塘边缘，场地被设计成多种综合用途，例如中央露台、步道系统以及球场、林荫道、池塘凉亭以及坐椅。与那些周围除了停车场没有其他景观的商业设计模式相比较，这样的景观设计提供了多种人与景观互动的模式。

七、 总结

一个项目的可持续发展在很大程度上取决于团队在完成以下三个任务中的设计方式：发展计划的目标设定，在场地评估中作出的选择，在评估背景中作出的选择。对本项目而言，以上三项任务与以下两个元素相关：环境因素和文化因素。在最初的设计中，并没有关注到场地的可持续发展以及场地与地区的生态、文化特征。景观设计师在方案细化阶段推出了可持续设计方案后，可持续设计方法成为项目的目标并贯穿始终。

初步设计阶段，设计师说服报业集团总部使项目选址由视野极佳的山顶改为曾作为垃圾填埋场的山下草地，设计师的自然使命感和业主的区域生态观令人叹服。这是对风景园林师使命的最佳注

to shaped woodland to pond edge to plaza –offer multiple uses of the site such as the central court terrace meander, walking paths, ball field, woodland walk, pond pavilion, and seating. In contrast with the surrounding business development where landscape are dominated by parking lots with no outdoor space except the sidewalk, this landscape provides multiple opportunities to engage with the landscape.

VII. Summary

The sustainable outcome of a project is largely determined by the design team's approach toward three tasks: the development of program goals,choices made in the assessment of the site,choises made in the assessment of the context. The key features of these three tasks for the Gannett/USA Today Headquarters project are listed in relation to two important sustainable factors:environmental factors,cultural factors. The relationship shows that the initial program goals of the project were not focused on sustainability,nor did the treatment and planning of the site and region have much ecological or cultural integrit.The landscape architects introduced sustainable site planning approaches later during the program refinement phase.Sustainable planning techniques were overlain with the basic program goals—and were not an integral part from the profect inception.

During the preliminary design phase, the architect persuade the The Gannett/USA Today Headquarters to move the building site from the healthy wooded hilltop to the meadow/fill site. The architect's mission

图5-2-8 水生植物
Fig.5-2-8 The aquatic plant

图5-2-9 人类活动场地
Fig.5-2-9 The human system design

脚——风景园林师作为土地的守望者、资源的守护者和景观环境的营造者，是有效协调人地作用机理与从事生态规划设计的主体之一。

设计的亮点之一是引入场地修复理念，倡导投入时间、成本、精力进行场地生态系统的修复。目前我国的相关研究主要聚焦于污染场地的修复，尚处于起步阶段。污染场地也称棕地，是指受到一种或多种有害物质污染但经过清理之后仍有潜力重新开发利用的土地。近年来，北京、沈阳、广州以及其他一些中国大城市将许多污染企业迁出城市中心，产生了大量棕地，这些棕地在经过清理达到适当的安全标准之后可以再开发利用。

亮点之二是设计中的一处区域性雨洪调蓄池。雨洪调蓄池的建设，不仅能消减下游管渠中的洪峰流量，对减小城市排水工程的造价也有良好作用，更能最大限度地实现雨水资源化利用。此外，它还可改善雨洪径流的水质，减轻对城市下游地区的洪水威胁。这一技术在我国也逐渐推广普及。

of natural sense and the regional ecological view of the owner are marvelous. This is the optimal footnote for the mission of the landscape architects——he landscape architects as the watchman of land resources, the guardian for resource and builder for landscape environment. They are effective mechanism and engaged to coordinator of the ecological planning and ecological design.

One of the highlights of the design is the concept of field repair, which advocated investing time, cost and energy to repair the ecological system of the field. At present, researches in China often focus on the restoration of the pollution sites,which is still in the initial stage. Pollution sites are also called brownfield. This kind of field are always polluted by one or more harmful material pollution but still has the potential for utilization after pollution control. In recent years, many enterprises which produces pollution in Beijing and shenyang, guangzhou and other Chinese cities are moved out from the city center.Then there are lots of brownfield in the center of these cities,which can be reused after cleaning and achieving adequate safety standards.

The other highlights of the design is the stormwater pond design.This stormwater pond design can not only reduce the peak flow in the downstream pipe canal,but also reduce urban drainage engineering cost.It also can maximally realize rainwater utilization. In addition, it also can improve the quality of the the rainfall flood runoff, easing the downstream flooding of the city. This technology also gradually popularization in our country.

思考题

1. 请简述甘尼特报业集团总部景观设计的项目概况。

2. 甘尼特报业集团总部景观设计的场地设计分为哪三个不同的景观空间？如何体现了场地修复的概念？

3. 本设计的种植设计原则是什么？

4. 雨水调蓄池的设计功能是什么？试举例其在我国的设计应用。

5. 人类活动场所设计原则是什么？

Questions

1. Please introduce the overview of Landscape Design for The Gannett/USA Today Headquarters.

2. What are the three distinct landscape spaces in the site design? How to embody the field repair concept in this Design?

3. What are the Principle of Planting Design?

4. What are the functions of the Stormwater Pond in this design? Please give some examples of application in China.

5. What are the principles of Human System Design in The Gannett/USA Today Headquarters?

注释/Note

图片来源/Image：本案例图5-2-1来自http://www.kpf.com/projects/Project2/11080 _ hr.jpg；其余图片来自“Sustainable Site Restoration-Landscape Design for Ganne”以及“Gannett/USA Toda Headquarters”

参考文献/Reference：本案例引自“Sustainable Site Restoration-Landscape Design for Ganne”以及“Gannett/USA Toda Headquarters”

推荐网站/Website：http://www.asla.org/awards/2008/08winners/415.html

第三节 克罗斯温茨湿地
Section III Crosswinds Marsh

一、 项目综述

作为底特律市韦恩县机场扩展项目的一部分，机场管理机构依照国家和联邦法律的要求，聘请史密斯建筑事务所来设计新的湿地以补偿那些在建设过程中损失掉的湿地。史密斯建筑事务将其环境保护专长运用到项目中，而且设计工作远远超出甲方的基本需求，创造了北美洲最大的自给型补偿性湿地，克罗斯温茨湿地还是游憩公园和野生动物保护区。湿地内部的自然系统循环良好，所以它根本不需要任何的水泵或者人工手段来稳固湿地群落。更重要的是，保护区内有木板路和平台组成的网络，以供公众使用。因此，克罗斯温茨湿地既是一个颇受欢迎的游憩目的地，也是一个极佳的学校学生和社区团体的教育基地（图5–3–1，图5–3–2）。

二、 场地及其背景介绍

克罗斯温茨湿地是一个1 400英亩（约566公顷）的多用途湿地补偿项目，位于密歇根州韦恩郡的西南部。该项目始于1986年，当时底特律韦恩机场正在更新它的总体规划，包括了一个主要跑道的扩张，这一扩张需要大范围的湿地改造和补偿。机场的补偿规模和空间限制促使机场管理者去寻求一个位于“机场外”同时又处于同一流域的补偿区域。史密斯建筑事务是当时该项目的首席顾问。最

I. Overview

As part of the expansion of the Detroit Metropolitan Wayne County Airport, the Airport Authority hired JJR to create new wetlands to replace those lost during construction in accordance with state and federal regulations. JJR put its environmental expertise to work and far exceeded the basic requirements, creating one of the largest self-sustaining wetland mitigation projects in North America,Crosswinds Marsh is also a recreational park and wildlife refuge. The natural systems function so well, Crosswinds Marsh requires no pumps or other artificial methods to maintain the wetland communities. What's more, the preserve includes a network of boardwalks and platforms that encourage public use. As a result, Crosswinds Marsh has become both a popular recreational destination, as well as an excellent educational tool for school children and community groups (Fig.5-3-1,Fig.5-3-2).

II. Introduction of the Site and Context

Crosswinds Marsh is a 1,400-acre (566 hectare) multipurpose wetland mitigation project located in southwestern Wayne County, Michigan. The project originated in 1986 when the Detroit Metropolitan Wayne County Airport updated its master plan, including a major runway expansion requiring extensive wetlands disturbance and mitigation. The scale of the mitigation and space limitations at the airport site led airport managers to seek an "off-airport" mitigation site within the same watershed. Smith Group JJR served as the prime consultant for the project.A location in Sumpter Township

图5-3-1 建设前的克罗斯温茨湿地
Fig.5-3-1 The Crosswinds Marsh before construction

图5-3-2 建设后的克罗斯温茨湿地
Fig.5-3-2 The Crosswinds Marsh as built

终选择桑普特镇的一片区域作为补偿区，因为它足够大，同时又位于韦恩郡的农村部分，并且可以满足补偿区的水文需求。

从桑普特镇韦恩郡私人土地所有者手中获得的土地已超过了1 400英亩（约567公顷）；其中约620英亩（约250公顷）的湿地已经营造或者恢复，成为克罗斯温茨湿地的一部分。在第一阶段，约有320英亩（约130公顷）的新湿地是由史密斯建筑事务所设计的。第二阶段，也就是最后一个阶段，于2000年完成。所有的工作都由韦恩郡机场部门管理。20世纪90年代，该项目历经计划、批准、建设，成立和监督。随着该项目的贯彻实施，它已经不仅仅是一个湿地补偿项目了，同时它还具有重要的生物多样性意义和公众使用功能。从那时起，它就成为一个备受赞誉的项目，1999年它获得了美国风景园林师协会颁发的主席奖。克罗斯温茨湿地现在是韦恩郡一个成功的公园和环境教育中心。从监测中获得的新数据使得研究者对密歇根州以及遍及美国中西部的其他湿地的恢复产生了新的认知。

克罗斯温茨湿地地势很低，这一区域是由艾瑞克古湖的湖底沉积物形成的，包括了滩脊和由一些蜿蜒水沟组成的水文模式。历史上，该地区曾经是一个由湿草甸、湿地和一个树木群落构成的湿地景观，后来它被人为地用农业沟渠以及排水管道抽干，用以种植粮食作物（玉米和大豆），放牧和干草耕地。在90年代早期，项目刚开始的时候，该位置的一大部分仍旧为农业所用。剩余的部分被作为居住用地或者正处于林地演替过程中。

湿地应该具有补偿作用这一目标是由美国环境保护机构、美国渔猎局、密歇根自然资源部门（现在被叫做密歇根州环境质量部）、韦恩郡其他不同

was selected for the mitigation site because it was sufficiently large, was located in a rural portion of Wayne County, and had the hydrological "potential" to support the required area needed for mitigation.

The property acquired from private landowners by Wayne County in Sumpter Township exceeds 1,400 acres (567 hectares); of this total about 620 acres (250 hectares) of wetlands have been created or restored at Crosswinds Marsh. In phase 1, approximately 320 acres (130 hectares) of new wetlands were designed by Smith Group JJR. Phase two, the final phase, was completed in 2000.All work was managed by the Wayne County Department of Airports. During the 1990s, the project proceeded through planning, permitting, construction, establishment, and monitoring. As the project was implemented, it grew well beyond a wetland mitigation project, adding significant biodiversity and public use components. It has since become an award-winning project, receiving the President's Award of Excellence from the American Society of Landscape Architects in 1999. Crosswinds Marsh is now a successful Wayne County Park and Environmental Education Center. New data from monitoring has produced new knowledge for wetland restoration elsewhere in Michigan and throughout the Midwestern United States.

Crosswinds Marsh is located in an area of low relief formed on ancient Lake Erie lakebed sediments, including beach ridges and a hydrological pattern composed of several meandering channels. Historically, the site supported a wet prairie, marshlands, and a wooded wetland landscape, which had been artificially drained with agricultural ditches and tile drains to support row crop (corn and soybeans), pasture, and hayfields. At the project's beginning in the early 1990s, a large proportion of the site was still used for agriculture. The remainder was in residential use or in the process of woodland succession.

Project mitigation goals were developed through consultation among the U.S. Environmental Protection Agency(EPA), the U.S. Fish and Wildlife Service (USFWS), the Michigan Department of Natural Resources (now the Michigan Department of Environmental Quality and referred to in the remainder of this chapter as MDEQ), various Wayne

部门、公民建议委员会和史密斯建筑事务共同磋商从而制定的。协商结果包括一系列湿地类型：森林、湿草甸、开阔浅水区和深水区。前三种类型可以补偿森林、湿草甸和机场湿地。深水湿地还可以用于建立温水渔场（图5-3-3）。

密歇根州环境质量部负责检查和许可认证，它要求311英亩（约125公顷）的湿地达到1.5:1的补偿比例，这些湿地由于跑道的扩张受到阻碍并且与底特律韦恩机场项目相关联。克罗斯温茨湿地的选址是由许可机构协商而定的，由补偿湿地的选址而得出了几项结论。首先，场地位于机场外，这样可以避免水鸟给飞机带来的不必要的风险。其次，流域外的位置需要提供足够的未开发土地并尽可能减少社区影响。最后，评估者认为一个整体的大区块相对于分散的几个小区块来说，对营造高价值的野生动物栖息地更加合适。

在建造过程中，克罗斯温茨湿地是美国最大的独立补偿项目。这一项目由大量的推土机作业，超过75万立方码（约573 450立方米）土方用于创造一个湿地流域，该流域范围在0～20英尺（约0～6米）深。克罗斯温茨湿地项目总花费约为1 200万美元，其中75%的费用是由美国联邦航空局的联邦跑道扩展基金支付的。这一项目是在机场和韦恩郡的大力支持下开始的。它的规模非常大，从而获得了公众的注意，所以它需要具有更广泛的功能并且使更多的人受益，包括生物多样性，公众可达性和娱乐性。史密斯建筑事务与各团体共同合作，从而将这些目标与功能整合到补偿项目的规划中。

三、 规划与设计介绍

1. 项目参与者

克罗斯温茨湿地的建设得到了联邦、州以及郡县各机构的大力支持。美国联邦航空局通过机场债券为其提供了基金。密歇根州环境质量部负责审查和批准颁发所有许可证。韦恩郡底特律机场的部门主管、工作人员和桑普特镇的官员都参与到整体的审查和项目批准中来。

史密斯建筑事务是规划，设计和施工等方面的首

County departments, a citizen's advisory committee, and Smith Group JJR. The resulting goals included a range of wetland types: forested, wet meadow, emergent, shallow open water, and deep water. The first three types were required as compensation fimpacts to forested, wet meadow, and emergent wetlands at the airport. The deep-water wetland was added to establish a warm-water fishery(Fig. 5-3-3).

MDEQ was responsible for reviewing and approving the permit, which required a 1.5:1 mitigation ratio for the 311 acres (125 hectares) of wetlands that were to be disturbed for the runway expansion and associated projects at the Detroit Metropolitan Wayne County Airport. The Crosswinds site was proposed in response to consultations with the permitting agencies (principally the MDEQ, EPA, and USFWS). At the time, several conclusions were reached regarding the mitigation site. First, an off-airport site would avoid unnecessary hazards to aircraft from waterfowl. Second, an out-of-basin location was required to provide sufficient undeveloped land and to minimize community impacts. Finally, a larger site was considered preferable to several smaller sites for creating more valuable wildlife habitat.

At the time it was constructed, Crosswinds Marsh was the largest single mitigation project in the United States. The project consisted of a massive earth-moving effort involving more than

750,000 cubic yards (573,450 cubic meters) of earth to create a wetland basin ranging from 0 to 20 feet (0 to

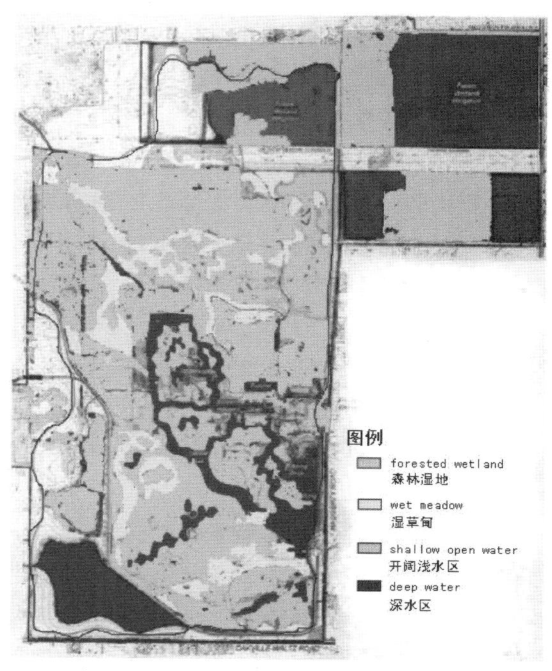

图5-3-3 克罗斯温茨湿地类型图
Fig.5-3-3 Wetland types of Crosswinds Marsh

席顾问。项目由史密斯建筑事务的环境工作室管理，其中就包括了风景园林师和环境科学家，他们在一个跨学科的体系下一同工作。其中专业教授包括盖瑞、凯瑟琳（他们负责水质和鱼类），还有威廉（负责植物）。顾问包括艾伦（来自东部密歇根大学，负责研究爬行动物，两栖动物和哺乳动物）和史蒂芬（来自密歇根大学，研究鸟类）。这些科学家进行基线数据搜集和执行后的监测，同时景观规划设计师做出设计决策和施工图，并监督项目建设（图5–3–4）。

史密斯建筑事务的项目团队根据该项目的补偿区的灵活性，共同建立了一个“生态框架”。风景园林师最初的工作是了解并确定沉降前的植物环境，并且运用严谨的科学原理对这一项目加以解释。整个项目中，园林设计师领导跨学科的团队研究栖息地的设计方法，并且对多变的土地环境和种植法进行试验（如，用大小不同的植物，在无表层沙土的上升犁田上种植）。同时，风景园林师还要对项目的公共使用部分进行规划，包括木制小道、徒步道、骑马道、独木舟游径（图5–3–5）。他们还成功的推广了木栈道和游径的使用材料，并且建立了湿地教育中心（图5–3–6）。

项目工程师负责建立水文模型并解释所有关于补偿克罗斯温茨湿地周边地区洪水的问题。迪斯布劳和克拉克–莫里是场地内的两条排水渠。项目工程师认为迪斯布劳排水渠可以持续地将水排入湿地而成为湿地的主要水源，并且可以改善下游水质。克拉克–莫里排水渠降低了水质，所以改变它的流向将其引到湿地外围而保持其水文上的独立性。工程师同样也建议用黏土隔离墙来隔离湿地与周边事物，其中就包括位于斜坡上的地区填埋场。工程师设计并设置了一套具体的外流控制结构来为整个项目建立一个基础水平面提升。这一控制结构可以阻止上游的鱼类迁徙，尤其是一些不良鱼种，其中包括鲤鱼类和黄鱼类。

一些团体作为公民建议委员会的一员参与到了项目中；另外一些更活跃的参与者包括密歇根联合保护协会，东密歇根环境行动委员会和密歇根大学。他们的兴趣多集中在为湿地补偿位置的类型和公共合理使用。他们更关注在湿地补偿区建设（项

6 meters) deep .The total cost of the Crosswinds Marsh project approximated $12 million, 75 percent of which was paid federal runway expansion funds from the Federal Aviation Administration. The project began with strong support from the airport and from Wayne County to enable the airport expansion. Its large scale garnered public attention and interest in providing additional functions and benefits to the project, including biodiversity, public access, and recreation. Smith Group JJR worked with all patties to integrate these objectives and functions within the mitigation project.

III. Site Plan and Design

1. Project Participants

Crosswinds Marsh involved a large complement of federal, state, and county agencies. The Federal Aviation Administration provided funding through airport bonds. MDEQ was responsible for reviewing and approving all permits. Wayne Country department supervisors, staff at Detroit Metropolitan Wayne County Airport, and Sumpter Township officials were involved in the overall review and permitting of the project.

Smith Group JJR was the prime consultant for planning, design, and engineering. The project was managed within the Smith Group JJR Environmental Studio, which includes landscape architects and environmental scientists working in an interdisciplinary setting. The professionals involved included Gary Crawford and Catherine Rising (aquatic acro-invertebrates, fish and water quality) and William Brodowicz (plants). Subconsultants included Allen Kurta (Eastern Michigan University; reptiles, amphibians ,mammals) and Stephen Hinshaw (University of Michigan; birds).The scientists conducted baseline data collection and post-implementation monitoring, while the landscape architects made the design decisions, produced the working drawings, and supervised project construction(Fig.5-3-4).

The SmithGroup JJR project team collaborated to establish an "ecological framework" for the project regarding how much mitigation flexibility was possible. The landscape architects' initial role was to provide an understanding of the presettlement vegetation context and to interpret and apply rigorous science to the project. Throughout the project, landscape architects led the interdisciplinary team, researching habitat design methods and experimenting with alternative soil mediums and planting methods (e.g. no topsoil, sand, planting in raised furrows, using different size plants).The landscape architects also led the planning of the public use

component of the project, including boardwalks and hiking, equestrian, and canoe trails(Fig.5-3-5). They also successfully advocated for the inclusion of boardwalks, trails, and an arrival educational wetland

图5–3–4 克罗斯温湿地总体规划图
Fig.5-3-4 Master plan of Crosswinds Marsh

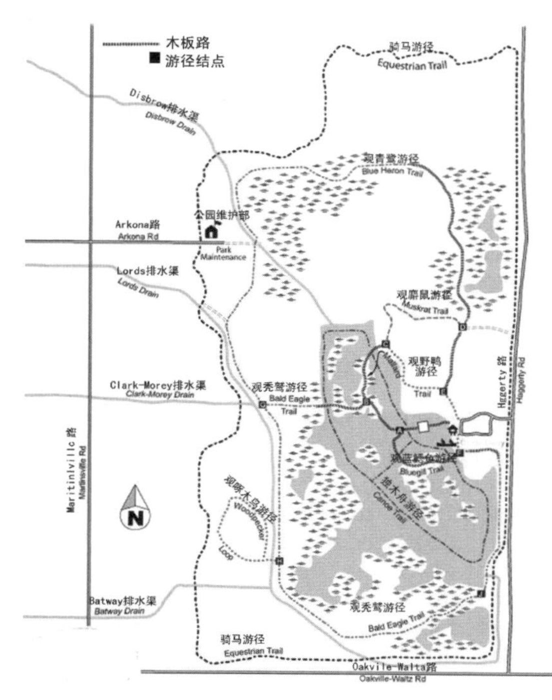

图5–3–5 克罗斯温湿地游径系统
Fig.5-3-5 Crosswinds Marsh trail system

图5–3–6 游客与环境教育中心
Fig.5-3-6 Visitor and environmental education center.jpg

center (Fig.5-3-6).

Project engineers were involved with the hydrologic modeling to address any concerns for post mitigation flooding of areas adjacent to the Crosswinds Marsh site. Two drainage channels existed at the site, the Disbrow and the Clark-Morey. The project engineers determined that the Disbrow Drain could continue to flow into and through the marsh to serve as the wetland's primary hydrological source and to improve downstream water quality. The Clark-Morey Drain, which had impaired water quality, was diverted around the perimeter of the marsh to maintain hydrological isolation. The engineers also recommended the use of clay cut-off walls to hydro logically isolate the marsh from its surroundings, which include a regional landfill situated down gradient. The engineers designed and set concrete outflow control structure to establish the base water elevation for the entire project. The control structure reinforced the decision to stop upstream fish migration because of concerns about immigration of undesirable fish species, including common carp (Cyprinus carpio) and gizzard shad (Dorosoma cepedianum).

Several groups participated as part of the citizen's advisory committee; among the more active participants were the Michigan United Conservation Club, the East Michigan Environmental Action Council, and the University of Michigan, Dearborn..Their interest focused on the type and location of public use appropriate for the site in the context of wetland mitigation—the primary goal—and associated habitat creation.

2. Project Goal

The Crosswind Marsh project began solely as a compensatory wetlands mitigation project. Thus the principal goal was to satisfy the requirements of the MDEQ permit to enable the runway expansion to be built, The permit specifically required that the mitigation be "sufficient in terms of physicals characteristics, water supply, and any other pertinent ecological factors to create the desired wetland types".

As the project evolved, Smith Group JJR, working in collaboration with Wayne County, advocated three

目主要目标）以及相关的生物栖息地建设的背景下，场地中有哪些区域适合建设公共活动区，其形式又是怎样的。

2. 项目目标

克罗斯温茨湿地项目最初只是一个补充性的湿地补偿项目。因此其主要目标是满足密歇根州环境质量部的需求从而使机场跑道扩展项目能够得到建设。密歇根州环境质量部要求补偿区应该有充足的物理特性，水供给以及其他相关的生态要素，从而创造一个理想类型的湿地。

随着项目的发展，史密斯集团与韦恩县共同合作为该项目提出三个额外目标：

（1）发展教育和消极娱乐使用；

（2）为公共利用提供机遇；

（3）把克罗斯温茨湿地建成一个县级支持的“保留性解释湿地”。

依据项目计划，这些额外的目标有利于创造“永久性的多样性的环境及社会功能”，这对韦恩郡也是十分重要的。同时这对于县内解决地方问题也是非常重要的，这些问题源于约30个人从家乡的迁离，并且失去了将近1 400英亩（约567公顷）的土地，这些土地曾经是私人所有的。

3. 公众、个人伙伴关系及合作

项目批准、设计和实施的过程涉及许多公共和个人的机构及其利益。当地社区也参与各种项目概念的讨论中。最初，该社区并不理解或重视如此大规模补偿项目的潜在利益。它曾经被简单地理解为是一个“大湿地”——它会替代农田和居民区，但是却没有什么积极的作用。管理者举办了数次的社区听证会，让公众更加了解这一项目的需求，并且意识到它将有潜力成为韦恩县的一项重要资产。

密歇根联合保护协会与其他镇一同参与到项目的发展中来。桑普特镇要求获得湿地的狩猎许可，但由于安全原因被否决。作为公众听证会和项目顺利实施的直接结果，克罗斯温茨湿地从社区居民眼中的大湿地逐渐转变成了一个成功地服务于当地居民的县城公园（图5-3-7）。在周围高地为骑马者

additional goals for the project:

(1) To develop educational and passive recreational use;

(2) To provide opportunity for public use;

(3) To establish Crosswinds Marsh as a county-sponsored "Wetland Interpretive Reserve".

According to the mitigation plan, these additional goals were added to provide "a variety of environmental and social functions in perpetuity". It was also important to the county address local concerns that arose from the displacement of about thirty individuals from their homes and the loss of near 1,400 acres (567 hectares) of land from the township tax base, which had previously been under private ownership.

3. Public/Private Partnership and Collaboration

The process of project permitting, design, and implementation involved multiple public and private organizations and interests. The local community was involved as various project concepts were developed. Initially, the community did not understand or appreciate the potential benefits of such a large-scale mitigation project. The plan was understood simply as a "big swamp"—displacing farmland and local residents while offering little or no positive attributes. Several community hearings were held to raise public awareness and understand of the project's requirements and its potential to become a significant asset for the country.

The Michigan United Conservation Clubs were involved during the evolution of the project along with the other members of the township and the country. Sumpter Township asked for hunting access to the wetland, but this was denied for safety concern reasons. As a direct result of the public hearings, and of the project's successful implementation, Crosswinds Marsh evolved in the community's eyes from a swamp into a successful country park for use by local residents (Fig.5-3-7). Perimeter upland trails were installed for equestrians, and an extensive system of drains, trails (accessible per the standard set forth in the Americans with Disabilities Act), and board walks were located throughout the site(Fig.5-3-8,Fig.5-3-9).

4. Biodiversity Data Issues and Planning Strategies

As in most mitigation and restoration projects, Crosswinds Marsh confronted a distinct lack of data regarding the specific ecosystems, plant communities, and biota involved. To overcome this lack of information, the project planners and designers conducted two complementary lines of research: a review of published literature and an inventory of adjacent wetlands. SmithGroup JJR researched vegetation by examining the only available source of

设置了游径，还设计了广泛的排水系统和道路系统（符合美国残疾人法案的标准），并且规划了遍布全区的木质道路（图5-3-8，图5-3-9）。

4. 生物多样性数据问题和规划战略

和大多数补偿及恢复项目一样，克罗斯温茨湿地也面临着特殊生态系统、植物群落和生物群系等方面的数据的缺失。为了克服信息缺失这一困难，该项目的策划者和设计者实行了两条互补的研究线路：已出版文献和邻近湿地的清单。史密斯建筑事务所检查了这一特定区域唯一可用的信息资源（移居后的测绘图和基于历史的测量图）来研究其植被特征。通过这一研究列出了韦恩县现存湿地生境类型，以确定规划区周边范围动植物多样性及分布情况。该清单记录了现存物种类型以及与湿地栖息地相关的特殊湿地物种，从而指导补偿规划的进一步发展。场地的植物学调查显示当地并没有珍稀的濒临灭绝物种。然而在机场却发现了受到威胁和濒临灭绝的植物物种；这些物种被记录在案并且以栅栏

图5-3-7 儿童、教师、大学新生们的户外学习基地
Fig.5-3-7 Scout groups, teachers, and elementary to university students use the preserve as an outdoor learning laboratory

图5-3-8 木板路
Fig.5-3-8 Boardwalks

this site-specific information: resettlement mapping and historical survey maps. An inventory was made of existing wetlands habitats in Wayne County to identify the diversity and distribution of plants and animals surrounding the project area. This inventory influenced the development mitigation plan by producing species lists and associating wetlands habitats with particular wetland species.

Botanical surveys at the mitigation site found no threatened or endangered wildlife species. However, threatened and endangered flora species were found at the airport; these species were documented and protected with fencing to avoid disturbing them during runway construction. Selected occurrences of at-risk endangered plant species were relocated with a tree spade to a suitable location within the mitigation site. The design of suitable habitat for the plant species was based largely on identifying soil types and hydrological conditions were established by using monitoring wells at both the airport and the mitigation area and then grading the mitigation site to achieve comparable hydrological conditions. The land is determined according to the inspection of existing conditions of the compensation site, and the construction of right hydrological conditions is through the supervision of airport and compensation site, then scoring the compensation site which have similar hydrological conditions .

The Disbrow Drain supported a warm-water fishery and macro-invertebrate population typically found in agricultural drains, freshwater marshes, and shallow water communities. Results of the initial aquatic surveys, however, indicated that large numbers of common carp (Cyprinus carpio) and gizzard shad (Dorosoma cepledignum) migrating from Lake Erie were disrupting in-stream habitat, dislodging macro-invertebrates, and displacing resident fish. This prompted the decision to "isolate" the hydrologic regime of the wetlands by cutting off upstream migration potential from Disbrow Drain.

The literature review, site investigation, and survey of nearby reference wetlands provided sufficient information to produce a comprehensive mitigation plan. The plan was developed in phases initially

图5-3-9 网络化的木板路与观望台保护了栖息地的敏感性并且鼓励公众使用
Fig.5-3-9 A network of boardwalks and observation platforms protects sensitive habitats and encourages public use

围住以保证其在跑道建设期间不受干扰。一些处境危险的濒危植物种类被重新移植到补偿区其他合适的位置。为濒危植物寻找适合的栖息地大多是依据土地类型和水文条件，工作人员同时对机场和补偿区进行监测，然后对补偿区进行等级划分并选择出与植物原栖息地水文条件类似的区域。土地的确定是根据补偿地点现存的条件的检查，而合适的水文条件是通过使用监督机场和补偿区域而建立的，而后给获得类似水文条件的补偿地点评分。

迪斯布劳排水渠可以用于温水鱼类的养殖，并且为在农田水渠，淡水湿地和浅水区发现的大型无脊椎动物提供生存环境。然而最初的水生生物调查结果显示，大量的鲤鱼和黄鱼从艾瑞湖迁徙而来，扰乱了河中的栖息环境，排挤走了大型无脊椎动物并取代了本土鱼类。因此管理人员决定切断迪斯布劳排水渠与上游的迁徙潜力，从而保持湿地水文上的“种群隔离性”。

文献资料总结，场地调研和附近湿地的调查为制定综合补偿规划提供了足够的信息。该规划的进展分为几个阶段，最初是由评分规划来决定的，因此划分了多种不同水位的湿地类型。植被规划为生物多样性的恢复或湿地、高地生境重建提供了一系列管理技术。

5. 项目评估

植物质量测评系统被用于场地的监测过程中。对20个试点地区超过5年的不断监测不仅记录了补偿区物种的构建率，同时也展示了在不同的季节性雨水量的条件下各种物种的比例。每个物种至少有两个试点区，沿着试点区每50英尺（约16米）会设置一个10.8平方英尺（约1平方米）的试验点。每个试验点的中心都用木桩标记，并且划分出了种植区和非种植区。按照规定在1994到1998年间对湿地进行了监测。1999年韦恩县主动采取了对地区子群的额外监测。除了拍照记录之外，每个试点的覆盖率、指示物种的出现频率、水位以及PH值等信息也都被记录了下来。

水文监测一年进行2次，分别在春季和夏末，记录下水深和地下水位。永久试验点的水质监测每年进行3次，分别在春末、夏季和秋季。温度、溶解

determined by the grading plan, which established a diversity of water depths correlated with specific wetlands types. The planting plan applied a range of management establishment techniques to restore diversity or wetland and upland habitats.

5. Project Evaluation

The Floristic Assessment system was used in the monitoring process at the site . Continual monitoring of twenty transects over five years has not only documented the establishment rates for mitigation species but showed how the proportion of species has changed in response to seasonal moisture conditions. A minimum of two transects was established in each habitat type, with 10.8-square-foot(1 square meter) plots established every 50 feet(16 meters) along the transect. The center of each plot was marked with stakes. Plots were selected to represent planted and unplanted areas. Permit-required wetland monitoring was conducted from 1994 to 1998. Additional monitoring of a subset of plots took place on the county's own initiative in 1999. In addition to photographs, the information recorded for each plot included percentage cover, frequency of indicator species, water level, and pH.

Hydrology was monitored twice a year in spring and late summer, recording readings from water-depth gauges and groundwater wells. Water quality was monitored three times a year, in late spring, summer, and fall, at permanent entering and existing sample points .Temperature, dissolved oxygen, pH, and conductivity were also measured. The hydrologic and water quality data collection has been discontinued. Wayne County does continue to collect water-quality data for bacteria because canoe rentals involve water contact.

Several classes of fauna were monitored: birds, herptiles, and mammals. Birds were monitored three times each year, during the spring migration and in summer and fall Specific stations were visited in the morning for a five-minute observation that included visual and audio

observations as well as callbacks (human-initiated birdcalls). The herptiles were monitored in conjunction with the vegetation sampling three times each year. Amphibians and reptiles were also monitored from March to June using night trapping, with drift fences and pitfall traps. Mammals were monitored with live traps set along the permanent transects. Muskrat houses and evidence of muskrat grazing were also noted.

Finally, aquatic biology was monitored three times a year in spring, summer, and fall using a stratified random sample. Macro-invertebrates were analyzed in July at seven locations for tolerance rating using the Shannon-Weaver diversity index (Shannon and Weaver 1949). Fisheries were monitored using seines

氧、pH值和导电性同样被监测着。水文和水质数据采集现如今已经停止了。由于有独木舟出租项目，所以韦恩县还在继续采集水质数据以监测细菌量。

鸟类、爬虫类和哺乳动物等动物种类也需要进行监测。鸟类每年被监测3次，分别在春季迁徙时节、夏季和秋季。特殊的监测站每天早上观察5分钟，观察包括视觉和听觉的观察并进行复查（通过人类模仿鸟鸣）。爬虫类与植被样本一同被监测，每年3次。每年3～6月监测两栖动物和爬行动物，运用放牧围栏及圈套陷阱的夜晚捕捉方法。哺乳类动物通过设置在永久断面的活兽陷阱进行监控，同样也会监测麝鼠的洞穴和麝鼠磨擦过的痕迹。

最后，水生生物每年监测3次，分别在春季、夏季和秋季，使用的是分层随机抽样的方法。每年的7月在7个不同的位置对大型无脊椎动物进行监测，并且运用莎伦——韦富多样性指标来进行分析。鱼类的种类和大小则通过使用浅水渔网和深水刺网来监测。

密歇根州环境质量部要求每5年进行一次大规模的整体监测，每10年对从机场迁徙到邻近克罗斯温茨湿地地区的受保护的动植物进行一次监测。这一要求并不为一些机场工作人员所认同，他们最初以为监测结果会引发更多的补偿工作。该监测要求同时也关系到整个项目的完成进度，因为在项目的其他部分都已经完成后，还需要5到10年的监测时间。监测时间远远超过了合同中签订的保证两年完成种植的协定。在该项目中，韦恩县和密歇根州之间的关系并不太稳定，而且一些联邦机构也参与其中。对于人员调整以及项目最终完成期限等问题的关注度也一再提升。

监测记录显示，有的区域已经被一些外来物种所侵占，其中包括芦苇和金钱草。根据条款规定，这些物种中的75%到80%被移除。该地区被重新种植并运用非化学手段监测控制着。

随着湿地植被的逐渐稳固，机场跑道扩建过程中暴露出来的大片泥土地就会导致湿地的开放水域变得浑浊。由于湿地的开阔水面的流向与盛行西风平行，从而加大了风力，这就导致了水的流动以及浊度升高。早些年，鲤鱼在湿地繁殖，不断地搅动水底的淤泥，从而使得水浑浊问题恶化。随着水生植物生长的

in shallow water and gill nets in deeper water to note species type and size.

MDEQ required five years of monitoring overall and ten years of monitoring for the

protected plant species moved from the airport to the site adjacent to Crosswinds Marsh. This requirement was not appreciated by the airport personnel who initially assumed that monitoring results could "trigger" additional mitigation work. The monitoring requirement also raised concern over project "closure" because it required a five- to ten-year wait after the project was otherwise complete. The monitoring requirement extended well beyond the project contractors' two-year guarantee period on planting. During the project, an "uneasy" relationship existed among Wayne County, the State of Michigan, and the several federal agencies involved. Concerns also arose about changing personnel in the regulatory agencies, and anxiety developed over the timing of the final project approval and sign-off.

Monitoring documented that several areas had been invaded by exotic species, including common reed (Phragmites australis) and loosestrife (Lythrum salicaria). These species were 75 to 80 percent eliminated through targeted treatments with the herbicide Rodeo. The sites were replanted and are now being monitored and controlled with nonchemical controls.

As the wetland vegetation was becoming established, large areas of clay exposed during construction caused high turbidity in the open-water area of the wetland. The long orientation of open water parallel to the prevailing westerly winds produced significant wind fetch, causing rates of water movement and increased turbidity. In the early years, carp proliferates in the wetland, exacerbating the turbidity problem by continually stirring up the bottom clay during feeding. The turbidity has lessened as aquatic vegetation has become established, stabilizing the clay substrate. The predominant plants established in the open water include Eurasion water-milfoil (Myriopbyllum spicatum), curly pondweed (Potamogetom crispus), and brittle water nymph. As the water quality has improved with project maturation, predatory fish species (such as bass) micropterus ssp., which feed on and limit the number of small carp, have become established.

To the surprise and delight of the project designers and the country park staff, bald eagles have moved into a stand adjacent to the wetland. The primary nest tree is located in a woodlot adjoining the main area of open water that was flooded by rising water levels. This location overlooks much of the open-water area at Crosswinds and is the last area to freeze in winter,

ensuring a nearly constant feeding habitat for the eagles. The eagles' presence gives the project high public visibility, lending credibility to Crosswinds

稳定，泥土的基质相应得到稳定，浑浊度有所减轻。在开敞水域的主要水生植物包括欧亚水生蓍草、菹草、睡莲。随着水质的改善，以小鲤鱼为食的食肉鱼类（比如鲈鱼）的数量将会更多并且更稳定。

令项目设计者和公园工作人员兴奋不已的是，秃鹫被吸引到了湿地附近的摊床上筑巢。它们筑巢的树木位于一片林地中，这片林地毗邻于被逐渐上升的水位所淹没的开阔湿地。秃鹫的栖息地可以俯瞰公园中的大部分开阔水面，这里也是唯一的冬季结冰的区域，这就为它们创造了稳定的摄食地环境。秃鹫的出现提升了公园的公众辨识度，使人们更加确信公园是真正的自然保护地和湿地研究中心。而公众的认可也在影响着公园的管理者——比如，他们不再使用会杀死湿地中鲤鱼的鱼藤酮（农药）。

外来物种的入侵就像项目中遇到的其他问题一样，在韦恩郡公园员工管理中是一个具有警示教育作用的问题。私人船只禁止在湿地内滑行，因为这有可能将斑马贝克带到湿地中。但是可以租赁独木舟以满足游憩需求（图5–3–10）。因此教育项目应该在生物多样性的大背景下讨论物种入侵的问题。

从这片区域管理中得到的经验教训已经应用到了湿地二期工程中，而且已经基于一期湿地建设的结果对设计做了相应的调整。二期建设强调湿地的植被茂盛，并且禁止任何公众进入。

持续监测游人使用情况以及相应的矛盾也是韦恩公园管理部门的工作之一。周边的小径主要用于骑马游览和保安巡逻。在余下的地区，骑马是要被监督的，而且骑马的游人和骑自行车的游人之间也存在着一些矛盾。此外，许多意想不到的矛盾还会不断出现，比如说，臭鼬、浣熊、水貂、还有负鼠已经学会了利用公园中的木板路去吃鸭子和鸟蛋。这个问题至今还没有解决办法。

四、 总结

克罗斯温茨湿地的产生是对机场建设中损失掉的湿地进行补偿，这是北美洲最大的自给补偿型湿地。这种补偿机制是发达国家对湿地保护和开发利用的实践经验，也是国际普遍性做法。具体来讲，

as a bona fide "nature preserve" and wetland study center. Their appearance has influenced project management—for example, eliminating the possible use of rotenone to kill the carp species in the wetlands.

The issue of invasive species, as well as other aspects of the project, has taken on an educational component under Wayne County Parks staff management. Private boats are excluded because of the risk of importing zebra mussels (Dreissena polymorpba) to the wetland. Instead, rental canoes meet the recreational demand(Fig.5-3-10). Education programs thus discuss invasive species in the broader context biodiversity.

Lessons learned at this site have been applied to the phase two mitigation project adjacent to Crosswinds Marsh, which was approved with certain modifications to the design based on the results of the phase one wetlands construction. This additional mitigation emphasized forested wetlands and restricted any additional public access.

Monitoring human use and related conflicts is part of ongoing management by the Wayne County Parks Department. The perimeter trail is used for equestrians and for security patrol. In other areas, horse use is monitored, and some conflicts exist between horse riders and bicycle riders who invent trails. Other, unexpected conflicts have emerged. For example, skunks, raccoons, minks, and opossums have learned to use the project's boardwalks to eat duck and bird eggs. No solution to this problem has developed to date.

IV. Summary

CW wetlands was built to compensate the lost wetlands during airport construction, this is the largest self compensation type wetland in North America. The compensation mechanism is the practical experience of wetland protection and development in developed country , it's also international universality practice. Specifically, When we must take up or damage the marsh inevitably in the urban construction or business development, the government or developers must bear the wetland ecosystem restoration and reconstruction task. And in order to prevent the continuous decrease of urban wetland area, the area should not less than the original wetland area. The United States just relied on the policy of wetland compensation in different place, then realized wetland "zero loss" goal around 425000 hm2 mainlands in the past ten years. It is urgent

The main function of CW wetland is as a recreational park and wildlife reserve, the so-called self-sufficient type means don't need any outside artificial intervention, the wetland itself has realized the whole process of the cycle. From the projects in

就是在城市建设或商业开发中，在不可避免而必须占用或损害湿地时，政府或开发商必须承担湿地生态系统的修复与重建任务，其面积不少于原有湿地面积，以防止城市湿地面积的不断减少。美国在过去10多年间，正是依靠这种异地补偿湿地的政策，才实现了大陆42.5万公顷左右湿地“零损失”的目标。我国亟待成立这样的行政规范来保护湿地。

克罗斯温茨湿地主要功能是游憩公园和野生动物保护区，所谓自给型是指不需要外界人工的任何干预，湿地本身已实现了生态循环的全过程。纵观国内外湿地的恢复与重建工程，发现它们大多采用建设湿地公园的模式，使之融入城市生态基础设施框架中，更好地发挥湿地对理想人居环境的综合服务功能。克罗斯温茨湿地不仅在建设和规划中有值得借鉴之处，在后期维护与管理中也有优秀经验可取。例如，每5年进行一次大规模的整体监测，包括动植物监测、水文监测等，以确保生物多样性要求，并控制外来物种入侵，保持地方性景观特色。

图5-3-10 独木舟
Fig.5-3-10 Rental canoes

the domestic and foreign wetland restoration and reconstruction, we find that most of them used the mode of wetland park construction in order to melt them into urban ecological infrastructure framework, then play the comprehensive service functions of wetland better to ideal living environment.CW wetland not only have reference in construction and planning, but also have advisable excellent experience in the later maintenance and management. For example, developing a mass overall monitoring in every five years, including the animal and plant monitoring, hydrology monitoring and so on to ensure biodiversity requirements, control alien species invasion, and keep the local landscape features.

思考题

1. 何为补偿性湿地? 补偿性湿地区位选择应该考虑哪些要素?
2. 在多学科组成的团队中，景观设计师应该起到怎样的作用?
3. 应该采取怎样的措施来保护被开发的湿地的物种多样性并且防止外来物种的入侵?
4. 项目监测与评估在补偿性湿地项目中的重要性?
5. 建设补偿性湿地首要考虑的要素应该是什么? 如何协调生态保护与公众利用之间的关系?

Questions

1. What is the meaning of wetland mitigation project? What are the main factors during the chosen of the location of the mitigating wetland?
2. What is the role of landscape architects in an interdisciplinary team?
3. How to protect the biodiversity of the wetland which is occupied, and how to stop the invasive species?
4. Try to tell the importance of project evaluation in the wetland mitigation project.
5. What is the main factor during the construction of mitigating wetland? How to harmonize ecological protection and public use?

注释/Note

图片来源/Image：图5-3-1，图5-3-2：http://lafoundation.org/research/ ；图5-3-3：Biodiversity Planning And Design: Sustainable Practices（ Author：John F. Ahern， Elisabeth Leduc， Mary Lee York） ；图5-3-4:：http://lafoundation.org/research/ ；图5-3-5～图5-3-6：http://www.google.com.hk ；图5-3-7：http://lafoundation.org/research/ ；图5-3-8～图5-3-10：http://www.google.com.hk

参考文献/Reference：Biodiversity Planning And Design: Sustainable Practices（ Author：John F. Ahern，Elisabeth Leduc， Mary Lee York） ；http://lafoundation.org/research/

推荐网站/Website：http://lafoundation.org/research/

第四节 乌兰野生动物园
Section IV Woodland Park Zoo

一、 项目综述

动物园在生物多样性的保护中扮演着非常重要的角色——并不仅仅是动物的收容所或者栖息地，它还是教育中心，让人类更多的了解我们的环境以及生活在其中的各种生物，同时与城市以外的物种建立情感维系。当人类看到动物们生活在与它们自然中的栖息地几乎一模一样的地方并且保有其自然习性时，他们开始尊重动物甚至开始保护野生栖息地。与此相反，那些生活在混凝土牢笼中并且被金属栅栏隔离起来的动物们却难以得到人们的同情和理解。就像大卫·汉考克斯在《不同的自然》一文中说的那样，在比喻和实践当中，动物园能够且必须成为野生环境的门户。景观规划设计师应该在美国乃至世界范围内的动物园规划发展中起到很重要的作用，他们可以促成多学科多团队的合作，从而设计出成功的，生态的动物园（图5-4-1，图5-4-2）。

从200年前动物园出现到如今，它的模式已经发生了很重大的改变。尽管动物园是以保护动物为主的，但是许多动物园的设计依然从人类的角度出发，而不是以动物的生存感受为主。华盛顿西雅图的规划设计公司琼斯&琼斯所设计的西雅图乌兰动物园却从动物的生存出发，对动物园规划设计做出了

I. Overviews

Zoos play an important role in the conservation of biodiversity—not only as refuges or places for captive breeding but as centers for education, raising humans' intellectual understanding of the world environment and its species, and building an emotional attachment with species other than our town. When people see animals living in an accurate replication of their natural habitat and engaging in natural social behaviors, they begin to respect animals as having a dignified existence in their own right, and they may become more interested in protecting wild habitats. On the contrary, animals confined to concrete cages and separated from the public by metal bars garner little sympathy or understanding. As David Hancocks argues in "A Different Nature", zoos can and must become gateways to the wild, metaphorically and practically. Landscape architects can play an important role in the future development of zoos in the United States and worldwide by bringing together interdisciplinary teams and coordinating the science necessary to create successful, biocentric zoos (Fig.5-4-1,Fig.5-4-2).

Zoos have evolved significantly since the first public zoos were created two hundred years ago. Despite their current roles as conservation centers, however, many zoos still base their design on an anthropocentric, rather than a biocentric, point of view. The architecture and landscape architecture firm of Jones & Jones, located in Seattle, Washington, succeeded in making the fundamental change to a biometric perspective at the Woodland Park Zoo in Seattle. Inventing the concept of landscape immersion, a term coined by Grant Jones, they

图5-4-1 乌兰野生动物园中爬树的金头狮狨
Fig.5-4-1 Golden-lion-tamarin climbing a tree

图5-4-2 乌兰野生动物园中的大猩猩
Fig.5-4-2 Gorillas in the Woodland Park Zoo

图5-4-3 乌兰动物园区位图
Fig.5-4-3 Location of Woodland Park Zoo

成功的根本性变化。设计师格兰特·琼斯创造出了"景观沁"这样一个新的规划概念，他成为第一个把动物当作客户的动物园设计师。

二、 场地及其背景介绍

1. 项目历史

1979年琼斯&琼斯建筑及景观规划有限公司完成了乌兰动物园的远期规划。这个占地90英亩（约36.4公顷）的动物园位于西雅图西北部，其东边界是奥罗拉大街，西边界是菲尼大街，北部和南部都是居民区（图5-4-3）。1968年，前景信托通过发行债券拨款450万美元用于特定的动物园整修，以此为基础成立了一个"市长动物园特别行动小组"，而该远期规划则是对1975年提出的目标的一个响应，由于此前并没有此类规划，一个咨询委员会将其称作1976项目，一旦这一规划完成并被批准入整体规划，计划内的实际工程也将随着资金的到位而逐步进行。尽管大部分资金来自前景信托债券，但是慈善事业的支持也帮助完成了大量的其他改善措施。

2. 项目参与者

这项规划是由西雅图市发起的，以完善FTBI在1975年所做的前期规划，该规划由建筑师G.R. 巴斯·里克负责，但是却遭到公众自发的反对，因为动物园占地面积太大，并且与周边的居民区非常不协调。琼斯&琼斯公司的许多景观规划设计及建筑

became the first zoo designers to describe the animals as clients.

II. Introduction of the Site and Context

1. Project History

The Woodland Park Zoo long-rage plan was completed in 1979 by Jones & Jones, Architects and Landscape Architects, Ltd. The ninety-acre (36.4-hectare) zoo is situated in the northwest sector of Seattle, Washington, and is bounded on the east by Aurora Avenue, on the west by Phinney Avenue, and on the north and south by residential communities (Fig.5-4-3). In 1968,$4.5 million was set aside by the Forward Thrust Bond Issue for specific zoo improvement, based on a Mayor's Zoo Action Task Force was formed, and the long-range plan was developed as a response to objectives set forth in

师都参与到了这个项目中，其中就包括格兰特·琼斯（主要负责人），乔恩·查尔斯，约翰保罗·琼斯，彼得·哈佛，约翰·安迪，大卫·沃尔特斯，约翰·斯旺森，埃里克·施密特和基斯·拉森。公司的合伙人之一吉姆·布莱顿指出规划师们95%的工作时间都用于领导和协调这个多学科合作的规划队伍。他强调景观规划师是非常重要的，他们会同心协力创造出概念性的规划理念并且将它付诸实践。琼斯&琼斯公司被选中来做这个项目主要是因为它的非常著名的“通过景色本身去发掘其自然过程和形式”的突破性的理念，还有其在规划中“自然第一”的承诺。在乌兰动物园的规划中，琼斯&琼斯公司再一次强化了他们的这一声誉，生态化的设计手段以及“景观沁”和“文化共鸣”的概念共同作用，吸引着游客融入动物们的自然生活环境中。

顾问团队也是设计组的重要组成部分，包括生物学家丹尼斯·鲍尔森；土木工程师唐纳德·霍根；还有地质学家菲利普·奥斯本。詹姆斯·福斯特是乌兰野生动物园的临时主管。最初担任设计协调员的大卫·汉考克斯是规划“生动的动物园”的专家，他确立了描述和展览主题，后来他成为新一任的动物园主管。汉考克斯是一名建筑师，他曾经与伊恩·麦克哈格合作设计过伊朗的帕蒂森项目。汉考克斯被琼斯&琼斯公司聘请来设计乌兰动物园，他努力探寻景观的内涵以确定动物园的建设可行性，从而也符合植物的气候学要求。丹尼斯·鲍尔森将这次的尝试描述为没有预定角色的“学科间团队合作”。此外，前景发展委员会，西雅图规划设计委员会以及许多的学者，还有自然主义者都在规划方案的制定过程中贡献了力量。社区、公众参与采取了前景发展委员会的形式，即充当“规划决心的责任性和连续性的媒介”。

三、 规划与设计介绍

1. 项目理念

“景观沁”的概念强调动物园中的动物们应该在其自然的生存环境中被观看（或者最大限度的复制其生存环境），这样游客就可以深入到观察区域

1975.Once the long-range plan was completed and approved in comprehensive plan, because no such plan existed, an advisory committee called the 1976, the actual projects contained within the plan were completed in phases as funding became available. While much funding has come from Forward Thrust monies, philanthropic support has helped complete numerous other improvements.

2. Project Participants

The long-range plan was initiated by the City of Seattle in response to the requirements of the Forward Thrust Bond Issue in 1975 after a previous plan, developed by architect G.R. Barth lick, was rejected by a public initiative because of its monumental scale and incompatibility with the neighborhood. Many landscape architects and architects at Jones &Jones were closel involved in developing the master plan, including Grant R.Jones (the principal-in-charge), Jon Charles Coe, Johnpaul Jones, Peter Harvard, John Andy, David Walters, John Swanson, Eric Schmidt, and Keith Larson. Jim Brighton, a partner in the firm, noted that the landscape architects spent 95 percent of their project time acting as leaders and coordinators within interdisciplinary teams. Landscape architects are needed, he says, to pull together a conceptual idea and to give it a physical form. Jones & Jones was chosen for this project because they were known for their “groundbreaking methodology for reading a landscape to determine natural process and form” as well as for their commitment to “placing nature first” in their designs. They further promoted this reputation with the Woodland Park Zoo long-range plan, which incorporated the concepts of landscape immersion and cultural resonance to fully engage visitors in the natural environments of the animals they came to observe.

Consultants were also an integral part of the team, including Dennis Paulson, biologist; Donald Hogan, civil engineer; and Philip Osborn, geologist. The interim director of the Woodland Park Zoo at the time was James Foster. The “living zoo” expert, David Hancocks, originally acted as design coordinator, establishing the presentation and exhibit themes, and later became the new zoo director. Trained as an architect, Hancock had previously consulted with Ian McHarg on the Pardisan project in Iran. Hancocks, which came to the Woodland Zoo project by retaining the service of Jones & Jones, sought to understand the landscape in order to determine the feasibility of organizing the zoo around the bioclimatology of plants. Dennis Paulson described the effort as “teamwork across discipline,” with no predefined roles. Additionally, members of the Forward Thrust Development Committee, the Seattle Design Commission, and many scholars and the naturalists provided their input as the plan was

去体验动物的生存环境而不仅仅是远远地看着。“文化共鸣”是格兰特·琼斯创造的另一个术语，它的含义就是超越景色本身去体验自然环境中人类与动物相互依赖的关系的方方面面。例如，在人们感受大象的生活方式的同时他们也在了解着东南亚的建筑，宗教和文化（图5-4-4，图5-4-5）。

2. 项目宗旨与目标

项目目标是由动物园特别行动委员会制定的：“乌兰野生森林动物园应该是一个显示动物的价值和美感并且展现动物行为和生理适应性的生命科学研究所”。因此，动物园应该着重于培养公众的理解，让游人理解野生动物以及它们与生态系统的关系。

根据长远规划的描述，动物园之所以有别于自然历史博物馆，就是因为它圈养着动物，因此，它必须供给动物必要的自然生存条件，这样既满足了动物的利益也同样满足了观赏者的需求。为了达成这个理想，大卫·汉考克斯按照社会生物学的原理改善了展览的模式，他选择更少的动物种类，但是每一种动物的自然种族规模更大，比如一群大猩猩或者一群羚羊（图5-4-6，图5-4-7）。这样游客就可以观赏到更多不同的动物群，并且了解它们不同面的生活，而不仅仅是观看它们的形态特征，同

图5-4-4 乌兰动物园鸟瞰图
Fig.5-4-4 Aerial model of Woodland Park Zoo

图5-4-5 乌兰动物园平面
Fig.5-4-5 Master plan of Woodland Park Zoo

developed. Community/public participation took the form of the Forward Thrust Development Committee, which acted as "the medium for accountability and continuity of intent".

III. Site Plan and Design

1. Project Concept

Landscape immersion emphasizes that animals in zoos should be exhibited in their natural environment (or the closest facsimile possible), which should reach out into the observation area so that visitors also experience the environment by standing in it rather than just looking at it .Cultural resonance, another term coined by Grant Jones, moves beyond landscape immersion to include aspects of the interdependent relationship between humans and animals in their native environment. For example, the architecture, religion, and culture of Southeast Asia is intimately tied to people's experience with elephants(Fig.5-4-3,Fig.5-4-4).

2. Project Goals and Objectives

The stated goal of this project was set forth by the Zoo Action Task Force:"The Woodland Park Zoological Garden should be a Life Science Institute demonstrating the value and beauty as well as

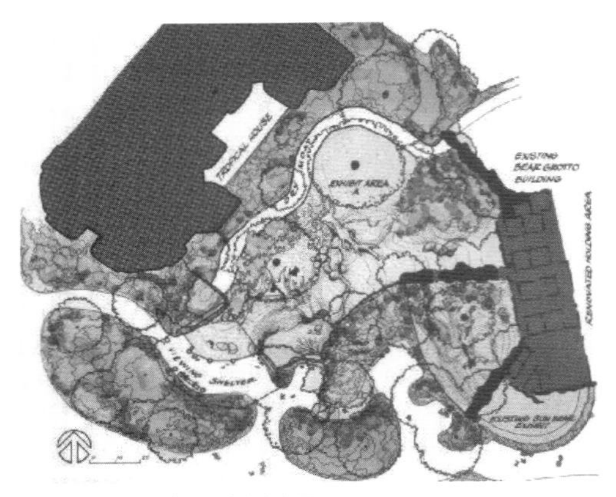

图5-4-6 重新设计的低地大猩猩展区
Fig.5-4-6 The redesign of the lowland gorilla exhibit

图5-4-7 群居的低地大猩猩
Fig.5-4-7 Social group of lowland gorillas

时还可以令人们反思人类自己的社会生活。动物园这样做的另一个意图是“为游客提供一个全面的环境体验”。为此，动物展示的不同主题区依照生物气候学的地带分区而划分，而没有按照动物来源的不同或者动物地理分区而进行划分，动物们生活的微气候复制于它们所生存的自然气候。

琼斯&琼斯在乌兰野生动物园中运用的设计途径标志着动物园设计手法的提升和颠覆，在传统的动物园中，动物无法在模拟其自然生存栖息地的环境中生存，甚至生活在有金属栏杆隔离的混凝土牢笼中；群居动物常常被单独拘禁，这导致了他们反常的行为。在这样的情况下，动物们不仅仅被剥夺了尊严，甚至丧失了博得游人同情的机会，因为游人往往只能看到动物的消极面——比如，认为它们是肮脏的并且具有攻击性。此外，大多数的动物园往往存在这样一种误区，那就是展示不成比例的大量的哺乳动物和外来物种。与此相反，乌兰野生动物园非常重视动物的区域重要性，不仅展示游客生活中可见的生命世界，也展示来自遥远的太平洋的动物。

动物园有几大主要的目的：游憩、教育、研究、还有保护。1993年，世界动物保护组织出版了世界动物保护战略，其中指出动物园和水族馆最主要的目的应该是贡献于动物保护运动，无论是直接的（通过圈养繁殖）还是间接的（通过提升游客意识）。再者，如上所述，非常必要的是着重于当地的或者局部范围的动物物种保护，因为糟糕的土地利用破坏了整个国家越来越多的动物生境。格兰特·琼斯说动物园的真正贡献以及正义之处体现在

behavioral and physical adaptations of animal life. As such, primary emphasis should be placed on fostering public understanding of animal life and its relationship to ecological systems"

As the long-range plan describes, a zoo differs from a natural history museum because it houses living animals; as such, it becomes "essential to give the animals all the necessary opportunities to engage in natural behavior, to the benefit of the animal and the zoo visitor. To achieve this end, David Hancocks developed the exhibit theme of social biology by selecting larger, natural group sizes of fewer animals species, such as a band of gorillas or a herd of antelope (Fig.5-4-6, Fig.5-4-7). This focus allows visitors to ob serve many different aspects of the animals, not just their morphological characteristics, and is intended to encourage people to reflect on their own social species. Another intention of the zoo was to provide visitors with a "total environmental experience. To this end, the presentation theme of bioclimatic zones moves away from conventional zoo design in that it present the animals within microclimates that replicate natural climates, rather than dividing the zoo into different continents or zoogeographic areas .

The approach by Jones & Jones for Woodland Park Zoo marked a progressive departure from the status quo in zoo design, in which animals are displayed in habitats that fail to resemble their natural environment, or even worse, display in concrete cages with metal bars separating them from zoo visitors. Often, social animals are exhibited in solitary confinement, which leads to unnatural behaviors. It these situations, the animals not only are stripped of their dignity but often lose the chance of gaining sympathy from zoo visitors, who may see the animals negatively—for example, describing them as dirty or aggressive. Additionally, most zoos advance a false image of animals, presenting a disproportionately large number of mammalian species and exotic species.

它可以唤起公众的环保意识，共同解决环境问题。同样的，汉考克斯说过“我们对动物园有着非常不同的需求，而观看熊猫或老虎已经不仅仅是动物园存在的理由了”。

乌兰动物园的主要责任很明确，那就是保护动物并且激发游客对于动物园的动物以及野生动物的同情心。这种融入进景观的理念可以激发人们的情感认知。格兰特·琼斯说他们试图唤起游人在旷野中的感觉：人类只是自然的入侵者，而且非常渺小，在乌兰动物园游览的游人们就像是来到了动物领地的客人们，并观察动物以它们自己的方式活动和交流（图5-4-8）。

3. 生物多样性问题以及规划战略

动物园可以通过许多途径来保护生物多样性。直接的方式，可以与其他动物园组成繁殖网络或者运用人工授精技术、低温贮藏技术来促进动物的繁衍，即使这些动物园在保护物种演化方面的作用有限。这样，动物园就变成了一个基因库。世界动物保护组织宣布，只有当对动物的长期的物种繁衍有益时，才能将其迁出野生环境，因此，为临灭绝物种创造生存空间以及建立繁殖网络就是至关重要的。

动物园还可以间接地保护生物多样性，通过教育游人并且鼓励环保行为，比如资助保护特殊生物栖息地或者生态系统的组织。最好的动物园设计就是让它能够使游人与动物心灵相通。就像大卫·汉考克斯在《一个不一样的自然》中所说的那样，参

图5-4-8 动物园中的狮子
Fig.5-4-8 Lion in the zoo

In contrast, Woodland Park Zoo includes animals of regional importance, educating visitors about the biotic world immediately surrounding them as well as about environment far from the Pacific Northwest.

Zoos have several main purposes: recreation, education, research, and conservation. In 1993, the World Zoo Organization published the World Zoo Conversation Strategy, Which notes that the greatest purpose of zoos and aquariums is to contribute the conservation movement, whether directly (through captive breeding)or indirectly(through visitor awareness).Furthermore, as noted above, a greater emphasis on local or regional species is needed, as poor land use practices destroy more and more local habitat across the country. Grant Jones says that the true contribution (and justification) of zoos is the way they increase public awareness and concern for environmental issues. Likewise, Hancocks writes:"There is a very different need for zoos , and going to a zoo to see pandas and tigers is no longer sufficient justification for its existence".

The priority at Woodland Park Zoo is clearly animal comfort and protection as well as in creating visitors' empathy for the animals they see and their wild cousins Landscape immersion can reinforce emotional and cognitive learning. Grant Jones explained how Jones & Jones attempted to re-create the feelings people have when they're in the wilderness: feelings of the intruder, of being very small in a big world. Woodland Park Zoo was designed to give visitors perspective of being a guest "in the animal's domain" and of observing animals that behave and interact on their own terms(Fig.5-4-8).

3. Biodiversity Data Issues and Planning Strategies

Zoos can project and conserve biodiversity in a number of ways. Directly, they can engage in captive breeding programs by participating in breeding networks with other zoos or in artificial breeding/cryopreservation efforts, though these efforts zoos play a limited role in protecting the survival of species .In this way, zoos can act as genetic reservoirs. The World Zoo Organization says that animals should only be removed from the wild if this action would benefit the long-term survival of the species; for this reason, giving space to endangered species and participating in breeding networks are vitally important activities.

Zoos can also conserve biodiversity indirectly, by educating visitors and encouraging future environmental action, such as contributing to an organization that protects a particular habitat or ecosystem. This is best accomplished when the zoo design evokes feelings of empathy among visitors. As David Hancocks describe in A Different nature, people surveyed after visiting a conventional zoo

观了传统的动物园后，人们看到的只是关在牢笼中的动物，它们肮脏而且好斗。而当人们进入到一个基于“景观沁”原理的动物园之后，他们却看到了动物们的美丽、强壮和有趣。而且后者都说参观过后他们更加关注环保事业了。格兰特·琼斯指出动物园的设计者一定要让游人感觉到动物们在这里生活要比在笼子中舒服得多。他在自己的文章中列出了一个成功的生态动物园的设计过程中该做与不该做的16项关键问题：

（1）游客不应该在高处观看动物，相反，动物们应该与视线水平或者高于视线；

（2）动物不应该被游人围绕着，因为在野外这种围绕有失尊严，相反，游人们应该在隐蔽的地方进行眺望；

（3）为动物提供本能的安全距离，在观赏区中设置不同的栖息地，使得动物可以选择他们想去的地方；

（4）不要将群居的动物单独拘禁或者设置过小的群体，只有在合适大小的群体中，动物才会表现出其本能的行为；

（5）不要展示那些残疾的、畸形的动物，而要为它们提供良好的遮蔽措施；

（6）不要运用人工产品来装饰动物，因为那样就会显示出以人类为中心的态度；

（7）不要设置“抢风头”的道具。相反的，忠实的再现动物生境，既不夸张也不变形；

（8）必要时才使用可见的屏障，不能呈现出动物被什么所包围的景象；

（9）不要让游人交叉观看，应该保持观景点的彼此隐蔽（最吸引游客注意的往往是其他的游人）；

（10）将观景区设置在二级道路周边，以防止主要道路分散游客或者动物的注意力；

（11）不要让一个展区一览无余；

（12）展区应该尽量再现动物们的野外生存环境；

（13）不要将游人安排在人工的环境中而将动物安置在自然的环境中，相反的，延伸自然环境到游客使用的区域中，以加深景观体验感；

（14）不要将生存环境不同的动物安置在同一

where animals are kept behind bars had only negative comments to make about the animals, describing them as dirty and aggressive. People surveyed after visiting a zoo based on landscape immersion, however, described the animals as beautiful, strong, and interesting. These latter individual also said they were more likely to contribute to an environmental cause after visiting the zoo. Grant Jones notes that it is crucial that zoo designers do not alienate the public by making them feel as if an animal would be better off extinct than in captivity. One of his article provides a list of sixteen key "do's and don'ts" for designing a successful biometric zoo:

(1) Views should not look down on the animals. Instead, animals should be or above eye level;

(2) Animals should not be surrounded by views, because this does not allow "the dignity apparent when encountered in the wild." Instead, provide small overlooks that are screened from one another;

(3) Keep the animals only as close as their natural flight distances allows. Provide "alternative locations" within the exhibit so animals may chose where they want to be;

(4) Do not keep social animals in solitary confinement or in too small groups . In natural-sized groups, natural behaviors will be evident;

(5) Do not display disfigured or deformed animals; up do provide them with good, off-exhibit facilities;

(6) Do not display animals with human artifacts, which promote anthropocentric attitudes;

(7) Do not allow props to "steal the show." Instead, reproduce natural habitat as faithfully as possible, without exaggeration or distortion;

(8) Use visible barriers only when absolutely necessary. Make it impossible to tell what contains the animals;

(9) Do not allow human cross viewing. Again, keep overlooks screened from one another.("Nothing attracts human attention more than other humans");

(10) Put viewing areas on secondary paths so that primary paths are not distracting to either people or animals;

(11) Do not allow an entire exhibit to be seen from one overlook;

(12) The exhibits should replicate the animal's natural environment;

(13) Do not put human in man-made settings and animals in natural settings. Instead, immerse the viewer by extending the natural setting into human-use areas;

(14) Do not display animals from different habitats together in one natural habitat setting;

(15) Do not display animals from strikingly different habitats in adjacent spaces;

(16) Plan all elements of exhibits concurrently to create an integrated whole.

类展区中；

（15）邻近的展区环境不应该差异过大；

（16）将所有的要素整合起来创造一个兼容完整的整体。

综上所述，乌兰野生动物园的展览主题非常符合社会生物学的原理。为了贯彻这一主旨，动物园将动物按照自然中的生活群体大小进行安放。这样动物就可以与其他的个体进行交流，就像它们在野外生活时一样。丹尼斯·鲍尔森是项目组中的生物专家，他认为动物园就像是一个展示生物多样性的“奇妙的切面”，我们可以看到大型动物也可以看到小型哺乳动物，还可以看到鸟类。与传统动物园还有一点不同的就是，乌兰动物园不仅仅考虑到动物的生存环境，它同样重视动物赖以生存的植物环境。事实上，在选择动物之前就已经对影响动物生活环境的植物进行了研究，最终选择了最利于创造动物原始生活气候带的植物。

生物气候学的概念是动物园展区设计的理论基础，它的核心理念就是“世界上的动物因为气候和植被的情况而分布于不同的栖息地”。这些栖息地分布在世界的不同区域，而且往往同一类型的栖息地分布在不同的大陆上。栖息地带的划分由三个相互作用又相互独立的因素决定：温度、降水和蒸发量。任何一个栖息地都可以按着这三个要素来划分，而且还可以确定与其相对应的植被覆盖情况，比如沙漠和温带雨林。

乌兰动物园的生物气候带的划分主要基于霍尔德里奇体系（图5-4-9）。西雅图处在体系中心周边，所以非常多种类的植物可以在这里生长。当设计师把场地的微气候与全球的气候带相匹配后，他们就可以营建只需少量甚至不需改造的展区，比如那些位于温带雨林区的区域。这个过程也使得动物园去关注本土的物种及其生活习性。其他的展区就需要做适度的调整（比如在岩石中置入加热圈以满足大猩猩的需求，或者最大限度的排水来创造沙漠环境），从而营造自然的生物气候带。

一旦生物气候带确定并且满足了相应的植被覆盖，动物物种就选定了。生态学家为动物园确定了选择动物时应该遵循的5个标准：

As mentioned above, the exhibition theme at the Woodland Park Zoo is social biology. To adhere to this theme, the zoo attempts to display natural-sized groups of social animals. Perhaps these results in fewer species than one might see in a conventional zoo, but also in a greater number of individuals within the species chosen so that they may interact with one another as they do in the wild. Dennis Paulson, the project's consulting biologist, noted that the zoo concept was to show a "wonderful slice" of biodiversity, from mega fauna to small mammals and birds. Also unlike conventional zoos, Woodland Park does not merely focus on charismatic mega fauna, nor does it forget about the plants that contribute to the animals' habitat. In fact, choosing the vegetation that best replicates the climatic zones occurred before the selection of animal species for the exhibits.

The bioclimatic concept, which forms the basis for the exhibits, is itself based on the idea that "animals are generally confined within certain parts of the world by the cause-and-effect relations of climate and vegetation". These habitats occur within different parts of the world and often on more than one continent. The habitat zones are classified as a function of three interacting, interpedently factors: temperature, precipitation, and evapotranspiration. Any habitat in the world can be classified based on these three parameters and can be characterized by a certain kind of vegetation expected in that life zone—for example, desert or temperate rain forest.

The bioclimatic zones at Woodland Park Zoo were determined primarily by using the Holdrege system (Fig.5-4-9). Seattle, when placed on the Holdrege system triangle, can be found near the center, which allows a wide variety of plants to survive there. By matching microclimates on site to global bioclimatic zones, the zoo designers were able to create exhibits that needed little or no modification, such as those found in temperate rain forest zones. This procedure also allowed the zoo to focus some attention on local habitats and species. Other exhibits needed only moderate modification (such as the inclusion of heat coils in rocks within the lowland gorilla exhibit, or maximizing drainage to simulate desert habitats) to duplicate natural bioclimatic zones.

Once the bioclimatic zones were established and the associated vegetation chosen, animal species were selected. The ecologist's report for the zoo (Paulson nd.) lists five attributes that should be considered when selecting animal species. These criteria are as follows :

Education—four factors of which are especially important: social behavior, evolutionary adaptations, convergent and parallel evolution, and adaptive radiation; Interest—similar to education, and the rationale for exhibiting social species;

驯化——有四要素是非常重要的：社会行为，适应进化需求，聚合平行的演进，适应性扩张；喜好——类似于驯化，还包括展示物种的基本原理；再现——再现物种的多样性；研究——一些物种，比如那些生活隐秘的动物，在动物园中进行研究往往更有效率；保护——保护稀有动物和濒临灭绝的物种。

之前提到过，琼斯&琼斯公司设计乌兰野生动物园是基于景观沁的理念，而在长远规划中，这一理念被做了如下解释：最理想的情况是游客们穿过特色的生物气候带，观看到美丽的景色并且沉浸到它动人的氛围中。我们知道这样震撼的景色里面居住着动物，但它们被看不见的栅栏阻隔着。营造成功的动物园景观完全依赖两个要素：

景观的完整性，只有完整，其景观特色才能被保护；见解、观点的准确性、隐藏栅栏、优化景观、适合的灯光与阴影，而最重要的是在视觉上统一的动物空间和视觉空间。

通过这种方式，动物与游客之间就不存在距离了。图5-4-10展示了隐蔽栅栏的具体方式以及景观沁原理的其他方面。这一理念通过物理形态表现出来，使得乌兰动物园真正的区别于传统的，以人为中心的动物园设计。琼斯&琼斯团队的一员乔恩·查尔斯·科认为融入到环境的过程中，公园最初吸引了游客的情感，而后又颠覆了他们的理解力。步行道路的循环性，游憩区域的设置，生物气候序列对于动物园设计的效果都有很重要的作用。分区被布置于动物园之内以保证自然生态系统的过渡明显可见，例如在草原周边种植落叶林和针叶林。这种渐变进一步引导游客们融入自然环境以展示物种。图5-4-11介绍了八种不同的观赏形式。左边是半开放半隐蔽形式的剖面

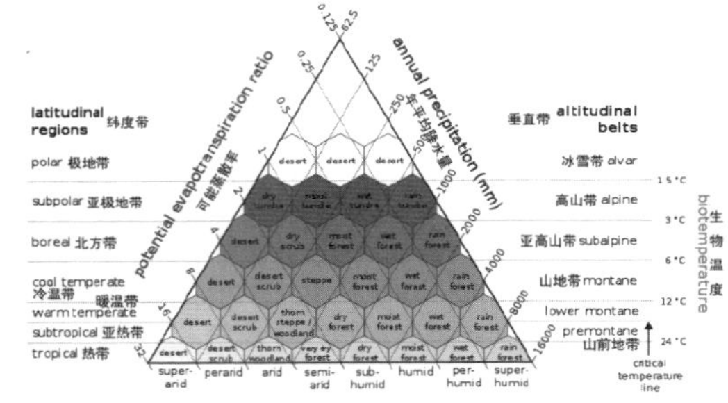

图5-4-9 霍尔里奇体系图
Fig.5-4-9 Holdridge System

Representation—of diversity; Research—some species. such as those with secretive habits, can be studied more effectively in a zoo environment; Conservation—of species considered rare or endangered.

As discussed earlier, Jones & Jones based the design of Woodland Park Zoo on the idea of landscape immersion, described in the long-range plan as follows: Ideally, the viewer should move through the characteristic landscape of the bioclimatic zone, seeing its sights and savoring its moods. Only then can we become aware that the landscape is also inhabited by animals, separated by unseen barriers. The success of this landscape immersion depends entirely upon two factors:

The completeness with which the characteristic landscape is projected; The care and accuracy which the viewpoints and views are located and composed.concealing barriers, enhancing perspectives, composing light and shadow, and, most importantly, visually unifying animal space and visual space.

In this way, no sense of separation between the animals and the visitors exists. Example of specific ways to achieve concealed barriers and other aspects of landscape immersion are shown in Fig.5-4-10.This philosophy, expressed in physical form, is what truly moves Woodland Park Zoo away from conventional, anthropocentric design. Jon Charles Coe, a member of the Jones & Jones team, notes that immersion first appeals to the visitors' emotions and then to their inflects. Careful attention to pedestrian circulation, the placement of recreation areas, and the bioclimate sequence all contribute the immersion effect. Zones are positioned within the zoo so that natural ecosystem transitions are evident, such as the placement of steppes exhibits next to deciduous forest and taiga exhibits. This attention to gradual transition further immerses visitors in the natural environments of the species display. There have eight viewing types to be used for all exhibits at the zoo. Open edge and partially screened edge types are shown on the left in plan and sections views. Mesh enclosure, shelter, animal day structure, and covered viewing into mesh enclosure type are shown on the right.Note that no direct views are allowed from the busy primary paths (Fig.5-4-11).

Conserving biodiversity in zoos seems to rely equally on which species are chosen as well as how those species are exhibited to the public. First and foremost, the comfort of the animals and their ability to engage in natural behavior must be achieved. Only then, through landscape immersion and the logical next step of cultural resonance, do visitors truly recognize animals as they exist in the wild, without bars(Fig.5-4-12).Cultural resonance is an idea that further shows visitor how animals and humans can do (or did)survive side by side. Grant Jones describes

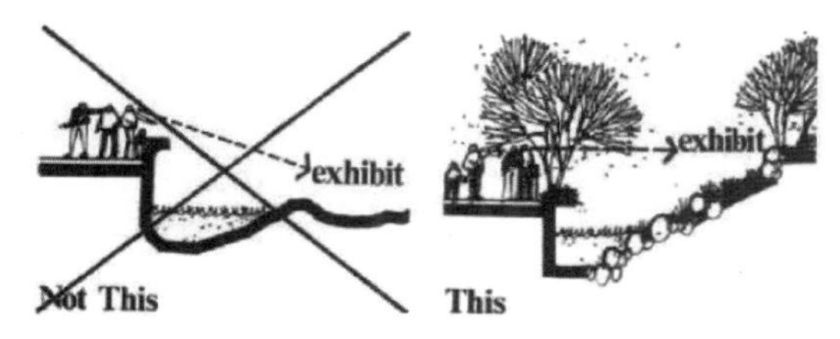

图5-4-10 展示景观沁原则的例子
Fig.5-4-10 Example of exhibit guidelines that illustrate the principles of landscape immersion

图。右边是通过网罩，遮盖物等遮蔽观赏的形式。其中主要道路上是不允许直接观看的。

生物多样性的保护依赖于物种的选择以及向公众展出的形式。首要的就是要保证动物们生存的舒适性和本能的保有。只有通过景观沁以及文化上的共鸣，游客才能真正的认识野生的动物，而不是牢笼中的动物（图5-4-12）。文化共鸣是人与动物共存的理念，他引导人们去思考当人与动物一起生活时他们各自会又怎样的反应。格兰特·琼斯是这样描述这一理念的：“我们坚信通过景观沁和文化共鸣的理念，游客们就会探寻到本质的层次，从而从相互关联的感受中去获益，最终真实地感受到人类对于地球的依赖性。”

产生文化共鸣的一个例子就发生在乌兰动物园中的亚洲象身上。在这里，人们不仅可以看大象，还可以看到反映大象与人类关系的人工产品，它们展示了泰国人对于本土大象的信赖。展览之路弯弯曲曲的穿过树木茂盛的山坡，最终到达了“Rong Chang”，或者到达了大象的居住地。动物园的解释体系中介绍，它们试图展示大象的多个面：野生状态的大象；作为役畜的泰国大象；泰国宗教文化中的大象。这里有一个简单的道理要让游客明白，那就是如果大象的栖息地没有了，那么最终，根植于这一栖息地的文化景观也会随之消失。有了这样的认知，游客们就会有更深层次的体验，他们会去思考动物在人类的生存环境中所扮演的角色。景观规划设计师可以通过设计动物园来对游客进行这种认知教育。

4. 项目评估

乌兰野生动物园的总体规划每4到5年就会进行一次更新。管理者会对存在的风险或问题进行检

this philosophy:"We believe that by combining the ideas of landscape immersion with those of cultural resonance, we have allowed the visitor to explore at the gut level, to reap the benefits of a feeling of interconnectedness, and to really sense our dependency on the planet for our humanness".

One example of the use of cultural resonance is the Asian elephant exhibit at Woodland Park Zoo. Here, human artifacts that played a part in the relationship between the humans and the elephants are included in the exhibit, illustrating reliance people in Thailand had on the native elephants. The exhibit winds its way through a wooded hillside, finally reaching the "Rong Chang," or the House of the Elephants. As the zoo's interpretive primer explains, the exhibit hopes to exhibit several "faces" of the elephants :"elephants in the wild; elephants as work animals in Thailand; elephants in Thai religious culture". There is a simple lesson to be learned: if elephant habitat is lost, so are the elephants—and, consequently, the culture rooted in this landscaped gone, too. With this realization, zoo visitors can experience a deeper appreciation for the intrinsic role animals play in human existence. Landscape architects can contribute to this sympathetic education of visitors by designing exhibits sensitive to the ecological and cultural history of the ecosystem that is their focus.

4. Project Evolution

The Woodland Park Zoo Master Plan has been updated every four to five years. Most monitoring takes places on projects where risks were taken or where problems have occurred—such as when an animal has escaped or when large amounts of vegetation have failed to thrive. Dennis Paulson explains that monitoring is the zoo's responsibility but that it usually isn't done. Grant Jones notes that zoo monitoring is often too simplistic and thus can reach the wrong conclusion. For example, if a tree dies in an exhibit, zoo staff may believe that it is best not to replace it. The correct response, however, might be to plant more in order to establish a microclimate more conducive to the tree's needs.

It was important for zoos to be willing to take risks and to move beyond conservative, conventional design in order to truly be successful in providing quality habitat for the resident animals. Monitoring must address both animal and human responses. A zoo's success has conventionally been defined in terms of maintenance ease first, visitor appreciation second, and animal comfort third..In the new mind-set, however, this order would be reversed. Creating and maintaining exhibits that cater to the animals' need and the immersion of viewers is not an easy task. It is, however, the only way that zoos will truly fulfill their obligations to the species they house.

测，比如动物逃脱或者大批的植被不能茁壮成长。丹尼斯·鲍尔森说监测工作是公园的职责，但是这项任务往往完成的并不好。格兰特·琼斯认为监测工作有时过于简单使得检测结果并不准确。例如，一棵树死了，工作人员可能认为不去替换它是最好的，但是正确的作法应该是种更多的树以稳定微气候，从而满足树木的需求。

动物园应该敢于冒险，打破传统，从而真正的为动物们提供高质量的生存环境。而监测必须估计到人类与动物的双重反应。传统的动物园规划理念往往是稳定第一，游人第二，动物第三。而如今这一理念应该被逆转。迎合动物的需求，满足游人体验不再是一个容易的任务了。只有这样，动物园才能真正履行它对于生活在其中的动物们的职责。

四、 总结

本案例不仅可以直接指导动物园的规划与设计，更引领所有参与此项工作的人员进行了一次思想的转变——尊重动物，将动物作为真正的“客户”。

回归自然，亲近动物是世界之潮流。上世纪末，我国各地动物园建设经历了第一次转型，大部分城市的动物园由市区搬至郊区，拆除圈养动物的铁栅栏、水泥地，取而代之的是生态化布置动物展示区和视线无障碍展示动物，由圈养笼舍式转变为散养开放式。如今，动物园把动物散养在一个区域面积较大的空间，以自然的方式展览。当我们徜徉

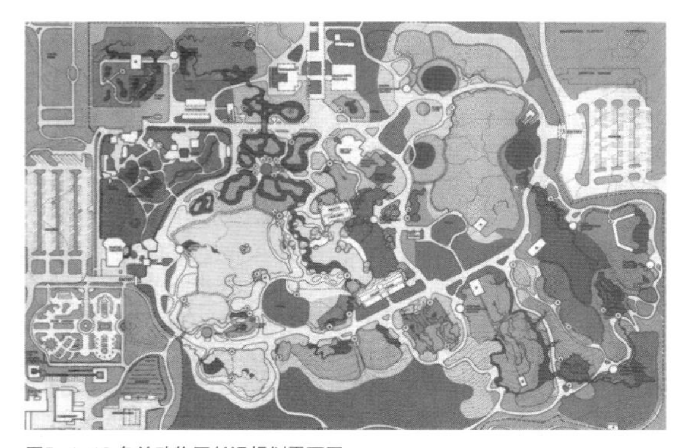

图5-4-12 乌兰动物园长远规划平面图
Fig.5-4-12 The long-range plan for Woodland Park Zoo

图5-4-11 动物园内八种不同的观赏形式
Fig.5-4-11 Eight viewing types to be used for all exhibits at the zoo

IV. Summary

This case is not only a direct guidance of zoo planning and design, but also led all the personnel involved in this work to make an ideological shift——respect animals, and treat them as true clients.

"Return to nature, close to animals" is a worldwide trend. At the end of last century, zoos construction around China have experienced the first transformation, most of the urban zoos have been moved to the suburbs, iron fence and concrete have been replaced by view barrier-free animal exhibition area with ecological layout, and change the way of keeping animals from captivity in cage to open free range. Nowadays, animals are kept in larger areas in the zoos, and exhibited in a natural way. When we wander in the animals' home which is more close to the nature, we can still find some problems, such as the lack of comprehensive consideration of the needs of both animals and people in exhibition design and construction. In addition, the development of zoos mostly stays in the display of the animals themselves, the construction and display of the overall

在这种与自然更亲近的动物家园时，仍能看到存在的一些问题，如动物园的观展设计和建设没有综合考虑到动物的需求和人的需求。另外，动物园发展多停留在展示动物本身上，对环境的展示和对动物园总体环境的营造做得还不够。

于是，我们期待第二次转型——浸入式动物园设计，这是目前国际上较为认可的一种展览理念。其特点是：把动物放置于一个模拟其原生环境的空间内，使动植物按最理想状态发展；力求重建自然栖息地，重视动物尤其是濒危动物栖息的舒适性；把参观者带入实地环境，使游客可以在更为自然的环境中欣赏野生动物。本项目就是典型的沉浸式动物园设计，其大量的设计细节和翔实的理论介绍，对我国第二次动物园转型具有启示意义。

environment is not enough.

Therefore, we look forward to the second round of transformation —— immersive zoo design, which is a more accepted international exhibition concept currently. Its characteristics include: put the animals in a simulated space like their native environment, so that the plants and animals live in an ideal condition; strive to rebuild their natural habitat, and attach importance to the comfort comfortableness of the animals' (especially endangered animals') habitats; lead the visitors into field environment, so that they can enjoy the wildlife in a more natural environment. This project is a typical immersive zoo design, the large number of design details and theoretical description have the meaning of enlightening on the second round of zoo transformation in China.

思考题

1. 在动物园规划设计中应该主要考虑的主体是人还是动物？怎样协调这二者之间的关系？

2. “景观沁”的设计理念是否适合动物园以外的景观规划设计？如果适合应该进行怎样的调整和改变？

3. 仔细研究不同的观赏方式，思考改善和提升的空间，尝试设计更加新颖合理的观赏方式。

4. 比较传统动物园与乌兰动物园的设计理念，找出乌兰动物园最创新的设计细节。

5. 思考动物园承担的社会责任有哪些，而针对这些不同的责任动物园设计又应该分为哪些层面？

Questions

1. What is the main part of a zoo during the process of zoo design, people or animals? How to harmonize these two factors?

2. Is the concept “landscape immersion” can be used in other landscape design projects other than the zoo design? Should there have some adjustments if it works?

3. Research different eight viewing types to find out the places that can be improved, and try to design new types that more suitable.

4. Think about the social responsibilities that zoos should have, and try to tell the different levels of zoo design that relative to the responsibilities.

5. Compare the traditional zoos with woodland park zoo and find out the most creative details of woodland park zoo.

注释/Note

图片来源/Image：图5-4-1，图5-4-2：http://woodlandparkzoojobs.silkroad.com/；图5-4-3：http://ditu.google.cn；图5-4-4：http://www.google.com.hk；图5-4-5：http://ditu.google.cn；图5-4-6～图5-4-11：Biodiversity Planning And Design: Sustainable Practices（Author：John F. Ahern，Elisabeth Leduc，Mary Lee York）；图5-4-12：http://www.jonesandjones.com/index.html

参考文献/Reference：Biodiversity Planning And Design: Sustainable Practices（Author：John F. Ahern，Elisabeth Leduc，Mary Lee York）；http://www.jonesandjones.com

推荐网站/Website：http://woodlandparkzoojobs.silkroad.com；htp://www.jonesandjones.com

第六章　城市公园景观生态规划设计

CHAPTER VI　CITY PARK ECOLOGICAL PLANNING & DESIGN

第一节　清水湾公园 Section I The Fresh Kill Park

第二节　安大略公园规划 Section II Lake Ontario Park Planning

第三节　橘子郡大公园总体规划 Section III The Planning of Orange County Great Park

第四节　伊利西安公园总体规划 Section IV Elysian Park Master Plan

第一节 清水湾公园

Section I The Fresh Kill Park

一、 场地简介

1. 项目简介

清水湾公园是由纽约市政府、纽约公园和游憩部门为主导而开发的一个公园，基地面积大约2163英亩（约8.8公顷），场地过去曾是纽约市重要的潮汐湿地及野生动物栖息地。主溪流、里士满溪流、大清水湾及小清水湾贯穿整个场地。自1948年之后，场地成为生活、商业、市政等固体垃圾的堆放地和填埋场，2001年纽约政府出台了一项法律，关闭了这个区域的垃圾场。场地43%的面积属于固体垃圾处理区，包括3/4、2/8、6/7和1/9四座垃圾山，如今垃圾山的覆盖工作已完成。

公园分为五大规划区域：北部公园、南部公园、西部公园、东部公园、中部汇流区公园。公园第一开发阶段为南北公园，包括周边道路系统的完善，以增加公园可达性。建成之后的公园将成为人们重要的游憩地，风景秀丽，栖息地得到保护，新的公园路系统将周边相联系，水上活动丰富，文化、商业、娱乐设施散布其间，垃圾场将摇身变为世界上最大、最生态、最创新的都市公园之一。

2. 场地历史

由于场地特殊的地理位置、冰积土及排水系统，场地过去生物种类繁多，生态系统类型丰富。

I. Project overview

1. Introduction

The City of New York, with the New York City Department of Parks and Recreation (DPR) as lead agency, is proposing the mapping and development of Fresh Kills Park. The project site is an approximately 2,163-acre property, it contains within its boundaries intact tidal wetlands and significant wildlife habitats. Main Creek, Richmond Creek, Great Fresh Kill, and Little Fresh Kill cross the project site. For many years, Fresh Kills Landfill operated as the City's principal municipal solid waste landfill, receiving household, commercial, and municipal solid waste and construction and demolition debris since 1948. Landfilling ended in 2001 because of a state law being passed. Approximately 43 percent of the project site is within an Solid Waste Management Unit(SWMU),in which there are four solid waste landfill sections—identified as 3/4,2/8,6/7and1/9.Now the work of closure construction has completed.

The proposed park has five key planning areas: North Park, South Park, East Park, West Park, and the Confluence. The first phases of the park implementation is North Park and South Park , along with proposed park roads to provide access. The park will feature recreational fields; landscaped areas and enhanced ecological habitats; new park roadways, including a new connection with the context; abundant water activities; cultural, entertainment and commercial facilities. The landfill will be become one of the world's largest and most ecologically innovative urban public parks.

2. Site History

清水湾也成为众多候鸟迁徙的目的地。随着斯塔滕岛城市化的发展及清水湾在20世纪后期作为垃圾场，整个岛屿的生态系统遭到严重破坏，环境恶化。现在随着垃圾场的关闭，清水湾将会展现一个新面貌（图6-1-1，图6-1-2）。

3. 设计主旨及思路

场地能否成功转变，更多是来自市民及利益相关者的需求，项目规划在充分尊重市民需求的基础上，提出了以下想法（图6-1-3，图6-1-4）：

1) 创造一个有活力的公园

提供一定的多样性的活动及服务设施，如游憩、运动、餐饮、教育与文化机构、滨水设施，保持公园的活力。这些活动区域应该集中起来，以维持场地大面积的静谧空间，保持环境质量；为大尺度游憩活动创造机会：公园将为游客提供便利的步行道、小径系统，规划组织马拉松跑道、自行车道及马道，同时结合垃圾山等有利地形，设计了数条山地自行车道，公园提供了大面积开放空间满足野餐、社区活动等。观察夜空等夜间活动也成为公园的特色活动。

2) 提供公园设施

在改善现有设施的基础上，增减新的公园设

Because of its geographical position, in combination with the glacial soils and drainage patterns, there were unusually rich ecosystems and plant communities to emerge decades ago. The Fresh kills is also a major destination of birds migrating. Urban development on Staten island has since destroyed much of the ecological richness originally found there, and certainly the use of the Fresh Kills marshes as landfill during the latter half of the 20th century further eroded the quality of the environment. Now, with the closure of the landfill, the site has a chance to present a new look (Fig.6-1-1, Fig.6-1-2).

3. Design Theme and Ideas

The transformation of the site consider the needs ,desires, dreams of citizens and the interests. The idea of the project includes (Fig.6-1-3,Fig.6-1-34):

Activate the park: Most people want to see a mix of active programming, including recreation and sports facilities, restaurants, educational and cultural institutions, and waterfront amenities and so on, The majority hope to retain large sections of quiet, scenic landscape, while offering concentrated areas for active programming.

1) Create Opportunities for Large-Scale Recreational Activities

The park will create extensive pathways and trails for walking, running, organized marathons, bicycling and horseback riding. It also would like an area dedicated to mountain biking trails. There are large-scale open spaces for picnicking, community events and so on. It also provides an a night-sky observatory.

2) Create Park Amenities

There has been strong support for improving

图6-1-1 场地区域鸟瞰图
Fig.6-1-1 Aerial view of Past Site

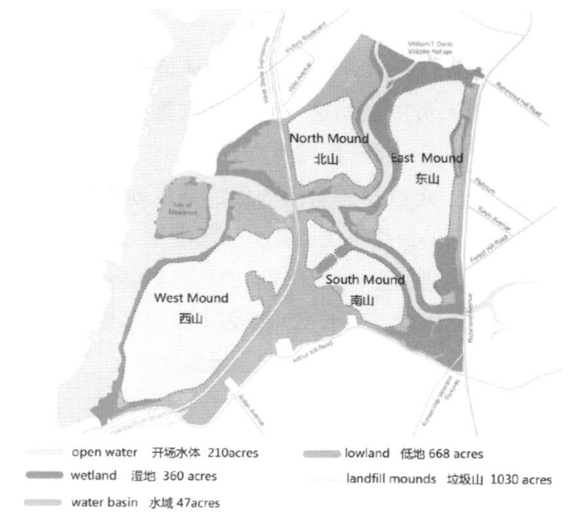

图6-1-2 场地土地类型图
Fig.6-1-2 Diagram of Past Distribution of Landtype

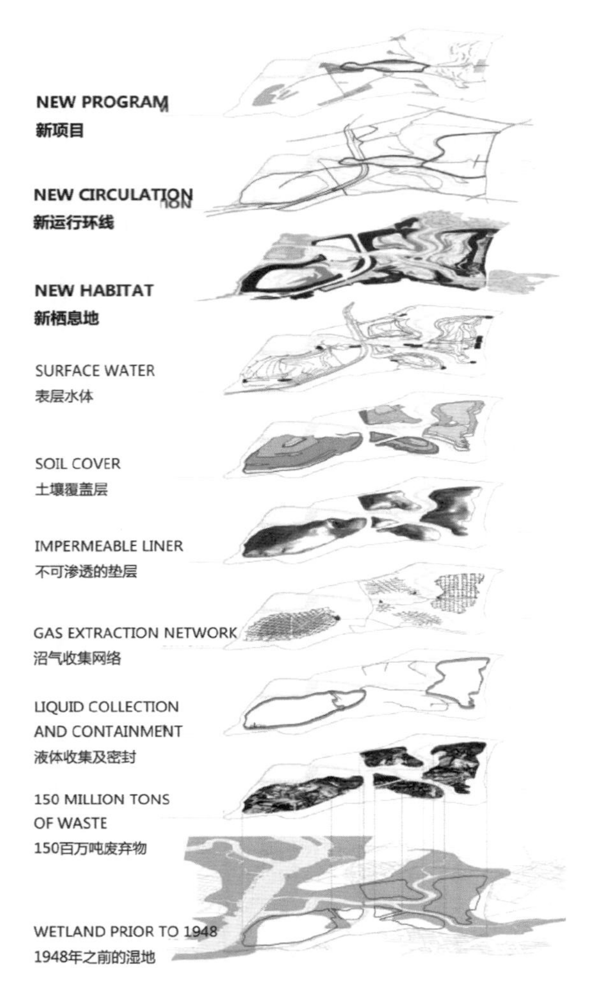

Fresh Kills Park: Lifescape
program+habitat+circulation

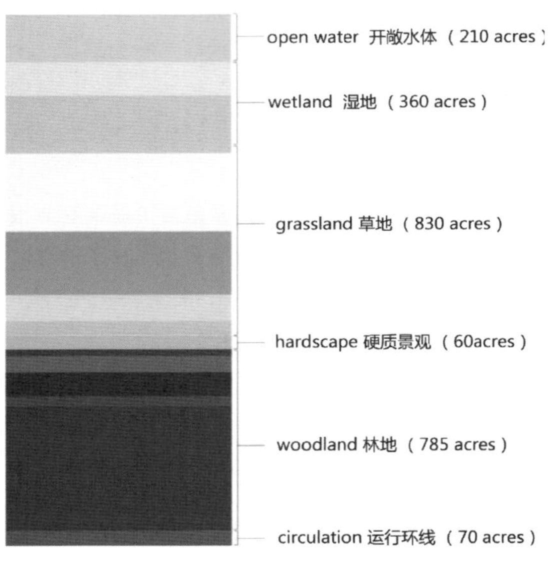

图6-1-4 清水湾公园主题
Fig.6-1-4 Fresh Kills Park theme

图6-1-3 清水湾历史及基础设施分解图层
Fig.6-1-3 Historical and infrastructural layers for The Fresh Kills Site

施，以满足周边社区市民的需求。

3) 借助清水湾公园提升区域自然资源

清水湾公园对于当地的自然环境有着重要意义，规划人员应借助这一机会来恢复自然系统，营造开放空间，提升区域生态系统，改善栖息地环境，创造一定数量的消极利用的野生动植物栖息地。规划新的道路系统以减少公园潜在人流及车流量对当地交通的影响：建成后的公园势必会对周边的交通产生影响，因而交通规划不仅要尽量减少对当地的影响同时还要助于提供当地的连通性。

4) 创造滨水游憩活动

利用现存的水资源为各种水上活动创造机会。

existing recreational and park amenities and building new ones, responsive to the needs of local residential communities.

Capitalize on Fresh Kills' vast scale to improve regional natural resources: Many people recognize Fresh Kills as a rare opportunity to restore natural systems, create wide-open spaces, and improve ecology and habitat in the region. A significant number envision extensive areas of passive park with diverse wildlife and plant habitats.

3) Build New Roadways to Mitigate the Impact of the Park on Local Congestion

There is widespread concern among Staten Islanders about the effect a destination park may have on traffic in already congested areas. Many people see transportation improvements as the key to managing anticipated impacts, and as an opportunity to help improve local connectivity.

5) 提供教育机会

许多市民意识到了教育机会的重要性，认识到清水湾转变为公园地的国际性重要意义。垃圾山的历史及作用可以通过教育展览的方式向人们展示，生态中心可以让当地的青少年能够切身体验生态科学。

6) 创造艺术及文化活动机会

场地独一无二的自然及尺度可以为环境艺术、表演活动及盛大文化活动提供机会。人们都很高兴能够在清水湾观赏一些艺术活动，如艺术装置、社区艺术工作坊、博物馆、画廊、露天剧场、盛大活动或展览等（图6–1–5）。

7) 展现可再生能源系统

公众希望公园的设计能够使用太阳能、风能、水能及沼气等可持续能源。而这些设计所带来的体验能够展示公园的前沿科技，增加其教育价值。

商业设施集中布置：当地居民认为大部分大型的商业中心应该集中布置在公园中心而不是在其边缘布置，但是商业的发展应该有一定的限制。商业的存在不仅能够增加公园的活力还能提升经济收益。

8) 增加青少年游憩活动

斯塔滕岛缺少运动设施，尤其是全年性设施。考虑到区域的不断发展及学龄儿童的数目不断增加，总体规划增加了一些活动设施，如室内田径训

图6–1–5 活动项目规划
Fig.6-1-5 Program plan

4) Create Opportunities for Waterfront Recreation

There is much existing water resources of Fresh Kills, seeing opportunities for different kinds of activities.

5) Create Educational Opportunities

Many people value the importance of educational opportunities and recognize the international significance of the conversion of Fresh Kills to parkland. The history and workings of the landfill can be explained through educational exhibits. An ecology center could involve local youth in ecological science experiments.

6) Create Opportunities for Art and Culture

The unique nature and scale of the site suggests opportunities for environmental art, performance art and cultural event programming. People are generally interested in seeing an arts program at Fresh Kills, including artwork installations, community and art workshops, a museum or gallery, an amphitheater, events and displays (Fig.6-1-5).

7) Demonstrate Renewable Energy Systems

The public is supportive of a park design that includes sustainable energy demonstrations harnessing solar, wind, water and methane power. Many persons feel these experiments would give the park a cutting-edge identity and augment its educational value.

Concentrate commercial facilities: Opinion is fairly consistent among local residents that large-scale, commercial programs should be located primarily in the center of the park, rather than along its edges, but that any such development be limited. Most people understand that commercial concessions are needed to activate the park and generate operating revenue.

8) Promote Youth Recreation

Many people feel that Staten Island has too few sports facilities, particularly year-round facilities. Noting that the borough is growing and there is an increasing number of school-age children, participants suggested that the Master Plan incorporate facilities such as an indoor track and field training center, indoor aquatic center and indoor tennis center.

9) Landfill Operations

There is a clear need to accommodate and avoid conflict with landfill operations is a major priority of the master planning process.

10) Environmental Health and Safety

Some people have voiced concern for health and safety at Fresh Kills, and have asked for the city's commitment to ensuring environmental regulatory standards are met and maintained. The city has in

练中心，室内水上运动中心，室内网球训练中心。

9）垃圾处理

总体规划的一项重点任务就是解决垃圾处理并避免其产生的冲突。

10）环境健康和安全

一些公众清水湾的环境健康和安全极为关注，希望政府能够确保清水湾能够达到环境监管标准。政府已经对此做出承诺，场地内的基础设施、监测及维护将会保证场地对公众使用的安全性。规划区域内的任何场地如果没有严格达到监测标准，政府将不会向公众开放使用。

11）清水湾规划过程

清水湾总体规划草案的完成是基于以下方面：场地的深入分析、社区需求评估及拓展、垃圾处理及其他垃圾处理模型的研究和总述。对于清水湾这样复杂且庞大的项目，基于主要代表中有识之士共识的决策是至关重要的。草案中所描述的场景和框架可以说是清水湾新的里程碑，并且这个草案将会是以后回顾、讨论、决策的基础（图6–1–6）。

fact made this commitment, and extensive on-site infrastructures, monitoring and maintenance will ensure the site is safe for public use. The city will not allow any part of the site to be opened for public use until regulatory standards have been clearly met.

11) Process at Fresh Kills

The Fresh Kills Park Draft Master Plan is the result of extensive site analysis, community needs assessment and outreach, landfill operations consideration, and studies and reviews of other landfill models. For a project as big and complicated as Fresh Kills, decision-making based upon informed consensus among the primary representatives is critical. The Draft Master Plan represents a milestone in terms of describing a vision and framework, and will serve as the basis for further review, discussion and decision-making (Fig.6-1-6).

II. The Master Plan

1. Circulation Plan

The site plan optimizes connectivity, access, and movement, the plan includes:Vehicular circulation plan, Non-vehicular circulation plan, Multi-use paths plan, Specially designated paths and trails, Waterfront access.

1) Parking Plan

Parking lots will be dispersed at appropriate

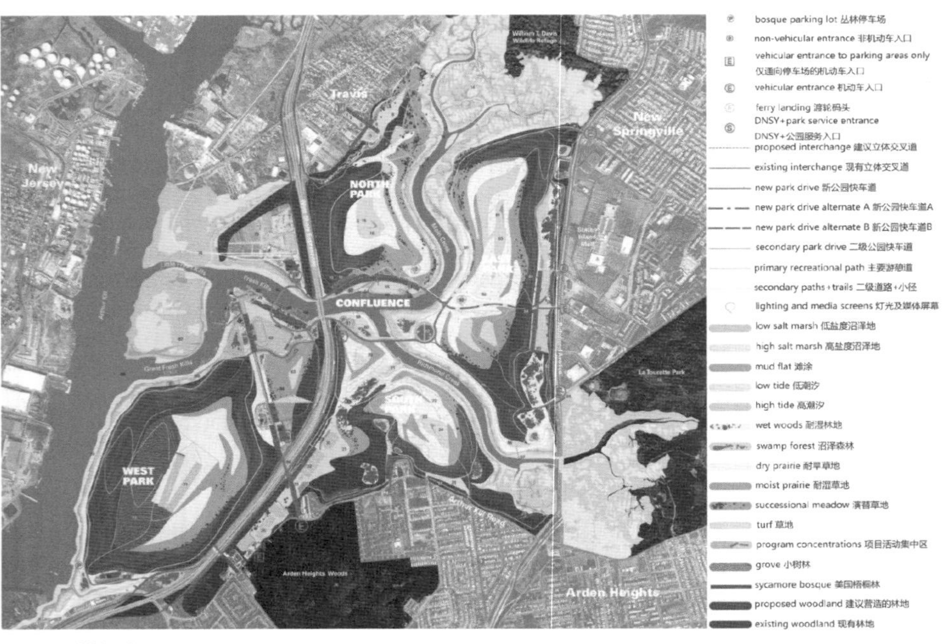

图6–1–6 总体规划图
Fig.6-1-6 Master plan

图6–1–8 丛林停车场鸟瞰图
Fig.6-1-8 Aerial view of the Parking Bosque

二、 总体规划

1. 交通规划

整个场地的交通规划强调连接性、可达性以及流动性。这个规划包括机动车、非机动车、多用途步行道、特殊道路及小径、滨水岛等方面的规划。

1) 停车场规划

规划并未将停车场集中布置，而是分散在公园内部的各个区域。周边社区居民可以通过公园的次级入口达到公园，他们可以通过步行或者自行车等交通形式到达这里。为了减少热岛效应及控制地表径流，停车场的铺装为透水铺装，并种植一些树木提供绿荫空间。由于这些停车场是一些主要的停车点，因而这些被树木环绕的停车区是场地重要的设计对象（图6–1–7，图6–1–8）。

2) 机动车交通规划

首先通过新建机动车道路系统以及入口来调整公园的机动车系统，所有道路在提供必要的连接性的同时还要尽可能保护开放空间及栖息地。

3) 非机动车交通规划

公园提供了一系列不同的道路及小径，为不同的使用者实行人车分离，在小道和机动车道交叉口，为游人的安全通行设计了特别的人行道（图6–1–9）。

4) 多用途小道

在垃圾山山脚设置了环形的小道，游客能够在此散步、慢跑、骑车甚至骑马。这些小道在引入照

locations throughout the site, allowing for localized or neighborhood access associated with the many secondary park entrances which are intended to provide local residents with access to the park by bicycle or on foot. In order to reduce heat island effect and control runoff, parking lots have permeable surfaces and trees surround them. At major gathering points, the tree-lined parking areas, will become major design features (Fig.6-1-7, Fig.6-1-8).

2) Vehicular Circulation Plan

The accommodation of vehicular circulation will first be finished through the construction of several miles of new park drives and new entrances. All the roads are designed not only to provide the needed connectivity but also to preserve large open spaces and habitat areas.

3) Non-vehicular Circulation Plan

There are a variety of paths and trails for multiple users. All paths are separated from roads, with special pedestrian crossings as needed to facilitate safe passage (Fig.6-1-9).

4) Multi-use paths

To create loops around the base of each of the landfill mounds for walking, running, cycling, and horseback riding. They will be the primary activity paths with infrastructure such as lighting, seating and so on (Fig.6-1-10).

图6–1–9 机动车运行环线规划示意图
Fig.6-1-9 Illustrative view of vehicular circulation plan

明、坐椅等基础设施后将成为公园内极具活力的道路（图6-1-10）。

5) 特殊小道及小径

这是设计人员为公园特殊人群设计的道路，如那些骑自行车的人，徒步旅行者等。这个规划为这类特殊群体提供了场地20英里（约32公里）的道路。

6) 滨水步道

能够让游客到达小溪周边的码头及游艇。

2. 建筑物规划及设计

无论场地内的新建筑还是旧建筑都必须和周边的生态和环境密切结合，并且能够适应场地来发展的需求，因而一些建筑具有创新性——拥有可移动的屋顶、可控制的表皮等。规划人员为所有建筑设计制定了一些原则：尽可能减少对环境的影响；尽可能利用被动式供热、散热和通风系统；尽可能利用多种的合适的具有可持续性的技术，如绿色屋顶；尽可能利用当地或循环利用的材料。

3. 景观及生态系统规划

基于场地的自然资源，总体规划期望能够培育一个多样的、具有恢复力的景观，并通过移除场地污染，提升其作为开放空间的价值。现有的场地既有人工景观也有自然景观，由于垃圾山的长期污染，四座垃圾山及周边去生态系统很脆弱，生态系统单一，生物种类不多。场地一些自然景观则拥有丰富的生态系统，例如数百英亩的盐沼地和潮溪网络（图6-1-11）。

基于现有的场地条件，规划人员在景观和生态系统建设方面设定了五个目标：营造一个多样化、有一定恢复力的景观系统，建成后的场地景观能够提升区域水体和环境质量，强化生态连续性，提升生物多样性及环境可持续性；建设生物廊道，连接现有的植物、动物、鸟类及为生物群，构建区域特有的生物栖息地及河口生态系统（图6-1-12）

根据现有的自然资源及新栖息地营造的潜在机

图6-1-10 非机动车运行环线规划示意图
Fig.6-1-10 Illustrative view of non-vehicular circulation plan

5) Specially Designated Paths and Trails

These are designed for special user group such as cyclists, hikers and so on. The plan provides more than 20 miles of such paths.

6) Waterfront

Waterfront access is accommodated by numerous docks and launches around the creeks.

2. Structure Plan and Design

Every structure, whether new or old, must be fully integrated with the local ecologies and site conditions and be able to adapt to site development in the future. So there are some innovations being applied to the architectures---- open roof structures, programmable surfaces and so on. The planners formulate some structural design principles that all structures 1.have a minimal impact on the local environment; 2.take advantage of opportunities for passive heating, cooling and ventilation systems; 3.maximize the use of appropriate, incorporate sustainable technologies ,for example greenroofs; 4 .utilize local or recycled materials wherever possible.

3. Landscape and Ecosystem Plan

The Master Plan proposes to build on these natural

会来组织公园的内部结构加强生态建设，保证公园每个阶段建设过程中（公园建设分为三个阶段）都能让游客明确地体会到“不断演替的景观”，提高生态系统演替的多样化。将生态建设规划与正在进行的垃圾山管理工作相结合，以减少费用，提升场地的可持续性。

在实现以上目标的同时，三个主要因素促使规划人员如何组织各类栖息地的结构，这三个因素为：现有自然资源的位置及栖息地营造的机会；实现与周边自然资源的相连；人们期望的公园空间结构及景观。建成后的公园主要包含三种类型的栖息地：湿地、草地、林地。现有的湿地以入侵物种为主，所以对其恢复的策略是除去入侵物种，重新创建一个物种丰富，能够自我维持的湿地群落。草地可以分为山丘草地（垃圾山表层草地）和非山丘草地。山丘草地是通过在垃圾山之前覆盖一些土壤而实现的，物种以适应能力强的本地物种为主，能够实现这个新生态系统的自我维持，也有利于减少长期维护的费用。非山丘草地的改造主要依据新的利用方式而定。林地面积将会适当扩大，这些林地主要是草地及新群落建设地等区域的周边林地，其目

assets to cultivate a diverse and resilient landscape, and to improve open-space asset through removing the pollution. The current site is a complex amalgam of artificial landscape and natural systems. Because of long-term contamination, off-mound ecosystem and on-mound ecosystem are very fragile and homogenous and only a few species survive. Yet some areas in Fresh Kills has abundant ecological assets, such as hundreds of acres of salt marsh, a network of tidal creeks (Fig.6-1-11).

Based on the current natural resources, The planner set five goals for landscape and ecosystem plan: Cultivate a diverse, resilient landscape that is a natural asset to the region in terms of ecological connectivity, water and air quality improvement, biodiversity and sustainability; Create meaningful habitat for the region and the estuary by building wildlife corridors linked to existing natural resources, taking into account not only plant life but also bird, mammal, fish, crustacean, insect and microbial communities(Fig.6-1-12); Organize the park internally around existing natural resources and local opportunities for enhanced habitat creation;Design and stage ecological improvements so that the parkland can be understood and enjoyed in each phase of its development as a legible "landscape in process," designed to promote successional diversification over time; Integrate ecological improvement plans with ongoing landfill management operations to increase benefits, reduce public expenditure and enhance site sustainability. In

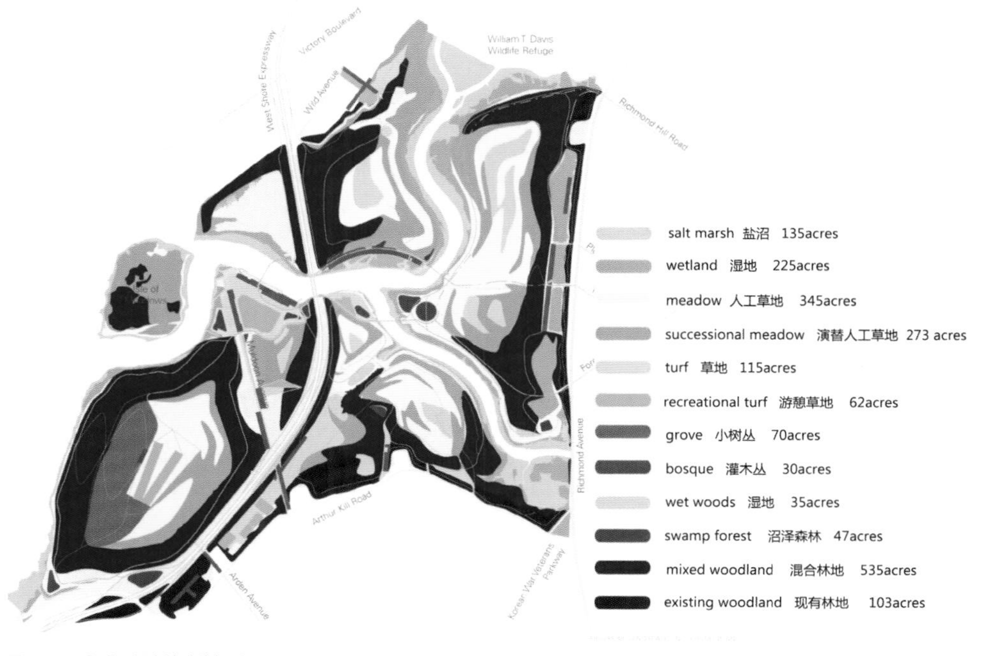

图6-1-11 景观及栖息地总体规划图
Fig.6-1-11 The master plan of landscape and habitat

图6-1-12 生命景观发展过程中生物多样性图表
Fig.6-1-12 Diversification chart of lifescape's development

的是便于为这些地区提供一个缓冲带，在公园内部形成一个林地廊道网络，将公园周边及其内部的自然系统联系起来。

整个场地植物群落建立在对场地实地分析及斯塔登岛健康自然生态系统研究基础上而确定的。物种主要是那些能够适应现在条件并能够改善场地不良条件且不需要较高维护的先锋物种。这些植物为后期公园物种多样化目标的实现创造了先决条件。植物种植采用了线、簇（孤岛）、面的模式。这种模式既有利于生物的迁移、种子的传播也能够保证人的移动。线模式的目的是引导场地物质流、水量、能量流，将新物种引入同质化区域，主要包括沼泽地边线、小径、林荫道、线性灌木丛等；簇（孤岛群）模式主要是为自然资源保护区及人类活动密集区等区域提供一系列的“保护巢”；面模式就是创造一个拥有大面积的类似于斑块的渗透层镶嵌体，目的就是创造一个自我维持的表层，控制土壤侵蚀，营造本土栖息地。它包括湿地、林地、运动场等。为了保证生态系统之间连续的物质流和能量流，规划人员强化了场地内外环境间的联系，公园中的种植模式、水流、小径、基础设施等为场地新的生态系统演替拉开了序幕。总而言之，随着时间的推移，规划将为公园培育不同景观类型，如湿地、林地、草地群落；提供多样的活动机会，成为纽约市的一处独一无二自然保护区（图6-1-13）。

三、 垃圾山修复规划

1. 现状及规划目标

规划之初，垃圾山表层土壤稀薄且贫瘠，土壤水分较低且变化很大，入侵种占主导地位，生物多样性基本不存在。如果能够创造性利用这些垃圾将

keeping with the goals mentioned above, three key factors direct the organization of the structure of the habitats, which include: location of existing natural resources and opportunities for habitat creation; connectivity with adjacent natural resources; desired spatial envelope and landscape setting for the park.

Three primary landscape types-wetlands, grasslands and woodlands- will coexist after the completion of the park.

The wetlands will be renovated to remove invasive species and create more diverse, self-sustaining wetland communities. Grasslands type includes off-mound grasslands and on-mound grasslands. After being covered with a layer of fresh soil, on-mound grasslands will be need to be cultivated over time, and the species are primary native plants which can be adapted to existing conditions and be helpful to construct self-sustainable ecosystems and reduce long-term maintenance. Off-mound grasslands will be shaped to accommodate new use in the park.

Woodlands would be expanded from where they are presently, as well as in significant areas of new planting, both on- and off-mound, to help create a large buffer rim around the site and enhance corridor connectivity with adjacent natural systems. The proposed palette of planted communities at Fresh Kills is based on site analysis and field study in healthy natural areas on Staten Island. The plan recommends that habitat creation efforts rely primarily on tough, "workhorse" species which will re-establish the prerequisite conditions for natural, successional processes to build diversity over time and do not require a high degree of maintenance. The masterplan propose a matrix of lines (threads), surfaces (mats), and clusters (islands) to maximize opportunities for access and movement-movement of seeds and biota as well as people and activities. Linear threads direct flows of water, energy and matter around the site, injecting new life into otherwise homogenous areas. Linear threads include edges of wetlands, pathways, avenues, linear shrubs and so on. Clusters of islands provide denser nests of protected habitat, seed source and program activity. Surface mats create a patch-like mosaic of mostly porous surfaces to provide self-sustainable coverage, erosion control and native habitat. Surface mats include wetlands, woodlands, playground and so on. In order to ensure material flow and energy flow between ecosystems, the planner strengthen the connection between the site and the context. Some things—the plant matrix, streams, pathways and infrastructure—will initiate new succession across the site. In a word, the landscape and habitat plan will create different landscape style, such as wetlands, grasslands and woodlands, and provide many opportunities for various actives and marks the site as a unique reserve in New York (Fig.6-1-13).

会极大改变现有场地的环境问题。垃圾山的土壤更新及植被覆盖不仅能够提升整个场地生态系统的健康性及多样性，也能够重塑垃圾堆表面的景观特征。重塑过程中要逐步减少局部覆盖的焚烧，控制侵蚀，改善土壤水文及排水状况，增加土壤厚度及减少长期维护成本。规划人员为垃圾山的恢复设定了以下几个目标：在保证垃圾山结构稳定的前提下增加土壤量并改善土壤质量；在避免积水的前提下通过植物的吸收作用增加土壤的蓄水量；减慢入侵物种的传播速度；重新引入本地物种，创建一个多样的种子库，营造一个繁茂的覆盖层；在满足监测要求前提下减少维护需求及成本。规划人员采用了一系列技术手段来实现这些目标，如场地的监测，垃圾山表层土壤的更新或者增加新鲜土壤，“耕种”垃圾现有土壤以改善土壤现状，营造新的草地覆盖层等。考虑到清水湾不同的现状条件及覆盖类型，有必要建立一套复合处理技术（图6–1–14，图6–1–15）。

2. 填埋场封场

封场就是将垃圾堆封存起来，以减少水体渗透，控制侵蚀，改善表面排水，为下面固体垃圾和土层之上环境之间建立一道屏障。封场是层体系的监测术语，这个体系一般由不渗漏层（2~2.5英尺[约0.6~0.76米]厚的保护物质层，具有屏障保护作用）和生长介质层（大约6英寸[约0.15米]后，为植物生长提供条件）。

清水湾封场由上至下有以下5个层组成：顶层

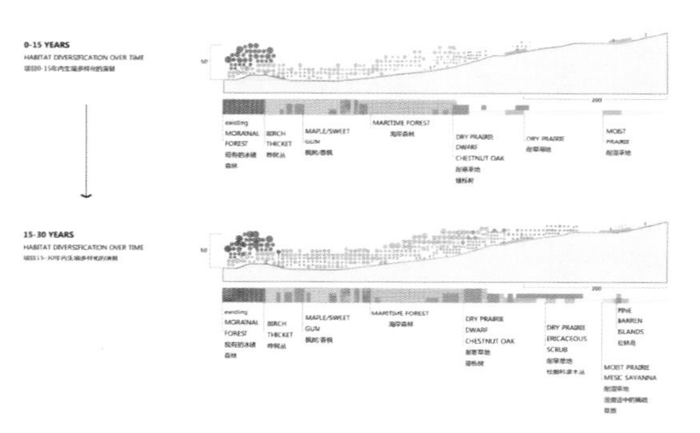

图6–1–13 栖息地多样性发展示意图
Fig.6-1-13 Habitat diversification development

III . Landfill Remediation Plan

1. Current Condition and Planning Goals

Before the plan, the current condition of the landfills is: soils are thin and of poor quality, moisture levels are generally low but also highly variable, invasive species dominate, and there is very little species diversity. it is of great significance to improve environmental condition if creative measures are taken to manage the landfills. The renovation of soils and vegetative cover on the mounds would not only improve the health and diversity of ecosystems across the site but would also improve the performance of the landfill caps by reducing localized cover burnout, minimizing erosion, improving soil hydrology and drainage, thickening soil depth and reducing long-term maintenance costs.

There are several goals for landfill mounds restoration: increase soil quality and quantity while ensuring structural stability; retain more water for plants while avoiding water accumulation; reduce the spread of invasive species; reintroduce native plant communities capable of building a diverse seed bank and establishing a robust cover; minimize maintenance requirements and costs, while complying with regulatory requirements.

There are a range of techniques for achieving these goals, ranging from in situ management over time to importing and/or manufacturing new soils for overlay on the cap and "farming" the slopes to renovate soils in situ, and then establishing new meadow cover. Given the range of different situations and cover types at Fresh Kills, it is likely that a combination of techniques will be necessary (Fig.6-1-14,Fig.6-1-15).

2. Landfill Sections Final Cover

Final cover is to be placed over all the landfill sections for the purposes of minimizing water infiltration, reducing erosion, promoting positive surface water drainage, and providing a physical barrier between the solid waste below and the above-ground environment. Final cover is the regulatory terminology for the system of layers that are generally comprised of an impermeable liner, 2 to 2.5 feet of barrier protection material and a roughly 6-inch thick layer of growing medium.

The five layers of final cover at fresh kills from the top to the bottom comprise the following.

Top Layer (Vegetation/Roadways/Surface)---- The top of the final cover is designed to direct runoff away from the landfill and control erosion of the cover. It typically consists of vegetative cover (including a 6-inch-thick layer of planting soil) but may also include asphalt or gravel road materials. The final cover vegetation includes warm season grasses. Annual rye grass is used initially to establish a vegetative cover to

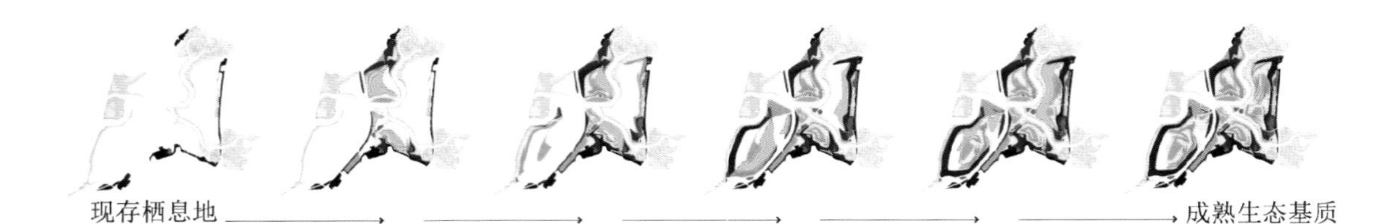

图6-1-14 新栖息地的培育
Fig.6-1-14 Cultivating new habitats overtime

图6-1-15 垃圾山恢复示意
Fig.6-1-15 Illustrative view of Landfill Mound restoration

（植被、道路、地表）：封场的顶层主要作用是引导表层的排水，控制覆盖层的侵蚀。一般包括植被覆盖层（包含有6英寸厚的种植土），也可能包括沥青层、碎石路等。封场植被一般为暖季型草。一年生黑麦草常常作为先锋种，以快速形成植被覆盖层控制水土流失，暖季型草然后逐步取代黑麦草。隔离保护层：它由2英尺厚的土壤组成，该层的作用是因极端天气而引起断裂或沉降的情况下，保护下面的水体阻挡层，存储多余雨水直至植被吸收完或者直至将水排出覆盖层，同时也可以防止穴居动物挖掘。排水层：由土壤或者地工合成物组成，这一层主要是将渗入到封场内部的水体排掉，以实现将上层土壤中过多的水分排出的目的，进而将封场的饱和度尽可能最小化。水体阻挡层：本层由土工膜或者压实黏土层组成，主要目的是阻止水体渗透到下层的固体垃圾中导致沼气的产生。同时它也能够密封垃圾所产生的沼气。沼气排放层：本层由土工复合材料或者沼气输送管组成，它可以将填埋场内部产生的沼气输送到沼气排放系统。

3. 渗透液控制系统

渗透液是由于雨水渗透到填埋场内部固体垃圾层后，一些可溶解的物质在垃圾层溶解而产生的液体。为了达到一些环境法律的要求，规划人员设计

prevent erosion until the warm season grasses take hold.

Barrier Protection layer — This layer is composed of a roughly 2-foot thick layer of soil. Its purpose is to protect the underlying hydraulic barrier layer from weather extremes that could cause cracking or heaving and to store excess water until plants uptake water or until the water drains from the cover, and protects against burrowing animals.

Drainage Layer — Made of either soil or a geosynthetic, this layer facilitates drainage of water that has infiltrated into the final cover, draining the overlying soils so as to minimize the saturation of the final cover.

Hydraulic Barrier Layer — This layer consists of a geomembrane, or compacted clay layer that prevents water from percolating into the underlying solid waste and creating leachate. It also functions to contain landfill gas.

Gas Venting Layer — This layer typically consists of a geocomposite layer or landfill gas ducts that allow landfill gas generated within the landfill to move toward landfill gas vents.

3. Leachate Control System

Leachate is created as percolating rainwater travels vertically through the solid waste layer and dissolves soluble substances in the solid waste layers into leachate. In accordance with the requirements of some environmental laws, models will be developed projecting a volume of flow and solute transport in order to evaluate the effects of the landfill on local groundwater and surface waters, identify the potential vertical leachate migration pathways, and to develop corrective measures, as necessary. Also, groundwater sampling and sensitivity analysis will be also performed. As a result of these modeling and monitoring investigations, a leachate control system will be selected by the city and approved for installation at the Fresh Kills Landfill. The objectives of the corrective measures system are to minimize leachate impacts to local groundwater or surface waters. The Fresh Kills Landfill Leachate Treatment Plant has a design capacity of 1,050,000 gallons per day and contains the following major unit processes: influent holding, sequencing batch reactors for

出了模拟模型，这些模型可以预测渗液的体积及流动方向，进而可以评估填埋场对当地地下水及地表水的影响，确定渗液垂直运动路径，进而制定出正确的必要的措施。同时规划人员也进行了地下水抽样分析及环境敏感性分析。结合模型及环境监测分析，政府部门为清水湾公园选用了一套渗滤液控制系统。这些措施的最终目的是尽可能减少渗滤液对地下水及地表水的不利影响。清水湾公园的渗滤液处理场可以每天处理1 050 000加仑（约3975立方米）废液，主要处理过程包括：渗流控制，生物治理的顺序分批式反应器，金属沉降，重力分离，砂石过滤，PH值调整，排污口扩散器排放。清水湾渗液控制措施包括：不渗透的封场，分隔墙，渗滤液回收井，渗滤液收集及运输系统以及处理场。这些措施能够尽量阻止渗滤液的流动以避免对清水湾地区的地下水及地表水的污染。此外填埋场下方的更新世地质构造可以作为不渗漏屏障的一部分用以隔离填埋场和深层地下水。总之，清水湾填埋场渗滤液控制系统就是在渗滤液排入水体之前将其收集和处理。公园将会对地下水和地表水持续监测以检测系统的性能，保证能够判定出渗滤液外泄所引起的环境污染以及渗滤液合适的移动路径以将其对环境不利影响最小化（图6–1–16）。

4. 垃圾山沼气管理系统

城市固体垃圾会包含一些可以焚烧的物质，这些物质由于生物或化学分解作用而产生废气。为了避免废气外泄而造成严重的环境的问题，填埋场设计了一套废气管理系统，这个系统主要包括废气收集系统，废气处理及燃烧系统、被动式通风系统。这些收集的废气可以通过前期处理之后在废弃处理厂转变为天然气亦或直接输送到燃烧平台。

5. 雨水径流管理系统

这个系统作用是在雨水进入当地水体之前，截留场地自我产生的雨水径流，除去水体中的悬浮物及污染物。它可以控制雨水对封场顶层的侵蚀，也可以阻止雨水对水体阻挡层的破坏。除了雨水控制系统之外，场地的各个部分都有对侵蚀及沉降物控

biological treatment, metals precipitation, gravity clarification, sand filtration, effluent pH adjustment, and discharge through diffuser outfall. These control measures at Fresh Kills, including the impermeable landfill covers, cut-off walls, leachate recovery wells, and a leachate collection and conveyance system and treatment plant, prevent the migration of leachate generated by the landfill sections from reaching local groundwater and surface waters in and around Fresh Kills. In addition, the mostly Pleistocene geological formation beneath the landfill forms part of an impermeable barrier between the landfill and deeper groundwater. In summary, the Fresh Kills Landfill leachate control system is designed to contain, collect, and treat leachate before it reaches surface waters. Moreover, groundwater and surface water monitoring programs are also used to measure the performance of the system and to ensure that any possible impacts to the environment due to leachate releases are identified, and appropriately mitigated, in order to minimize any negative impacts to the environment (Fig.6-1-16).

4. Landfill Gas Management System

Municipal solid waste landfills contain buried wastes that produce gaseous compounds through biological and chemical decomposition. In order to avoid significant impacts to the environment from air emissions at the landfill, The Fresh Kills landfill has an gas emissions management system which is generally composed of an active landfill gas collection system, a landfill gas recovery and flaring system, and a passive venting system. the collected gases are either sent to a gas recovery plant for re-use as pipeline natural gas after pre-treatment or directed to a flaring station.

5. Storm Water Management System

The system has been designed to detain all site-generated stormwater runoff on the site and to facilitate the removal of suspended sediments and any adhered pollutants prior to any discharges to local waterbodies. It is designed to prevent erosion of the top layer of the final cover, thereby preventing damage to the underlying hydraulic barrier layer and thus minimizing infiltration and leachate generation. In addition to the stormwater management system, erosion and sediment control practices across the site, but particularly on the landfill sections, greatly reduce the potential for water quality impacts from the landfill on the surface waters flowing through Fresh Kills. In summary, site drainage and runoff at Fresh Kills is controlled through vegetative cover, grading, and stormwater collection and control systems. To that end, final cover plays an important role in stabilizing and protecting the soil from erosion during rainfall events.

制的措施，这些措施在填埋场尤为重要。它们大大降低了流经填埋场地表水在流经清水湾时对水体质量造成的影响。总之，场地排水及雨水径流的控制是通过植被覆盖层、分类处理、雨水收集及控制系统等完成的。在降雨过程中，封场在土壤固定控制侵蚀方面发挥了重要的作用。

四、 汇流区设计

公园包括五大分区：汇流区、北部公园、南部公园、东部公园和西部公园。每一个公园都有自己独特的属性、栖息地和设施。所有的公园都通过环形交通网络与中部汇流区相连接，这个交通网把整个场地联为一体，将不同区域相联系（图6–1–17）。

汇流区：

汇流区位于这个巨大公园的中部。溪湾区和尖点区是中部区重要的开发区。尖点区将规划成为公园最大的人流集聚地，通过跨越清水湾标志性大桥将其打造成公园的门户区尖点区是公园滨水活动、文化商业活动等最佳区域。这些活动依赖于完善的基础设施、适合高人流量的构筑物、足够大的停车区域。本区域为人们提供游憩活动、综合运动设施及场地，可以满足看台甚至体育场的需求。随水岸线延展的漫步串联起餐馆、聚餐区、露天市场、码头、渡轮站点。溪湾区将规划成为环线北部区域重要的滨水和文化活动区。对公园权益人来说，这一区域是个非常重要的规划区，是到达滨水区参与滨水活动的重要战略点。溪湾区的服务对象主要为家庭经济社区，强调生态、教育价值及与水相关的活动。此区域最有可能最早为公众提供游憩活动、商业及文化项目，为人们提供独具特色的公园体验。区域硬质和软质边界的结合有利于露天平台及船坞的设置，这些设施均临近得到恢复和维护的潮汐湿地（图6–1–18~图6–1–22）。

五、 总结

滨水公园是城市中自然因素最为密集、自然过程最为丰富、生态环境最为敏感的地域，是最重要

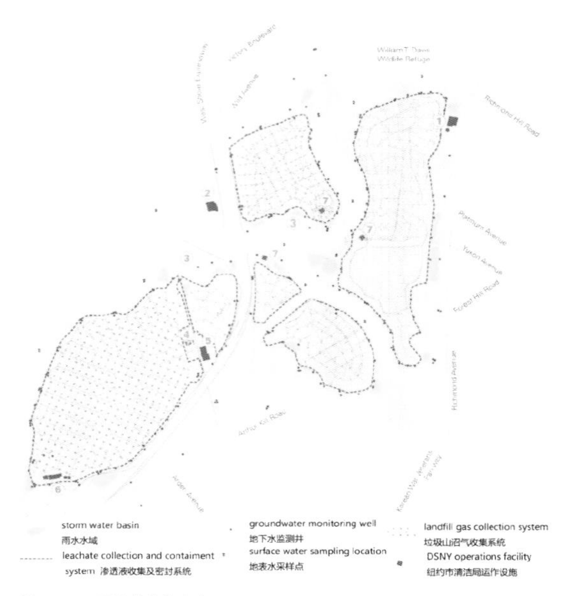

图6–1–16 环境健康和安全
Fig.6-1-16 Environmental health and safety plan

IV. The Confluence Design

The park has five main areas: the Confluence, North Park, South Park, East Park and West Park, each with its own unique attributes, habitats and amenities. All park areas are linked to the central confluence circulation network that organizes the site as a whole and establishes connectivity between the different areas of the park (Fig.6-1-17).

The Confluence:

The Confluence is at the physical center of the vast expanse of park. The Creek Landing and the Point are its major development locations. The Point is planned as the largest concentration of destination programs in Fresh Kills Park and will serve as a gateway destination marked by a signature bridge crossing Fresh Kills Creek. The location is optimal for iconic, waterfront programs and cultural and commercial uses that depend on a high degree of finish, visibility and proximity to other amenities, structures with large footprints and ample parking areas. The Point offers opportunities to accommodate active recreation programs and multi-use sports facilities and fields with the ability to include stands and even a stadium. A long promenade along the water's edge can support restaurants, a banquet facility, an open-air market roof, a boat marina and a ferry or boat-taxi launch. Creek Landing is planned as a concentration of waterfront and cultural activity on the northern side of the loop drive. It will be a key location for

图6-1-17 公园五个分区
Fig.6-1-17 Five areas of the park

的生态廊道之一。清水湾公园作为城市滨水公园，展示了纽约城市滨水景观的美学价值。项目场地曾是纽约市潮汐湿地和野生动物栖息地，一度作为垃圾场使这里的生态系统受到破坏。场地修复是关键的第一步，通过填埋场封场、渗透液控制系统和沼气管理系统的设计等措施，恢复了场地的安全性。规划重视挖掘城市滨水公园景观的美学价值，保留地域历史文化的多样性，从而创造城市和谐的生态环境。

清水湾公园设计中，注重公众共享性，提升滨水区活力，保护自然资源，激发艺术及文化创造活动，展现生态循环系统，配套商业设施，同时结合湿地植物景观设计将湿地生态教育、旅游及生物多样性保护等功能综合起来，成为公众参与性很高的活动场所，为纽约市的城市生活提供了丰富的物质和精神资源。

总体规划的生命景观包含三项功能：活动项目＋栖息地＋循环。生命景观是大尺度环境再利用及更新的生态过程，不仅仅恢复场地生态系统的健康及生物多样性，也能够为潜在的游客提供精神享受及想象力。生命景观也是一系列动态培育过程——不仅土壤、空气、动植物等生态系统要素的培育，还包括人类活动、经营模式、管理行为等方面的动态策划，也包括环境技术及再生能源的开发、教育功能的创造（图6-1-23）。

access to and interaction with the waterfront, a programming goal of particular importance to Fresh Kills Park stakeholders. The Creek Landing is scaled and oriented primarily toward family and community use, with an emphasis on ecological, educational and participatory water-related programs. the site offers the greatest potential for early public access to recreational, commercial and cultural programming and to the creeks that provide a unique way to experience the park. The combination of hard and soft edges makes the site an advantageous location for decks and docks adjacent to restored and well-maintained tidal wetland (Fig.6-1-18~Fig.6-1-22).

V. Summary

Waterfront Park is the area in a city that with the most concentrated natural factors, the most abundant natural process, the most ecological environment sensitivity, is one of the most important ecological corridor. Qingshui Bay Park as the city waterfront park, shows us the city waterfront landscape aesthetic value of New York. The project site was once a tidal wetland and wild animal habitat in New York, has been used for long time as a garbage dump, which destructed the ecosystem. And the site remediation is a the first key step, by closing the landfill, and with the help of osmotic fluid control system and methane management system design measures, restored the site safety. The planning of it focus on searching for city waterfront park landscape aesthetic value, retention of regional history and culture diversity, so as to create a harmonious ecological environment of city. Qingshui Bay park design pays attention to public sharing, developing waterfront activities, protecting natural resources, and stimulate artistic and creative activities, presenting ecological circulation system, business facilities, and combined with the landscape design of wetland plants to the wetland ecological education, tourism and biodiversity protection functions, becoming public participation with high activities, contributing to New York city as wealth of material and spiritual resources.

Lifescape of the masterplan is defined by three functional layers: program, landscape, and circulation. Lifescape is an ecological process of environmental reclamation and renewal on a vast scale, recovering not only the health and biodiversity of ecosystems across the site, but also the spirit and imagination of people who will use the new park. Lifescape is about the dynamic cultivation of new ecologies at Fresh Kills over time—ecologies of soil, air and water; of vegetation and wildlife; of program and human activity; of financing, stewardship and adaptive management; of environmental technology (Fig.6-1-23).

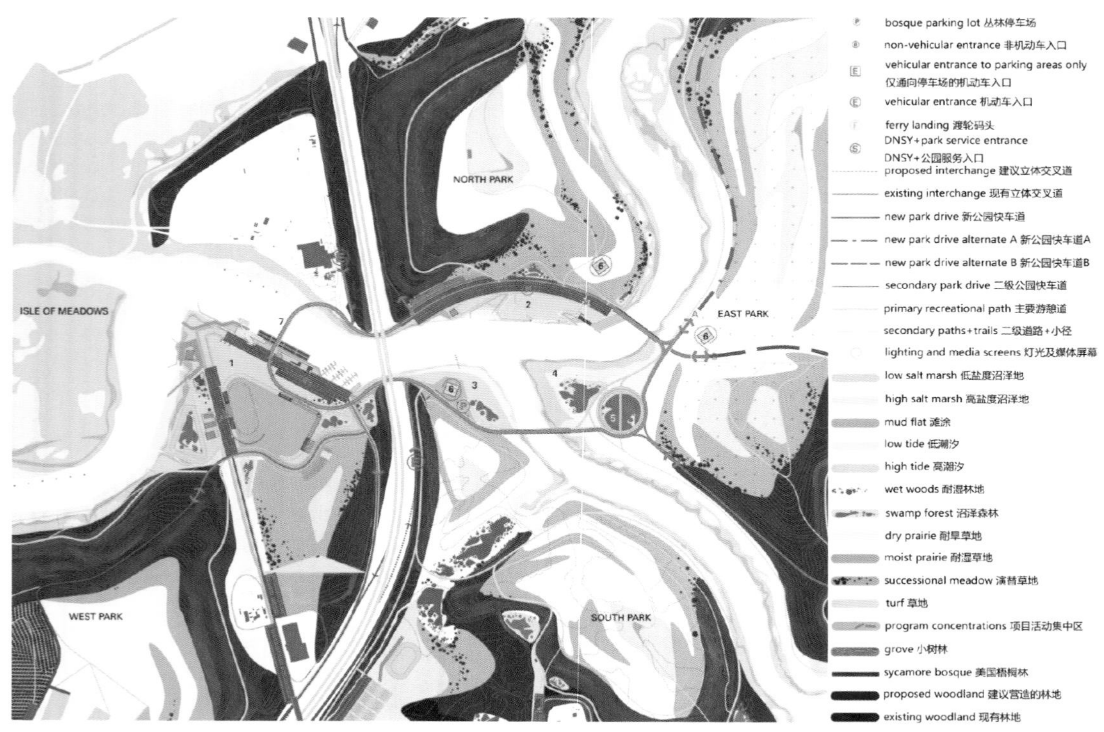

图6-1-18 汇流区总体规划
Fig.6-1-18 The master plan of the confluence

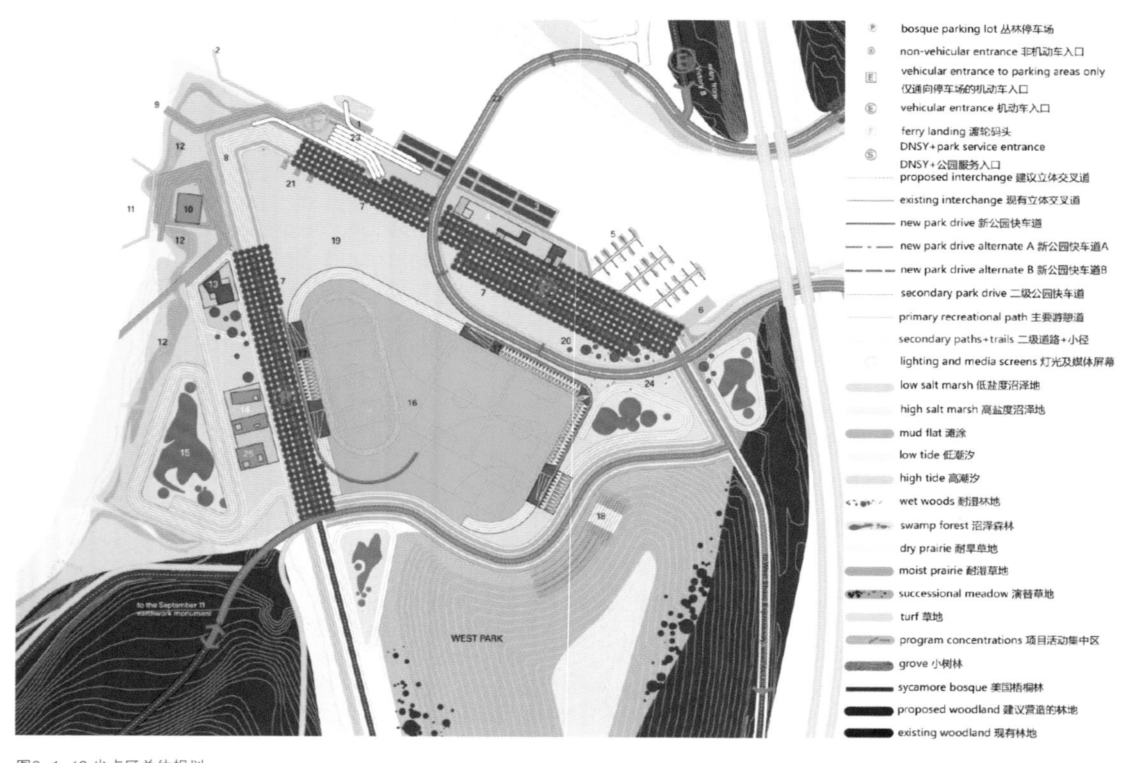

图6-1-19 尖点区总体规划
Fig.6-1-19 The master plan of the Point

第二节 安大略公园规划
Section II Lake Ontario Park Planning

一、 项目概况

安大略公园将成为多伦多市的一个地标性公园和典型的目的地。它将成为多伦多市及区域内一处代表性场所——一个典型的滨湖景观，一处充满活力的场所。这个公园将特色鲜明的社区与滨水区相连接，它也是一处避难所、受保护的生态区、城市荒野地。公园总体规划制定过程中的一个重要挑战是理解场地现有的各类生态系统以及场地利用——从东部少见的三叶杨林到汤米汤姆森公园中游艇俱乐部社区及风筝放飞场地。总体规划团队咨询了城市官员、社区居民、公园使用者及公众，期望规划一个优秀的公园：各种强度的活动在此发生，为社区居民及长期栖居于此的动物提供适宜的微气候条件（图6-2-1）。

I . Project Introduction

Lake Ontario Park will be a landmark park and defining destination for Toronto. It will be a place with great presence in the life of the city of Toronto and the region—an iconic lakefront landscape and a place full of life. The park will knit together the distinct communities and territories of the waterfront, but it will also allow for points of refuge, protected niches and urban wilds.

A key challenge of the Lake Ontario Park master planning process was to understand the diverse existing ecologies and uses of the site—from the rare stands of Eastern Cottonwoods to the community boating clubs and kiteboard launch sites protected by Tommy Thompson Park. the Master Plan team has consulted with city agencies, residents and park users, as well as the wider public, to plan a park with different intensities of activity, with 'microclimates' for both the human communities and the animal species that have long inhabited the site (Fig.6-2-1).

二、 安大略湖场地特征

安大略湖公园集聚了各类滨水场地：现存的公园、深受欢迎的海滩、未被充分利用的工业用地、私人租赁区及区域重要生态区。这个场地具有以下特征：

1. 大型荒野

整个场地的大部分区域是自然演替的景观——

II . The Feature of Lake Ontario Park Site

The Lake Ontario Park site is an assemblage of various waterfront places: existing parks, beloved beaches, under-utilized industrial sites, privately-leased areas and regionally important ecological territories. The site has some characteristics:

1. Big and Wild

Large portions of the existing site are defined by spontaneous, successional landscapes – dense, extensive and thriving communities of pioneering

图6-2-1 公园场地
Fig.6-2-1 The park site

稠密茂盛的先锋植物与野生动物群落。茂密的漆树林、成熟的三叶杨林地、广阔的湿地及草地、混合林地、不断变化的鹅卵石铺设的海岸线,结合场地曾经的工业背景形成了既具有美学价值又具有“荒野”功能的广阔景观。由于活动场地的偏僻及场地主要区域处于未管理状态，大量的水生生物物种和陆生生物物种栖居于这些荒芜的城市景观中。当人们意识到这块场地靠近城市中心区时，这块大尺度的自然区给人以深刻的印象：它们是多伦多市（北美中心大城市之一）的大型自然地块（图6-2-2）。

2. 环境问题突出

安大略湖公园景色极为优美，而这些荒芜的基本场地曾经是人造场地，曾有一段时期用于工业和港口。过去的湖滨填埋及工业利用遗留了大量的环境问题，这些问题将在公园的发展中得到解决。过去的研究表明场地的土壤及地下水已经受到污染。

3. 公园套公园

新规划的公园将吸纳湖岸线的各个部分，并将其整合成一个巨大的、内部相互联系的整体——一个由连续公园地及超过37公里长的水岸边界组成的925英亩（约374公顷）的新公园。规划这个新公园的最大挑战之一就是显示出那些水边界上的深受人们喜爱的场所：外部港口航海联盟、风帆冲浪俱乐部、风筝冲浪海滩、狗儿戏耍区、基地区、汤米汤姆森公园、活拜海滩及东海滩公园这些空间是多

plants and wildlife. Sumac thickets and mature Cottonwood stands, broad wetlands and meadows, mixed woodlands and dynamic cobble shorelines combine with the industrial context to create an expansive landscape that is both aesthetically and functionally 'wild'. Because of the physically remote setting and unmanaged character of major areas of the site – these wild, urban landscapes have indeed become populated with a relatively broad range of terrestrial and aquatic wildlife species. When understood the site's proximity to downtown, the scale of the existing natural areas is impressive: these are 'big nature' sites in one of North America's great urban centers (Fig.6-2-2).

2. Variable Environmental Conditions

Lake Ontario Park is an Unusually Beautiful Landscape. Paradoxically, this wild, elemental site exists upon mostly man-made land, with a history of past industrial and port uses. Historic lake-filling and industrial uses have left a legacy of variable environmental conditions that will be addressed. Not surprisingly, past studies indicate that the quality of soils and groundwater has been affected.

3. Parks with the Parkland

The new park takes previously disparate parts of the lake shore edge and assembles them into one large inter-connected whole – a new park comprised of 925 acres of contiguous parkland and more than 37 kilometers of waters edge. One of the great challenges in planning a new park at this site is the presence of the many loved 'places' that exist along the water's edge: the Outer Harbour Sailing Federation, the windsurfing club, the kiteboarding beach, the off-leash dog areas, the Base Lands, Tommy Thompson Park, Woodbine Beach and the parks of the Eastern Beaches. These spaces are indeed some of Toronto's most cherished sites, and all design efforts need to balance the critical need to unify the vast park while preserving the elements and character that make the site truly remarkable. The Master Plan attempts to

图6-2-2 不断变化场地历史
Fig.6-2-2 A dynamic site history

伦多市极为珍贵的场地，所有场地的设计应该寻求一个平衡：在保存彰显场地内涵的要素与特征的同时，保证这个巨大公园的一体化。总体规划认为应尽可能地记录、提升、利用现有场地的美学价值及特有的游憩功能（图6–2–3）。

三、 规划框架

规划框架提供了一个灵活的、具有弹性的组织框架，包括以下几个思路：

1. 一个连续的绿色滨水线

安大略湖公园具有前所未有的潜力，成为一个连续的绿色湖滨地带，并联结城市重要的生态资产：从堂河下游及其绿道到汤米汤姆森公园。多伦多市将为创新性公园的规划及设计提供一套新的标准，以营造影响深远、活跃的、给予精神享受的自然湖滨公园（图6–2–4）。

2. 37公里的湖岸线

安大略公园大量广阔的、多样的、自然状态的环境条件提供了多种的布局可能性——不同的场地、建造条件、场地特征、栖息地及利用方式。公园大型的人工陆地所形成的之字形的湖岸线提供了滨水体验，栖息地及活动的多样化的布置组合。这个湖岸线与多种自然过程相互作用：潮汐漂流物、沉积物、侵蚀、湖水的凝固和解冻、波浪运动、冰块冲刷以及先锋物种及野生群落的出现等。这些作用使得未被利用的场地在提供新的公众利用的同时，为场地创造了一系列动态环境条件和潜力，实现海滩及沙丘区域植物的多样化。总体规划基于这

图6–2–3 汤米汤姆森公园鸟瞰图
Fig.6-2-3 Aerial view of Tommy Thompson Park

record, enhance and exploit the unique beauty and distinctive recreational settings of the existing site (Fig.6-2-3).

III . Planning Framework

Planning framework promotes a flexible and resilient organizational framework which includes several ideas as following:

1. A Connective, Green Waterfront

The Lake Ontario Park project has the unprecedented potential to assemble a continuous green lakefront that links the City's great ecological assets: the lower Don River and Greenway to Tommy Thompson Park. Toronto will set a new standard for innovative park planning and design with its vibrant, active and refreshingly wild lakefront park (Fig.6-2-4).

2. 37 Kilometers of Shoreline

Vast, varied and rough, the extensive edge conditions of Lake Ontario Park support an incredibly diverse array of places, found conditions, features, habitats and uses. The long, crenelated shoreline of the largely man-made landmass of the park creates opportunities for a uniquely diverse array of waterfront experiences, habitats and activities. The shoreline interacts with natural processes of littoral drift, deposition, erosion, freeze and thaw of the lake, wave action, ice scour, and the emergence of pioneering plant and wildlife communities. These interactions produce a dynamic set of conditions and potentials, supporting beach and dune plant diversity, while prompting new public uses for otherwise under-utilized sites. The master plan builds on this dynamism to create a park that is alive, changing and multivariate (Fig.6-2-5).

3. 3 Transects

Three primary path armatures create a single

图6–2–4 连续的绿色网络
Fig.6-2-04 Continuous green network

个动态变化营造了一个充满活力、多元变化的公园（图6–2–5）。

3. 3条横断面

三条主要路径系统创造了一个连贯的独立公园框架，游客可以沿着每条路线体验多种特色鲜明的景观。这三条主路线是循环道路网络的主要要素，这个相互联系的网络将多样的公园整合为一个整体，同时将公园与周边的社区及城市相联系。作为连续的、顺畅的干道，主路线串联不同的景观进而创造了一个统一的整体，同时也保留了大量不同的特色景观及活动项目（图6–2–6）。

4. 40个“空间”

925公顷的场地具有丰富的特色景观，这些景观被分为不同的分区，这些分区为公园中的项目提供了不同的体验及机会。“横断面”、“场所”及“基地”将不同的场地整合为一个统一的整体，每个场地拥有不同的景观、体验及利用方式（图

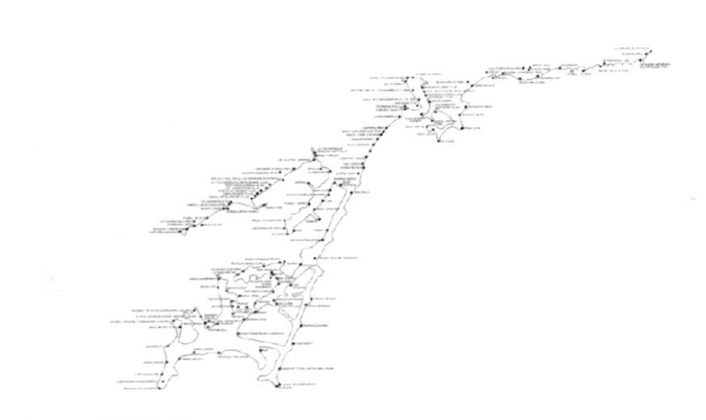

图6–2–5 岸线特征
Fig.6-2-5 Shoreline feature

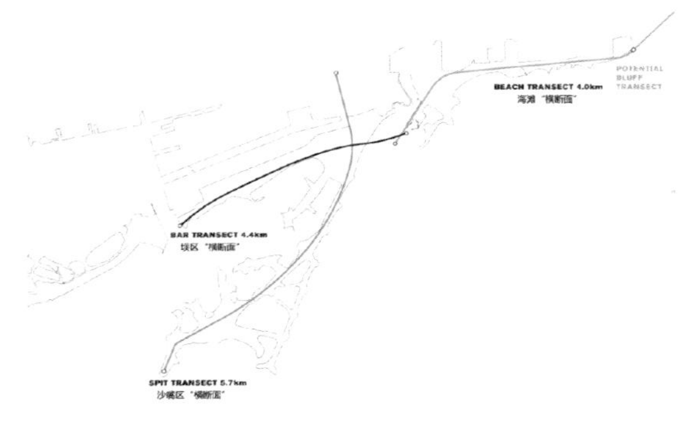

图6–2–6 三条横断面
Fig.6-2-6 Tree connecting transects

coherent park framework and introduce visitors to a variety of distinctive landscapes along each trajectory. The transects represent the principle elements in a 'connective network' of circulation paths that unify the diverse park and link to the adjacent neighborhoods and city. As continuous, unobstructed pathways,the transects pull together disparate landscapes to create a unified ensemble, while retaining a great deal of variety in the character and programming (Fig.6-2-6).

4. 40 Rooms

Varied zones of distinct character within the 925-acre site offer dramatically different landscape experiences and opportunities for park programming. The transects, rooms and outposts pull together disparate sites to create a unified ensemble of many different landscapes, experiences and uses (Fig.6-2-7).

5. 400 Outposts

The Master Plan amplifies the site's numerous points of prospect and refuge as defining elements of the Park's design. The 'outposts' are both new and existing elements of the Lake Ontario Park site, some built and some wild: overlooks, frog ponds, picnic promontories, earthwork outlooks, unique plant colonies, water landings, seating and program amenities. These places are intended to lend a more intimate scale of experiences to the otherwise large-scale park (Fig.6-2-8).

6. The Historic Shoreline

The Master Plan looks to reinvent and amplify the historic dynamic character of the original shoreline through a vibrant mix of active and passive landscapes, the cultivation of vast areas of 'urban wilderness', and an understanding of the park's evolving form over time. Specifically, the bar and the bay form two historical landscapes that the Master Plan seeks to recover, or at least reinterpret.

IV. Planning Overview

1. The Master Plan

The Lake Ontario Park Master Plan capitalizes upon and amplifies the existing "wild" characteristics of this unique site – the rough and remote landscapes, the open horizons and vistas, the active Outer Harbour, the exposure to the elements and to the vast scale of the lake. It creates a new active parkland, a place with extensive opportunities for physical exertion, play and passive use, while at the same time diversifying the range, extent and health of its various ecosystems (Fig.6-2-9).

6-2-7）。

5. 400个“基地”

总体规划强调了众多具有发展潜力或避难性场地作为公园设计中的明确要素，这些“基地”既包括安大略公园场地新的和现有的要素，也包括人工建造的或者野生的要素：俯瞰点、蛙池、野餐岬、大地景观、特殊植物栖息地、水上标志物、座位及活动项目等，这些场所的主要目的是区别于其他大尺度公园，为本公园游客提供更为亲密的尺度感（图6-2-8）。

6. 历史海岸线

通过积极和消极景观的结合；大面积城市荒野地培育；对公园演变历史的理解等措施，总体规划再次彰显和强化了原有海岸线的历史变化特征。更为具体地讲，公园现存的坝和海湾是其两大历史景观，而这两者也是总体规划试图再次发现，进而重新诠释的对象。

四、 规划概况

1. 总体规划

安大略公园总体规划紧紧抓住了场地现有的独特“野生”特征——未开发的野生景观；开阔的水平线；狭长的景观；充满活力的外海港口；显露的要素；显著的巨大尺度的湖面。规划设计了一个新的活跃的公园，一个提供身体锻炼、游戏及消极利用等多种机会的场所；同时它也保证了多样性生态

Lake Ontario Park is intended to become a regional park that will be a signature destination and amenity for the Greater Toronto Area. It will also serve the adjacent neighborhoods, which are expected to change dramatically over the next 30 years. A new residential community of over 40,000 people is planned for the Port Lands. The retail, commercial and office development will bring many more people to the area and the park over time. At the same time, major planning projects for the Central Waterfront, West Don lands and Lower Don lands are paving the way towards achieving a connective network of open space and park amenities.

In the context of the changing waterfront and city, it must be understood that the greatest efficacy of the Master Plan is in providing a series of strategic recommendations to guide organization of land-use principles, key program sites and features, circulation networks and new and expanded natural areas. Although some of the park features and places currently exist, the Master Plan largely reflects a long-term vision that will require continued consultation and design prior to implementation.

A project of this scale and complexity requires determined leadership at the civic level in addition to a clear set of guiding principles that will ensure that the park grows into a coherent, signature park for the city. To this end, the Master Plan is intended to be both a bold, big picture vision for the future park, as well as a grounded, practical framework of realistic opportunities. To a large extent, the Master Plan proposals are based on the challenges and opportunities identified throughout the research and consultation phases of the project. Information on the project site, however, is still being compiled by a range of consultants at varying capacities. The information from these findings - soils investigations, tree inventories, bathymetry soundings, Environmental Assessments - will be used in the process of translating the following principles into the design of physical park.

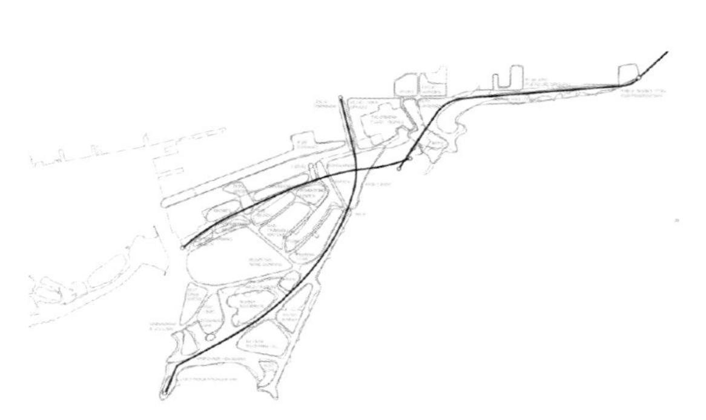

图6-2-7 公园40个特色鲜明的空间
Fig.6-2-7 40 distinct rooms of the park

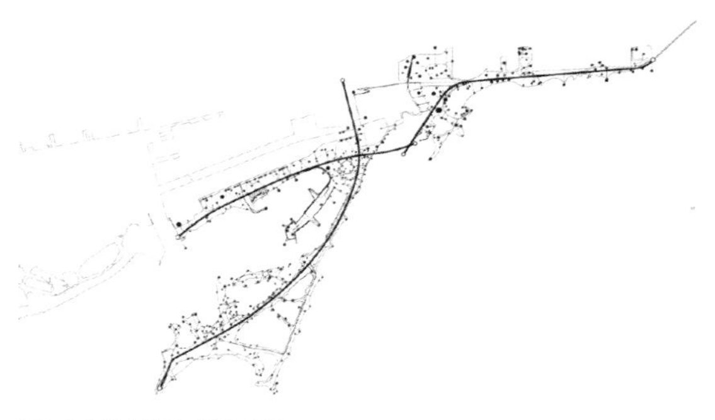

图6-2-8 安大略公园总体规划
Fig.6-2-8 Lake Ontario Park master plan

系统的范围、广度及健康性（图6–2–9）。

安大略公园将会成为一个区域级别的公园，将成为大多伦多区重要的目的地和基础设施。同时它还会服务毗邻的社区，这些社区将在未来30年中发生巨大的变化。一个容纳40 000人的新社区将在港口地建成。零售业、商业和办公等行业的发展将会逐渐吸引更多人到此区域及公园。同时，中央滨水区、丹河西部段和堂河下游段等重要规划项目极大地促进多伦多市建设一个相互联系的、由开放空间和公园组成的网络。

在不断变化的滨水区和城市背景下，总体规划的最大效力就是提供一系列战略性建议，以指导土地利用原则、重点项目地址及特征、交通网络、新的及拓展的自然区。尽管公园现今已保留一些特色景观及场所，总体规划更多地反映了公园长期发展的愿景，这个愿景实现的要点并不是实施过程而是更多地依赖于连续不断的咨询及设计。

如此巨大尺度及复杂的公园不仅需要一套能够保证其成为一个连贯的重要的城市公园的指导原则，也需要坚定的领导机制。总而言之，总体规划不仅是公园未来大胆的宏伟的预想，也是实际可行的基本的实践框架。很大程度上，总体规划的提议均是基于项目调研和咨询阶段所界定的挑战与机遇。项目场地信息由顾问在不同承载力条件下编纂而成。调研所搜集的信息包括：土壤调研、场地现有树木信息、水深测深等信息，环境评估应用于将指导原则转译成自然公园设计的过程中。

安大略湖公园总体规划包括六个主要目标：

（1）提出一个大胆、灵活、弹性的景观框架。总体规划是战略建议和自然景观提升的框架。尽管总体框架显示出一个大胆、连贯、明智的公园组织结构，但是景点的具体结构形态、尺度、位置、数量及特征将会依据未来规划及设计复审、环境评估流程及多伦多社区变化的需要而进行适当的调整。

（2）创造一个与众不同的滨湖区公园，包含有特色的活动及独一无二的以水为导向的活动项目。安大略公园为一些活动及项目提供了一个特别的大尺度结构形态，它在多伦多市的独一无二性体现在：可以容纳各种积极活动及消极游憩活动、多

There are the six primary goals of the Lake Ontario Park Master Plan:

(1) Promote a bold, flexible and resilient landscape framework. The Master Plan is a framework of strategic recommendations and physical landscape improvements. While the Master Plan demonstrates a bold, coherent and sensible organization of the park, the precise configuration, scale, location, quantity and character of its elements are adjustable according to future planning and design review, environmental assessment processes and the changing needs of the Toronto community.

(2) Create a vibrant lakefront of distinctive activities and unique water-oriented programming. Lake Ontario Park offers extraordinary large-scale settings for a range of activities and programs that are unique in the city, allowing for extensive active and passive recreation, diverse watersports, educational amenities and cultural enrichment (Fig.6-2-10).

(3) Build a connective system to unify the large park. Create a network of pedestrian and multi-use trails that link together and unify the diverse parcels of the existing site with the surrounding city.

(4) Cultivate a rich mosaic of terrestrial and aquatic life that becomes the center of a continuous regional greenway. Recognizing the presence of extraordinary ecological resources, the Master Plan's ecological goal is to preserve, restore and diversify the range, extent and health of its various ecosystems.

(5) Create a realistic and sustainable landscape. By promoting flexible environmental site management responses, natural succession and a sensitive implementation timeline, the Master Plan offers a realistic and sustainable park framework.

(6) Preserve and amplify the remarkably unique settings. Responding to its historical formation, its large-scale and its elemental character, the Master Plan seeks to capitalize on the existing "wild" characteristics of this unique site.

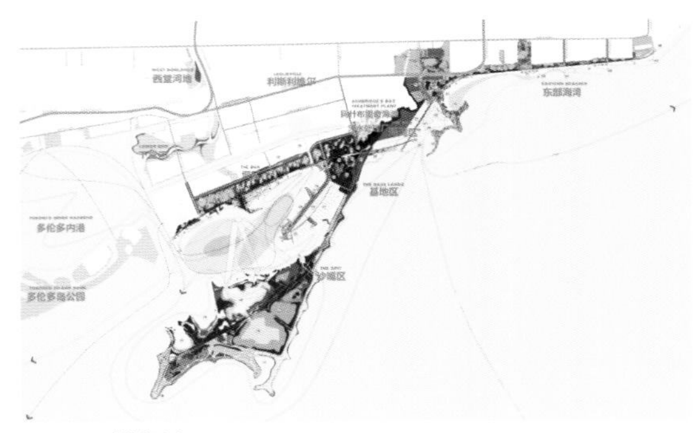

图6–2–9总体规划图
Fig.6-2-9 The master plan

样的水上运动、教育场所及丰富的文化大餐（图6-2-10）；

（3）构建一个相互联系的体系，实现公园的整体性。总体规划构建了一个步行道及多用途道路网络，连接及整合场地各个分区及周边的城市用地；

（4）将公园培育成富含陆生生物及水生生物的镶嵌体，成为区域连续绿道的中心。基于对场地现有的极为丰富的生态资源，总体规划的生态目标就是保存、恢复、丰富各种类型生态系统的范围及健康；

（5）营造切实可行的、可持续的公园景观。通过提升场地环境管理响应的灵活性、促进自然更替及合理的实施期限，总体规划提供了一个切实可行的、可持续的公园框架；

（6）保存及强化场地的特色方面。针对场地历史形态、巨大尺度及景观元素特色，总体规划紧紧扣住了这块场地现有的“荒野”特性。

2. 公园土地利用

总体规划的这一部分界定了公园的土地利用类型及分区，为整个公园景观提供了可以评价的土地利用分类，包括积极利用、消极利用、栖息地及自然区、划船区等；每种土地利用类型仅仅是为了描述的目的，而并不在于土地利用是如何在安大略公园确定的。我们也要注意到这些描述并不是排他性的，这是由于各种土地利用类型的边界常常是模糊不清的（图6-2-11）。

积极利用：积极土地利用类型主要包括能量消耗较高的区域，如活跃的运动场地，也包括正式或非正式利用而指定的或者多用途区域、各种多用途道路等；消极利用：消极的土地利用类型包括一般的正式和非正式公园利用区域，也包括开放草坪、野餐去、运动场、探险场、步行道、垂钓码头、木板人行道、海滩等。消极利用类型不同于栖息地及自然区利用类型，消极利用区域的主要目的是公园的一般利用及游憩活动。尽管也是由软质的绿色景观及海滩组成，消极利用区仍然允许大量的公园日常利用及游客活动，其景观维护理念也不同于栖息地及自然区；栖息地及自然区：这类土地

2. Park Land-use

This section of the Master Plan identifies land-use categories and areas of the park site, and provides an estimated land-use breakdown for the full park landscape, as per the categories of active; passive; habitat & natural areas; and boating. The land area estimates and park uses for each category are provided for descriptive purposes only, and are not intended to be prescriptive of how land-use is to be defined in Lake Ontario Park. It is important to note that these descriptions are not exclusive, as the boundaries between the various land-use categories are often blurred (Fig.6-2-11).

ACTIVE: The 'ACTIVE' land-use category refers to those park areas designated for high energy; active sports play, including customized and multi-use fields for formal and informal use, and all multi-use trails.

PASSIVE: The 'PASSIVE' land-use category refers to those park areas designated for general park use, formal and informal, including open lawn areas, picnic areas, playgrounds, youth pods, pedestrian trails, fishing piers, boardwalks, and beaches. The 'PASSIVE' land-use category differs from the 'HABITAT & NATURAL AREAS' category in that these areas are intended

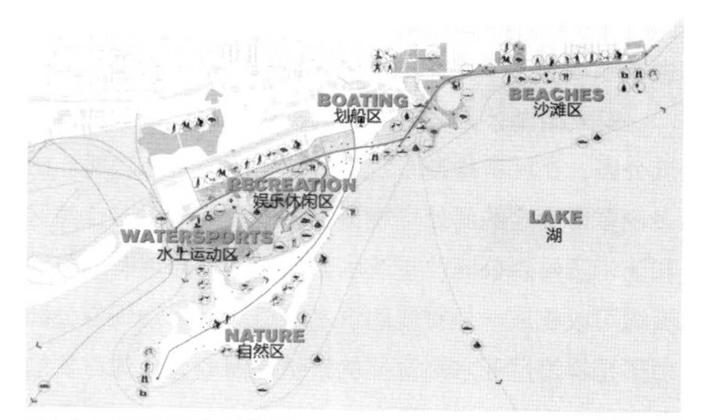

图6-2-10 公园活动总体规划图
Fig.6-2-10 Park activities master plan

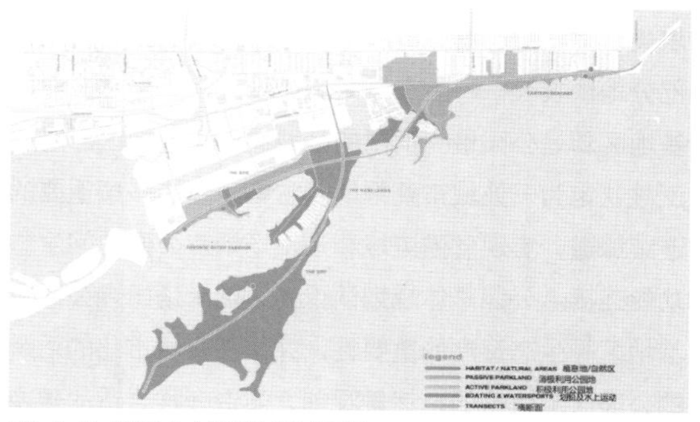
图6-2-11 公园公共土地利用总体规划图
Fig.6-2-11 Park land-use master plan

利用类型主要涉及的区域是那些人工管理下景观日益自然化的区域，包括恢复区，需要人工管理的野生林地区、草地及湿地（主要管理雨水及栖息地）。尽管这些公园区在管理、恢复或者“筑巢”期严格限制公众的进入，但是通过步行道、慢跑道、远足道、教育基地及山丘俯瞰点来保证其一定的可达性；划船：这类土地利用方式涉及主要用于划船或者与划船相关的活动的区域，也包括公共和私人游艇俱乐部和组织机构、水上运动场地以及游船始发站。

五、 公园生态

1. 生态资源

由于位于唐河流域出口附近，如果能够增强安大略公园与顺沿安大略湖岸线的休莫河流域和罗杰河流域的连接程度，公园将具有重要的区域生态功能，这三个流域是大多伦多区的重要组成部分，并且向北延伸至橡树岭冰碛区。滨水区是水生生境与陆生生境之间重要的交界面，因而对于安大略湖的生态系统健康及物种多样性具有重要的影响。

安大略湖公园具有很好的机遇成为一处重要的湖滨公园，在城市绿道系统中发挥重要的生态功能。由多伦多市确定的两处环境重要区——樱桃海滩自然区和汤米汤姆森公园位于滨水区附近。公园现有的景观资源可以建成一处重要的滨水绿道。汤米汤姆森公园由于是科洛尼亚水鸟重要的繁殖地而成为全球重要的鸟类区，也是北大西洋鸟类迁徙路线上的一处重要停留站，还是蝶类重要的迁移停留站；在汤米公园发现了重要的稀缺植物物种。38公顷大小的基地区是一块具有重要生态潜力的地块。尽管安大略湖公园的大部分地块狭长——生态学术语中称之为“边界”——基地区却是公园最为宽阔的区域之一。基地区已经广泛被认定为一处城市荒野地，它构建起了一条重要的连续绿道，实现堂河流域和汤米汤姆森公园之间生态功能连接。公园总体规划认识到了汤米汤姆森公园、基地区等生态资产的重要性，并试图在城市化的活跃的公园区与敏感的自然景观创之间造一种合适的转换（图6-2-12）。

to be the primary areas for general use and leisure activities. Although consisting principally of soft, green landscape and beaches, the 'PASSIVE' areas will receive the bulk of daily park use and visitor activity, and will be maintained in a manner that that differs from the 'HABITAT & NATURAL AREAS'.

HABITAT & NATURAL AREAS: The 'HABITAT & NATURAL AREAS' land-use category refers to those park areas that are intended to evolve into managed and increasingly 'naturalized' landscapes, including regeneration areas, managed native woodlands, managed meadows, and managed wetlands (for stormwater and habitat). Although these park areas may have restricted or limited public access during management, regeneration, or 'nesting' periods, the intention is to maintain their accessibility through walking, jogging, and hiking trails, educational outposts, and hillform overlooks.

BOATING: The 'BOATING' land-use category refers to those park areas that are principally used for boating and boating-related activities, including public and private boating clubs and organizations, water-sport sites (z), and boat launches.

V. Park Ecology

1. Ecological Resources

Situated near the mouth of the Don River watershed, Lake Ontario Park will fulfill an important regional ecological function by enhancing connectivity to the watersheds of the Humber and the Rouge Rivers along the Lake Ontario shoreline. These three watersheds occupy most of the Greater Toronto Area and extend north to the Oak Ridges Moraine. The waterfront provides a crucial interface between aquatic and terrestrial habitats, and is thus critical to ecosystem health and the diversity of species in Lake Ontario.

Lake Ontario Park offers a unique opportunity to create a signature lakefront park that provides an important ecological function, within a continuous urban greenway. Two Environmentally Significant Areas (ESAs) designated by the City of Toronto are situated along the waterfront, encompassing the natural areas of Cherry Beach and Tommy Thompson Park. The Lake Ontario Park site already includes landscapes that are important resources that contribute to building a significant waterfront greenway. Tommy Thompson Park is recognized as a globally Important Bird Area (IBA) because of its concentration of breeding colonial waterbirds. It is also a key stopover site on the North Atlantic Flyway and an important migratory stopover for butterflies. Provincially rare and significant flora species have also been identified at Tommy Thompson Park. The 38 hectares of Base Lands represents a site

2. 连接性

安大略湖公园构建了一条可持续的绿色湖滨，实现了堂河绿道与汤米汤姆森公园之间的连接。多伦多市正在努力营造一个具有极高生态价值的连续的绿道网络，而安大略公园包含了这座城市几处重要的生态资产：汤米汤姆森公园和基地区、坝区、堂河绿道等。

总体规划的主要目标是实现安大略湖公园文化、生态及经济三方面的协同发展。

公园的一个重要挑战是在维持和提升公园水生及陆生栖息地生态质量的同时，改善公园公众游憩景观。由于基地区能够满足公园使用者观察体验自然的需求，因而这个区域提供了一个良好的机遇实现上述的平衡，进而也减少了公众涌向汤米汤姆森公园而造成的生态压力。尽管具体的土地利用规则将会依据公园未来分区的布置而制定，但总体规划为了实现堂河与汤米汤姆森公园之间连接而提出了一些建议：

1）活动强度的递减

在堂河与汤米汤姆森公园之间连接通道内，总体规划期望公众活动强度不断递减。总体上来说，坝区西部区域高速、正式、积极地利用逐步转变为东部的消极的公园利用，更为东部的基地区和汤米汤姆森公园提供小径、教育与解说，这些活动均要符合城市GR分区及多伦多区域保护局的总体规划要求。最终，活动利用强度的递减与这些区域荒野特征及功能的递增相协调。

2）减少破碎化

由于场地存在大量的碎石及湿地仅限于坝区和

图6-2-12 自然资源
Fig.6-2-12 Natural resource

with significant ecological potential. While most of Lake Ontario Park is long and narrow - all "edge" in ecological terms - the Base Lands is one of the widest zones. Already recognized as an 'urban wilderness', the Base Lands is a critical link in building a continuous greenway between the Don River watershed and Tommy Thompson Park. The Master Plan recognizes the importance of these ecological assets - Tommy Thompson Park and the Base Lands - and attempts to create an appropriate transition between the urban and 'active' park areas, and sensitive, natural landscapes (Fig.6-2-12).

2. Connectivity

The Lake Ontario Park project has the potential to assemble a substantial green lakefront that links the Don River Greenway to Tommy Thompson Park. In considering Toronto's opportunities for creating a continuous, green network with high ecological value, the Lake Ontario Park site includes some of the City's great ecological assets: TOMMY THOMPSON PARK & THE BASE LANDS, THE BAR, THE DON GREENWAY.

The primary goals of the Master Plan incorporate cultural, ecological and financial directives for the future of Lake Ontario Park. A significant challenge for the Lake Ontario Park project is to simultaneously promote the enhancement of public recreational landscapes while maintaining and improving the ecological quality of aquatic and terrestrial habitats. The location of the Base Lands provides an ideal opportunity to achieve the required balance as sufficient nature viewing opportunities can be created to satisfy many Park users, thereby reducing the pressure for more public access to Tommy Thompson Park. Although specific land-use regulations will be set by the park's future zoning designations, the Master Plan has the following recommendations for the function of the 'Don to Tommy Thompson park' connection:

1) Gradient of Activity

The Master Plan envisions a gradient of public, human activity across the 'Don to Tommy Thompson park' connection. In a general sense, the high speed, formal and active uses proposed for the western portion of the Bar will transition into passive park uses in the eastern portion. Moving further east, the Base Lands and Tommy Thompson Park will offer trails, education and interpretive opportunities consistent with the City's GR zoning and the TRCA (The Toronto and Region Conservation Authority) Master Plan, respectively. As a result, the decrease in active uses will correspond with an increase in 'wilderness' character and function.

2) Reducing Fragmentation

基地区，因而公园原有的水文遭到严重的干扰。总体规划最大程度利用场地环境管理及公园战略措施的实施来改善周边水文现状，以实现堂河与汤米汤姆森公园之间的连接通道能够成为具有较高价值的野生生物通道网络。

Current hydrology in the park site is interrupted by piles of rubble and conditions for wetland establishment are limited in the Bar and Base Lands. To the greatest extent possible, the Master Plan will utilize environmental site management and strategic park implementation practices to promote contiguous hydrologic conditions to ensure that the 'Don to Tommy Thompson Park' connection performs as a high-value wildlife network.

3. 涵养的栖息地及自然区

安大略湖公园场地不仅巨大而且类型丰富，包含多种类型的陆生和水生栖息地。尽管原有的场地主要由工业景观、人造景观等组成，但数量众多的新形成的栖息地和生态系统也存在于场地内，主要包括汤米汤姆森公园数百公顷的繁茂的林地、草地、湿地及池塘；西岸线的成熟三叶杨地；坝区及基地区的初期野生地；近堂河河谷区的峡谷区域，堂河谷包含了丰富的迁徙物种。针对这些现有的自然资产，总体规划建议将这些区域培育成为种类丰富且可迅速恢复的景观，提升其作为区域重要生态绿道的公园功能（图6–2–13）。

针对栖息地及自然区，总体规划确定了五个主要目标：栖息地及自然区域内的景观不仅种类丰富而且可迅速恢复，成为一处重要的自然资产，公园在区域生态连接性、水质量改善、生物多样性及可持续发展方面承担相应的功能；不仅考虑到植物，也应顾及到鸟类、哺乳类、鱼类及昆虫等生物，构建野生生物廊道连接现有的自然资源，最终实现公园成为区域及湖区重要的栖息地的目的；设计一个持久的景观框架，这个景观框架要具有充足的灵活性，能够正确调整现有的自然资源，促进栖息地的创造；构建生态改善计划，便于公众能够清楚地了解公园发展的每个阶段都是“不断变化的景观”，促进生态系统结构和功能的改善；确保公园早期的成功和长期可持续发展能够精明实施和经济地规划，以保证公众能够在各个季节探索和享受公园。

在实现总体规划中的生态目标的同时，三个因素对所建议栖息地景观的组织起着重要的推动力：现有自然资源的位置及关键栖息地营造的机遇；由东至西的生态连接，包括堂河下游段与汤米汤姆森公园之间连续的绿色通道；公园所期望的景观布局。

3. Cultivating Habitat & Natural Areas

The Lake Ontario Park site is vast and varied, with a wide range of terrestrial and aquatic habitats. Whereas the site today comprises a complex amalgam of industrial, man-made landscapes, there are a multitude of newly emergent habitats and ecosystems. These include the hundreds of acres of thriving woodlands, meadows, wetlands and ponds at Tommy Thompson Park, the magnificent stands of mature cottonwoods in the western shore, the diverse patches of nascent wildlands in the Bar and Base Lands, and the proximity to the ravines of the Don River valley from which a rich mix of species migrate. The Master Plan proposes to build on these natural assets to cultivate a diverse and resilient landscape, one that should improve the performance of the park as a regionally important ecological greenway (Fig.6-2-13).

The Master Plan has five main objectives for habitats and natural areas in Lake Ontario Park: Cultivate a diverse, resilient landscape that is a natural asset to the region in terms of ecological connectivity, water quality improvement, biodiversity and sustainability; Create meaningful habitat for the region and the lake by building wildlife corridors linked to existing natural resources, taking into account not only plant life but also bird, mammal, fish, and insect communities; Design a durable landscape framework that is flexible enough to accommodate existing natural resources and local opportunities for enhanced habitat creation; Design and stage ecological improvements so that

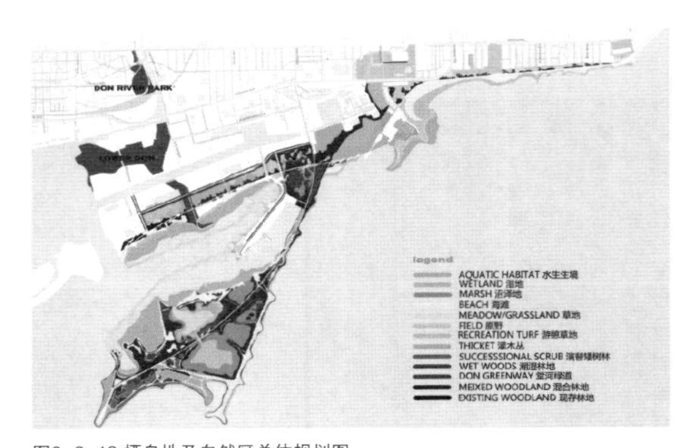

图6–2–13 栖息地及自然区总体规划图
Fig.6-2-13 Habitat & natural areas master plan

4. 增加植被覆盖

除了加强堂绿道与汤米汤姆森公园之间的生态绿道建设，安大略湖公园将有助于实现多伦多市所确定的植被覆盖率由17%增加到35~40%的目标。

为了支持多伦多市确定的绿化覆盖目标，总体规划在公园植被覆盖方面确定了四个主要目标：尽可能地保留现有的植被物种；保留重要文化点，如樱桃海滩的三叶杨；在东海滩栽植新树种，保留此处的混合树林样本；种植适应不断变化湖岸线的本土物种；在详细规划阶段需要咨询森林管理人员，以确保公园中绿化植物物种满足城市关于绿化植物种类、多样性、恢复力等方面的要求（图6-2-14）。

六、 主要区域规划

1. 坝区

1) 场地介绍

由于樱桃海滩和北河岸的140英亩（约56.7公顷）开放地域，形成充分利用的景观及特色鲜明的水上活动，坝区被设想为逐步转变的修复性景观，不断扩大的游憩区，与公园西部底端相连的绿道连接点。

坝区由东部裂缝一直延伸至基地区，成为整个公园区转变最为显著的区域，大片荒废的工业土地曾经占据坝区，现今这个区域的景观焕然一新，尺度巨大并且处于不断修复的过程中。坝区被设想为积极利用、消极利用和自然区相混合的景观区——这些利用方式由海滩、河岸线、田地及林地而界定。坝区为游客提供了一系列的道路及小径、数量众多的游憩地以及大幅度改善的湖滨区及设施。借助于风、海浪和微气候等条件，坝区的海岸线集聚了大量的划船、风帆冲浪活动、滑翔冲浪俱乐部。这些使用者及以水为导向的生活方式很好地界定了安大略湖的特征。

除了两处场地用于积雪堆放和木材刨切，大部分的坝区工业场地将处于一个粗放的、任其自然发展的状态。这些历史工业场地在不断变化的环境条件下将存在遗留物，这些遗留物将在公园建造期间

the parkland can be understood and enjoyed in each phase of its development as a legible "landscape in process," designed to promote ecosystem structure and function; Ensure both early successes and long-term sustainability with smart phasing and economic planning to create a park that will be explored and enjoyed in all seasons.

In keeping with the ecological goals of the Master Plan, three primary factors drive the organization of the proposed habitat landscapes: location of existing natural features and opportunities for significant habitat creation; east-west ecological connections across the waterfront, including a strong, continuous green connection between the lower Don River and the Tommy Thompson Park; desired landscape setting for the park.

4. Increasing Canopy Cover

In addition to promoting a green connection between the Don Greenway and the Tommy Thompson Park, Lake Ontario Park will contribute to the City of Toronto's stated goal of increasing the overall urban tree canopy from 17% to 35~40%.

To support Toronto's canopy goals, the Master Plan offers four primary objectives for canopy cover at Lake Ontario Park: 1 Preserve existing canopy species to the greatest extent possible; 2 Ensure the preservation of significant cultural stands, i.e. the Cottonwoods at Cherry Beach and the mixed specimens along the Eastern Beaches, through strategic infill of young tree species; 3 Promote the use of species native and well-suited to the dynamic shorelines of Lake Ontario; and 4 Consult with Forestry staff in detail planning stages to ensure that canopy species proposed at Lake Ontario Park meet the City's criteria for species type, diversity and resilience (Fig.6-2-14).

VI. Major Areas Plan

1. The Bar

1) Area Description

With 140 acres of open, underutilized landscape and the distinctive wateroriented activities of Cherry Beach and the North Shore, the Bar is imagined as a transformed remediated landscape, an expansive recreation area and a greenway connector that ties together the western end of park.

Stretching from the Eastern Gap to the Base Lands, the Bar will be one of the most transformed in all of Lake Ontario Park, taking mostly derelict postindustrial lands and creating a new, large-scale remediated landscape. The Bar landscape is envisioned as a diverse mixture of active, passive, and natural areas - defined by beaches and shoreline, and fields and woodlands. This landscape will offer a series of paths and trails,

得到处理。总体规 划对坝区的构想体现了针对一系列环境问题所采取的灵活的、有效的策略措施。

2) 设计目标

作为海港生态和游憩背景下的主要焦点，总体规划关于坝区未来生活及景观提出了6个主要目标：基于渔人岛沙坝的历史活力营造一个充满活力的、不断变化的景观：①一个持久的、可迅速恢复的框架，这个框架对新的野生物种群落及游憩地点具有开放性；②保存及恢复成熟的三叶杨林地及重要的河滩植物群落，进而营造樱桃海滩所特有的乡村风格的林地环境；③在北河岸建造特色鲜明的别墅环境，服务于水上运动及游憩；④为游憩、游戏、教育与学习等目的而布置的设施应具有一定的灵活性，便于正在进行的及将来进行的土壤调查、场地管理、环境评估及避免对生态敏感区的干扰；⑤通过保存和设计自然区、消极利用公园区等平衡公园活动项目，能够实现堂河绿道与汤米汤姆森公园之间的生态绿道连接；⑥在为消极利用、多重的或者特别目的而提供各种类型道路系统的同时，应该建立一个灵活的交通网络，连接坝区的不同场所；⑦提升昂温大道设计，将其作为公园快速路而非主干道，车道尽可能窄小同时提供街道停车场以满足骑自行车人员及其他道路人员的需求（图6–2–15）。

3) 场地环境管理的适应性设计

坝区景观要满足增长的游憩及生态需求。为了创造变化的环境条件，坝区的主要自然设计策略是

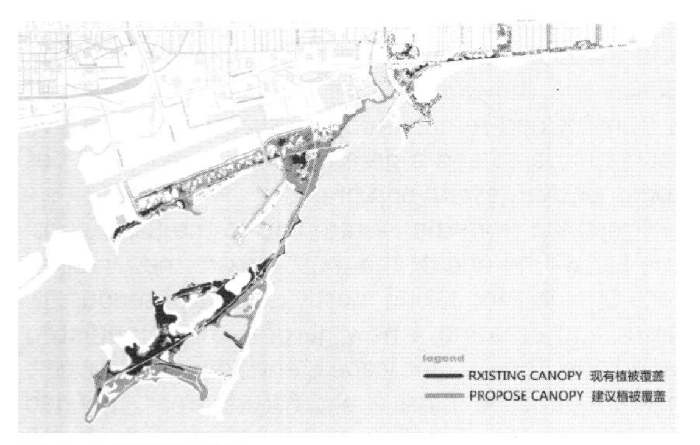

图6–2–14 植被覆盖规划图
Fig.6-2-14 Canopy cover master plan

numerous recreational settings and much-improved lakeshore edges and amenities.

With wind, wave and microclimatic conditions, the shoreline of the Bar has become an extraordinarily unique collection of boating, windsurfing and kite boarding clubs. These users and their water-oriented lifestyles contribute to the defining character of Lake Ontario

With the exception of two sites for snow dumping and wood chipping, the industrial sites along the Bar have largely been left to develop as a rough, unmanaged thicket landscapes. These historic industrial lands have a legacy of variable environmental conditions that will be addressed in the creation of Lake Ontario Park. The Master Plan's recommendations for the Bar represent a flexible and effective strategy for responding to a range of environmental issues.

2) Design Objectives

Seen as a major focal point in the context of harbour ecology and recreation, the Master Plan has six primary objectives for the future life and landscape of the Bar: 1.Build on the historic dynamism of the Fisherman's Island sand bar to promote a landscape that is alive and changing: ①A durable, resilient framework that is open to new types of wildlife communities and recreational settlements; ②Cultivate the rustic, wooded setting of Cherry Beach through the preservation and restoration of the mature Cottonwood stands and significant beach plant communities; ③Cultivate the distinctive "cottage" setting for water sports and wateroriented recreation at the North Shore; ④Create a flexible distribution of new facilities for multi-purpose recreation, play, education and learning that can adapt to ongoing and future soils investigations, site management, environmental assessments and ecologically sensitive areas; ⑤Encourage a green 'connection' between the Don Greenway and the Tommy Thompson Park by balancing park programs and uses with the preservation (where possible) and creation of natural, passive park areas; ⑥Develop a flexible circulation framework that connects the disparate 'places' in the Bar while providing a varied range of path treatments for passive, multi- and specialty-purpose use; ⑦Promote the design of Unwin Avenue as a park drive and not an arterial road, with the narrowest possible lane widths plus intermittent on-street parallel parking, subject to accommodating the needs of cyclists and other road users (Fig.6-2-15).

3) Adaptive Design for Environmental Site Management

The landscape of the Bar is one of expansive recreation and ecology. In response to variable environmental conditions, the primary physical design

大地艺术覆盖及土壤填运的组织框架。最终，坝区营造成由小山丘及山谷构成的新景观——这与渔人岛沙坝过去历史中的景观不同。连绵起伏的山丘景观结合草地及林地可以为场地营造各种类型的自然区（茂密混合林地、草地及泽地森林），同时为游客提供不受风影响的运动、全年开放的游憩、野餐、游戏及室外开阔区。

设计建议的主要目的是提出合适的方法或要点，指导坝区景观基于未来土壤信息及现有环境资产的条件下如何营造。山丘的具体位置、尺度、形态结构并不详细规定，但需要能够因未来决策的变化而进行适当的调整，这些决策往往受以下要素的影响：环境敏感区、成熟树林区、视线通道等潜在“特色”机遇、绿道节点等。

在制定场地环境管理策略时，总体规划人员研究了已经建造好的相似场地。

旧金山的克里斯场公园和帕洛阿尔托的拜斯比公园就是两个范例，它们通过在旧金山海湾的废弃地上布置大地艺术作品为游客提供游憩娱乐。在这两个项目中，游客可以看到在连绵起伏的山丘、野草及海岸植被的覆盖下——创造了适合场地气候条

strategy in the Bar is an adaptive organizational framework of earthwork capping and soil excavation. The result will be a re-formed landscape of shallow hills and hollows - not unlike the historical landscape of Fisherman's Island sand bar. The resulting landscape of undulating hills with meadow and woodland planting will support a variety of natural area types (thicket, mixed woodland, meadow and wetwoods), as well as wind-protected areas for sports, year-round recreation, picnicking, play and exposure to the environment.

This design recommendation is intended to offer an adaptive approach, or outline, for how the Bar landscape might be planned in relation to future soil information and analysis of existing environmental assets. The precise location, size, configuration of the hillforms are not fixed, but rather are adjustable according to future findings on areas identified as environmentally sensitive, areas of mature tree stands and potential 'feature' opportunities such as view corridors, greenway connections and the like.

In developing the environmental site management strategy of The Bar, the Master Plan team has researched comparative sites that have been designed and constructed along similar parameters.

Crissy Field in San Francisco and Byxbee Park in Palo Alto are two examples of public recreational landscapes constructed with topographic earthwork caps above historic waste sites along the San Francisco Bay. In both projects, the visible landscape is one of rolling hills covered in native grasses and

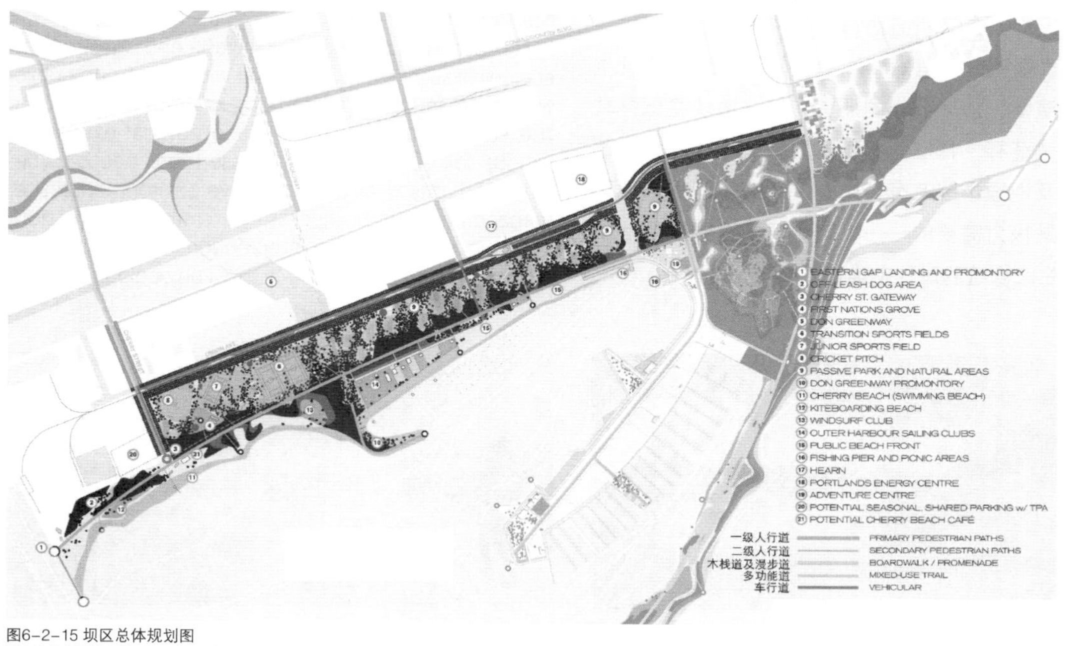

图6-2-15 坝区总体规划图
Fig.6-2-15 The Bar master plan

件的特色公共景观（图6–2–16）。

2. 基地区

1）场地简介

由于基地区的位置及大小，它成了安大略公园最为重要的生态构成要素。尽管公园的大部分区域较为狭长——生态学称之为“边界”——但基地区是整个公园最为宽阔和野生的分区。由于位于汤米汤姆森公园与坝区相交的区域，基地区成为重要的生态连接点：一条便于陆生生物迁入或迁出汤米汤姆森公园栖息地的宽阔绿道。

基地区过去曾是一处开放的水体。这个区域曾是一个垃圾堆，它大部分区域还遗留下来了一些土壤及环境问题。尽管存在一些遗留的环境问题，但是基地区为多种野生生物提供了生存的栖息地，尤其是那些鸣禽类。由于位于中间位置，且部分区域已经野生化，因而这块土地具有很大的潜力成为真正意义上的城市野生地。相对于坝区规划建议，总体规划对于基地区的建议更为简单。现在进行的和未来的土壤及水质调查将决定栖息地的可能性、具体位置、尺度及混合性，这种思路或许更适合这个关键生态场地的条件（图6–2–17）。

2）场地环境管理的适应性设计

由于基地区遗留的湖滨垃圾问题，之后所发生的生态演替，加之公众的期望，这一地块将保存其现有的“荒野”特征。尽管基地区区域内各类土壤的组成及地理学范畴有待确定，但是广泛的场地环境管理是极为需要的——在管理及培育野生自然及高价值生态系统的过程中确定人类干扰的水平。

coastal vegetation - producing distinctive and green public landscapes that are well-suited to the climatic conditions of the sites (Fig.6-2-16).

2. The Base Lands

1) Area Description

The Base Lands is the most important ecological component of Lake Ontario Park because of its location and size. While most of Lake Ontario Park is long and narrow — all "edge" in ecological terms — the Base Lands is one of the widest and wildest zones. Located at the point which the Tommy Thompson Park intersects the landscape of the Bar,the Base Lands represents an essential ecological connector: a broad gateway for terrestrial wildlife migrating to and from the remote habitats of the Tommy Thompson Park.

The Base Lands were historically open water. The area was a dump site and the vast extent of the Base Lands are have a legacy of variable soil conditions and environmental issues. Notwithstanding this legacy, the site has evolved to support a diversity of habitats for wildlife, most notably songbirds. With its central location, vast scale and legacy of wild colonization, the Base Lands offer a unique opportunity for a truly immersive "urban wilderness". As with the recommendations for the Bar, the Master Plan's proposals for the Base Lands are preliminary. Ongoing and future investigations on soil and water quality will determine the feasibility and precise location, scale and mix of the habitats that might be suitable to this critical ecological site (Fig.6-2-17).

2) Adaptive Design for Environmental Site Management

The Base Lands' legacy of lakefill and subsequent ecological succession, combined with the strong public will to preserve its 'wilderness' character. Although the specific make-up and geographical extent of various soil conditions in the Base Lands needs to be identified, it is clear that extensive environmental site management will be necessary - creating a question as to the level of human intervention in managing and cultivating wild nature and high-value ecologies in the

图6–2–16 坝区效果图
Fig.6-2-16 View of The Bar

在为基地区制定一个灵活的规划时，总体规划团队、多伦多及区域保护局以及五大湖区域内的重要生态学家担任同等重要的角色。通过生态调查及咨询，总体规划团队认为：如果一些环境条件的改善能够保证大面积精心布置的湿地、湿草甸及湿林地生长繁茂，那么基地区作为栖息地的价值将会得到极大的提升。这个观点也与多伦多及区域保护局的建议相一致，保护局认为基地区生态栖息地的类型应该是湿草甸、茂密的林地及沼泽地。场地地形现状及水文条件以及基地区所应该具有的高价值生态功能成为确定这些栖息地类型的基本原则（图6-2-18）。

总体规划中关于基地区环境管理的一条建议是对特别地块的重塑，这些重塑的场地将会营造一系列的小山谷或者湿地的水体——从阿什布里奇海湾延伸至外港口——它们将提供水生生物栖息地，同时为公众提供教育及消极的游憩体验。这些湿地水道的具体位置、大小以形态结构以及整体可行性将需要进一步的环境评估及利益相关者的复审而确定。

Base Lands.

In developing a flexible plan for the Base Lands site, the Master Plan team has coordinated with TRCA and lead ecologists in the Great Lakes Region. Through ecological research and consultation, the Master Plan team understands that the habitat value of the Base Lands would be greatly enhanced by the creation of conditions in which large, well-configured wetlands, wet meadows and wet woods can flourish. This notion is consistent with TRCA recommendations that the most ecologically appropriate habitat types for the Base Lands are wet meadow, woodland thicket and swamp. The rationale for targeting these types of habitats is based upon the existing physiographic and hydrological conditions, as well as the potential high-value ecological function of the Base Lands (Fig.6-2-18).

One of the Master Plan's proposals for environmental site management in the Base Lands is the excavation of targeted soil conditions. The excavation sites would be re-created as a series of lowland hollows, or wetland waterbodies -stretching approximately from Ashbridge's Bay to the Outer Harbour - that will provide aquatic habitat as well as an educational and passive recreation resource for public experience. The precise location, size, configuration of this 'wetland waterway', as well as its overall feasibility will be determined through further environmental and stakeholder review.

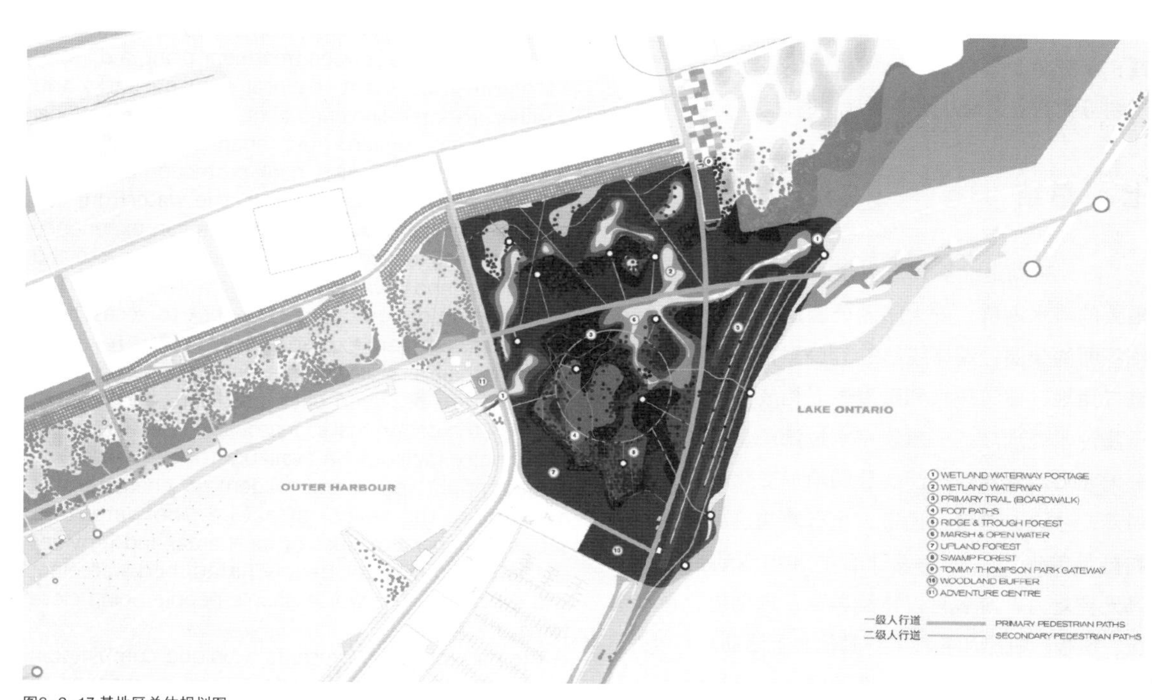

图6-2-17 基地区总体规划图
Fig.6-2-17 The base lands master plan

图6-2-18 基地区效果图
Fig.6-2-18 View of the base lands

3) 湿地水道

湿地水道被视为：一个解决环境问题的创新方法；一个改善水生及陆生栖息地（湿草甸和沼泽地）的方法；轻舟和小艇到达阿什布里奇海湾与外港口之间区域的一种方法。水道的特征与皮利角公园的泰森航道很相似——一条人工挖掘的、管理良好的水道，会成为航游整个公园的主要方式。由于未来的环境评估还未确定，规划人员认为湿地水道的特征应该满足下列条件：应该是较浅的非线性的水道，能够到达各个湿地水体；应该是植物并列排布、水面开阔的湿地和沼泽地；水体的位置及尺度依赖于现有的环境问题及生态重要性的区域；路桥（水陆联运）界定了水道的东西端，同时限制可达性；水道为游客提供了一条新的体验路线；水道的规划与实施应该遵从改善栖息地的目标。

七、 总结

安大略湖是北美洲五大淡水湖之一，属于世界最大的淡水湖群。安大略湖公园是多伦多市一条连绵不断的公园，湖岸线超过37公里，公园将多伦多港口用地、建筑垃圾和疏浚淤泥构成的狭长土地、一座污水处理厂、一座饮用水过滤厂、几个现存的公园和海滩连接在一起。公园地形变化丰富，既有海湾、沙滩、港口、通道，也有海沟、湖床、护岸和陆地等形态。规划设计后的滨水活动也同样充满无穷魅力，游客可以体验游泳、皮划艇、风筝滑板、帆板、帆船和各种各样其他水上活动。

规划设计中坚持可达性原则，积极探索主要道路与大尺度开放空间的设计，强调各个分区间的协

3) The Wetland Waterway

The wetland waterway is envisioned as 1) a creative means for handling environmental issues, 2) a means for enhancing aquatic and terrestrial habitat (wet meadows and swamp), and 3) a means for canoe and kayak access between Ashbridge's Bay and the Outer Harbour. The character of the waterway is envisioned much like Theissen's Channel in Point Pelee Park – a man-made and actively managed waterway that is a primary means for navigating the park. Pending future environmental assessments, the primary characteristics of the wetland waterway are proposed as follows: It will be a shallow, non-linear waterway achieved through a series of wetland water bodies; It is envisioned as plant-lined and open-water wetland and marsh; The location and scale of water bodies will be dependent on existing environmental issues and areas of ecological significance; Land bridges (portages) will define the east and west ends of the waterway, will limit craft access; The waterway will provide a new interpretive route for guided tours; and Planning and implementation of the waterway will respect the objective of enhancement of habitats.

VII. Summary

Lake Ontario (Ontario, Lake) is one of North America's five largest freshwater lake, belonging to world's largest freshwater lake group. Lake Park is a park in Toronto city with long lakeshore line, which longer than 37 kilometers. The park connects Toronto port land, building garbage, a strip land constituted by sludge dredging, a sewage treatment plant, a drinking water filtration plant, several existing parks and beaches. Park terrain varied a lot, including functional parts like bay, beach, port, channel, and also has land form of trench, lake, bank protection and earth. After the design of the project, the waterfront still possesses amazing activities, visitors can experience the swimming, canoeing, sailing, windsurfing, kite and skateboard and a variety of other water sports. The planning and design of the park stick to accessibility principles, actively exploring the design for main roads and large scale open space, emphasizing the coordination between each partition, and in each partition the design makes reasonable open space and community facilities be available for residents. The biggest bright spot is the "wilderness" characteristic, manifesting the beauty of nature. The Waterfront shows the characteristics of local area, and provides spacious wild land for people hiking, bird watching, and other activities which enable people being close to nature.

The park planning presents a unique combination of opportunities and challenges with respect to design, ecology, programming and sense of place.

调，并在每个分区内给居民提供合理的开放空间和社区设施。规划最大的亮点是保持“荒野”特征，体现自然之美。水岸的乡土气息和广袤的荒野特色能满足人们远足、观鸟、亲近自然、返璞归真的活动需求。

在设计、生态、项目活动及场所感等方面，公园规划独具一格地将机遇与挑战结合起来。被设想为一处野生生物保护区域文化生活、社会生活及游憩的发生地，公园为多伦多居民提供前所未有的体验及机会。在一定程度上，这个巨大的城市公园在吸引居民及游客并为他们提供一系列游憩体验、促进经济发展的同时，也界定了周边社区的特征。同时，通过自身丰富的空间、尺度及栖息地布置以及在城市宜居性方面的贡献，公园对城市可持续发展作出了重要贡献。

Conceived as a diverse reserve for wildlife, cultural and social life, and active recreation, the park offers an unprecedented array of experiences and opportunities for the residents of Toronto. In some way, the large urban park defines the character of the neighborhoods around them, drawing both residents and visitors for a full range of experiences and catalyzing economic development. At the same time, it makes a fundamental contribution to sustainability of the city through its generosity of space, scale and range of habitat opportunities, and its contribution to urban livability.

思考题

1. 安大略湖项目中，场地的功能布局与场地特征如何相呼应的？
2. 规划框架对于总体规划有何作用？
3. 总体规划强调了公园的哪些特征？对于场地的未来发展有何影响？
4. 土地利用规划是如何解决因为土地利用类型边界模糊不清而造成的规划难题？
5. 公园生态规划强调了哪些内容？

Questions

1. In the project, how does the function be assigned to echo the characteristics of the site?
2. What effect does the planning framework to the master plan?
3. What characteristics are strengthened in the master plan? What effect do they have to the site development in the future?
4. In the park land-use plan, how does the plan deal with the problems which are result from blurred boundaries between the various land-use categories?
5. What are be emphasized in the park ecology plan?

注释/Note

图片来源/Image：http://www.waterfrontoronto.ca/explore _ projects2/port _ lands/lake _ ontario _ park

参考文献/Reference：http://www.waterfrontoronto.ca/explore _ projects2/port _ lands/lake _ ontario _ park；http://www.cityofkingston.ca/residents/recreation/parks/lop/phase1/index.asp

推荐网站/Website：http://www.waterfrontoronto.ca/explore _ projects2/port _ lands/lake _ ontario _ park

第三节 橘子郡大公园总体规划

Section III The Planning of Orange County Great Park

一、 项目综述

这块占地1 347英亩（约5.45平方公里）的项目被称为“大公园”，它的前身是加利福尼亚州橘子郡艾尔托洛海洋空军基地，未来将会成为该地区居住、商业以及学习的中心。橘子郡大公园总体规划描绘了对一种新型公园的展望，在这里对有关社会和环境可持续性的新观点进行了调查和测试，并且橘子郡市民参与其中，对这些新概念大胆设想，建设更加健康和可持续的未来。大公园将不同社区和南加州的文化结合起来，同时对当地生态进行修复（图6–3–1）。

这种长期的项目将需要许多年才能够建立与发展起来。这个发展计划的其中一部分，是一个占地27英亩（约10.9公顷）的预览公园，它将作为大公园的访客中心、观测区以及设计元素和功能的原型区域。2007年夏天，预览公园的第一阶段以一个大型橘色氦气观测气球的升空拉开序幕。其他的设施将于2008年夏天建成，同时还有更多的计划中的功能。橘色观测气球泊靠在上空500英尺（154.2米）的高度俯瞰整个预览园，柑橘园、石雕以及植物的

I. Introduction

Known regionally as "The Great Park," this 1,347 acre project will become the heart of future districts for living, commerce, and life-long learning on the former El Toro Marine Air Station in Orange County, California. The Comprehensive Master Plan for the Orange County Great Park outlines a vision for a new kind of park. Here, new ideas for social and environmental sustainability are investigated and tested, and the citizens of Orange County become key participants in imagining these new ideas to create a healthier and more sustainable future. The Great Park knits together the diverse communities and cultures of Southern California while restoring the region's natural heritage(Fig. 6-3-1).

This long-term project will take many years to build and grow. Part of its development plan is a 27-acre Preview Park that serves as a visitor center, observation area and prototyping area for elements and features being designed for the Great Park. In Summer 2007 the first phase of the Preview Park opened with a large orange helium filled observation balloon. Other facilities were added in summer of 2008, and more features are planned. Anchored by an orange observation balloon that rises up to a 500-foot height for a commanding view, the Preview Park introduces several prototypes for the orange groves, stonework, and plantings that will eventually fill the entire park(Fig. 6-3-2).

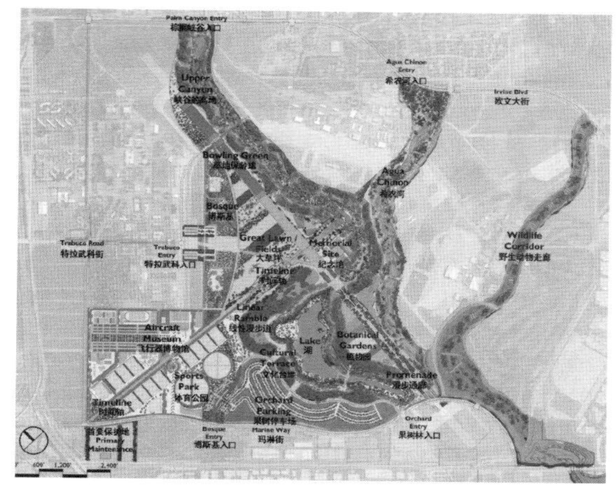

图6-3-1 橘子郡大公园总平
Fig.6-3-1 The master plan of Orange County Great Park

设计原型会出现在预览园当中，而这些最后也将组成整个公园（图6-3-2）。

二、 规划设计

该公园坐落在加州欧文前El托罗海军航空站（于1999年关闭），幅员辽阔。大公园占地1 347英亩，比圣地亚哥巴尔波亚公园还要大。大公园主要包括一个人工构建的2.5英里（约4公里）长的峡谷，一条改造后由地下露出地表的天然溪流、大湖、文化梯田、大草坪，航空博物馆、温室、植物园、步道和一个运动公园。尽管此类设施类似典型的公园设施，但是每个园区的功能可以多元化地发展。例如温室，是一个横跨湖中央的桥型结构，外部气孔结构既能保持空气流通，同时还能为内部植物提供遮挡（图6-3-3）。

另一个公园创新是大公园气球（目前已开放），能将游客带上400英尺（约122米）的高空，使人们可以在建设期间见证“公园的成长”。博物馆、农作物和具有教育意义的公园项目使当地居民和过去在橘子郡工作过的人感到很自豪。通过历史博物馆展览，冥思纪念碑以及在这工作过的人们的故事来纪念El托罗空军基地。同时许多公园项目，从小径漫步到公众节日都褒扬了该地丰富的社区和文化多样性。

大公园是三种公园体验的集合。峡谷是一个美丽的绿洲——一个漫步与休闲的乐园——是人们野

II. Planning and Design

Located in Irvine, California on the site of the former El Toro Marine Corps Air Station—which closed in 1999—the park is a project of immense scale. At 1,347 acres, the Great Park will be larger than San Diego's Balboa Park. The design for the Great Park features a constructed two-and-a-half mile canyon, a daylighted existing stream, a large lake, a cultural terrace, a great lawn, an aviation museum, a conservatory/botanical garden, a promenade, and a sports park. And while these may seem like typical park amenities, the vision for each park feature is anything but. The Conservatory, for example, is a bridge structure that will span the central lake and "breathe" through a porous skin while providing shade for the collections(Fig. 6-3-3).

Another park innovation, the Great Park Balloon(currently in operation), takes visitors 400 feet into the sky, allowing people to witness the "growing of the park" during construction. Museums, agricultural plantings, and educational programs honor the people who have lived and worked on Orange County's land in the past. The El Toro military airbase is commemorated with historic museum exhibits, a contemplative memorial, and the stories of the men and women who once served here. And a variety of park programs, from nature trails to public festivals, celebrate the area's rich diversity of communities and cultures.

The Great Park will actually be three park experiences in one. The Canyon is a beautiful oasis—a

图6-3-2 园内的景点与活动
Fig.6-3-2 The view spots and activities in the park

图6-3-3 温室植物园
Fig.6-3-3 Greenhouse botanical garden

餐和孩子探险的理想之地。生态公园是生态支柱，它提供了物种多样性，原生群落和野生动物。而农田和纪念公园纪念了该地作为物产丰富的农田和近些年作为空军基地的历史。

1. 大峡谷

大峡谷是大公园中最鲜明的元素。两公里长，深60英尺（约18.3米），长满了郁郁葱葱植物的大峡谷比周边地区要凉爽许多。大峡谷可以称得上是一座绿洲，游客们在这里可以漫步在穿越原生棕榈与林地期间的小道，并带有地中海风格的装饰。一条常年流淌的小溪串联着大大小小的水池，贯穿了整个大峡谷。一系列引人注目的小桥可以让游客从许多不同的地点穿越峡谷，营造了许多欣赏峡谷美景的机会。这其中要数“双塔桥”最为引人注目，它连接了一条锯齿形的小路，横跨了大峡谷最宽的部分（图6–3–4）。

2. 退伍军人纪念园

在超过50年的时间里，E1托罗空军基地在和平时期担任国家军事训练设施的角色，在冲突时期则成为海外军事行动的指挥地。橘子郡大公园记录了E1托罗空军基地的历史，在公园内老式战机被展示在先前的跑道残骸上。退伍军人纪念园将荣耀那些为国奉献的军士（图6–3–5）。

3. 植物生态园

植物园处在橘子郡大公园的中心位置。参观者可以近距离观看南加州植物，也可以以一种全新的方式体验人与植物、食物与健康、社会与环境之间

place to wander and daydream—a place for families to picnic and for children to explore. The Habitat Park is an ecological backbone that provides species diversity, native communities and wildlife. Finally, the Fields and Memorial Park commemorates the history of the site as both a productive agricultural landscape and, more recently, a military base.

1. Great Canyon

The Canyon is the Great Park's most distinctive element. Two miles long and up to 60 feet deep, the heavily planted Canyon is dramatically cooler than the surrounding urban areas. The Great Canyon is an oasis where visitors can stroll along paths and trails bordered by native palms, woodlands and Mediterranean ornamentals. A perennial stream with a string of small pools will run the length of the Great Canyon.A series of dramatic bridges will allow visitors to cross the canyon at various points, creating many opportunities to enjoy the view. The most dramatic of these bridges is the "Bridge of 2 Towers," which connects a zig zag path across one of the widest parts of the Canyon(Fig. 6-3-4).

2. Veterans Memorial

For over 50 years, MCAS El Toro served the country as a training facility in peacetime and a staging area for support of overseas military operations in times of conflict. The history of El Toro will be remembered at the Great Park Air Museum, where vintage aircraft will be displayed on remnants of the former runways. A veterans memorial will honor those who served our country(Fig. 6-3-5).

3. Botanical Gardens

The Botanic Garden is the heart of the Great Park. Visitors will be able to observe Southern California's plants in habitats up close and in detail. They will be able to experience, in a totally new way, the relationships between people and plants, food and health, society and setting. A "garden bridge" will link the garden to the cultural terrace (Fig. 6-3-3).

图6–3–4 大峡谷
Fig.6-3-4 The Great Canyon

图6–3–5 退伍军人纪念园
Fig.6-3-5 The Veterans Memorial

的关联。一座景观桥将花园与自然地形巧妙的连接起来（图6–3–3）。

三、 规划设计要素

公众在大公园建设中起了关键作用。在总体规划阶段，设计方与相关人士，如退伍老兵、环境组织以及艺术家进行了多次会议，这样的结果是大公园真正成了每个人的公园，反映了橘子郡居民的不同利益、不同价值以及不同背景。更重要的是，大公园为可持续性、生态责任和南加州公共空间建立了新的标准，提供了一个独特的场所供人们游览、放松、娱乐和交流。

1. 开放空间

随着时间流逝，橘子郡正在经历巨变。过去的开阔地、果园和远景正变成郊区景观。总体规划中，在存留下来的基地南北跑道上设计了一个开放空间的走廊，因此在公园中保留了一种延伸感，而这种延伸感曾在橘子郡盛极一时。游客在各端都能看到许多远景；轴的北端连接圣安娜山，而南端朝向海岸，可以远眺圣洁奎因山（图6–3–6）。

2. 历史

总体规划从几个重要方面歌颂了该地的历史以及橘子郡的区域性格。大公园的柑橘、坚果和鳄梨果园，农田以及社区公园都将反映该地过去的农业盛况。例如柑橘的历史重要性在公园中的果园区会显得活灵活现。规划同样讴歌了海军常年的精忠报国。例如该地的一份纪念表记录了基地历史上的重要时刻（图6–3–7）。

3. 联系

总体上该规划着眼于若干因素以发展联系。从生态上看，该公园是连接山脉和海洋间保护区的重要一环。公园还将为该郡的各个社区建立社会联系，具体方式为通过在遍布该郡的小径上进行自行车骑行、远足来将所有临近地区与公园以及较远的社区联系起来。在文化上，公园联系并纪念了该地的历

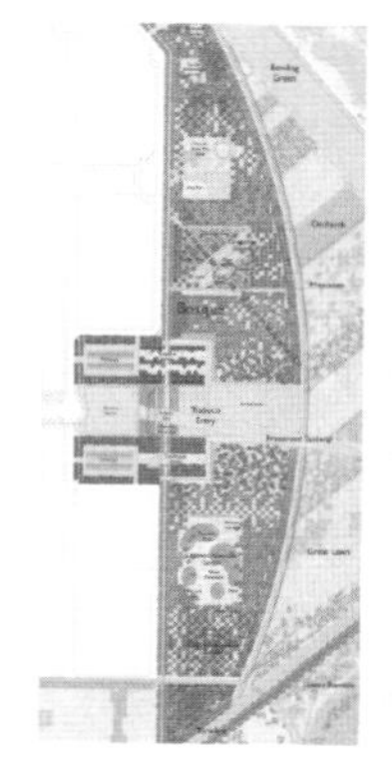

图6–3–6 开放式走廊与尽端的邻里公园
Fig.6-3-6 The open space corridor and neighborhood pocket park

III. Planning and Design Elements

The public has played a major role in creating the Great Park. During the master plan phase, intensive sessions were held with stakeholders groups, including veterans, environmental groups, and artists. As a result, the Great Park truly is a park for everyone, reflecting the interests, values, and backgrounds of all people of Orange County. More importantly, the Great Park will set new standards for sustainability, ecological responsibility and public space in Southern California, providing a unique place to visit, relax, recreate and meet people.

1. Open Space

Over time, Orange County is undergoing a profound change. What was once a landscape of open fields, orchards and vistas is being transformed into suburban development. The Master Plan includes an open space corridor in place of one of the existing colossal north-south runways of the airbase, thus preserving in the park a sense of the expansiveness that was once predominant in Orange County. Visitors will be treated to vistas at each end; the north of this axis delivers the Santa Ana Mountains, while the Southern end looks toward the coast, framing the San Joaquin Hills(Fig. 6-3-6).

2. History

The Master Plan celebrates the history and regional character of Orange County in several important ways. The county's agricultural past will be reflected in the Great Park's citrus, nut and avocado orchards, agricultural fields and community gardens. The historic importance of citrus, for instance, will "come to life" in the park's Orchard Parking area. The master plan also honors the decades of service to the nation by the Marine Corps. For example, a timeline bisecting the site commemorates major moments in the airbase's history(Fig. 6-3-7).

史感——眼前的一切。田地、大事年表以及纪念碑都将游客与该地的历史联系了起来（图6–3–8）。

4. 地形

在接下来的几年，大公园造地工程将启动；这包括进行大公园峡谷和湖泊的开挖，相关地貌的建设，比如文化梯田，以及大量的室外区域。届时也将完成对现有的阿瓜奇诺溪流（目前仍在地下）的改造，使之可以露出地表。这些地貌改造将有效改变现有的平坦平庸的地貌，形成具有小气候，同时也能供各类动植物栖息生长的新景观（图6–3–9）。

5. 文化

各种公园项目，从教育类到公众节日，都赞颂了那些为如今的橘子郡作出贡献的人们。公园里的各种设施，如可供行走观赏的大事年表、农田、冥思纪念碑、文化梯田以及博物馆都将使文化变得生机勃勃并且真实具体，同时还将那些偏远的社区与公园联系起来。

6. 生态

大公园的目标就是将贫瘠而广阔的El托罗航空站变成活的景观——这对于大公园的构想是至关重要的。当地的综合生态重建栖息地位于公园的三个主要地点——野生动物长廊，希农河和峡谷。在其他地区，包括植物园也同样能看到当地植物群落。这些栖息地将遵循以下生态原则和重建策略，而且随

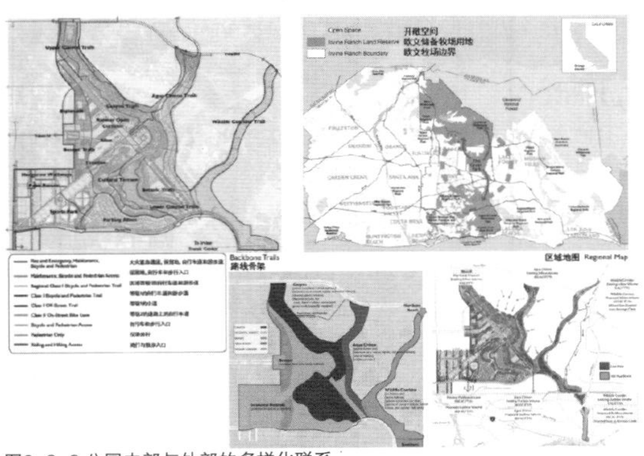

图6–3–8 公园内部与外部的多样化联系
Fig.6-3-8 Diversified linkage between the inner and outer parts of the park

3. Connections

Overall, the master plan focuses on several opportunities to develop connections. Ecologically, the park is a vital link in the chain of land reserves stretching from the mountains to the sea. The park will also create social connections to the communities throughout the county by knitting together riding, hiking, and multiuse trails from all parts of the region, linking all neighborhoods to the park and communities beyond. Culturally, the park connects and celebrates the sense of history here—what came before us. The fields, the timeline, and the memorial all deal with connecting visitors to the site's history(Fig. 6-3-8).

4. Topography

Over the next several years, terra-forming of the Great Park site will begin; this includes digging the Great Park canyon and lake, and creating associated landforms, such as the Cultural Terrace, and a number of outdoor areas. Daylighting of the existing Agua Chinon stream (now buried underground) will also be accomplished. These landform changes will essentially change the now mostly flat, featureless land into a new landscape complete with microclimates and capable of sustaining a wide variety of plant and animal life(Fig. 6-3-9).

5. Culture

A variety of park programs, from educational opportunities to public festivals, celebrate the diverse populations that make up present-day Orange County. And park features, such as the walkable Timeline, the agricultural fields, the contemplative memorial, and the Cultural Terrace and its museums will make culture vibrant and tangible and help link the park to

图6–3–7 公园内的历史时间轴
Fig.6-3-7 Historical timeline in the park

图6–3–9 公园内丰富的地形
Fig.6-3-9 Undulating terrain in the park

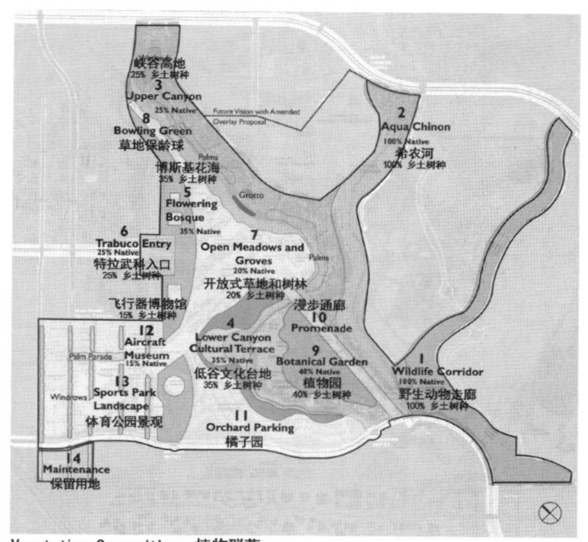

Vegetation Communities 植物群落

California Natives
加利福尼亚州乡土植物
- Coastal Sage Scrub 滨海的鼠尾草灌木丛
- Grasslands and Meadows Environment 草原及牧场环境
- Woodland Environment 林地环境
- Streamside Environments 细流环境

Xeric Landscapes / Cultral Landscapes
旱生景观 / 文化景观
- Southern California 南加利福尼亚
- Parkscape 公园景观
- Orchards, Fields and Community Gardens 果园农田和社区花园
- Open Meadows and Groves 开敞的草坪和果林

图6-3-10 公园内的植物的种类、比例与分布
Fig.6-3-10 The species, proportion and distribution of plant within the park

着时间的流逝，仅需少量维护（图6-3-10）。

四、 可持续设计

规划中有一些具体的且已经证实的生态手段，用以确保大公园自身的可持续性，并能产生真正的环境利益，具体将从五个方面衡量（图6-3-11）。

1. 能源

当地能源的产生与存储依靠一个占地1英亩（约4047平方米）的光电阵列实现，它将覆盖大草坪附近的一座户外亭子，是大公园中最大的再生能源装置。此外，在能源生产区将种植40英亩（约16.2公顷）的柳枝稷或其他具有生物能的作物用以进行现场能源生产。

2. 水

水的可循环性衡量包括水质的保持，天然湿地内

communities far beyond the park's boundaries.

6. Ecology

Our goal for the Great Park is to transform the sterile expanse of the El Toro airbase into a living landscape—that is fundamental to the vision of the Great Park. Ecologically complex restored native habitats will be found in three major sections of the park—the Wildlife Corridor, Agua Chinon, and the Canyon. Native plants communities are found in other areas as well, including the Botanic Garden. These habitats will be constructed by following ecological principles and restoration protocols, and will require only modest management as time passes(Fig. 6-3-10).

IV. Sustainability of Design

The Master Plan includes tangible, proven ecological techniques that will help ensure the Great Park can sustain itself and produce real environmental benefits. These efforts will be measured in five major categories(Fig. 6-3-11).

1. Energy

Conservation of energy and on-site energy generation include a one-acre photovoltaic array that

Sustainability Goals
可持续发展的目标

Biodiversity
生物多样性
Provide ecological habitats and connections to reflect the local natural heritage, and to enhance biodiversity in the region

Connection to Nature
自然连接
Bring nature and environmental education opportunities into the greater Orange County area

Land
土地
The US Navy is performing remediation of the land generally through mechanical means. The Great Park will develop healthy, living soil through natural soil amendments and other means as necessary - perhaps with phytoremediation

Air Quality
空气质量
Improve air quality of both internal and external environments

Water
水
Protect and conserve both natural and potable water resources

Well Being
福利
Protect and improve the health and productivity of those who visit and work there

Energy
能源
Reduce the use of fossil fuels and emissions of greenhouse gases

Materials
物料
Minimize the impact of construction materials and the generation of waste

Inclusion
容纳
Encourage community participation and civic engagement so that all visitors can obtain an equivalent experience in the park

Heritage
传统
Instill a sense of place that references the history of the site and the region

Transit-Oriented
公共交通导向
Provide a transit oriented development for the surrounding community with less-polluting transportation choices and connections within and beyond the park

Monitoring
监测
Incorporate ongoing measurement and monitoring of key sustainability metrics

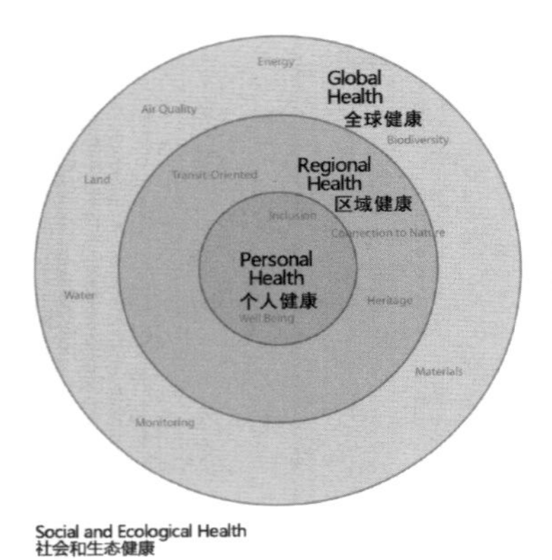

Social and Ecological Health
社会和生态健康

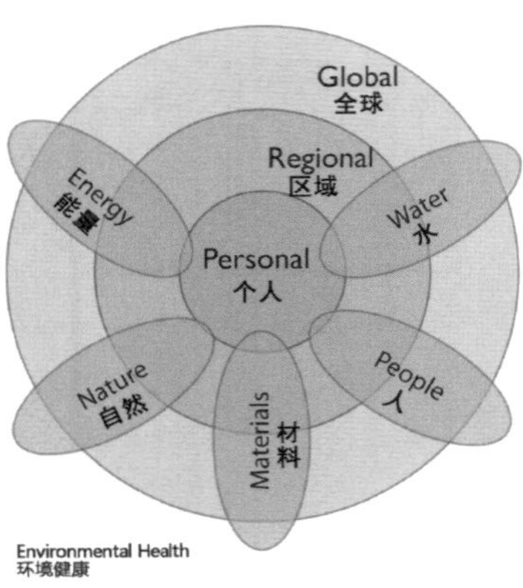

Environmental Health
环境健康

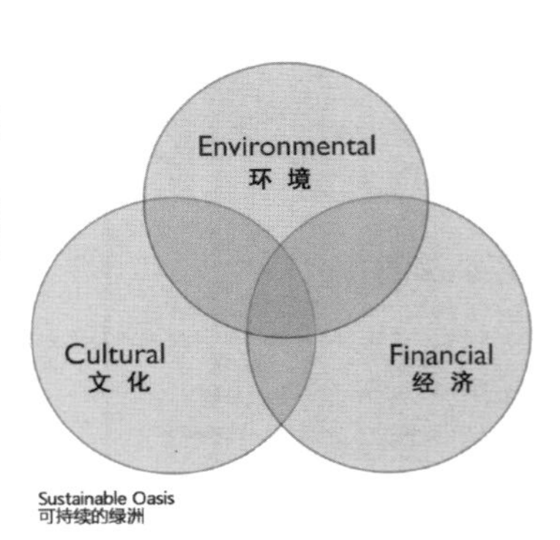

Sustainable Oasis
可持续的绿洲

图6-3-11 可持续设计的目标
Fig.6-3-11 The goals of sustainable design

的水循环和径流保持。一个开放的沼泽地和雨水道网络能减少洪水的流出，同时增加地下水蓄水层对水的存留。此外，公园还拥有天然净水系统阵列，通过三个方式净水。第一种是利用多孔行道和渗透装置。第二种是利用生态沼泽和集成在景观区的渗透、析出介质。第三种是对净水湿地下游水的存留。

3. 物料

建造大公园的材料将被打捞起来循环使用，进行生物工程处理，不会产生废物。大部分拆除的跑道将原地使用，现有建筑的红木支架将用作桥梁支架。总计超过350万吨水泥和钢铁将循环使用。拆除和土方工程中产生的所有绿色垃圾都将运往一个永久堆肥设施，该设施能生产营养丰富的有机肥料用以修复土壤。

4. 自然

重建当地栖息地，增强了大公园中的自然过程，强化了生物多样性，并在公园内部与外部形成了生

will cover an outdoor pavilion near the great lawn and will be the largest of many renewable energy installations in the Great Park. In addition, over 40 acres of switch grass or other biomass crops will be grown in the production zone for energy generation on site.

2. Water

Sustainable water measures include water conservation, water recycling in natural treatment wetlands and runoff capturing. An open swale and storm drain network will minimize flood flows exiting the site and maximize re-capture to the groundwater aquifers. In addition, the Park will have an array of natural treatment systems implemented in a three-stage treatment process. Stage one includes use of porous pavement and infiltration devices. Stage two features bioswales and infiltration/exfiltration media integrated into landscape zones. Stage three involves capturing water downstream in treatment wetlands.

3. Materials

The materials used to create the Great Park will be salvaged, recycled, ecologically engineered, and waste neutral. Portions of the demolished runway will be used on site, and redwood planks from existing on-site buildings will be used as bridge planks. In all, more than 3.5 million tons of concrete and steel will

态联系。野生动物走廊不对公众开放，而将仅为动物的活动保留，它为公园提供了重要的生态支柱，并且是海岸和橘子郡中心地区现有自然区域的重要生物连接通道；这条连接通道在近150年来首次实现动物迁徙穿越鞍峰山谷。希农河是一条天然水道，被埋在水泥管道中60年后，将重见天日，并重建为一个有效的南加州水滨生态系统。在峡谷内有一条常流溪河和许多池塘，反映了南加州丘陵和低地水生栖息地地貌，将为当地大量的植物和动物提供生存条件。峡谷中也有许多特殊栖息地，包括春季水塘、岩礁和长满蕨类植物的洞穴。总体上，大公园规划中包括了75%的本地及外来植物；这些植物生长无需大量水分，且不对加州生态产生侵略性。

5. 人

大公园的目标就是使该地的可持续性变成一种对游客来说看得见、摸得着的体验。在橘子郡骑自行车能给游客提供一种有趣、无污染以及健康的方式探索公园的各种元素。在植物园或者作物产地种植新鲜的有机食物，在公园的咖啡馆或者农贸市场上都能买到。此外，关于环境可持续性的新观点和元素都将进行研究、测试，并应用于在建的公园，使之成为艺术家、科学家和游客共同合作的研究拓展中心，在这里，游客可以与专家一道为我们的时代问题寻求解决办法。

五、 总结

橘子郡大公园设计中，多处景点独具特色，标识性很强，令人过目不忘。大峡谷是最鲜明的地形要素，橘色氦气观测气球是最大的设计亮点，而横跨湖中央的桥型结构温室也同样匠心独具，这些景点共同构建了公园的独特形象。

场地前身是海军航空站，公园规划设计中注重历史景观文脉的延续，规划保留了大公园的柑橘、坚果和鳄梨果园，来反映曾经的农业盛况；设计了退伍军人纪念园缅怀为国奉献的军人；设计了可行走观赏的大事年表、农田、冥思纪念碑、文化梯田以及博物馆，让游人能够深刻体验历史演绎中沉淀

be recycled. All green waste from demolition and earthworks on the site will be brought to a permanent composting facility which will provide rich organic supplements, high in nutrients, for soil amendment.

4. Nature

Natural processes will be enhanced in the Great Park by restoring native habitats, enhancing biodiversity and creating ecological connections within and outside the Park. The Wildlife Corridor, off limits to the general public, will be reserved for animal movement, providing an essential ecological backbone for the Park and a critical biotic link between existing natural areas in coastal and central Orange County; this wildlife connection will allow animals to migrate across the Saddleback Valley for the first time in nearly 150 years. The Agua Chinon, a natural waterway buried in a concrete channel for 60 years, will be daylighted and reestablished as a functioning southern California riparian ecosystem. Within the Canyon, a perennial stream and ponds, reflective of southern California's foothill and lowland aquatic habitats, will support a wide variety of native plants and animals. The Canyon will also showcase unusual habitats, including vernal pools, rock outcrops, and fern grottoes. Overall, the Great Park Master Plan includes 75 percent native and "California friendly" non-native plants; these plants have modest water needs and do not appear on any of the state's invasive plant lists.

5. People

The goal of the Great Park is to create a place where sustainability becomes a tangible experience for visitors. Orange bikes will offer visitors a fun, non-polluting and healthy way to explore all elements of the Park. Fresh organic food grown at the Botanic Garden and, perhaps, in the Production Fields, will be available in park cafes and at the Farmer's Market. In addition, new ideas and opportunities for environmental sustainability will be investigated, tested and built through the Living Park, a research and outreach center for collaboration between artists, scientists and visitors. Here, visitors can participate with experts in finding solutions to problems of our time.

V. Summary

In the design of Orange County Great Park, many view spots are very impressive given its unique features. The grand canyon is the most obvious geographic factor; the orange helium view balloon is regarded as the biggest light spot in the design, as well as the bridge-typed greenhouse across the lake, which all establish the unique image of the park.

The former building of the park is the navy airport, so the designer pays much attention to saving the history and culture within the spot. They have preserved the citrus, nut and avocado orchards to reflect the once

下来的文化传统，真正感悟大公园独有的景观、气质、特色，从而建立持久的吸引力。

场地记忆的延续，对于景观设计师来说是创作中富有挑战意义的命题。这项工作不仅意味着要探究场地历史而且意味着在历史环境中注入新的生命。历史延续性是橘子郡大公园的创新源泉，这一原则不仅从宏观层面引导了公园景观结构的发展，还通过对历史价值深层次的挖掘，不断实现新的突破。由此可见，文脉的传承不仅是为了保留场地记忆，更是为了构建场地深厚的传统文化底蕴，从而营造出高品位的、人性化的景观空间，满足游者多样化的精神需要。

prosperous agriculture; they have built a demobilized-soldier monument park in memory of the servicemen; they have put up great event chronology, farm land, meditation monument, culture terrace and museum for visitors to expose the culture and tradition passed from history to them and let them feel the unique landscape, temperament, and characteristic of the park, which can set up lasting attraction.

The extension of the spot memory can be a challenging proposition in the design for landscape designers, because it not only calls for research of the spot history but also summons new charm injected into the spot. The extension of history is the innovation origin of Orange County Great Park, which not only guides the landscape structure development but also makes new breakthroughs by exploring deeper value of its history. Therefore the extension of culture and history serves for not only preserving spot memory, but also passing deep cultural legacy to establish high-quality and humanizing landscape space to meet various spiritual needs of the visitors.

思考题

1. 分析橘子郡公园中，峡谷公园、生态植物园与纪念园三者之间的联系。

2. 选取规划设计要素中你最感兴趣的一个进行深入了解，对其进行介绍与分析并说出你感兴趣的原因。

3. 分析了解橘子郡公园可持续设计五个部分之间联系，并论证通过这些措施能否达到其最初做出的可持续设计的目的。

4. 橘子郡公园在历史延续与创新的联系上处理得如何？好是好在哪里，差是差在哪里？

5. 尝试提出一些你所知道生态措施，将其规划到公园中去，并论证你的措施将如何与已有的措施进行联系从而起到应有的作用。

Questions

1. Discuss the link between Canyon Park, Botanical Garden and Veterans Memorial in Orange County Great Park.

2. Describe one planning and design element that you find interesting. Introduce, discuss and describe the reason why you find it interesting.

3. Discuss the link between the five aspects of sustainable design. Demonstrate how these measures achieve the sustainable design goal.

4. How did Orange County Great Park balance the relationship between adoption and creation? Discuss the pros and cons.

5. Try to list some ecological protection measures you know. Apply these measures into the design of the park and demonstrate how they can work with the existing measures to achieve their goals.

注释/Note

图片来源/Image：文中图片均来自http://www.asla.org/awards/2008/08winners/127.html

参考文献/Reference：本案例引自ASLA2008年荣誉奖项目详细描述

推荐网站/Website：http://www.asla.org/awards/2008/08winners/127.html ；http://www.ocgp.org/ ；http://www.jrbobshawdesigns.info/#! _ _ken-smith-workshop-west

第四节 伊利西安公园总体规划

Section IV Elysian Park Master Plan

一、 项目概述

伊利西安公园总体规划是一个由社区和城市居民共同参与决定的行动条款的工作列表，这一规划对于公园的整体提升和作为公园使用者的洛杉矶居民而言都是至关重要的。规划编制的决定来源于洛杉矶市民对于保留和保护伊利西安公园这一巨大资源的强烈要求。

规划通过借鉴历史信息形成一系列行动建议以改善公园现状条件及解决长久以来妨碍公园正常使用的问题。这些建议以行动条款的形式服务于伊利西安公园肌理的重新编制，并使整个公园重新焕发活力。

公园里的每一个区域都不是独立存在的，他们之间通过社区、游憩道路网、自然区域，及它们内部被积极或消极使用的设施等连接。因此，公园的规划不能采用分区设计而是以主题的形式进行。总体规划由四部分组成，它们分别是：公园的娱乐和游憩、游赏公园、公园场地、公园的养护。

在洛杉矶，新的公园和休憩用地的开发在伊利西安公园的四周正在进行（图6-4-1），此外，城市社区发展规划对于与伊利西安公园的连接也是一个良好的机遇，因此，重新焕发伊利西安公园活力

I. Overview

This Master Plan is a working list of action items that were determined by the community and the City as critical to the on-going improvement of the Park and to the citizens of greater Los Angeles who use and enjoy the Park. The decision to develop a new master plan grew out of the desire of local citizens to preserve and protect this great resource called Elysian Park.

This Master Plan draws on historic information to form a list of recommendations to improve existing conditions in the Park and to solve persistent problems that continue to deter Park use. These recommendations in the form of action items serve to re-weave the fabric of Elysian Park; to strength the pieces in order to revitalize the whole.

Not one area is perceived as its own place in Elysian Park. All areas are connected by the communities, the network of trails, the precious natural areas, and the active and passive recreational uses set among them.Therefore the Park is not planned by area but by theme. there are four chapters to this Master Plan. These are: FUN AND RECREATION IN THE PARK ,GETTING AROUND THE PARK ,THE PARKLAND,TAKING CARE OF THE PARK

New park and open space development in Los Angeles is occurring all around Elysian Park(Fig.6-4-1). In addition, future City plans of new community developments are positive opportunities to connect into the resources of Elysian Park. Revitalization of Elysian Park is a key component in the city's strategy

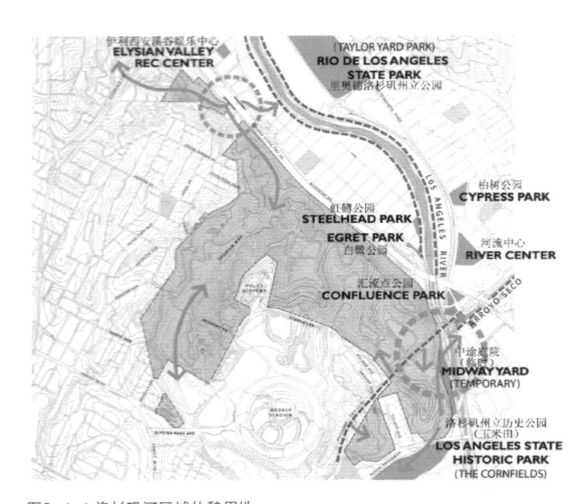

图6-4-1 洛杉矶河区域休憩用地
Fig.6-4-1 Los Angeles river regional open space

就成为以创造和维持洛杉矶城市高品质生活这一城市战略的关键组成部分。

二、 场地介绍

伊利西安公园对洛杉矶城而言，是一个充满生机的历史肌理的遗存。当波托拉探险队在1769年到达过去的洛杉矶古城所在地时，他们在靠近百老汇附近的布埃纳比斯塔山的山脚下扎营，而那里就是现在伊利西安公园的所在地。最初西班牙卡洛斯国王授予印第安人村落的土地中就包含这个公园的范围。

今天，洛杉矶已发展成为一个多文化交融的特大都市，伊利西安公园作为这个城市最古老的公园，依旧默默地为人们提供着一个逃离拥挤的城市生活，体验自然的庇护所。

伊利西安公园环绕着道奇体育场在的查韦斯山谷，几乎是一个山坡上的社区，同时也是洛杉矶警察学院的家。公园西边毗邻回音公园，南面是中国城，北面是5号快速路、柏树公园和伊利西安溪谷，菲格罗亚大街从110州级公路（帕萨迪纳快速路）向北穿过公园（图6-4-2）。

现在，伊利西安公园所拥有604英亩（约2.4平方公里）的绿色空间的重要性比以往要大得多。洛杉矶的人均休憩用地占有率在全美的主要城市中是

to create and maintain a desirable quality of life in Los Angeles.

II. Introduction of the Site

Elysian Park is a vibrant, living remnant of our City's historic fabric. When the Portola expedition arrived at the future site of the Pueblo de Los Angeles in 1769, they camped at the foot of Buena Vista Hill near Broadway in what is now Elysian Park.The original Spanish land grant from King Carlos of Spain to the Pueblo included part of the Park.

As Los Angeles has matured into the multi-cultural megalopolis it is today, Elysian Park, the city's oldest Park, has quietly persisted as a refuge where a crowded populous can experience nature within an urban environment.

Encompassing Chavez Ravine where Dodger Stadium is located, Elysian Park is mostly a hillside community that is also home to the Los Angeles Police Academy. Elysian Park is bordered by Echo Park on the west, Chinatown on the south and the 5-Freeway, Cypress Park and Elysian Valley on the north.The Figueroa Street Tunnels take northbound State Route 110 (the Pasadena Freeway) through the park(Fig.6-4-2).

Today, the significance of the Park's 604 green acres is greater than ever. Los Angeles has the lowest number of open space acres per resident compared to any other major American city. Conserving and

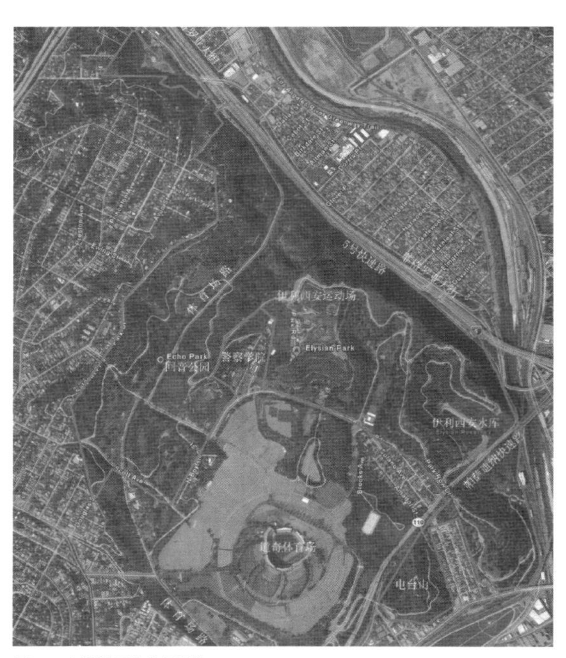

图6-4-2 伊利西安公园区位
Fig.6-4-2 The location of Elysian Park

最低的。保护和强化城市公园及休憩用地对于为城市居民提供高品质生活来说是至关重要的。城市游步道、充满灵感的鸟瞰场所、野餐地点和绿色游憩区域，这些都可以为包括居住在市中心区和伊利西安公园附近社区的居民提供舒适的空间。在1886年，伊利西安公园掀起了一波要求保持开敞空间和使其正式成为一个公园的活动。在2005年的今天，这个公园依旧处在保持自然和城市化的冲突夹缝中，现在的决策将世世代代影响着洛杉矶人。

enhancing City parks and open space is critical to providing an improved quality of life for millions. The urban trails, inspirational overlooks, picnic spots and green recreation areas provide an oasis for both downtown residents and the numerous neighborhoods surrounding Elysian Park.In 1886 Elysian Park rode the wave of the American desire to maintain open space and officially became a park. Today, in the year 2005, the Park is still in the middle of the collision between nature and urbanization. The decisions made now will impact the citizens of Los Angles for generations to come.

三、 相关规划简介

1. 1971年的总体规划

1) 规划目标

伊利西安公园最重要的特征就是荫庇溪谷的温和气候，这使得多样外来物种的引进成为可能，同时也为长远的规划提供了条件。规划建议将现有的公园打造成为动植物群落及人类活动的天堂，而不仅仅是为体育活动等提供场地。

在伊利西安公园与格里菲斯公园之间应该存在一条明显的廊道，人们可以通过自行车或者步行的方式穿梭于两者之间，因此，规划应该将两个重要的设施连接起来。这样就为廊道两侧的美化和土地利用带了许多机遇。

完全开发后的公园最重要的作用体现在以下几类群体中：在双休日时，空旷的草坪上的野餐者，独自或结伴步行在公园中，或在规划线路上骑车的人，在公园的美丽景色中享受一两个小时的人。伊利西安公园将会很好的服务于市民，利用轨道交通将这公园周边整个区域的人流与公园联系起来，这将会使得公园及其内部各个部分具有充分的可达性（图6–4–3）。

2) 道路体系

规划需要一个广泛的道路系统贯穿整个公园。这些道路应该有两种。一种是具有足够宽度，服务于越野跑步及维护车辆的道路，另一种服务于徒步旅行。这些道路的位置在逻辑上应该合理，能够连接所有主要的有趣的地方。沿路应该有有趣的惊喜，例如不寻常的植物组合，鸟类喂食器或类似的

III. Introduction of Related Planning

1. 1971 Master Plan

1) Planning Goals

The most significant feature of the park is the mild climate to be found in the sheltered valleys. It is this condition which has made it possible to introduce a wide variety of exotic plant materials and suggests a plan for long range development.it is the reconmmendation of this plan that the direction indicated by existing uses and the park be a haven for flora ,fauna and those people who are seeking forms of recreation other than highly organized sports or concessions.

A significant corridor should exist between the two parks to be used for bike riding,hiking or just walking from the park to Griffith park,thus tying these two facilities together.There could be many opportunities for beautification and use areas along the way.

The most important aspect of the fully developed park will be its attraction for the traditional Saturday and sunday picnickers in the open meadows and the single individual or small group who wish to walk the trails,ride the proposed tram or simply enjoy an hour or two in the quiet beauty of the park.Elysian park would serve people well and with the installation of proposed people movers throughout the area connecting to the park tramway,there would be adequate transportation to the park and to all parts within it(Fig.6-4-3).

2) Circulation

The park demands an extensive trails system throughout the entire park.These trails should be of two kinds.those which are wide enough for park maintenance vechicles and cross country runners and smaller trails for single file hiking.These trails should be logical in their location and should connection all main points of interest. These should be interesting surprises along the way such as unusual plant groupings,bird feeders or similar features.rest points should have rustic benches or logs for the convenience of hikers and ,where practical,domestic water should be

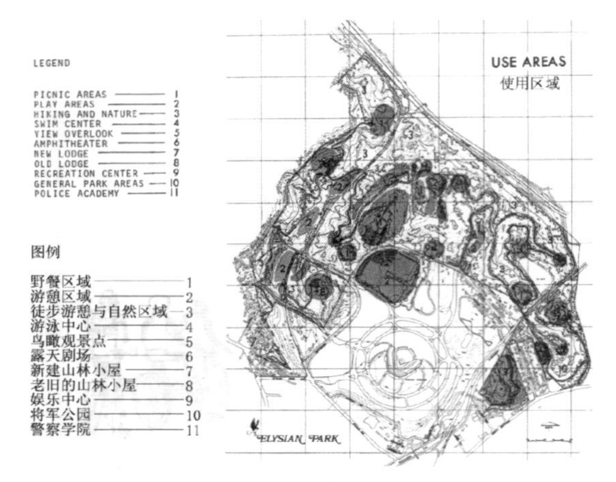

图6-4-3 伊利西安公园1971年总体规划的使用区域划分
Fig.6-4-3 Use areas of Elysian ParkmMaster plan in 1971

特征。休息点应该有风格淳朴的长凳或方便旅行者的日志，在合适的位置，应当将公园内的水体与饮水设施连接起来（图6-4-4）。

3）植物群落

规划将公园进行分区，并对每个分区进行特性分析，以确定哪些物种是不适宜种植的（图6-4-5）。公园的优势在于之前形成的生态环境，尽管看起来这一环境是无计划的，但是它却为植物群落的扩展提供了出发点。在许多区域，规划修复了原有的植物群落。我们所处时代的社会氛围要求那些动植物群落丰富区域的健康存在，并且被很好地维护。对大多数区域来说，所选择种植的植物

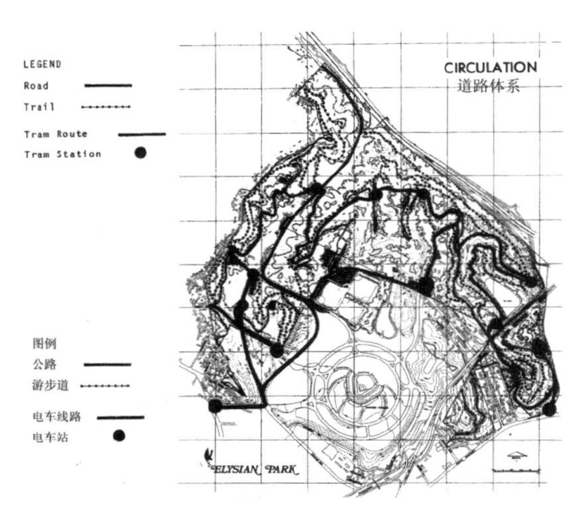

图6-4-4 伊利西安公园1971年总体规划的道路规划
Fig.6-4-4 Circulation of Elysian Park master plan in 1971

extended for drinking fountains(Fig.6-4-4).

3) Plant Communities

Planning divided the park into zones,each of which will be analyzed for its own merit with planting generally restricted to related species(Fig.6-4-5). The real advantage now existing in the park is the previous development of an ecological environment which,although seemingly haphazard,has nevertheless generated the springboard from which we can begin our expansion of the plant communities.In many areas we will only be renovating that which already is there.The very social climate of our times cries out for places where rich experiences in flora and fauna can be found in pleasant and well maintained.For the most part the plants selected here have been given careful consideration as far as maintenance is concerned,but is is not fair to the majority who will come to this park to be deprived of beauty because of maintenance difficulties or the vandalism of the few.Therefor,when we have named a few things which will not stand up under trampling,we are recommending a number of different kinds of protection to preserve them (Fig.6-4-6).

It will be necessary to prepare the soil properly by deep ripping in areas where shallow soil is encountered and when it is possible.downed trees and brush should be put through a chiper and the results returned to the soil as mulch,thus building an even deeper humus covering for the great variety of plants which will grow here.

4) Irrigation

The function of the water plan are to supply the water necessary to sustain planting in all areas of the park and to provide adequate water supplies for fire hydrants throughout the park.

The fist stage is a new water pumping station to be installed near the riverside drive entrance to provide

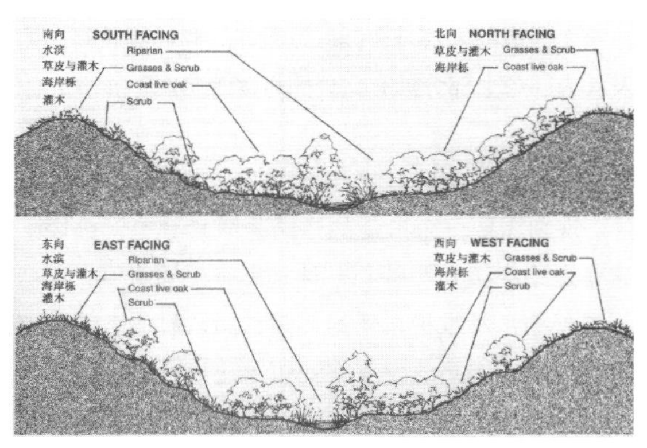

图6-4-5 对海岸栎的生境调查
Fig.6-4-5 Investigation to the habitat coast live oak

已经充分考虑了养护的问题，但是由于养护的困难和故意破坏而剥夺大多数人观赏美丽景色的权利是不公平的。因此，规划明确了一些不允许被践踏的植物，同时也提出一系列保护措施来保护它们（图6-4-6）。

当土层稀薄时，通过深层撷取的方式准备土壤是必要的。腐朽的树木和灌木应该被收集作为土壤的覆盖物，这样可以为大量多样的植物生长区域提供较厚的腐殖质覆盖层。

4）灌溉系统

用水规划的功能在于维持整个区域植物的生长需水，同时也用于提供公园抵御火灾的消防用水。

第一阶段是滨水车行道入口附近安装一个新的水泵站，可以提供750 000加仑（约2839立方米）的水量。高度开发区域的新的灌溉或自动喷水灭火系统应实现完全自动化。灌木区域应该利用灌木层顶的自动喷淋系统进行灌溉。公园的其他区域应被作为低强度开发区域和农业类型的灌溉体系。本土植物范围内只在夏天和发生特大旱灾进行灌溉，主要为保护重要的树或灌木覆盖。此外，灌溉系统不应该被频繁使用，除非干旱已久（图6-4-7）。

2. 2003年伊利西安公园扩展地块规划设计

1971年的总体规划提及扩展伊利西安公园的范围及申请扩展用地的内容，因此在2003年，规划了位于伊利西安公园北侧，滨河车行道西侧的扩展用地（图6-4-8，图6-4-9）。规划在尊重现状场地条件的基础上，分析了该地块的地质条件、植被覆盖情况，设计了将该公园与伊利西安公园主体区域联系起来的道路系统，使其成为伊利西安公园的重要组成部分（图6-4-10）。

四、 伊利西安公园总体规划

1. 公园的娱乐和游憩

伊利西安公园是当地社区的后花园，其对于周边社区而言也是一个区域性公园，同时它也是洛杉矶市的观光胜地。公园复杂多样的自然地势加剧了公园内

water to the 750000 gallon tank .New irrigation or sprinkler systems in the more highly developed areas should be fully automated .Shrub areas should be irrigated by standard shrub head sprinkler system. All other parts of the park should be considered low development areas and an agricultural type irrigation system .The native areas should be watered in summer only during times of extreme drought and then primarily to save important tree groves or heavy shrub cover.Then the system should not be used again unless the drought has been long(Fig.6-4-7).

2. 2003 Elysian Expansion Parcel on Riverside Drive

In 1971,The master plan refer to the content of extending the boundary of elysian park and applying for land.Thus,in 2003,the expansion parcel was planned which was in the riverside driver west and in the elysian north(Fig.6-4-8,Fig.6-4-9).Basing on respecting for the site conditions,planning analysed

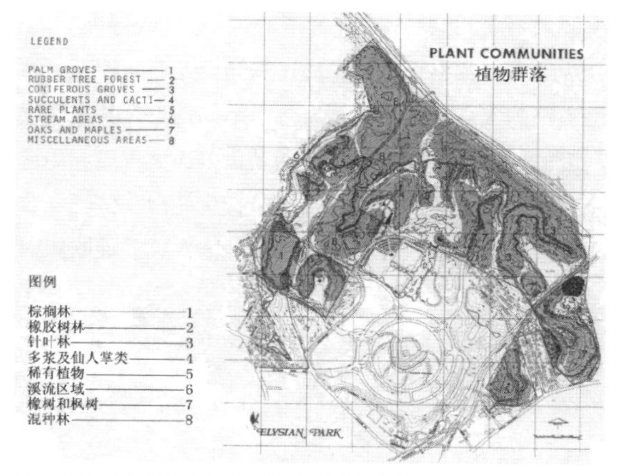

图6-4-6 伊利西安公园1971年总体规划的植物群落设计
Fig.6-4-6 Plant communities of Elysian Park master plan in 1971

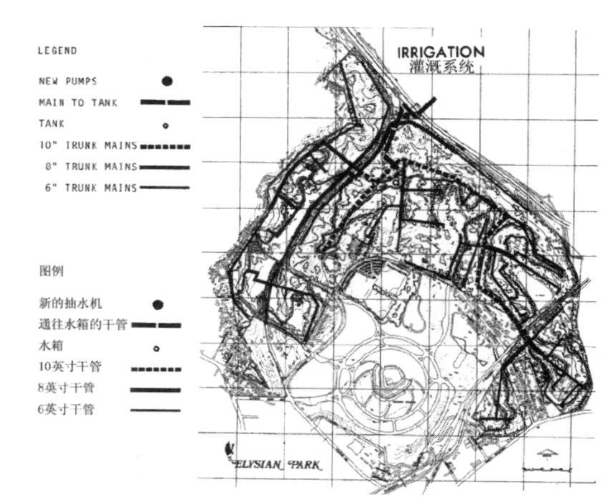

图6-4-7 伊利西安公园1971年总体规划的灌溉体系规划
Fig.6-4-7 Irrigation plan of Elysian Park master plan in 1971

部更多的消极场地与更少的积极场地的发展。时至今日，公园提供了四个球类活动场地，其中两个主要用于东北部小型联赛，此外还有一个篮球场、三个网球场、七个锻炼设施场地、众多的野餐区域和一段大约八公里的多样化的游憩道路（图6-4-12），考虑到在公园内有限的动态游憩活动的可能性，对现有设施的维护和利用最大化就变得十分关键，规划的这一部分回顾了当下正被使用的公园设施的情况，并从以下方面对未来的改进提出建议。

未充分利用区域：改进提高公园中里欧波利蒂野餐区、伊利西安水库、天使峰野餐区和电台山的利用率。运动场地：更新现存位于里欧波利蒂和索拉诺峡谷的垒球场。野餐区域：在伊利西安运动场上方开辟新的野餐区域。适应性游憩中心：构建新的露天休息室，翻新现有的凉亭以适应各种活动的需求（图6-4-11），提供团队野餐用的烧烤架。观景点：设计观景点以突出场所的特质，重新设计鸟瞰视点。游憩道路的端点：重新设置游憩道路的端点，并提供便利设施。慢跑道路：设计活动项目以最大程度利用慢跑道路，设计环形路线，提供多样化的景观体验。公园中的狗：在新设置的道路端点处提供狗粪便收集器

the situations of Geological conditions and vegetation coverage and designed the trails connecting this park and the main part of elysian park,making the park an important part of elysian park(Fig.6-4-10).

IV. Elysian Park Master Plan

1. Fun and Recreation in the Park

Elysian Park acts as a backyard to local neighborhoods, a regional park to surrounding communities and a tourist attraction for the City of Los Angeles.The complexity and variety of the Park's natural topography has led to the development of more passive play areas and fewer active ones. As it exists today, Elysian Park offers: four public ball fields, two of which are used primarily by Northeast Little League, one basketball court, three tennis courts, seven play equipment structures, numerous picnic areas and approximately eight miles of diverse trails(Fig.6-4-12).Given that active recreation possibilities are limited within Elysian Park, it becomes critical that existing facilities are maintained and used to the fullest potential.This chapter reviews the Park's existing facilities "pinpoints" areas of under use, and makes recommedations for future improvements for the following.

UNDERUTILIZED AREAS:Apply improvements to increase Park usage of Leo Politi Picnic Area, Elysian Reservoir, Angels Point Picnic Area, and Radio Hill. SPORTS FIELDS:Renovate existing softball field at

图6-4-8 滨河车行道及体育场路公园区位
Fig.6-4-8 The location of Riverside Drive Stadium Way Park

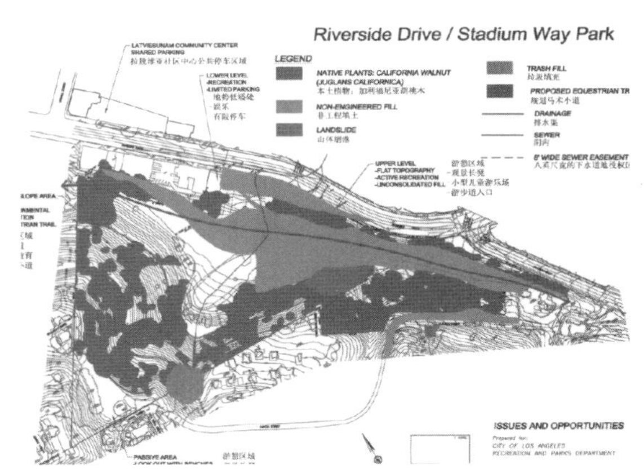

图6-4-10 滨河车行道及体育场路公园的问题与机遇
Fig.6-4-9 The issues and opportunities of Riverside Drive Stadium Way Park

图6-4-9 向西北方向看的公园斜视图
Fig.6-4-9 Oblique view to park site, looking northwest

图6-4-11 伊利西安公园的凉亭
Fig.6-4-11 Arbor in Elysian Park

具。活动项目：作为对社区需求的回应，在利用现有公园设施的基础上开展活动项目。通过公众参与的形式开展新活动项目，以充分利用设施及游憩道路。获取用于活动开展的土地。

2. 游赏公园

对于那些熟悉公园情况的人而言，在公园周边找路也是一件很具挑战性的事情，原因主要在于公园的位置、公园内部用地结构及环绕公园的高速路及铁路的混乱。每条进入公园的道路都具备不同的外观，公园周边的地形比较复杂，同时环绕在公园周边的社区也具有不同的特征，此外，道路设计、高速路出入口及相关的标示也都侧重于指示前往道奇体育场，而不是伊利西安公园（图6–4–13）。

所有这些因素使得公园道路畅通对于行人和车辆的重要性凸显出来。所有前往伊利西安公园的游客应当能够轻松得知各种公园设施的方位，并进入到自然环境中去，同时在公园中感到安全和放松。因此，游赏公园的过程应当成为整个公园体验的重要组成部分。考虑到公园的交通是当地社区居民所关心的主要问题，同时也是其他活动正常开展的基

Leo Politi.Renovate existing softball field at Solano Canyon. PICNIC AREAS:Develop and implement a new picnic area above Elysian Fields. ADAPTIVE RECREATION CENTER:Construct a new exterior restroom. Reconstruct the existing arbor to accommodate program activities.Provide group picnic barbeque(Fig.6-4-11). VIEWPOINTS:Design viewpoints to enhance the special and unique quality of each site.Re-design Point Grand View overlook.TRAILHEADS: Establish trailhead locations. Add amenities at each trailhead.RUNNING/JOGGING TRAILS:Create programs and events that utilize the trail system to the fullest extent possible.Create new trail loops and a variety of trail experiences.DOGS IN THE PARK:Provide dog waste dispensers at established trailheads.PROGRAMMING: Respond to community needs. Expand programs that will be supported by the existing facilities offered in Elysian Park. Implement public partnerships to establish new programs that would enhance the usage of existing facilities and trails. Acquire land to support expanded programming.

2. Getting around the Park

Finding one's way around Elysian Park is challenging even to those who know the Park well. The reasons why are primarily due to the Park's location, the fragmentation of it s lands, and the tangle of freeway and railways that wrap the Park. Each major entryway into Elysian Park has a completely different look and feel due to the complexity of the topography and the

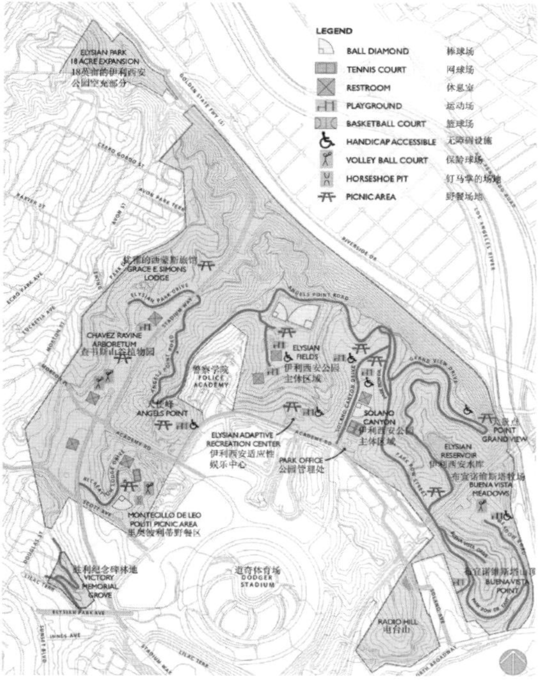

图6–4–12 现有的娱乐设施
Fig.6-4-12 Existing recreation facilities

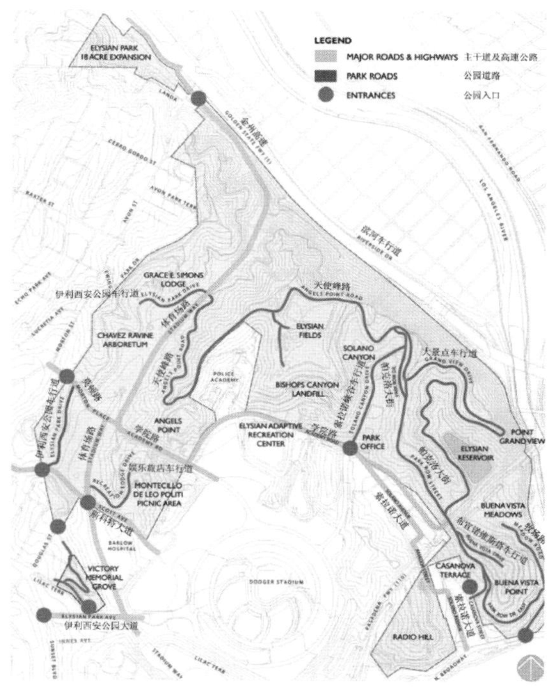

图6–4–13 现状道路
Fig.6-4-13 Existing circulation

础，这一部分所包含的行动方案和建议如下。

道路交通：修缮从滨河道路到学院路之间的体育场路，进而提供一个从伊利西安山谷进入公园的人行入口，增加公园使用的安全性，降低行车速度，同时提供一条具备审美情趣的园路。公园道路的车道宽度标准化，以减少除设计区域以外路段的路边停车（图6–4–14）。

停车：在设计区域内，沿着靠近游路端点、野餐区域和观景点的公园道路设置大量的小型豆荚式停车场（图6–4–15）。为方便使用公园设施，在公园内提供停车的机会。游路、交叉口和阶梯：构建一个初步的游路体系，为公园的多元化体验提供服务。通过利用桥、道路交叉口、步行道、阶梯和附加游路将现有的游路体系联系起来。使公园内部道路与外部的现有几规划的道路和出入口联系起来。标示系统—寻路、定位、识别和道路标志：详细叙述现有由洛杉矶游憩和公园管理部门所批准的寻路和说明性标示。在整个公园内部提供游路端点标示和游路距离标示。进入公园的通道：在已经被大众认同的公园入口处提供入口标示。

3. 公用场地

基于对加利福尼亚乡土景观的初步描述，伊利西安公园已经被密林、沿海低矮的灌木和海岸

differing characters of the communities that border the Park. In addition, the road design, entrances and exits from the freeways, and their associated signage are geared to accessing Dodger Stadium rather than identifying the Park(Fig.6-4-13).

All these factors elevate the importance of Park circulation to both pedestrian and motorist. All visitors arriving at Elysian Park should be able to easily locate park facilities and access the natural environment. While in Elysian Park, people need to feel safe and be able to relax in the park surroundings. The process of getting around the Park should be an integral part of a great park experience.Given that circulation is of primary concern to the local community and is the basic infrastructure that supports all Park activities, this chapter includes plans of actions and recommendations for:

ROADS AND TRAFFIC:Modify Stadium Way from the Riverside Drive to Academy Road to provide pedestrian access from Elysian Valley, increase park user safety, reduce speeds, and provide a park road aesthetic. Standardize park road lane width. Reduce roadside parking except at designated areas (Fig.6-4-14).

PARKING:Pod Parking Develop small parking lots (pod parking) in designated areas along park roads adjacent to trailheads, picnic areas, and viewpoints.(Fig.6-4-15)Develop parking opportunities throughout the park that will support full utilization of all park facilities. TRAILS, CROSSINGS & STAIRWAYS:Develop a premier system of trails for a variety of experiences that utilizes the entire Park.Connect the existing trail system together within the Park with the addition of bridges, road crossings, walkways, stairways, and additional trails.Connect trails within the Park to existing or future trails and access points outside the

图6–4–14 设计后的道路
Fig.6-4-14 Designed trail

图6–4–15 豆荚式停车场
Fig.6-4-15 Pod parking

栎和胡桃林所覆盖。在1796年，这些植被的分布由坡度、海拔、地形和土壤类型综合决定（图6-4-17）。

一个世纪之后，放牧和资源开采使得这一地区的景观呈现出裸露和剥蚀的外观。1886年，将伊利西安公园作为一个公众享有的公园的决定以树木的种植和大量桉树的引进为开端。1893年，洛杉矶园艺界在伊利西安公园构建起了南加利福尼亚州的第一个植物园。在查韦斯山谷的植物园中，最初的种植主要是一些栗树和橡胶树。两排并行的棕榈树种植形成了现在的棕榈树林荫大道，这些棕榈树是在1895年到1900年期间种植的。

在之后的几十年中，大量的外来树种被种植在公园内，这包括了喜马拉雅雪杉、橄榄树和大量的桉树。近期，大火破坏了位于加州高速路上方的具有历史意义的喜马拉雅雪杉林。大火之后，乡土景观的遗存重新显现出来，最突出的就是位于面向北方的斜坡上的加利福尼亚胡桃树。通过改造铁路站场，在伊利西安公园周边出现了新的休憩用地，新的州立和市立公园、新的游路和洛杉矶河流廊道复

Park. SIGNS—WAYFINDING, LOCATION ,IDENTIFICATION AND TRAIL MARKERS: Expand the existing wayfinding and interpretive signage plans approved by Los Angeles Recreation and Parks Department for Elysian Park.Provide trailhead signage and trail mileage markers throughout the Park trail system. GATEWAYS INTO THE PARK:Provide entry signage at identified locations to Elysian Park.

3. THE PARKLAND

Based on the first written descriptions of native California landscapes, the slopes of Elysian Park must have been clothed in chaparral, coastal sage scrub, and Coast Live Oak and Walnut woodlands. In 1796 their distribution was determined by combinations of slope aspect, elevation, topography, and soil type(Fig. 6-4-17).

A century later the influences of grazing and resource extraction had left the landscape bare and denuded. In 1886, once Elysian Park was dedicated as a Park for public enjoyment, beautification began with tree planting and the introduction of several thousand Eucalyptus trees. The Los Angeles Horticultural Society established the first botanical garden in Southern California in Elysian Park in 1893. Among the original plantings in the Chavez Ravine Arboretum were a magnificent cape chestnut, several expansive Tipu trees and a grove of exotic Rubber trees. The double row of Palms along what has become known as the

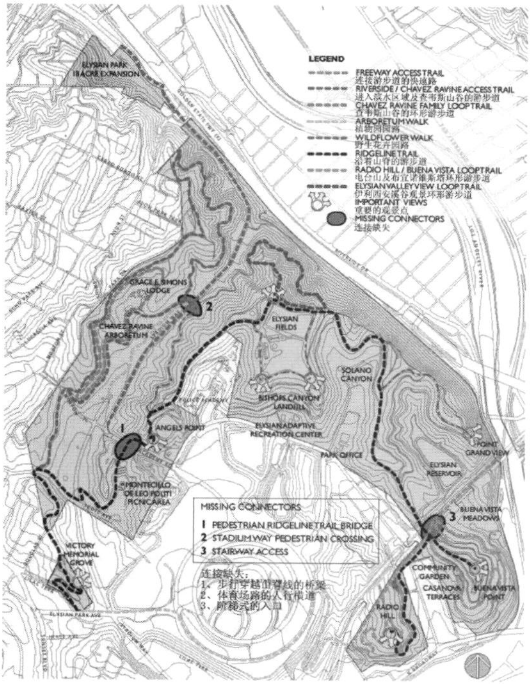

图6-4-16 规划道路体系
Fig.6-4-16 Proposed trail system

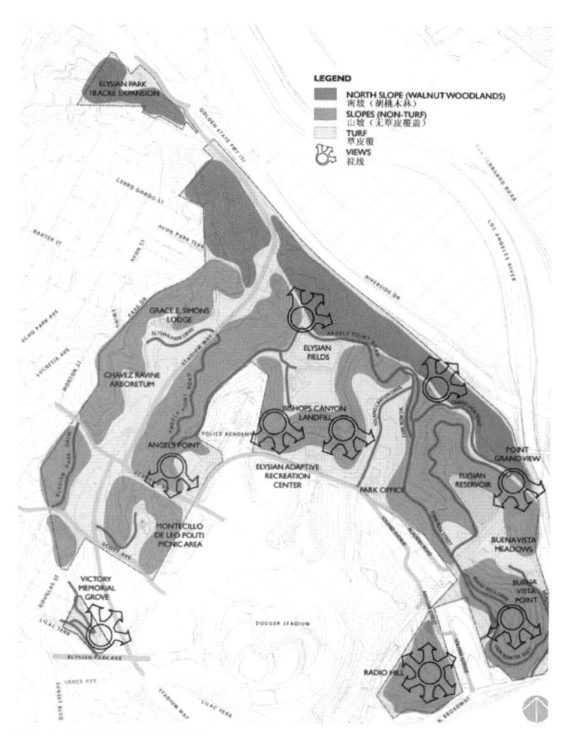

图6-4-17 现有植被及俯瞰点
Fig.6-4-17 Existing vegetation and overlooks

兴规划与周边的社区复兴协同实施。公园的容量和公园周边休憩用地的规划使得伊利西安总体规划必须抓住机遇融入更大的区域环境中。这部分内容主要包括：公园景观、动植物群落、本土植被群落构建和恢复、伊利西安公园周边土地的利用、伊利西安公园与周边环境的联系。

1) 行动方案

我们在现实中将会发现一些很讽刺的事情，在过去的一个世纪里，公园内的乡土植物与外来植物相比，变得越来越稀少。行动方案的目标是构建一个乡土植物占主体的公园，这样可以使公园在很长一段时间内维持自身的发展，同时可以在提供一个能够展示历史文化和景观特征的环境的基础上减少养护成本。查韦斯山谷植物园和棕榈树林荫大道构成了最重要的具有历史纪念意义的植物资源，这些资源的价值不仅仅在于其壮硕的大树，同时它也描述了在洛杉矶城出现之初的文化视角。因此，保留外来树种的精华，恢复和改善乡土植被群落将会强化公园的历史精髓（图6–4–18）。

2) 行动款项

强化公园本土植被群落的生物学价值；引导市民参与到本土植被群落恢复知识的学习和实施过程中；创造一个能够长期提供生物栖息地的自我维持的乡土景观；保护和强化所选择由外来树种构成的具有历史意义的林地；将伊利西安公园与其他城市公园和道路及洛杉矶河的区域野生动物廊道连接起来；支持土地征用，以完善公园功能，满足最适宜使用与享受为目的的审美活动需求。

4. 公园养护

伊利西安公园提供了一个直接临近洛杉矶的巨大的多样的休憩用地。随着选择在伊利西安公园周边区域居住和工作的人口的增多，对在公园内休憩放松的需求也势必会持续增长。使用需求的增加要求类似灌溉系统之类的养护体系实现充分的自动化，这样工作人员才能够更好地操作公园的维护工作。同时，公园养护也意味着保护那些对于洛杉矶

Avenue of the Palms, were planted sometime between 1895-1900.

Over subsequent decades an array of other exotic species was planted throughout the Park, including groves of Deodar Cedars, Pines, Olives and more Eucalyptus. Most recently, fire has transformed the historic bank of Deodar Cedars above the Golden State Freeway. In the wake of the fire, remnants of the native landscape have re-emerged most notably the California Walnuts characteristic of such north-facing slopes.Also re-emerging around Elysian Park are new open spaces reclaimed from railyards, new State and City parklands, new trails, and plans for a Los Angeles River corridor revitalization, developed in conjunction with the revitalization of adjacent communities. The sheer volume of park and open space planning surrounding Elysian Park necessitates that this Master Plan look at opportunities to connect Elysian Park to a greater regional context. This chapter addresses: Park landscapes,flora and fauna, native plant communities and restoration,land outside elysian park, connecting park lands.

1)Action Plan

Some irony may be found in the fact that native vegetation has become more rare and desired in the Park than the exotic species imported there a century ago. The objective of the Plan is to establish a unified native plant parkland sustaining itself over time,reducing maintenance costs while providing the context in which to show off the historic and iconic landscape features. Chavez Ravine Arboretum and the Avenue of the Palms constitute the most prominent and valuable historic vegetation resources in the Park. These resources are valuable not just for the impressive maturity of the trees, but for what they tell us about cultural perspectives at a time when the City of Los Angeles was emerging. Saving the best of exotic ornamentals and restoring and improving the native plant communities will strengthen the best of the Park's history (Fig.6-4-18).

2)Action Items

Enhance the biological value of the native plant communities in the Park;Include citizens in restoration knowledge and implementation:Create a self sustaining native landscape over time that supports habitat;Preserve and enhance historic groves of selected exotic trees;Link Elysian Park to other City parks and trails, and to the regional wildlife corridor of the Los Angeles River; Support land acquisition to complete the Park function and aesthetic for optimum use and enjoyment.

4. Taking Care of the Park

Elysian Park provides a very large and very diverse open space directly adjacent to downtown Los

市民具有重要价值的历史。

植物园、卡萨诺瓦平台和里欧波利蒂野餐区都是重要的历史遗存。在人们俯瞰洛杉矶城时，观景点提供了一个解读城市的首选位置。

老旧的墙体、长凳和构筑物为未来的设计和相同要素在公园内的应用提供了动力来源。对公园的保护远远重要于对公园的再发掘和周边土地的增值，养护公园关系到未来。这部分涉及以下内容：

1）维护公园。

构建新的公园配水系统。利用先进的技术建造自动化的灌溉系统。保持一个干净的，管理良好的公园。保护资源来支持预期的改善。

2)保留历史遗存。

强调历史特点和独特的公园资产。在观景点继续开展诠释工作。为支持植物园的发展提供资源。

3)制定设计标准。

将林带扩展到公园的边界，扩展包含诠释要素

Angeles. Greater demands on the Park for relaxation and recreation are inevitable as more people choose to live and work in the near vicinity. As usage increases the need for Park upkeep and systems that can be automated such as irrigation should be fully implemented. Park staff will then be better positioned to handle the maintenance of a premier Park. Taking care of the Park also means taking care of the history that is so valuable to Los Angeles residents.

The Arboretum, Casanova Terrace, and the Leo Politi Picnic Area provide graceful remnants of past history.Scenic viewpoints provide prime locations for interpretive signage as many overlook the urban patterns of greater Los Angeles.

Examples of old walls, benches, and structures provide the impetus for future design and implementation of those same elements to be applied through out the Park. Protection of existing Park land is more important than ever as Elysian Park gets re-discovered and its value increases along with the lands around it. Taking care of the Park is about the future. This chapter addresses:

1) Maintaining the Park

Construct a new water distribution system for Park

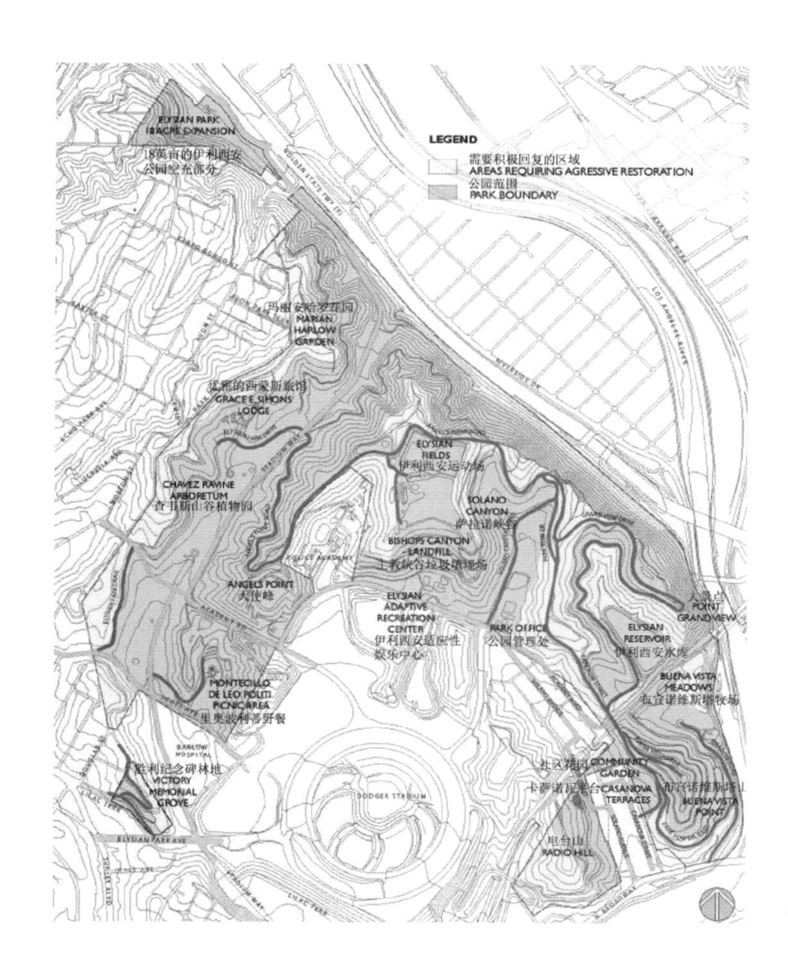

图6-4-18 植被优先恢复区域
Fig.6-4-18 Restoration priorities

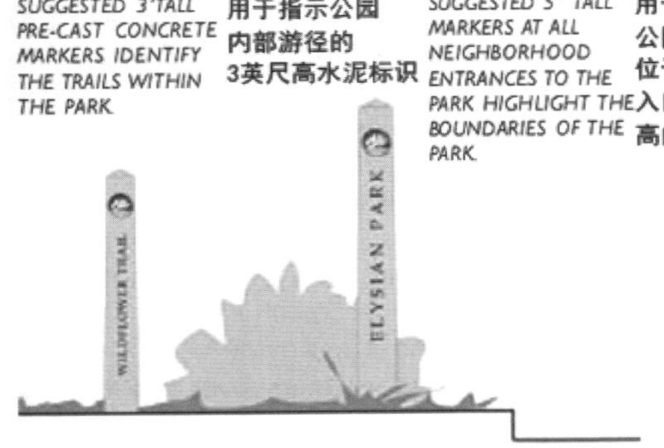

图6-4-19 公园中的标示及标记
Fig.6-4-19 Sighs and markers in the park

的标示计划（图6-4-19）。为野生动物维持一个相对荫庇的空间，照明的运用主要考虑安全因素。在适当的情况下运用复古风格的设计和材料。

五、 主要场地设计

1. 天使峰

天使峰野餐区是一块位于垃圾填埋地顶端，拥有大量舒适餐桌，相对平整的草坪。覆盖垃圾填埋地的填充土壤使得该地区的植被种植面临不确定性。树木及草皮的维护变得困难。附近的路边停车的不便使得游人需要带着沉重的冷藏器、野餐篮和孩子走很长一段距离才能使用该区域的服务设施。这些固有的问题使得该区域难以被成功利用（图6-4-20）。

use.Build a new automated irrigation system that takes advantage of new technologies.Maintain a clean, well-managed park. Secure resources to support the desired improvements.

2) Saving Places with History

Emphasize the historic features and unique assets of the Park. Continue to develop interpretive opportunities at scenic viewpoints. Provide resources to support the Arboretum.

3) Setting Design Standards

Expand tree lined arterials to Park boundaries. Expand sign program including interpretive elements(Fig.6-4-19). Maintain a "dark Park" for the benefit of wildlife. Lighting should be primarily for security. Apply historic design and materials where appropriate.

V. Major Site Plan and Design

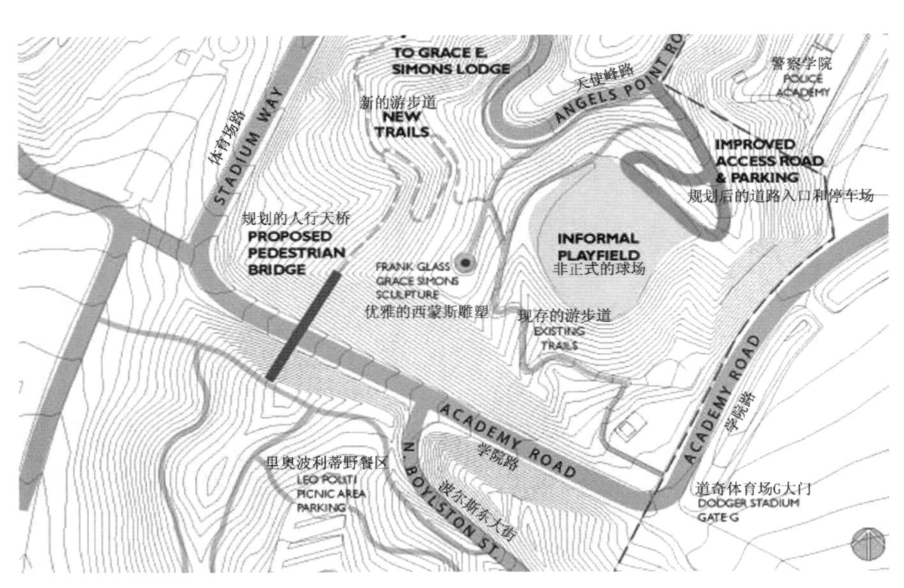

图6-4-20 天使峰野餐区提升规划
Fig.6-4-20 Angels Point Picnic Area improvement plan

建议如下：改善路况及邻近野餐区域的停车设施。创造非正式的游憩场地。修复垃圾填埋地的表层覆盖土壤以服务于植被的恢复。使用公园的绿色废弃物及山体崩坏所导致的道路修复工程中所获得的公园表层土壤，重新构建服务设施周边的土壤结构。

2. 战争纪念地和卡萨诺瓦平台

现存的战争纪念碑被浓密的植被所遮挡，或位于缺乏吸引游客视线的区域。历史公园剩余的部分也渐渐被灌木丛所覆盖，正慢慢从人们的视线和记忆中消失。实质情况是年久失修使得公园变得缺乏自身特色和公众意识。

建议如下：对现有的纪念碑和公园进行场地规划，以突出它们的特色；重新设计胜利纪念地的空间布局以更好的突出纪念碑。修建通往琼斯纪念地的道路，提供坐椅。修复卡萨诺瓦平台现存的墙体。在卡萨诺瓦平台提供标示解释；突出伊利西安公园的历史特征；改善维斯塔点的寻路标识；提供描述性标识和观景望远镜（图6–4–21）。

3. 棕榈大道

在周末和节假日，公园中大量的游人选择在树荫下聚集。这对树木根系及树干底部的影响十分明显。

建议如下：委任一名知名的植物学专家对树木的健康状况进行评估，并提供必要的措施维护这些树木；制定停车位置与树木根部的安全距离，对树木进行特征描述以突出其生物及历史特性（图6–4–22）；补种缺失的树木，沿着体育场路补种树

1. Angels Point

The Angels Point Picnic Area is a relatively flat meadow on top of a landfill with a large number of tables suitable for large group picnics. The fill soils capping the landfill have made plant establishment problematic. Efforts to maintain tree and meadow areas have been difficult and often unsuccessful.The available parking is limited to the roadside, requiring visitors to lug heavy coolers, picnic baskets,and small children a long distance to use the amenity.These inherent challenges have kept Angels Point Picnic Area from being fully successful(Fig.6-4-20).

Recommendations: Improve access road and small parking lot adjacent to the picnic area. Create informal playfield. Remediate disturbed landfill cap soil for improved restoration planting opportunity. Use Park green waste and any available Park topsoil from trail landslide repair to build new soil structure around new facilities.

2. War Memorials and Casanova Terrace

Existing war monuments are hidden behind overgrown plantings or sited in areas without strong sightlines that draw visitors to them. Remnants of historic Park structures are slowly being covered with bush and allowed to disappear from view and memory. The net effect is that years of deferred maintenance have created a park with few distinguishing features and diminished public awareness.

Recommendations: Develop site plans for existing monuments and park features that will give them more prominence:Redesign Victory Memorial site layout to better celebrate the monument.Construct a pathway to Jones Memorial and provide seating. Restore the retaining walls of Casanova Terrace. Provide interpretive signage at Casanova Terrace;Increase the prominence of Elysian Park's history;Improve way-finding signage to vista points;Install interpretive signage, view tubes(Fig.6-4-21).

3. Avenue of the Palms

On weekends and holidays throngs of visitors

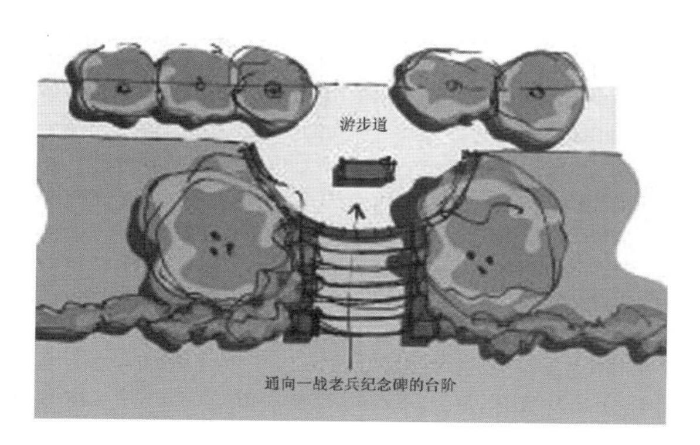

图6–4–21 胜利纪念碑的改进设计
Fig.6-4-21 Victory Memorial improvement

图6–4–22 对棕榈树大道的保护
Fig.6-4-22 Protection for the Avenue of the Palms

图6-4-23 游步道端点的设施
Fig.6-4-23 Trailhead amenities

木，以强化公园特性；种植新的棕榈树以培育新一代树木。

4. 游步道端点

伊利西安公园缺乏标示清晰的游步道端点。相交游步道缺乏指示性的标志和标识，而这些是游步道设施的基本组成（图6-4-23）。

建议如下：在靠近豆荚式停车场或其他停车设施的位置设置游步道端点； 在重要的游步道端点处设置包括垃圾箱、饮水设施、标识系统、道路名称标志和狗粪便收集器在内的设施；在植物园道路沿线提供对植物历史和习性进行说明的标示。

六、 总结

伊利西安公园总体规划强调公众参与，重视“足下文化”，关注普通市民的生活需求。建成后的公园，将成为洛杉矶市绿地系统中最大的绿色生态斑块之一，动植物资源丰富，是“城市绿肺”和“都市氧吧”。同时，也扮演着保护生物多样性的角色。伊利西安公园为城市居民提供户外休闲、观赏、游戏、运动、娱乐的空间，使每个人都有机会享受自然的生活，陶冶精神。

场地本身历史底蕴深厚，文化渊源可追溯至18世纪，有丰富的文脉线索可挖掘，这也为设计师的创意提供了永不枯竭的源泉。总体规划重视公众可达性，设计游步道、自行车道等满足不同功能的游览线路，妥善解决与外围城市道路的衔接，合理布置入口及停车场地。规划重视标示系统的设计，为

looking for parking choose to pack together under the trees. Significant impacts to the roots and the base of trunks have occurred.

Recommendations: Commission a certified arborist to evaluate the health of the trees and the measures needed to sustain them;Define designated parking a safe distance from the root zones Design and implement interpretive features that will highlight the significance of these trees and their history(Fig.6-4-22); Replace missing trees and extend the allee along Stadium Way to reinforce the identity of the Park;Plant new row of Date Palms to start next generation of trees.

4. Trailhead

Well marked trailheads have not been developed in Elysian Park.. Markers and/or signage that would connect one trail to another are missing, as are basic trail amenities(Fig.6-4-23).

Recommendations:Establish trailhead locations adjacent to pod parking or parking facilities wherever possible;Include amenities such as trash receptacle, drinking fountain, signage, trail name marker, and dog waste dispenser at significant trailhead locations;Provide interpretive signage on the Arboretum trail specific to the history and plantings along the trail.

VI. Summary

Elysian park master plan emphasize the public participation, and pay attention to the "the usual civilization" and citizens' lives. When completed the park will become the biggest green patch in the green space system of Los Angeles where animal and plant resources are rich, it also will be the "city green lung" and "urban oxygen bar". At the same time, Elysian park also plays an important role in protecting biodiversity. Elysian park provides outdoor recreation, sightseeing, games, sports and entertainment space for residents , so that each person has a chance to enjoy the natural life and edify spirit.

Site itself has profound historical details , its cultural origin can be traced back to the eighteenth century,and there are context clues that can be excavated,all of those provide exhaustible source of ideas for designers.Master plan pays attention to public accessibility, properly tackling the connection between peripheral city roads and internal traffic in the park by designing landscape swim trails and bikeway which can fulfill different functions and reasonablely decorating entrance and car parking. Elysian park master paln alos pay attention to the design of the sign system which provide clear direction instructions for tourists.

In the planning scheme,controling invasion of exotic

游者提供清晰、明了的方向指示。

规划中，控制外来植物入侵也是难点之一，生物入侵是造成场地生物多样性丧失的主要因素。外来入侵植物一般具有很强的适应能力，生长迅速，抗逆性强，生物量巨大，控制的难度较大。本项目行动方案目标之一，就是构建乡土植物占主体的公园，唯此，才能实现公园的自维持和可持续发展。

plants is a difficulty,biological invasion is the main factors causing the loss of biodiversity. Alien plants generally have strong adaptability,rapid growth, strong resistance, huge biomass and is hard to be controlled. One of the goals of action plan is to build a park mainly covered by the local plants,only in this way can we realize the self-sustaining park and the sustainable development of the park.

思考题

1. 伊利西安公园总体规划自始至终强调公园与外部交通、公园与城市绿地系统及公园与居民社区之间的协调与连接，试思考这三种衔接如何通过规划设计得以实现。

2. 伊利西安公园总体规划是一个持续的衔接的过程性规划，从第一次总体规划开始先后经过多次提升规划的修编，这种编制方式和程序大大不同于国内的公园总体规划编制情形。试思考这种编制方式和编制类型对于国内类似公园规划的指导价值。

3. 伊利西安公园总体规划中对于植物群落的在科学的本土植物群落调研的基础上恢复了本土植物群落，同时保留了部分宝贵的外来物种。在国内的公园规划中，对于植物群落的构建缺乏科学合理的本土植物调查，植物配置规划也就过于随意，结合伊利西安公园总体规划的内容，思考如何真正做到生态的植物配置规划设计。

4. 随着国内人们生活水平的提高，饲养宠物并且带宠物散步成为越来越多人的休憩选择，国外公共空间的规划设计很好的考虑人们的这一需求，而在此方面，国内的大部分公共空间设计还很缺乏此类考虑，试考虑如何更好地适应人们的这种休憩需求。

5. 公共场所的大喇叭变成了易于辨识的标示系统，这成为社会文明进步的标志，但在国内，这种进步还仅仅局限于部分场所和空间，试考虑如何将标示系统与风景园林学科的众多实践类型进行融合。

Questions

1. From beginning to end,elysian park master plan emphasizes contacts between the park traffic and external traffic, the park and urban green space system , park and community.Please try thinking about how to realize the three cohesion through the planning and design.

2. Elysian park master plan is a continuous plan, from the first overall plan to successive promotion plan, the establishment ways and procedures have greatly differences to the domestic park master plan.Try considering the guidance value of this way and the type of planning formulation to domestic similar park planning.

3. Based on the research of native plant community,elysian park master plan restored the native plant community and retain part of precious exotic species.The park's planning in China lacks scientific and reasonable native plants survey for the construction of the plant communities, plant configuration planning is too casual. Refer to the content of the elysian park,think about how to make the ecological planning and design for plant disposition.

4. With the improvement of domestic people's living standard, keeping pets and sneaking pets gradually become an very common recreational choice.Foreign public space planning and design have good consideration on this demand, and in this respect, most of the domestic public space designs lack such considerations.Please try to consider how to adapt to this demand.

5. Public trumpets are replaced by the sign system,this becomes the sign of social civilization progress, but in our country, this progress is limited in some places and spaces. Please try to consider how to combine the sign system and landscape practices better.

注释/Note

图片来源/Image：图6-4-1，图6-4-11~图6-4-23来自《伊利西安公园总体规划文本》/Elysian Park Master Plan ；图6-4-2来自谷歌地图截图/Google Map；图6-4-2~图6-4-7来自《1971年伊利西安公园总体规划文本》/Elysian Park Master Plan in 1971 ；图08-10来自《滨河车行道、体育场路公园规划》 Riverside Drive/Stadium Way Park Pllan。

参考文献/Reference：www.laparks.org

第七章 生态技术的应用

CHAPTER VII APPLICATION OF ECOLOGICAL TECHNOLOGY

第一节 纽约总督岛公园及公共空间总体规划 Section I Governor Island Park and Public Space Master Plan
第二节 布鲁克林桥公园规划 Section II Brooklyn Bridge Park Planning
第三节 碳技术、浮桥藻类公园案例 Section III Carbon T.A.P.//Tunnel Algae Park
第四节 生态城——阿科桑底 Section IV Arcology — Arcosanti

第一节 纽约总督岛公园及公共空间总体规划

Section I Governor Island Park and Public Space Master Plan

一、 项目概述

总督岛保护教育公司的任务就是通过组织居住用地、休憩用地和公共设施来重新开发总督岛。在对民意代表及公共官员进行咨询之后，总督岛保护教育公司提出了一个多目标的策略，通过这一策略可以使总督岛获得新生。这包括：拓展公众使用的途径和土地的特色利用，投资于历史遗产的保护和岛上基础设施的建设，实行一个多阶段、功能复合的发展策略，打造一个世界级的公园和公共空间（图7–1–1）。

岛上公共空间的游览和使用将岛屿与城市生活紧密联系起来，而对岛上建设的持续投资对于岛屿的保护和开发是不可或缺的。打造一个世界级的公园和公共空间，是实现该岛屿从一个废弃的只能通过水路进入的军事基地，转变成一个区域性的充满活力的、功能复合的旅游目的地的必然途径。岛屿的开发和租赁将会分期进行，而第一阶段的开发主要着眼于教育和休憩使用。未来的租赁对象在较长时间内将由商业和非营利性的公司组成。新的用途和空间的开发将会依托岛屿自身的优势，这包括岛屿上极好的视野，港湾的特殊优势和现有的景观。

I. Overview

The mission of GIPEC(Governors Island Preservation & Education Corporation) is to redevelop Governors Island with a mix of tenants, open space, and public amenities. Working in consultation with civic leaders and public officials, GIPEC has articulated a multi-pronged strategy to bring Governors Island back to life: Expand public access and early signature uses of the Island;Invest in historic stabilization and Island infrastructure; Execute a multi-phase, mixed-use development strategy;Create a world-class park and public spaces (Fig.7-1-1).

Visitation and use as a public space connects the Island to the life of the City while ongoing investment in the physical fabric of the Island is essential for both preservation and new uses. The creation of a world-

图7–1–1 总督岛鸟瞰
Fig.7-1-1 Bird's-eye view of Governors Island

但是项目的开发也存在着巨大的挑战，无论是公共空间还是居住用地都必须满足多用途的使用。

岛屿实际上是由两部分组成的，南侧部分荒凉且平坦，拥有大片的沥青路面、稀疏的植被和发育不良的树木。这一区域缺乏公众活动所需要的设施。然而许多现存的或将来的建筑已经被规划好，潜在的租赁者期盼着岛上环境的转变。这一公园和公共空间总体规划将是释放总督岛发展潜能的多期计划中的第一步。

二、 场地介绍

总督岛是一个位于纽约港中心的172英亩的岛屿。这个小岛坐落在距离曼哈顿南部800码（约731.5米），距离布鲁克林区400码的位置。这里最初是德拉瓦人的定居点，之后变成了荷兰在纽约最早的殖民地，在殖民时期，它成为英国总督的驻地。在作为聚居地存在了大约两个世纪之后，总督岛变成了一个活跃的军事基地，在1800年到1960年间，这个基地归属于美国陆军，之后为美国海岸警卫队使用。1996年，美国海岸警卫队撤离了这个岛屿。

这个岛的北部拥有建造于1810年至1940年间的房屋、砖结构的构筑物和炮台，因而这一区域在1985年被宣布成为一个国家历史地标。在1996年，这一总面积为92英亩（约37.2公顷）的土地包括了当地的地标保护区域。

岛的南部是在1900年通过垃圾填埋形成的，总面积80英亩（约32.37公顷），在这一区域有着大量的非历史性建筑。在2003年，布什总统授权了包括两座建于1812年的炮台在内的22英亩（约8.9公顷）土地为岛上的历史保护区，其现在作为一个国家纪念地由国家公园管理处管理。在2003年，联邦政府将岛上剩余的150英亩（约60.7公顷）土地转让给了总督岛保护教育公司。转让契约规定该岛屿的开发主要用于为公众提供可进入和使用的公共空间，服务于教育和文化用途，并要求限制其他将来的开发利用。同时，契约还要求开发在经济上是可持续的。总督岛保护教育公司作为公共实体负责整个项目的操作、规划及开发。这个岛屿在2003首次对公众开放。

class park and public space on Governors Island is essential to the transformation of the Island from an abandoned military base, accessible only by boat and closed to the public, into a vibrant, thriving mixed-use destination for the region. Development and tenancy of the Island will proceed in multiple phases, with the first phase bringing a discrete group of educational and recreational uses. Future tenancies will include a mix of commercial and not-for-profit enterprises over time. The creation of new uses and spaces takes advantage of the Island's attributes, including its extraordinary views, unique vantage point on the Harbor and established landscapes. But it must also address the challenges of the Island, both as a public space and future home for a broad array of uses.

The Island is in effect now two islands—the south is desolate and flat, with acres of asphalt and sparsely planted, stunted trees. It lacks many of the public amenities necessary for year-round public usage. And, while many possible uses for the Island's existing and future buildings have been contemplated, potential tenants await the Island's transformation. The Park and Public Space Master Plan is the first phase of a multi-phase process to unlock the potential of Governors Island.

II. Introduction of the Site

Governors Island is a 172-acre island in the heart of New York Harbor. It is located 800 yards from Lower Manhattan and 400 yards from Brooklyn. First settled by the Lenape, then the site of the first Dutch settlement, the Island was home to the British royal governor during the colonial era. For almost two centuries after these initial settlements, Governors Island was an active military base, first home to the U. S. Army from 1800 to the 1960s and later the U.S. Coast Guard. The U.S. Coast Guard left the Island in 1996.

The Island's northern portion, which includes houses, brick structure and forts dating from 1810 to 1940, was declared a National Historic Landmark in 1985. This district, totaling 92 acres, received local landmark protection in 1996.

The southern part of the Island, created with landfill in the 1900s, totals 80 acres and features numerous non-historic buildings. In 2003, President Bush designated 22 acres of the Island's historic district, including two 1812-era forts, as a National Monument that is now managed by the National Park Service (NPS). In 2003, the Federal government deeded the remaining 150 acres of the Island to the Governors Island Preservation and Education Corporation (GIPEC). The Federal deed stipulated requirements for public access and public space, educational and cultural uses and restricted other future uses. It also established the expectation that the Island would become economically

三、 地形设计

总督岛保护教育公司意图通过构建公园与公共空间，实现岛屿的转变。为了达到这个目的，同时使这个岛获得新生，西8城市规划和景观设计事务所在总体规划中的关键概念是通过吸引眼球的方式塑造岛屿南部的地形，最终创造一个具有丰富体验、视野和茂密植被的真实的、长久的景观。

景观的重塑将之前被遗弃的南部区域与现在正被使用的北部历史保护区结合起来。这一思路对于构建一个复合功能的游憩目的地是必不可少的，同时其对于公共空间的使用和岛屿的未来发展也至关重要。

在20世纪之交，这个岛屿的面积只有现在的一半那么大，在1901年和1911年，超过四百万立方码的填充材料被对方到岛的南端，岛的面积从92英亩翻倍增长至172英亩（图7–1–2）。这些人造景观区域平坦，而且几乎没有树木覆盖，而岛的北部却呈现出相反的绿树成荫的景象。这一戏剧性的差异一直延续到今天。当人们参观这个岛时，可以漫步在绿树成荫的历史保护区。走到南端时，将会经历一种极具吸引力的不同体验，人们会看到一个80英亩的平坦、荒芜的景观。在这里只有很少的树木和绿荫为你遮挡太阳和海风，大部分区域处于到2100年为止的百年一遇的洪水水位。

在岛屿实体外观第一次转变的基础上，西8团队的设计提供了二代地形转变的建议。数量众多的填充材料将被用于在公园和公共空间内塑造空间，这些材料来自对岛上拆迁物的再利用，也有一部分来自场地之外（图7–1–3）地形的转变为进行多种多样的活动提供了条件，同时也在实体外观和可塑造性上提供了机遇，这种转变创造了丢失已久的场

sustainable. GIPEC is a public entity established to operate, plan and develop the Island. The Island was open to the public for the first time in 2003.

III. Topographical design

GIPEC seeks to catalyze the overall transformation of the Island beginning with the creation of the park and public space. To support this aim and bring the Island back to life, the West 8 Team's key concept for the Master Plan design is to dramatically sculpt the southern part of the Island —transformation through topography —to create a true and lasting landscape with a rich array of experiences, views, and settings for trees and plants.

The re-shaping of the landscape integrates the abandoned southern part of the Island with the northern Historic District that many visitors already use today. This deliberate approach is essential to create a thriving mixed-use destination, both for public space uses and the future of the Island.

At the turn of the 20th century, the Island was half the size it is today. Between 1901 and 1911, more than four million cubic yards of fill material were added to the Island's southern end, nearly doubling its size from92 acres to 172 acres(Fig.7-1-2). These man-made landscapes were flat and treeless, in stark contrast to the lush rolling topography of the northern portion of the Island.

The dramatic difference between these two areas remains today. When visiting the Island, you can stroll through the Historic District and enjoy the shade of mature trees and verdant open spaces. As you continue to the Island's southern end, you are presented with a dramatically difference experience—a 80 acres of a flat, barren landscape. It is harsh and uninviting, with few trees and little shade to protect you from the hot sun or wind. Much of the area is below the projected 100-year flood line by 2100.

Building upon the Island''s first major physical transformation, the West 8 Team's design proposes a second-generation of topography transformation. Hundreds of thousands of cubic yards of fill material, reused from Island demolition and brought in from off-site, will be used to shape spaces within the park

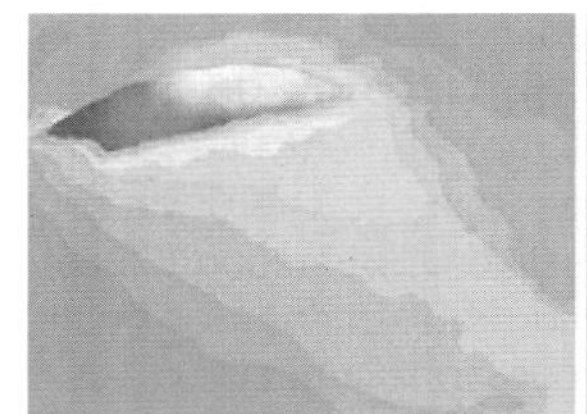
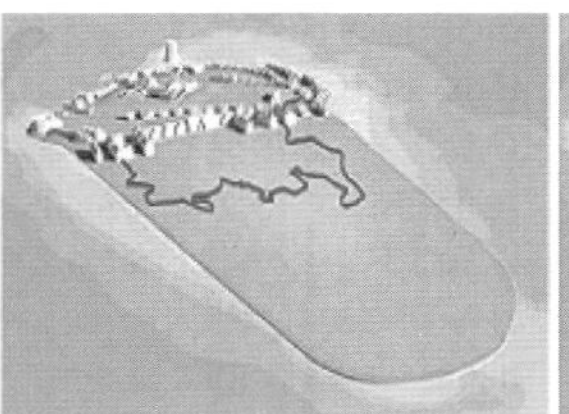
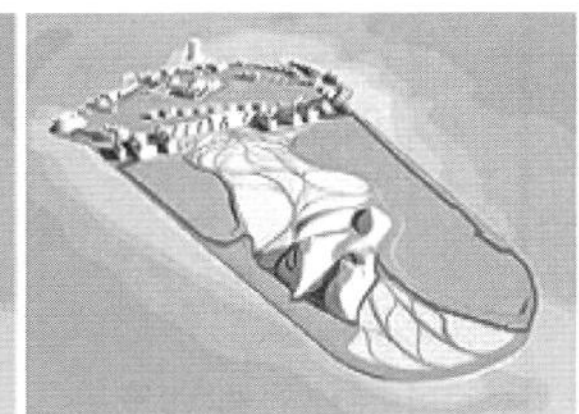
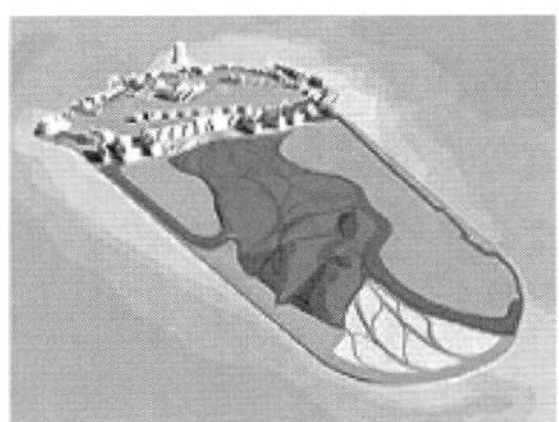

图7–1–2 设计后的到2100年的百年一遇洪水水位变化
Fig.7-1-2 Change of the projected 100-year flood line by 2100 after design

所的人文意义。步移景异的景观吸引前行探索。岛屿南部提升和重新塑造的地形将整个岛屿联系为一体。当你从岛的北部行至南部，你将会途经历史性城区、格罗夫起伏的吊床，然后从低缓、柔和的草坪行至群山之中的低山坡上。

地形的重塑也考虑到了洪水和气候变化的影响，为公园和公共空间的长久持续作出了考虑。一些地区，如湿地花园和部分伟大长廊，经过设计能耐得住洪水泛滥。公园的其他地区，如吊床林地，通过抬高地势确保树木的茁壮成长。

四、 视线设计

公园和公共空间的设计提供了陆上360度全景体验：一种是沿着宏伟的长廊所产生的缓缓的视线变化（图7–1–4），另一种是位于山顶的令人惊喜的观景点（图7–1–5）。 在接近总督岛的渡船上，游客仿佛第一次看到曼哈顿南部低矮的天际线。顺时针走向岛东部边缘，依次看到的是绿色的布鲁克林大桥公园和滨水办公区的码头。当看到巴特米尔克通道时，风、声音和气味在渐渐发生变化，直到自由女神像出现在眼前。从岛屿的西侧望去，艾里斯岛依稀可见，新泽西城的摩天大楼、哈德逊河与东河汇流点以及曼哈顿南部的天际线成为以威廉姆斯城堡为景框中的的前景（图7–1–6~图7–1–8）。

在沿着逆时针方向游览的过程中，游客可以观

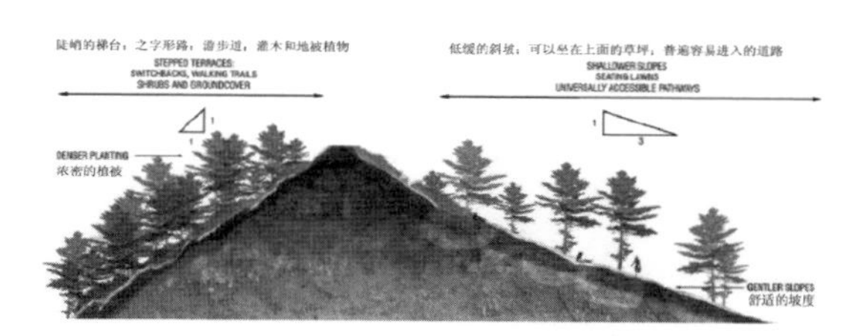

图7–1–3 山体的坡度创造了多样的植物种植环境、生物栖息地和活动项目
Fig.7-1-3 The slopes of the Hills create varied planting, habitat and program opportunities

and public spaces (Fig.7-1-3).

Shifts in topography give opportunities for different activities, transforming the Island physically and programmatically, and creating a humane sense of place that was missing before. Views shift as you move through one area to the next, drawing you in to explore. Raising and sculpting the topography in the southern part of the Island unifies and knits together the entire Island. As you travel from the northern part of the Island toward the south, you experience the original rises and hills of the Historic District, then the undulating Hammock Grove, then the lightly sloping Play Lawn which eases into the gentle lower slopes of the Hills.

Grading and the topographic changes also address the predicted effects of flooding and climate change so that the park and public spaces will last for generations. Some areas, such as the Wetland Gardens and parts of the Great Promenade, are designed to withstand flooding. Other areas of the park, such as the Hammock Grove, are raised to ensure the trees'long term health.

IV. Sightline Design

The park and public space design offers two 360°experiences on the land: one more slowly shifting along the Great Promenade (Fig.7-1-4) and the other

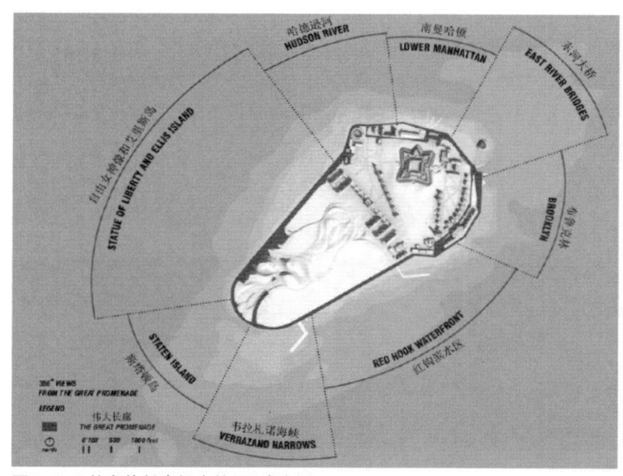

图7–1–4 从宏伟长廊望去的360度全景
Fig.7-1-4 360 View from the Great Promenade

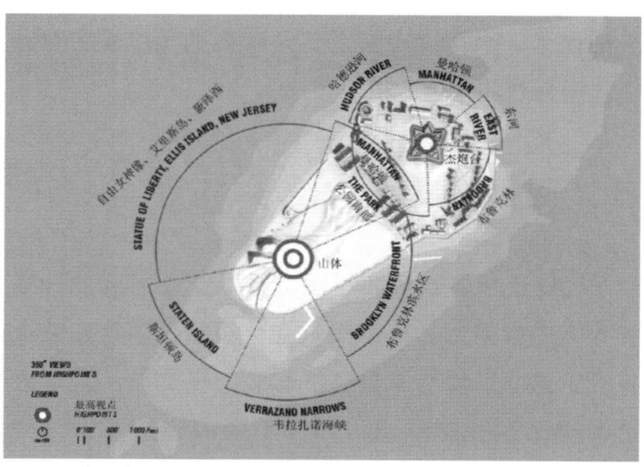

图7–1–5 从最高视点望去的360度全景
Fig.7-1-5 360 View from the highpoint

图7-1-6 从总督岛眺望曼哈顿的建筑群
Fig.7-1-6 Architectural Complex of Manhattan seen from Governors Island

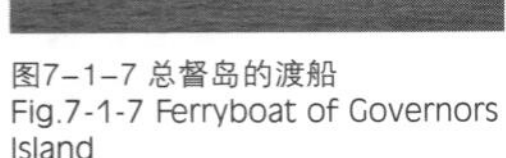

图7-1-7 总督岛的渡船
Fig.7-1-7 Ferryboat of Governors Island

图7-1-8 从总督岛眺望自由女神像
Fig.7-1-8 The statue of Liberty seen from island

赏不同的风景和拥有不同的体验，甚至可以骑自行车、独自或者结伴出行（图7-1-9）。随着季节变化，甚至是一天内的时间变化，视觉和感觉都是不同的。

西8团队的设计使宏伟的长廊成为一个能够进行全景体验的场所。在长廊中有马赛克铺装，宽度可以使人们在此会面，停车休息或流连。长廊的栏杆方便游客在接近水的地方依靠和赏景，在位于岛屿南端的两层游步道处可以俯瞰景色，而岛屿西侧在数量上有比东侧多两倍的场所供游人驻足观景。

岛屿南侧的山体为游人欣赏港湾的全景提供了场地。当你爬上山体四分之一高度时，岛上的休憩用地、建筑、港湾和天际线都浮现在眼底。再向上爬仅仅82英尺（约25米）的高度，山体最高峰为游人提供了一个可以观赏之前只有乘坐飞机或者直升机才能看到的景色的机会。

五、 可持续设计

气候变化导致了海平面上升、全球气温的升高和更强的暴风雨，纽约气候变化小组认为这些变化将会给这一区域带来具体的影响，岛屿新的地形改造考虑了这些影响。山体和地形的塑造考虑了到2100年时可能出现的百年一遇的防洪线。这些区域

providing surprise and drama from viewpoints on top of the Hills (Fig.7-1-5). Disembarking from the ferry at Governors Island, a visitor sees the Lower Manhattan skyline as if for the first time. Walking clockwise towards the Island's eastern edge, the green of Brooklyn Bridge Park is followed by the piers of the working waterfront . As Buttermilk Channel meets the Harbor, the wind, sounds and smells change until the Statue of Liberty faces you beyond. From the Island's western edge, Ellis Island is visible, then the skyscrapers of Jersey City, the confluence of the Hudson and East rivers and the Lower Manhattan skyline again, now framed by Castle Williams in the foreground(Fig.7-1-6~Fig.7-1-8).

This journey offers different views and experiences when undertaken counter clockwise, or alternatively on a bicycle, alone or with others(Fig.7-1-9). Views and sensations change in different seasons, days or even different hours in the same day.

The West 8 Team's design for the Great Promenade enhances the experience of this panorama. The promenade pavement widens into mosaic paving at places to meet, disembark or linger. The promenade railing invites visitors to come to the water, look, and lean. Two levels of pathways created at the South Prow Overlook and western Promenade double the number of places to hang out and enjoy the views.

The Hills offer a more dramatic 360°panorama of the Harbor. As you make your way up one of four hills, the Island's open spaces and buildings fall away, and the Harbor and skyline reveal themselves. While just 82 feet high above grade, the tallest hill affords views previously only experienced from a plane or helicopter.

V. Sustainable Design

Climate change will bring rising sea levels and temperatures and intensified storms, with specific impacts for the region as identified by the New York City Panel on Climate Change (NPCC). The creation of new topography on the Island specifically addresses these effects. The Hills and topography are shaped around the elevation of the projected 100-year flood in 2100. These raised areas, and the enrichment of soil

图7-1-9 岛上的非机动交通工具
Fig.7-1-9 Non maneuvering traffic tools on the island

的抬升和土壤条件的改善为树木及其他植被提供了一个可持续的环境。

超过1 300棵新的树木将会被补充栽种到已经种植的1 700棵树中。植被及树种的选择注重乡土化，同时注重是否能够适应当地的环境，以保证植被在全球温度变化的情况下能健康生长。

新地形的塑造，再利用了建筑和停车场的拆迁材料，将这些材料与垃圾区别开来，并作为循环利用的标志。通过对之前为沥青表面的区域进行绿化和布置可渗透性的铺装，降低降水的流失和热岛效应。

降水对于场地的影响将会被极大的改善。超过19英亩（约7.7公顷）的停车和道路铺装将会被移除并进行绿化种植，暴风雨所带来的降水将会被定点收集和处理，以使流入河口和城市下水道的水量最小化。

对雨水系统的进一步的改进措施如下：收集雨水以再利用于灌溉；从山体及自由平台收集雨水，并导引至岛屿南端的湿地公园；改善宏伟长廊的雨水收集系统，以减少排入港口的水。

conditions, provide sustainable habitat for trees and other plantings.

More than 1,300 new trees will be planted adding to the 1,700 trees already on the Island. Plant and tree selection will focus on native, as well as locally-adapted, species expected to thrive as temperature ranges in the region change.

The topography also reuses materials from demolished buildings and parking lots, keeping material out of landfills and serving as an icon of recycling. The replacement of acres of these impervious asphalt lots with lawn, plantings and permeable paving also reduces stormwater run-off and decreases the urban island heat effect.

Stormwater impacts on the site will be greatly improved from the existing Island conditions. More than 19 acres of paved parking lots and roadways will be removed and planted. Stormwater will be collected and managed on-site to minimize the amount of runoff into the estuary and City sewers.

Further improvements to the stormwater system include: Collecting stormwater for reuse in irrigation;Collecting water from the Hills and Liberty Terrace and outfalling it to the Wetland Gardens at the South Prow;Improving stormwater collection along the Great Promenade to lessen drainage into the Harbor.

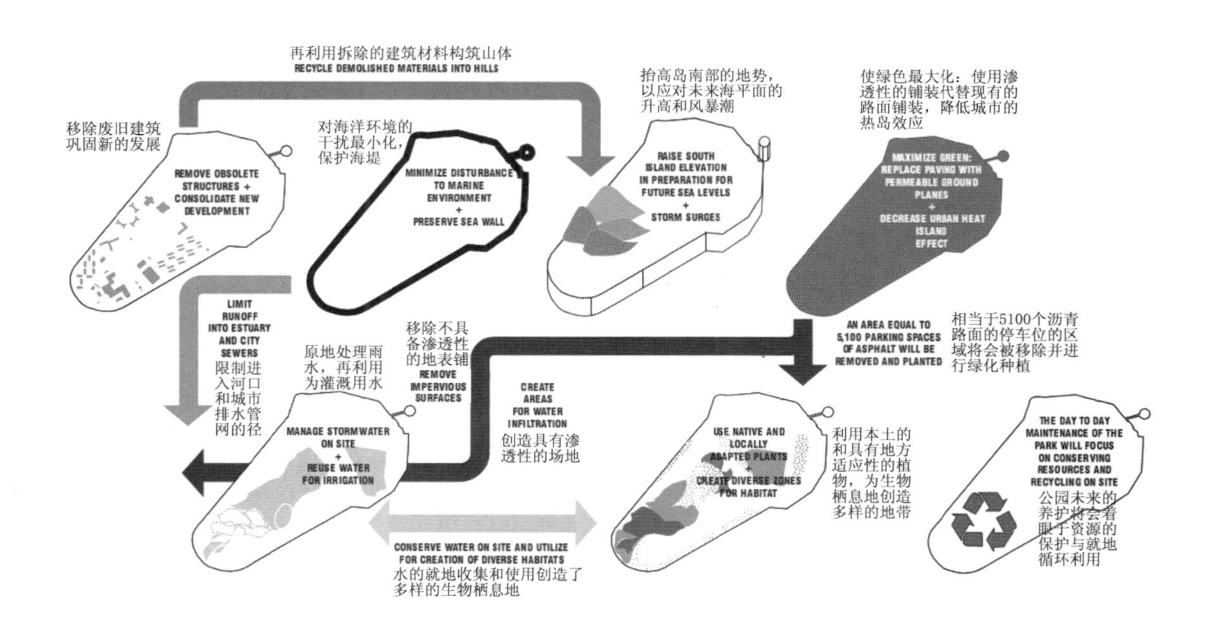

图7-1-10 可持续与环境保护的理念渗透到设计、建设及养护的每一个部分
Fig.7-1-10 Sustainability and environmental performance are integrated into every aspect of design, construction, and maintenance

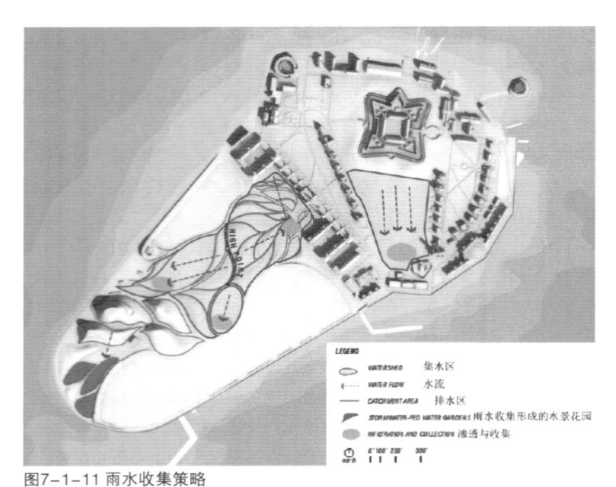

图7-1-11 雨水收集策略
Fig.7-1-11 Stormwater collection strategy

图7-1-12 总督岛场地规划
Fig.7-1-12 The site plan of the island

随着公园和公共空间设计的进展，随后工作计划将解决废弃物最少化和回收利用、可持续发展战略的具体措施、施工、运营和维护公共空间和公园的建筑等问题（图7-1-10，图7-1-11）。

As the park and public space design progresses, subsequent phases will address minimizing waste, recycling, and sustainability strategies for the specification, construction, operations and maintenance of the public spaces and park buildings (Fig.7-1-10,Fig.7-1-11).

六、主要场地设计

1. 宏伟长廊

环绕在总督岛四周的伟大长廊全长2.2公里，走完全程需要45分钟，如果骑自行车则需要的时间要少得多，在这一过程中，游客可以在靠近水边的地方体验到纽约港的全景：曼哈顿南部及布鲁克林区低矮的天际线、布鲁克林大桥、布鲁克林滨水区、开放的港口、新泽西城、自由女神像和艾里斯岛、流淌着的哈德逊河和东河。

伟大长廊是为步行者、自行车驾驶者、跑步爱好者、轮滑及滑板爱好者等使用群体共同设计的。长廊的宽度足以使两辆自行车并行、交谈，还可以和步行者安全的共享道路。宏伟长廊在全岛范围内公共的、禁止机动车通行的，这使得近、远期的游人及租赁者能够利用岛上建筑提供住宿，其中一些建筑是面向宏伟长廊的。

2. 历史保护区

在92英亩（37.2公顷）的总督岛历史保护区中，有22英亩（约8.9公顷）是国家纪念地。历史保护区的景观和公共空间保持了其原有的特色，同

VI. Major Site Plan and Design

1. The Great Promenade

The Great Promenade runs for 2.2 miles along Governors Island's perimeter. In a 45-minute walk, or much less time on a bicycle, visitors experience a 360° view of New York Harbor from the water's edge: the skylines of Lower Manhattan and Brooklyn, the Brooklyn Bridge, Brooklyn''s waterfront, the open Harbor, New Jersey, Liberty and Ellis Island, and the confluence of the Hudson and East Rivers.

The Great Promenade is designed for both walkers and bicyclists, as well as runners, strollers, and roller bladers. It is wide enough for two cyclists to bike and chat and to share the path safely with pedestrians. It is a public, car-free thoroughfare for the entire Island which accommodates visitors as well as future visitors and tenants for the Island's buildings. Some of these buildings front the Great Promenade.

2. Historic District

The 92-acre Governors Island , including the 22-acre Governors Island National Monument, is a nationally and locally designated historic district. The landscapes and public spaces of the Historic District retain their historic character, with limited and selective planting of trees and foundation plantings, restoration and repair of brick pathways, and rejuvenation of key spaces (Fig.7-1-13). The Island's two halves are truly joined together for the first time. The Great Promenade wraps the Island's perimeter. Liggett Terrace and Yankee Landing serve as transitions between the Historic District and the new

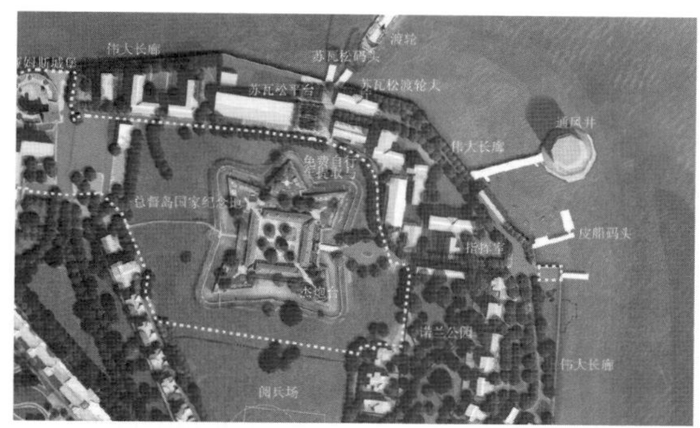

图7-1-13 历史保护区场地规划
Fig.7-1-13 The site plan of the historic district

图7-1-14 岛上的历史遗迹
Fig.7-1-14 The island's historic sites

时有限制和有选择地进行绿化和基础种植，保护和恢复砖石道路，同时恢复了一些关键的空间（图7-1-13）。岛屿的两个部分首次真正的联系起来。宏伟长廊环绕在岛的四周，赖继特平台和洋基码头则作为介于历史保护区和新开发公园和公共空间之间的过渡区域（图7-1-14）。整个岛上的装饰物、灯光器具、标示及栏杆的颜色保持协调统一，这样有助于将岛屿的南北、新旧两部分联系为一体。

3. 赖继特平台

赖继特会堂是一个横亘在岛屿中部的U型建筑物。其最具吸引力的就是它的拱廊，这些具有纪念意义的拱廊将历史保护区的公共空间与赖继特会堂联系起来（图7-1-15）。这些拱廊为公共艺术提供了一个展示的场地。拱廊的开口朝向赖继特平台，这是一个有花床、马赛克铺装、种植艺术品、坐椅、餐饮亭和娱乐设施构成的巨大的城市广场。从平台眺望，吊床林和嬉戏草坪尽收眼底，同时还能依稀看到远处的山体。

4. 吊床林

吊床林提供了一个光影婆娑的空间，它位于赖

park and public spaces(Fig.7-1-14). A common palette of park elements— furnishings,lighting, signage and railing—is provided throughout the Island to integrate the old and the new.

3. Liggett Terrace

Liggett Hall is a massive U-shaped structure that transverses the Island. Its most distinctive feature is the Arch, the monumental archway that connects the Historic District public spaces to Liggett Terrace (Fig.7-1-15). The Arch itself provides a wonderful opportunity for public art to accentuate this moment. The Arch opens into Liggett Terrace, a great urban plaza made welcoming by flower beds and mosaics, plantings, art, seating, cafe carts and play structures. From the Terrace, the Hammock Grove and the Play Lawn unfold beyond, with the promise of the Hills in the distance.

4. Hammock Grove

The Hammock Grove provides an area of filtered light and shade between the cultivated sunny space of Liggett Terrace and the open expanse of the Play Lawn and the Hills and Harbor beyond. Visitors come to the seven-acre grove to enjoy the seasons(Fig.7-1-17. Fig.7-1-18).They walk, bicycle, sit under a tree or nap in a hammock. Hammock Grove offers a wonderful respite and chapter in the park's unfolding topography and views. Hammock Grove contains 300 new trees representing 55 species.The Grove's filtered light creates different moods in different weather, but

图7-1-15 赖继特平台及周边区域场地规划
Fig.7-1-15 The site plan around Liggett Terrace

图7-1-16 赖继特平台的门厅
Fig.7-1-16 The hall of Liggett Terrace

继特平台、嬉戏草坪、群山和港湾之间。游人在这个樱木树林内感受季节的变化（图7–1–17，图7–1–18）。在这里可以散步、骑车，或者在吊床上小憩。这一区域在公园起伏舒展的地形和风景中是一个精彩的过渡区域。吊床林内有300棵新种植的包含55个种类的树木。虑过果林的光线在不同的季节产生了不同的气氛，也创造了一个令人难忘的公园体验。

5. 山体设计

岛屿的南半部分从赖继特平台平坦的植被和铺装开始向南延伸，经过吊床林舒缓的斜坡路和嬉戏草坪后进入到四座高度在46～82英尺（约14～25米）之间的山体中（相对高度）。对于熟悉这座岛屿历史的游客而言，这些山体的存在是令人震惊的。岛屿的南部曾经是垃圾填充地，地市平坦、单调且被建筑物环绕。现在这些南向的山体提供了一个可供探索的地形，同时也为鸟瞰全岛提供了一个较高的视点。

通过多种多样的路径，这些山体吸引游人南行。他们为草坪、树林和园径提供了一个大的背景，并且使对自由女神像的探索的吸引力最大化。每一座山体都绿树成荫，游人可以坐在舒缓的草坡上，探索不同的路径，沿着山脊散步。山体上丰富的植被、树木，为鸟类、昆虫及其他野生动物提供了生境。对于游人而言，上山的道路一条是大众化的舒缓的道路，另一条则是由陡坡和阶梯的颇具挑战性的道路。沿着道路设置了供人们休息的坐椅和观景点。树林及灌木丛过滤着人们的视线，使人不断地探索前方（图7–1–19）。

6. 自由平台

自由平台将纽约城与自由女神像联系起来。站在

图7–1–17 吊床林
Fig.7-1-17 The Hammock Grove

always provides a memorable park experience.

5. The Hills

The southern half of the Island stretches from the flat plantings and paving of Liggett Terrace through the gentle sloping paths of the Hammock Grove and Play Lawn to four hills rising in height from 46 feet to 82 feet (relative to grade). For the visitor familiar with the Island's history, the Hills are a surprise. The Island's southern half had been landfill, utterly flat, sterile, devoid of topography, crowded with buildings. Now, a grouping of hills frames views looking south, providing interesting terrain for exploration, and culminating in an overlook high above the Harbor.

From various approaches, the Hills lure visitors south. They provide a backdrop to the lawns, trees and paths of the park and are sited to maximize the impact of the moment of discovery of the Statue of Liberty. Each hill is lush and green, inviting visitors to sit on its grassy lower slopes, explore its pathways, walk along a ridge. Planted with a broad array of trees, shrub sand vegetation, the Hills too invite birds, insects and other wildlife to make their home. For visitors, One pathway has a gentle grade for universal access; another is more challenging with steeper slopes and stairs. Along the way, benches provide resting and viewing points. Trees and shrubs filter views and create anticipation for what lies ahead (Fig.7-1-19).

6. Liberty Terrace

Liberty Terrace connects New York City to the Statue of Liberty. Standing on the Terrace, the visitor sees Lady Liberty's face. As visitors tour the Great Promenade, from either direction, the Statue comes in view, as a surprise when rounding the South Prow, or first in profile on the Island's western Promenade. As cyclists head south or pedestrians meander through the park, the trees, hills and lawns frame views of the Statue, all culminating at Liberty Terrace. Here, the Statue is a natural magnet to draw people to a terrace designed for gathering, views and people-watching (Fig.7-1-20).

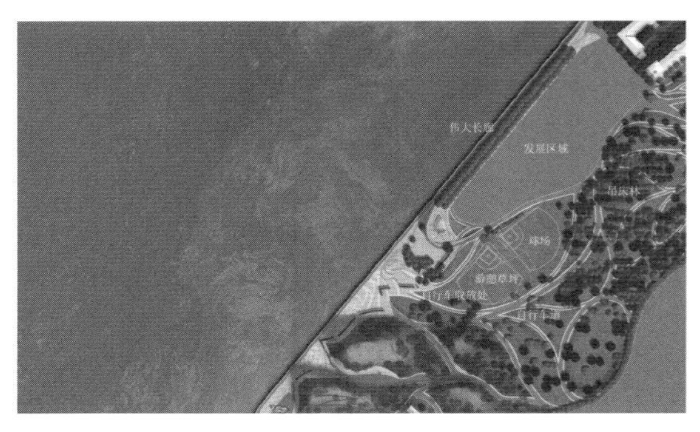

图7–1–18 吊床林及周边区域场地规划
Fig.7-1-18 The site plan around the Hammock Grove

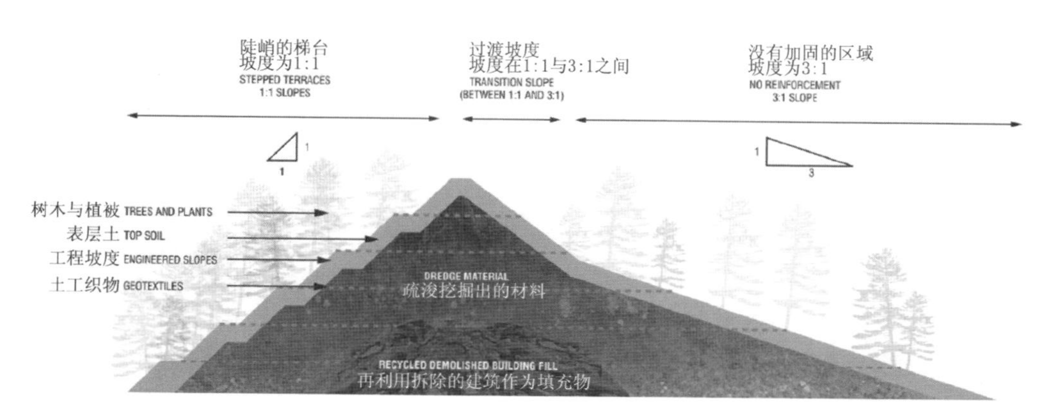

图7-1-19 设计团队分析了山体的工程构建途径
Fig.7-1-19 The team analyzed the engineering approach to building the hills

自由平台上，游人可以看到自由女神像。当游人行进在宏伟长廊上，转过岛屿的南端或者在伟大长廊的西段时，自由女神像令人惊喜的从另一个方向进入人们的视野。当骑车或步行穿过公园时，树木、群山和草坪映衬在自由女神像周边，所有这些景色在自由平台达到高潮。在这里，自由女神像是一个天然的磁石，吸引着人们前往设计好的地形聚会、观景图7-1-20）。

7. 岛屿南端

当游人转过岛屿的最南端时，远处的航标和乌云在这里都是可以看得到的，温度骤降、微风迎面而来，曼哈顿的天际线和布鲁克林滨水区渐渐隐去。

就好像是从一艘船的船首望去，面前是宽广的大海和蓝天。驶往斯坦恩岛的渡船和作业船只穿梭在港口内。费雷泽诺大桥延伸向远方。岛屿南端的设计向人们展示了自然气候变化的景观和处于淡水

图7-1-20 自由平台的露天座椅
Fig.7-1-20 Open air seating of Liberty Terrace

7. The South Prow

The bell of a buoy sounds, storm clouds are visible in the distance, the temperature drops and the breeze picks up as the visitor rounds the Island's southern tip. Gone are the Manhattan skyline and the Brooklyn waterfront.

Ahead, as if viewed from a vantage of a ship's prow, is a vast expanse of sea and sky. Ferries to Staten Island and working ships and tugboats travel through the Harbor. The Verrazano-Narrows Bridge stretches in the distance. The design of the South Prow showcases the view, the natural changes of weather and the exploration of a habitat at the intersection of saltwater and freshwater (Fig.7-1-21).

图7-1-21 岛屿南端及周边区域场地规划
Fig.7-1-21 The site plan around the south prow

河咸水交界处的生物栖息地（图7-1-21）。

七、总结

本项目是城市滨水区生态设计的又一例证，也是城市更新的成功案例。昔日废弃的仅靠水路才能进入的军事基地转变为功能复合的、充满活力的现代公园。这种城市更新方法体现了一种更加审慎、明智与和谐的发展过程，案例中采取了一系列更新途径，诸如保护、修复、再利用以及再开发等。

总督岛公园的规划与设计提供了360度全景体验，注重视线的转换和步移景易的观赏效果。结合现场地形，为游人提供了丰富的视觉体验。可持续性设计和弹性设计也是本项目的特色。

纵观全文，最突出的亮点是规划设计中对于现场地形条件的详细踏勘与缜密分析，重视调查研究，“先诊断，后治疗”，即贯彻了格迪斯提出的“调查—分析—规划”的科学方法。这一点在目前也被视为城市规划的一般程序，因为只有通过认真的现场踏勘与调查，才能得到宝贵的基础数据和资料，这是规划的前提，然后才能在分析这些资料和数据的基础上选择规划体制。

这位苏格兰生物学家、城市问题理论家还特别强调注重保护地方特色，认为每一个地方都有一个真正的个性，这种个性或许沉睡已久，规划师的任务就是把它唤醒。

VII. Summary

This project is another instance of the urban waterfront ecological design,and it is also a successful case of urban renewal.Former abandoned military base only can be entered by boats is changed into a compositely functional,lifeful and modern park.This method of urban renewal reflected a kind of more prudent,wiser and more harmonious development process, cases adopted a series of renewal way, such as protection, restoration, reuse, and redevelopment and so on.

The planning and design of governors island park provides a 360-degree panorama experience, pay attention to sight conversion and the ornamental effect that each movement of step will bring the different sceneries. Combining with the terrain,the planning scheme provides a rich visual experience. Sustainable design and elastic design are also the project characteristics.

Throughout the full text, the most prominent window is the detailed inspection and careful analysis of the site conditions in planning and design,"diagnosis first, treatment after", namely take the scientific method "survey-analysis-planning"put forward by Gerdes. This point is also seen as the general procedure of the urban planning, for it is only through the investigation and inspection earnestly that the planners can obtain the valuable basic data and material which is the premise of the planning, and then based on the data,the planners can decide to choose which kind of planning techniques.

The Scottish biologists, urban issues theorists also laid special emphasis on protecting local characteristics and thought that each place has a real personality which perhaps is sleeping deeply, the task of planners is to make it awakened.

思考题

1. 总督岛公园与公共空间总体规划是一个在内容上相对完整的美国公园总体规划案例，通过与我国当前公园总体规划的比较，分析它们在编制内容、编制深度及编制程序上的优缺点。

2. 低碳、绿色、可持续、应对气候变化等已经成为风景园林行业的一系列标签，本案中通过对地形的利用及塑造实现了可持续与弹性设计的目标，试想在其他类型的公园规划中如何体现这一系列新时代行业目标。

3. 视线的设计是总督岛公园规划的一大亮点，结合中国传

Questions

1. The governor island park and public space master plan is a relatively complete planning case of American parks in the content.By comparing the park planning in China,analyse the advantages and disadvantages of planning contents, planning depth, and planning formulation process.

2. Low carbon, green, sustainable, and climate change have become the industry label of landscape architecture.In this case, through the use and shaping of terrain, the design team realized the sustainable and flexible goals.Please imagine that how to embody the new industry target in other types of park planning.

3. The sight design is the highlight of the governor island park and public space plan. Please refer to "Each

统园林中对“步移景异”的理解，探讨如何使之与现代风景园林规划和设计更好的融合在一起。

4. 当前湿地公园的建设在类型、规模上多种多样，结合总督岛中湿地公园的选址、定性，思考在何种情况下，湿地可以作为风景园林项目构成要素，而这些湿地的功能定位需要依据哪些客观条件。

5. 人文资源是人类重要的财富，同时也是风景园林规划的重要对象，结合总督岛公园及公共空间总体规划中对人文资源的保护与利用，思考如何在调查踏勘、方案构思、施工、运营及维护等不同阶段协调好对人文资源的保护与利用。

movement of step will bring the different sceneries" in the Chinese traditional garden design,discuss how to combine it with modern landscape planning and design.

4. The current wetland parks have many kinds in the types and size.Please refer to site selection and qualitative analysis of the wetland park on the governor island,think about in what circumstances, wetland can become the element of landscape architecture projects, and what kind of objective conditions should the wetland functional orientation follow.

5. Humanistic resources is an important human wealth,it is also an important object of landscape architecture.Please combine the content of humanistic resources protection and utilization in the governor island park and public space master plan,think about how to coordinate conservation and utilization from survey,design, construction, operation and maintenance and so on.

注释/Note

图片来源/Image：图7-1-1，图7-1-6～图7-1-9来自www.sinovision.net ，其余图片均来自《纽约总督岛公园及公共空间总体规划》/Governor Island Park and Public Space Master Plan

推荐网站/Website：本案例引自www.govislandpark.com

第二节 布鲁克林桥公园规划

Section II Brooklyn Bridge Park Planning

一、 项目简介

布鲁克林桥公园面积约占地85英亩（约34.4公顷），包括码头、高地和周边水域。预计将拥有1.3公里长的水岸线，以杰伊街北、大西洋大道东南弗曼街以东以及东河的西面为界，包括了1到6号码头（图7–2–1）。

该项目沿着东河从大西洋大道的尾端到杰伊大道，也就是曼哈顿大桥的北面。这些地段分别隶属于纽约州、纽约市、纽约港务局、新泽西港务局和一些私营部门。这个项目建成后能够为周边的社区、区域乃至整个城市提供一个全新的室外公共娱乐场所，并且可以让市民直接体验到滨海的乐趣。该项目为公众提供了全新的亲水体验，创新的利用木板、浮桥和管道，大大延长了水岸线。该项目基

图7–2–1 基址位置
Fig.7-2-1 The location of the site

I. Project Description

The approximately 85 acre Project, consisting of piers, upland and water area, would stretch along 1.3 miles of Brooklyn waterfront approximately bounded by Jay Street on the north, Atlantic Avenue on the south, Furman Street on the east, and the East River on the west, and would include Piers 1 through 6 (Fig. 7-2-1).

The proposed Project would extend along the East River from the foot of Atlantic Avenue to Jay Street, north of the Manhattan Bridge. It would comprise parcels currently owned by the State of New York, the City of New York, the Port Authority of New York and New Jersey, and private sector entities. The Project would provide the surrounding communities with a major new precinct of outdoor public recreation and the opportunity to experience the waterfront directly, while also serving both the larger City and the region. The proposed Project would offer the public unparalleled access to the water, making innovative use of boardwalks, floating bridges, and canals that would wind along the water's edge. It would also include rolling hills, marshland, and abundant recreational opportunities with multipurpose playing fields, playgrounds, shaded ball courts, open lawns, and 12 acres of safe paddling waters. There would be pockets of natural landscape recreated on some of the parkland to attract birds and other wildlife. The Project's pathways would increase the water's edge from 2.4 miles to 4 miles and provide pedestrian connections both to the water and to the full range of the Project's experiences.

The entrances to the proposed Project at Atlantic

地内拥有起伏的地形、湿地、多功能运动场、操场、顶棚球场、开阔的草坪和12英亩的安全划水区域等丰富的娱乐场所。在原有场地上将重新建立一些与周围环境不同的小的景观区域来吸引鸟类和其他野生动物。通过该项目的建设，将原有水岸线从2.4英里（约3.9公里）增加的4英里（约6.4公里），同时增加行人与水的互动以及与整个项目范围内的联系。

项目的主入口设置在大西洋大道的富尔顿码头，位于“曼哈顿高架桥下的社区”，这样的设置可以使得附近的居民与游客最大限度的使用到每个入口的便利设施，同时不需要走很久便能够到达操场与草坪区。布鲁克林公园总共有五个分区，从南边开始依次为：大西洋大道，包括6号码头与其高地；2、3、4、5号码头和其高地；1号码头和富尔顿码头；间桥区域；以及北面的曼哈顿大桥和曼哈顿桥大门。项目计划囊括基地范围内的所有可用元素以及对于五个分区来说的特殊组件（图7-2-2）。

二、 规划目标和指导准则

项目规划的目标之一是让公园能够融入整个城市结构中，主要是通过园路与现有的城市道路结合的手段。因为目前基地很大一部分是被高速路从城市中分割开来的，而把现有道路延伸至公园的做法会让居民进入公园时有一种亲密感。通过这种方式再辅以欢迎式的入口能够促进游客对公园的熟悉程度。

布鲁克林桥公园的几个主要入口分别设置在大西洋大道、老富尔顿街以及约翰街的“曼哈顿高架桥下的社区”。这些入口均设置在城市的主要路口

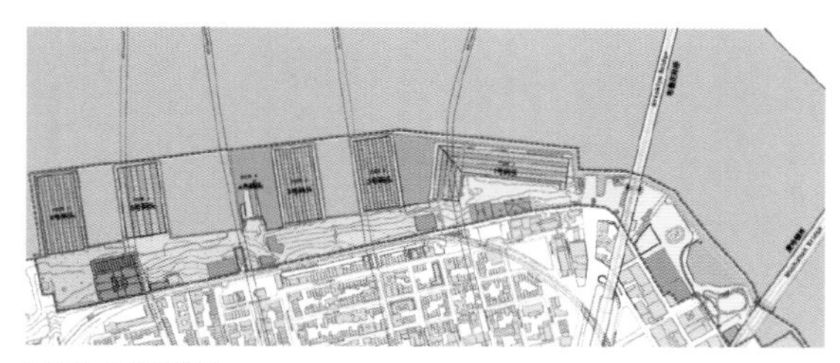

图7-2-2 基址总平
Fig.7-2-2 The master plan of the site

Avenue, at Fulton Ferry Landing, and in D.U.M.B.O. ("Down Under the Manhattan Bridge Overpass", the acronym for the neighborhood located in that vicinity) are designed so that at each entrance the visitor would have access to a wide range of amenities nearby and would not have to walk long distances to arrive at playgrounds and lawn areas. The proposed Project is envisioned as five interconnected areas. From the south, these are: the Atlantic Avenue Gateway including Pier 6 and its upland; Piers 5, 4, 3, and 2 and their uplands; Pier 1 and Fulton Ferry Landing; the Interbridge Area; and North of the Manhattan Bridge and the Manhattan Bridge Gateway. The Project program contains elements available throughout the Project ("parkwide elements") and specific components for each of the five subareas (Fig. 7-2-2).

II. Project Objectives and Guidelines

One of the goals of the proposed plan is to integrate the site into the fabric of the city by creating a network of roads and sidewalks that weave the site into the existing grid. Because a large portion of the site is separated from the existing city by highway alignments, the extension of existing streets into the park would give the pedestrian a sense of familiarity and context when entering the park. The goal of promoting a sense of familiarity is coupled with another goal of the park, which is to provide a welcoming entrance for its visitors.

Generous park gateways would be created at the main entrances to Brooklyn Bridge Park, ie. Atlantic Avenue, Old Fulton Street, and in DUMBO at John Street. These gateways are located at major urban junctions and would serve as visual beacons to regional park visitors and local residents alike. Each gateway would provide generous, clear and safe pedestrian access, street tree planting and seating. Additional park entrances would be created at the termination of existing city streets, defining a continuation of the city and its pedestrian network to the edge of the park. These include: Joralemon Street, Middagh Street, Doughty Street, New Dock Street, Main Street, Washington Street, Adams Street, Pearl Street, Jay Street, Water Street and Plymouth Street. It is a goal of the proposed plan to ensure that from all vantage points, views from these connections will be unobstructed, visually interesting and welcoming (Fig. 7-2-3).

An additional goal of the road system in the park is to define the extent of the development parcels and delineate the line between the development and the park. The placement of the buildings at the edges of the park and at the gateways reflects three goals related to views: Respect the mandated view plane from the Brooklyn Heights Promenade; Respect the existing view corridors of streets that terminate in the park; To the maximum extent practicable, site the

处，它们将与这些枢纽作为一种视觉信号来引导游客与附近的居民。在每个入口附近都设置有宽阔、清晰安全的人行通道、行道树和坐椅。其他的次入口将设置在现有城市街道的尾端，来承接城市道路以及步行网络与公园边界的连续性。这些次入口分别位于：尤拉勒蒙街、麦道街、道蒂街、新码头街、梅茵街、华盛顿街、亚当街、珍珠街、杰伊街、水街、普利街。这个目标达成后，游客可以在这些枢纽处看到畅通、直观、有趣且开放式的视觉通廊（图7–2–3）。

公园道路系统建成的另外一个目的是能够划定新增地块的范围同时区分开公园与新增地块。在公园边缘与入口处的建筑必须要遵循下面三个准则：

不阻碍布鲁克林高地处人行道的视平线；不阻碍通往公园的现有的视觉通廊；为了最大限度地提高可行性，需要合理地为每一个发展地块选址，以期能够在每个入口处对公众最大限度地发

proposed developments so there is a maximum amount of park provided to the public at each entrance.

III. Project Planning and Design

1. Atlantic Avenue Gateway and Pier 6

Atlantic Avenue is the southern gateway to the park, providing pedestrian and mass transit connections to adjacent neighborhoods and regional access. Atlantic Avenue slopes down towards the water, affording a clear view of the water as one approach the Project. A burst of green would be visible in the distance as one passes beneath the Brooklyn-Queens Expressway drawing visitors into the Atlantic Avenue gateway. The upland area of Pier 6 and the edge of thepier, located at the foot of Atlantic Avenue, would provide all the amenities of a neighborhood park including playgrounds, lawns, access to the waterfront, and recreational opportunities. The sidewalks approaching the park along the north side of Atlantic Avenue would be tree-lined and ample, sufficient to accommodate large groups of visitors at one time. Safe crosswalks will be established from the upland areas and between the development sites to the park. The Bikeway would be

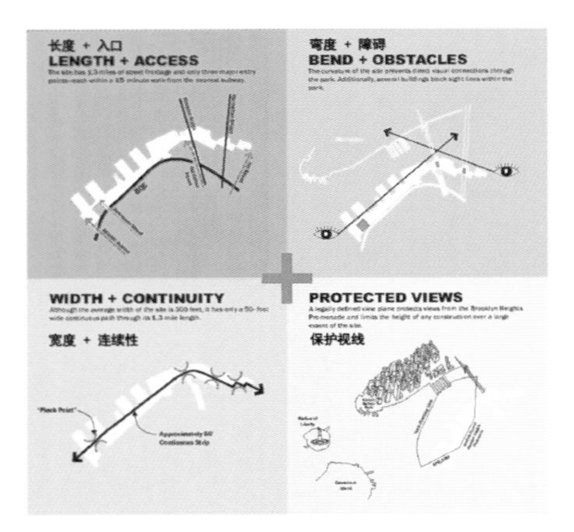

URBAN JUNCTIONS
after more than a century of the site supporting port related uses and being cut off from the city fabric, the idea of urban junctions was posed as a means to concentrate development at the park entries – connecting the city with the park.

城市交点
基址在过去一个世纪里都是被用于与港口相关的产业，而且与整个城市结构相剥离，城市交点的理念主要是作为一种手段来集中发展园区入口，从而将城市与公园

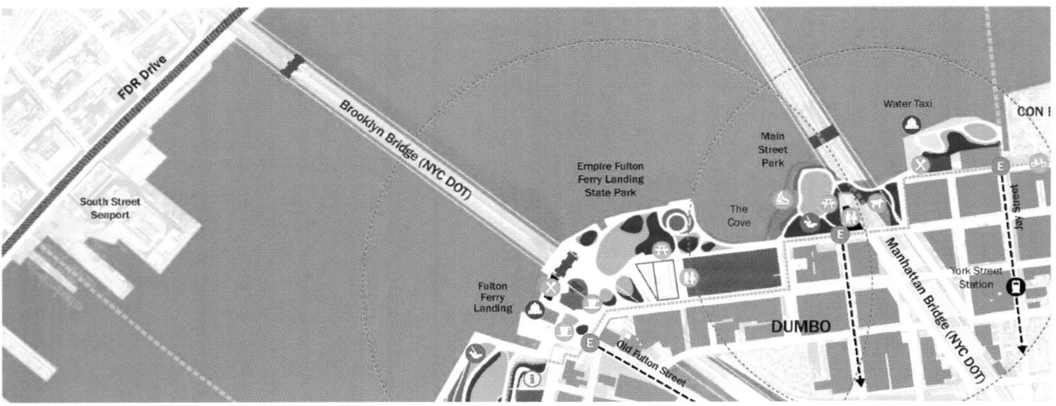

图7–2–3 城市交点的组织
Fig.7-2-3 Organization of urban junctions

挥公园的功能。

三、项目规划设计

1. 大西洋大道入口和6号码头

大西洋大道是公园的南大门，连接了行人、公共交通与周边的社区与街区入口。大西洋大道以缓坡的形式逐渐靠近水面，明晰的视觉通廊与入口通道合为一体。具有强烈视觉冲击的绿色能够将布鲁克林–皇后区高速路上游客的视线吸引到大西洋大道入口上。6号码头与其丘陵地区位于大西洋大道的尾端，为周边邻里社区提供休闲娱乐设施：操场、草坪以及海滨。大西洋大道北侧的人行道绿树成荫，可以容纳大量的游客。在高地和园区发展用地之间将建立安全的人行道。单独设计的自行车道会避免与行人冲突。

在6号码头远离高地的地方，有着充足的光线、和煦的微风和良好的排水，利用这些优势设置一些海滩沙堤，同时也会在这里设置一些操场、草坪和沙滩排球。此处将保留许多原有的原生滨海植物，并利用地形保护它们免受海风的侵袭。在码头比邻的丘陵地区边缘，将保留一个原有建筑作为游客中心和公厕。

2. 2号、3号、4号、5号码头和相邻高地

1) 人行步道与高地

如上所述，海滨长廊将从大西洋大道入口一直延伸到富尔顿码头。而南边，海滨长廊将横跨一个新建的潮汐汊道。一些现有的建筑，可继续作为公园维护和管理用房。2号和3号码头上的棚子可以用来停靠非机动船。

2号和5号码头之间的高地将被抬高，用来减弱布鲁克林–皇后区高速公路所带来的噪声影响，同时还能够提供观赏海港景色的地点。

2) 5号码头

5号码头设计有三块户外场地，市民可以在此参与下列的体育活动：足球、橄榄球、板球、曲棍球。此外，在5号码头的西边，设置有一个室内建筑。可

designed to avoid conflicts between pedestrians and roadways.

Moving farther away from the upland, a "beach barrier" with dunes is proposed on Pier 6 to take advantage of this location's sunny, windy environment and well-drained structure. Active programming would be inserted, such as playgrounds, lawn areas, and beach volleyball could be located on Pier 6. Vegetation, including a variety of native shoreline plants, and topographical forms would provide shelter from the wind. An existing concrete masonry building at the edge of the pier that abuts the upland area could be preserved and used as a visitor's center and comfort station.

2. Piers 5, 4, 3, and 2 and Related Uplands

1) Promenade and Uplands

As described above, a waterfront promenade would extend from the Atlantic Avenue entrance all the way to the Fulton Ferry Landing. In the south, the promenade would bridge over a newly created tidal inlet. Some of the existing upland buildings could be reused for park maintenance and operations. A shed on Pier 2 or 3 would be used to store non-motorized boats.

The upland area between Piers 2 and 5 would have an elevated and sloped topography that would reduce noise from the Brooklyn-Queens Expressway and provide views of the harbor from the uplands.

2) Pier 5

The proposed plan includes three outdoor fields on Pier 5, on which any of the following sports could be played: soccer, football, rugby, cricket, lacrosse, or field hockey. In addition, it is contemplated that the field at the western edge of Pier 5 could be housed in an indoor structure. This structure would provide year-round sports courts. Pier 5's perimeter would provide a continuous water's edge esplanade, from which park patrons might fish or sit or walk along.

3) Pier 4

A shallow water habitat area would be created in the vicinity of Pier 4 and the adjacent railroad float transfer bridge, which would remain intact. In the area of Pier 4, a new floating

boardwalk would be created that would connect to the larger circuit of walkways and provide a place for launching kayaks and bird-watching. On the upland area adjacent to Pier 4, there could be a new beach that would connect to the larger circuit of walkways. The beach would provide direct access to the water and serve as a launching point for non-motorized boats, but swimming would not be permitted (Fig. 7-2-4).

4) Piers 2 and 3

Portions of the warehouse sheds on Pier 3 could

以提供全年的室内运动场地。5号码头的外围，有一块滨水的空地，游客可以在此钓鱼、休憩或者散步。

3）4号码头

4号码头的附近将设计一个自然浅水栖息地，同时其附近铁路浮桥将保持不变。4号码头还将设置一个新的浮桥，它将与更大的人行构筑物相连，并提供皮艇码头与观鸟点。四号码头比邻的高地区域有可能改建成一个新的海滩，也会与人行构筑物相连。游客可以在海滩处戏水、租赁非机动船只，但不能够在此处游泳（图7-2-4）。

4）2号和3号码头

3号码头原有的仓库大棚可以改建成室内娱乐场地，同时还能够提供必要的阴凉区域。2号码头将被改造成一个大型的市民草坪，这个草坪还可以灵活地作为海滨长廊使用。

be reused to house active recreation courts and also provide essential shading. The Pier 2 area would be transformed into a large civic lawn that would lend itself to programmatic flexibility and waterfront promenades.

3. Pier 1 and Old Fulton Street Gateway

1) Old Fulton Street Gateway

Old Fulton Street at Fulton Ferry Landing is conceived as the primary gateway entrance to the park with direct access to borough and regional roadways. A large civic plaza is proposed at the base of Fulton Ferry Landing to provide a generous public gathering space at the Project entrance. Pedestrian connections to Fulton Ferry Landing and Pier 1 would be improved with ample sidewalks, designated safe crossings and street trees. In order to create a scenic Fulton Ferry gateway and improve physical and visual connections within the proposed Project, the Purchase Building would be removed.

2) Pier 1

A hill on Pier 1 would be created with views into the park and out towards the harbor, Governor's Island,

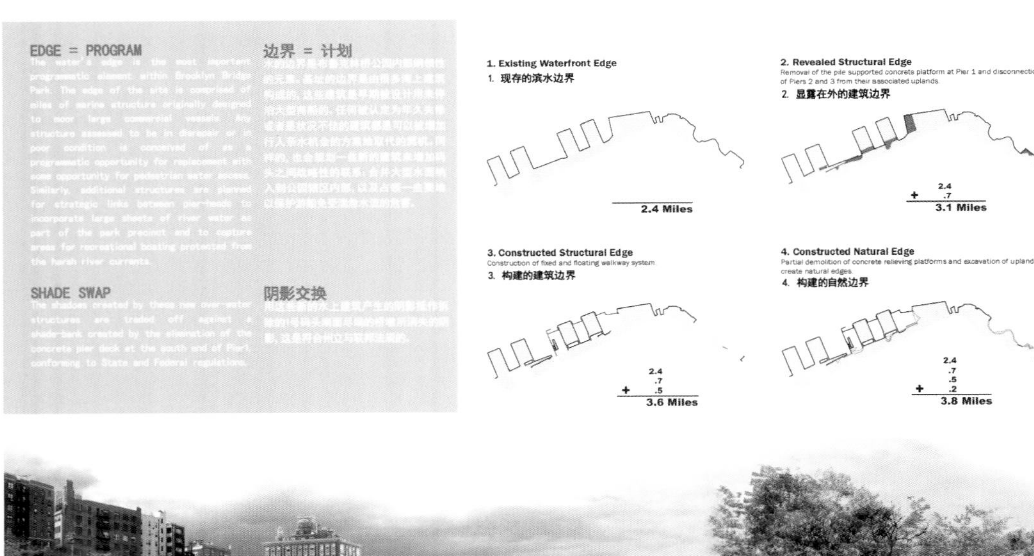

图7-2-4 滨水边界的改造
Fig.7-2-4 Transformation of the waterfront edge

3. 1号码头和老富尔顿大道入口

1) 老富尔顿大道入口

在富尔顿码头的老富尔顿大道被看作是公园的主入口，它与城市与区域道路相连接。在富尔顿码头的基础上设计了一个市民广场，给周边的居民提供了充足的公共集散空间。富尔顿码头与1号码头有足够的人行道连接，并对原有人行道进行了改善。为了创造出一个风景优美的富尔顿入口并改善入口处的视觉与触觉感官，因此附近的商业建筑都将被拆除。

2) 1号码头

1号码头处将堆起一个小土丘，在这里可以内看公园，远眺港口、总督岛、自由女神像、曼哈顿和布鲁克林大桥。横跨弗曼街的人行天桥连接起1号码头和施贵宝公园。这将有助于施贵宝公园的重建，并且能够在临近社区提供一个进入到布鲁克林桥公园的次入口。在码头正对东河的一边将辟出一块空地，沿码头边缘建立起浅水栖息地。通过拆除支撑1号码头的木桩来增加额外的人行步道，在该区域还会设计一个新的双向海滨长廊，围合出一个供皮划艇使用的水湾。

4. 间桥区域

布鲁克林与曼哈顿大桥之间的区域很大程度已经开发成了公园用地，其中包含了帝国–富尔顿公园和梅茵公园。原有的开放水湾将予以保留，让游客们可以直接到达海岸。原有的烟草仓库将被改建成一个花园、咖啡厅或者租给艺术团体。在曼哈顿桥附近的现存的纽约市环境保护局将被改建成具有社会、文化、教育或其他用途的公共建筑（图7–2–5）。

四、 规划设计要素

1. 滨水入口与水系循环

布鲁克林桥公园的主要资源之一就是环绕周围的水。因此该项目中的几个组成要素就是在视觉和触觉上增加与水的互动。4号码头对面的海滨区域，在未来有可能转化为一个海滩并开展皮划艇和独木

the Statue of Liberty, Manhattan, and the Brooklyn Bridge. A pedestrian bridge across Furman Street that would link the hill on Pier 1 to Squibb Park is also proposed. This would serve to reactivate Squibb Park as well as provide an additional entrance into the Brooklyn Bridge Park from the adjacent community. An esplanade would be created along the portion of the pier fronting on the East River, and shallow water habitat zones would be established along the pier edge. By removing the pile-supported deck portion of Pier 1 and providing additional walkways, the Project could create a new two-sided waterfront promenade and provide a large basin for kayaking.

4. Interbridge Area

The area between the Brooklyn and Manhattan Bridges is already largely developed as parkland, containing both Empire-Fulton Ferry State Park and the new Main Street Park at the foot of Main Street. At the water's edge, the existing open water cove would be retained to allow park visitors to reach the shoreline. The restored exterior shell of the former Tobacco Warehouse may be used to house a walled garden, cafe, or space for arts groups. The existing New York City Department of Environmental Protection building adjacent to the Manhattan Bridge at Washington Street may be reused for community, cultural, educational, or other uses (Fig. 7-2-5).

IV. Planning Elements

1. Waterfront Access and Circulation

One of the primary assets of Brooklyn Bridge Park is its proximity to the water. Several elements of the proposed plan encourage interaction with the water, both visually and physically.The waterfront area across from Pier 4 could be transformed into a beach for launching kayaks and canoes. The area between Piers 1 and 6 would feature a waterfront promenade extending roughly along the bulkhead line. This paved promenade would serve as a main pedestrian thoroughfare running through the Project and would allow views of the water, piers, harbor, and the Manhattan skyline. Through a series of sloping ramps and floating and fixed walkways, park users would also be able to experience the water at sea level. This water-level access would allow for kayak launching in certain areas, as well as fishing, additional park circulation, and other water-dependent activities. These walkways would provide for an entirely different experience of the park, offering dramatic views of the columnar forest of piles that support the pier deck.

In the interbridge area, existing access to the water (in Empire-Fulton Ferry State Park, the Main Street Park, and Fulton Ferry Landing) would be extended to connect with the areas to the north and south. North

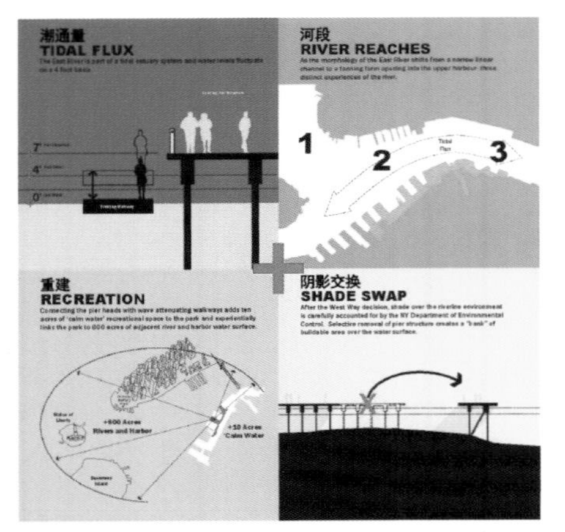

=

EDGE COMPLEXITY

The park master plan was built around the idea of edge complexity in literal and scientific ways (edge habitats on the site), in spatial ways (range and types of edges that people can occupy), and in metaphoric ways, creating a language of edges throughout.

边界复杂性

公园的总体规划是围绕边缘复杂性的理念建立的，分别通过文字描述的方法，科学的方法（边缘生境），空间的方法（游客可以在不同边界类型和边界范围内活动），和隐喻的方法创造了一种贯穿基址的边缘语言。

图7-2-5 间桥区域的边界复杂性
Fig 7-2-5 Edge complexity of

舟项目。1号和6号码头之间的区域将形成一条沿海岸线的海滨走廊。这条走廊的铺设将成为公园内部一条主要的步行道，并在视线上将水、码头、港口和曼哈顿的天际线串联起来。游客能够借助一系列的坡道、浮板和固定的人行道得到最佳的亲水体验。现在的水位已经在某些区域开放独木舟、捕鱼和其他水上活动。这些临水的步道展现出来的景致完全不同于公园的其他区域，游客可以在这里看到初具规模的支撑码头甲板的木桩。

在联桥区域，现有的亲水区（帝国–富尔顿码头州立公园、梅茵街公园和富尔顿码头）将向北部和南部区域扩大。北面的梅茵街公园、新的人行道、游憩场和水上平台将把城市公园和亚当街的北面地区联系在一起（图7–2–6）。

2. “安全水”区域和水系利用

从4号码头的南部边缘到1号码头的南部边缘之

of the existing Main Street Park, new walkways and an esplanade would be created, as well as an overwater platform linking the City park to the area north of Adams Street (Fig. 7-2-6).

2. “Safe Water” Zones and Water-Dependent Uses

From the southern edge of Pier 4 to the southern edge of Pier 1 two connected “safe water zones” are planned. These would provide approximately 12 acres of secure water area for non-motorized boats, including kayaks, canoes, and paddle boats. Marine structures would define the area, serving to attenuate waves from passing boats. Floating boardwalks would be arranged to contain the boaters and kayakers and provide additional wave attenuation within the safe water area. A channel or “canal” would connect the two safe water zones, between Piers 3 and 4 and between Piers 1 and 2. This canal would allow kayaks or other small non-motorized craft to navigate from the area between Piers 1 and 2 to the area south of Pier 3. Piers 2 and 3 would be connected across the canal to the upland areas of the Project by overhead pedestrian walkways that would also provide access

PARK PROGRAMMING

The location of the park recreational program will be in direct response to varying structural capacities of the site. Activities that do not add significant weight loads, such as field sports, will be located on the piers. Activities that add considerable loads to the ground will be located on terra firma.

园区规划

公园游憩方案的位置，将直接受到基址不同结构承载力的影响。不显著增加负载的活动，如田径将被放置在码头上。而显著增加负载的活动将被放置在地面上。

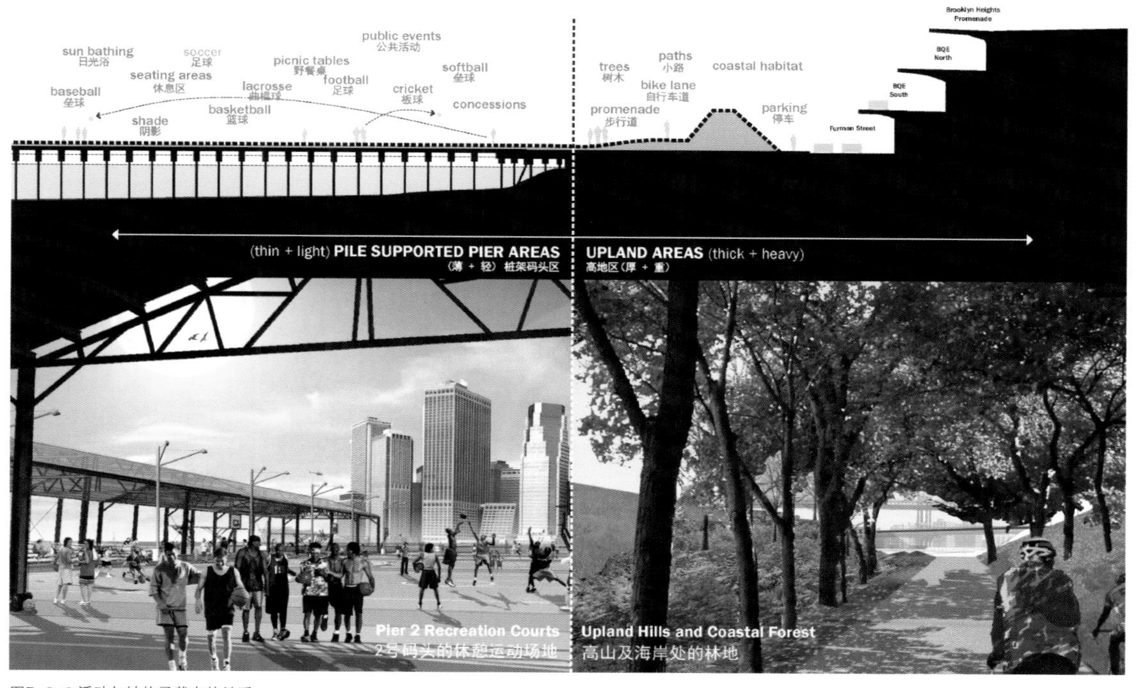

图7-2-6 活动与结构承载力的关系

Fig.7-2-6 The relationship between activities and structural capacities

间的“安全水域”已经纳入计划。这将为皮划艇、独木舟、桨船等水上项目提供12英亩（约4.86公顷）的安全水域。这个区域会根据海洋结构来设计，用来减弱经过船只所产生的波浪的影响。人们会用浮板来联结皮划艇和船员，并能够在安全水域中划出额外的波浪衰减区。3号和4号码头以及1号和2号码头之间的安全水域将通过水渠和“运河”连接。这条运河可以允许皮艇和其他小型非机动船在此通行，可以从1号和2号码头之间的区域到3号码头南面的区域。2号和3号码头会通过横跨运河的行人天桥连接公园内部的高地，这些天桥也将作为紧急车辆通道。4号码头将被保留作为自然保护区。1号码头则保留有一些木材，这些木材将作为过去工业的一种见证而存在。5号和6号码头之间的区域将作为历史性或有教育意义的船只停泊点。4号和5号码头之间是帆船和机动船的码头。游艇码头仅提供一

for emergency vehicles. The remnants of Pier 4 would be left in place and used as a nature preserve. At Pier 1, some timber piles would be left in place following removal of the pier deck to preserve a physical record of the site's industrial past. Outside the safe water zones, the area between Piers 5 and 6 would provide slips for the mooring of historic or educational vessels. Between Piers 5 and 4 would be a marina for sailboats and powerboats. The marina would provide limited boating services, including utility hookups and fueling.

Water taxi stops would be located at Pier 6, at the slip between Piers 2 and 3, at the north side of Pier 1, and near the John Street site, allowing waterborne transportation options for Project users and others coming to the Project site (Fig. 7-2-7).

3. Bikeways

A designated bikeway, coordinated with the Greenway Initiative effort, a local greenway advocacy and planning organization established to create a continuous greenway bicycle path from Greenpoint to Bay Ridge, would be integrated into the Project from Pier 1 to Pier 6. Entry for cyclists to the Project would primarily be at

些有限的船舶服务，比如提供工具和加油。

在6号码头、2号和3号码头之间的停泊点、1号码头北侧、约翰街设置了水上的士站点，给当地居民和游客提供到达公园的水路选择（图7-2-7）。

3. 自行车专用道

在一个地方林荫道宣传和策划组织提倡和努力下，在绿点与湾岭之间架起了一条特定的自行车道与林荫道相结合自行车林荫道，同时它也连接了1号和6号码头。自行车道的入口主要设置在老富尔顿大街和大西洋大街。自行车道的设计将最大限度上避免与人行道冲突。

4. 车辆的出入与停靠

车辆的出入与停靠点需要服务项目基地周围两个公园的游客以及附近商业和住宅用户。将会有一条新的街道建立在公园内部，接纳从弗曼街进入到酒店、宾馆和住宅的车辆。这将作为公园空间与发展区域之间的界线。这些道路就意味着这里有着不同的土地性质，例如减少酒店服务与住宅单元与开展公园活动相对应。

在寒冷的天气，5号码头将开通直通车服务，游客、运动员和粉丝们可以直接开车到达运动场，为了便利园务还会设置缓坡并提供小卡车。将分别在临近的街道、路边设置停车设施，同时还将在基地内设置停车场。

Old Fulton Street and Atlantic Avenue. To the maximum extent possible, the bikeways would be designed to avoid conflicts between pedestrians and roadways.

4. Vehicular Access and Parking

Vehicular access and parking would be needed for both park visitors and for users of commercial and residential buildings on the Project site. New streets would be created within the park to allow access from Furman Street into the hotel, restaurant and residential uses in the Project. These roadways would provide a clear boundary between park spaces and development parcels. These streets would define where the different activities would take place, such as drop-off at hotels and service to residential units versus the beginning of park activities.

In the cold weather months, park users may be permitted to drive onto the Project at Pier 5, which would allow field sports teams and their supporters to reach the playing fields directly by vehicle and provide for drop-offs and pickups.

Parking would be provided on adjacent local streets, in nearby off-street parking facilities, and within the Project boundaries at the proposed development sites.

5. Renewable Energy

The Project design would incorporate new technology to provide renewable energy, such as solar energy, to the extent practicable. Photovoltaic cell installations could provide a significant amount of the energy demand of Brooklyn Bridge Park, so consideration is being given to utilizing photovoltaic cells, and, possibly, hybrid streetlights. Photovoltaic cells could be mounted on the roofs of the remaining pier sheds.

6. Habitat

An important design goal is to establish the maximum number of sustainable, functioning habitats

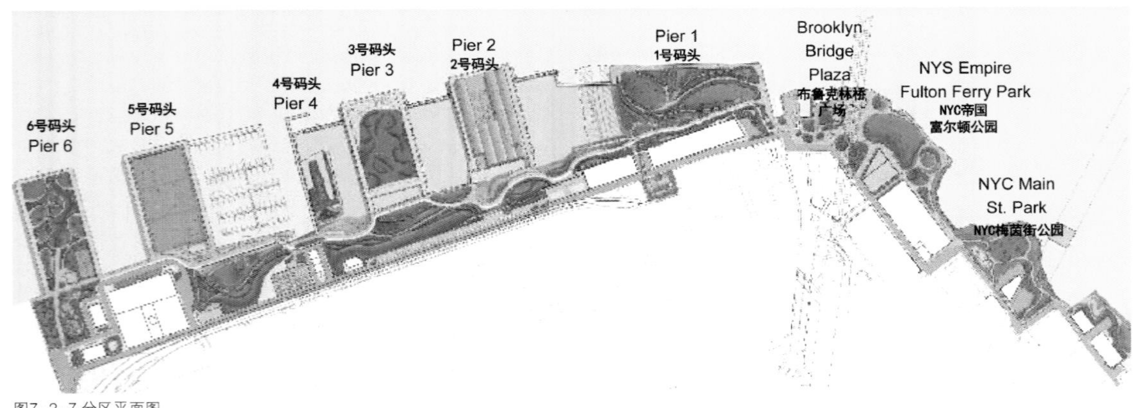

图7-2-7 分区平面图
Fig.7-207 The partition plan of the site

5. 可再生能源

公园将采用新技术，在切实可行的范围内提供诸如太阳能之类的新能源。光伏电池装置也在考虑之中，因为它能够满足布鲁克林桥公园大部分的能源需求，但是也有可能采用混合路灯。而在其他的屋顶和码头将采用光伏电池。

6. 自然栖息地

布鲁克林桥一个重要的设计目标是在其中尽可能多地建立可持续运作的自然栖息地。这包括6号码头的沙丘；1号码头的沿海地段与1号、4号和5号码头的高地森林；1号、2号、3号、5号和6号码头的高地灌木林；2号和3号码头的高地草坪；1号码头、4号码头以及布鲁克林和曼哈顿大桥之间的山地、淡水沼泽与潜水栖息地；2号与3号码头附近的湿地。

7. 生态恢复

本着尊重生态的原则，设计人员设计了几组“后

in the Brooklyn Bridge Park. Natural habitats may include dunes on Pier 6; coastal forest on Pier 1 and its upland, and the uplands of Piers 4 and 5; shrubland on Piers 1, 2, and 3 and on the uplands of Piers 5 and 6; a wildflower meadow on the uplands of Piers 2 and 3; marsh and shallow water habitats on and adjacent to Piers 1 and 4 and between the Brooklyn and Manhattan Bridges; and freshwater swale and wetlands near Piers 2 and 3.

7. Ecological Restoration

With respect to site ecology, the park planners explored a number of initiatives grouped under the term of "post-industrial nature" that were aimed at re-establishing a series of functioning ecosystems on the currently lifeless site. The combined planting and stormwater treatment strategy for Brooklyn Bridge Park is founded on four guiding principles: to create many different natural areas that serve individually as gardens but work together to establish a new site ecology; to treat as much stormwater as possible onsite; to maximize area of shade and cover from the wind, and to preserve open space. Natural habitats being reintroduced to the site include coastal shrublands, freshwater wetlands, coastal forest, a wildflower meadow, a marsh, and shallow water habitats.

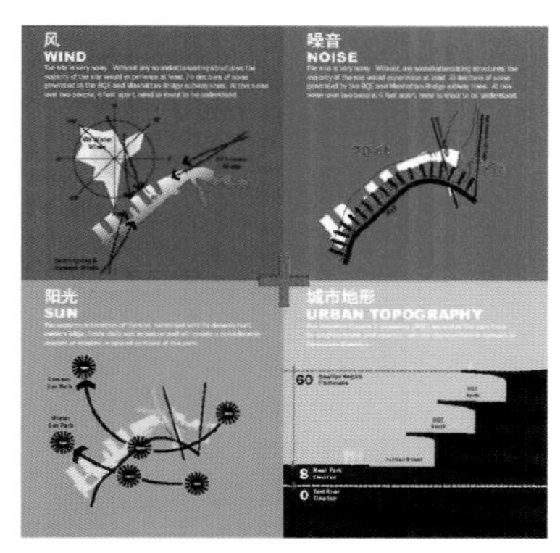

MICRO-CLIMATE and FORM

The site is exposed to a great deal of environmental forces: high solar and wind exposure, uncomfortable noise levels from the adjacent expressway, river currents, wakes, and sedimentation. The physical composition of the park proposes an inseparable link between micro-climate and form.

微气候和形式

该处基址所受到的环境力量很大：强烈的日光和风蚀，临近高速路产生的噪音，水流冲击，尾波冲击，泥土沉淀。园内实体的组成在微气候和形式之间提出一种不可分割的联系。

图7-2-8 微气候和形式
Fig.7-2-8 Micro-climate and form

工业自然”，希望通过这种手段在原本死气沉沉的基址上重新建立起生态系统。布鲁克林桥公园的混合种植和雨洪管理系统的建立遵循下面三项指导原则：建立许多不同的自然区域，散能够独树一帜，合则能共同组成一个新的生态点；尽可能多地收集雨水；多设置荫蔽的区域，能够给开敞的空间遮风避雨。一些沿海的灌木丛、浅水湿地、沿海的森林、野花草地等自然栖息地也会重新在这里出现。

8. 微气候与人体舒适度设计

自然植物可以使得基址更适宜人们休闲娱乐。设计师统筹考虑园内的种植分布、日照方向与风向因素，最大限度地提高人体的舒适度（图7–2–8）。依据不同的气候，基址内不同的相关问题所产生的需求，都分别通过不同的元素来解决：贯通南北的树篱不仅可以提供阴影，还可以遮蔽夏日午后水面反射的阳光；地形与建筑通过不同形式和不同方向的结合，既保证冬季充足的阳光，又可以遮挡寒风；地形较为高的山丘能够减弱布鲁克林–皇后区高速公路所产生的巨大噪音，使得公园更加安静与舒适，同时还能够提供望向海港的绝佳视野（图7–2–9）。

五、 总结

公园因布鲁克林大桥而得名，这座赫赫有名的大桥建于1883年，是当年世界上最长的悬索桥。公园所在地曾是废弃货船厂，是曾繁荣一时而今衰落的城市滨水区。规划从生态建设入手，旨在提高城市滨水空间的景观质量。建成后的布鲁克林桥公园，是一处独特的公共滨水公园，它不仅提供了舒适宜人的公共休闲空间，而且给人们提供了观赏纽约港迷人景致的绝佳视角。本项目设计中，充分利用滨水景观的综合功能优势，开拓城市人居、游憩、旅游、经济发展与生态保护于一体的现代城市滨水景观，使枯燥乏味的工业码头和高地重新焕发生机。

根据人类聚居环境学理论，人类环境分为居住和聚集两类，滨水地区更适合后者，所以现代滨水区在开发使用中，优先考虑社会公众的需求，强调

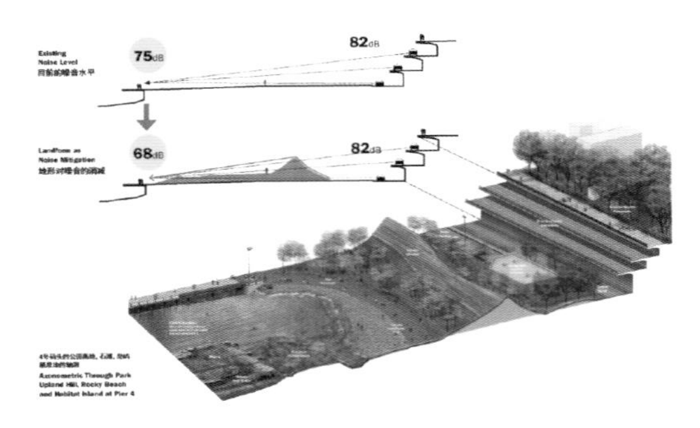

图7–2–9 噪音的衰减
Fig.7-2-9 Noise attenuation

8. Microclimate and Human Comfort

Natural plantings also have the potential to make the site more hospitable as human habitat. The dispersal of planting that is suggested throughout the site will be coordinated with solar orientation and wind protection to maximize human comfort (Fig. 7-2-8). Depending on the season, or the particular site-related issue, different needs are accommodated through different park elements: a nearly continuous meander and hedgerow of trees running north-south will provide shade and relief from the intense summer afternoon sun that reflects off the water; the topography and buildings provide protected sun exposure in the winter and shelter from the wind in different ways and from different directions; and tall hills absorb the tremendous noise generated by the Brooklyn-Queens Expressway, making the park quieter and more pleasant while also providing new elevated views to the harbor (Fig. 7-2-9).

V. Summary

The park is named after the Brooklyn Bridge which was built in 1883, the longest suspension bridge at that time, and the location of the park was once a deserted cargo factory, an urban front that prospered and then declined. The planning starts with biological construction and aims at promoting landscape quality of the urban water front. The completed Brooklyn Bridge Park is quite unique. It not only creates comfortable public relaxation space, but also provides a wonderful view for the charming scenery of New York habour. The planning has taken full advantages of the comprehensive function of water front landscape, and has developed contemporary urban water front landscape, the combination of residence, refreshment, travel, economic growth and biological protection, injecting new vigour to the boring industrial port and highland.

According to Sciences of Human Settlements, human environment is divided into living and gathering, and the water front is suitable for the

大众化与公共性，力求创造符合时代需求的城市滨水景观。本设计中，清晰安全的人行通道、自行车专用道的合理规划及车辆出入与停靠的完善措施保障了公园良好的可达性和安全性；开放式的视觉通廊、具有亲和力的滨水空间、安全水域的划分体现了公共性作为城市滨水区发展的基本取向；生态修复、微气候与人体舒适度设计、可再生能源的利用凸显了设计的人文观与自然观。

later. So in modern water front development, public needs are the first thing taken into consideration, which emphasizes popularity and publicity. In the planning, reasonable planning for the distinct and safe pavement and bicycle lanes and perfect measures for vehicle entering and parking secure easy access and safety of the park; the division of open vision corridor, favorable water front space and safety water area has fully reflected publicity as basic orientation for urban water front development; design for biology restore, microclimate and human comfort, and the use of recycling energy expose the humanity and natural views of the designer.

思考题

1. 布鲁克林桥公园的设计事务所MVVA（Michael Van Valkenburgh and Associates）界定的“后工业自然”是“能快速激活场地生态性；能在密集使用的城市中发展；是不加修饰的人工环境”，其是如何在这个设计中得到体现的?

2. 尝试分析布鲁克林桥公园的城市交点的设置与分布对于其与城市街道以及周边社区联系所起到的作用。

3. 公园38%的面积是由码头支撑的，如果是你将如何通过分区与活动的设置来解决这个问题，又或是你可以提出更好的解决办法。

4. 布鲁克林桥公园的另一个主要特征是多样组合的边缘状况，这不仅为人，还为其他生物提供了机会。尝试从各个方面来分析公园的边缘组合所带来的优势。

5. 布鲁克林大桥公园的成功在于它体验的多样性、创造可重复使用的材料，以及保持一种充满生机的历史感。试论证MVVA是如何从这几个方面获得成功的。

Questions

1. The Brooklyn Bridge Park Design Firm MVVA (Michael Van Valkenburgh and Associates) defined “The Post Industrial Nature” as “capable of quick inducing land ecological characters, capable of developing in highly populated city area, non edited artificial environment ”, try to discuss how these concept been reflected in this case?

2. Try to discuss the set up and distribution of the urban junctions in Brooklyn Bridge Park though how are they connected to the city streets and surrounding blocks.

3. 38% of the park area is supported by the piers, if you will be the designer how would you solve this problem , or you can put forward a better solution.

4. Another major feature of the Brooklyn Bridge Park is a diverse portfolio of the edge, which not only for the man, but also provides an opportunity for other organisms. Try to analyze all aspects of the advantages brought about by the park's edge combination.

5. Brooklyn Bridge Park's success is the diversity of the experience, reusable materials, and a vibrant sense of history. Try to demonstrate how MVVA succeeded from several aspects.

注释/Note

图片来源/Image：图7-2-1，图7-2-3~图7-2-6，图7-2-8，图7-2-9来自http://www.asla.org/2009awards/011.html ；图7-2-2，图7-2-7来自http://www.brooklynbridgepark.org/

参考文献/Reference：本案例引自ASLA2009年荣誉奖项目详细描述

推荐网站/Website：http://www.asla.org/2009awards/011.html ；http://www.brooklynbridgepark.org/ ；http://www.onebrooklyn.com/

第三节 碳技术、浮桥藻类公园案例
Section III Carbon T.A.P.//Tunnel Algae Park

一、 项目综述

碳技术、浮桥藻类公园（图7-3-1）是一个概念设计公园，它由PORT建筑设计与城市规划咨询公司设计完成。该公园是一座连接了布鲁克林和曼哈顿的可以生产生物燃料的浮桥式藻类公园。公园可通过其中涵养的藻类将纽约市区车水马龙的行车隧道中产生的二氧化碳气体转化为生物燃料。这座连接了布鲁克林和曼哈顿的具有碳捕获功能的临时性浮桥，由精心设计的湿地、休闲娱乐设施以及高速自行车道组成，其主路则是横跨河流的码头状结构。

藻类公园不仅能够实现碳封存和碳消耗，同时也将生产氧气、生物塑料、保健食品、农业饲料以及生

图7-3-1布鲁克林-炮台公园隧道碳技术浮桥藻类公园
Fig.7-3-1 The Carbon T.A.P.Tunnel Algae Park over the Brooklyn-Battery Tunnel

I. Overview

Carbon T.A.P.//Tunnel Algae Park (Fig.7-3-1) is the concept-park design competition which was proposed by PORT Architecture + Urbanism. It is a Floating Algae Park between Brooklyn and Manhattan to produce bio-fuel. It will tout a series of algae pontoons to convert carbon-dioxide emissions from NYC's busy streets into bio-fuel. The temporary floating carbon-capturing bridge between Brooklyn and Manhattan will house well thought-out wetlands, recreation facilities, and high-speed bike lanes. The parkway will keep on moving across the river on pier-like armatures.

The algal park won't only sequester and consume carbon, but will also produce oxygen, bioplastics, nutraceuticals and agricultural feeds, along with bio-fuel of course. In a way, it will be an one-stop solution to problems relating to fuel, agriculture, ecology (includes avian and aquatic habitats too), and new public spaces (construction of swimming pools, boardwalks, and plazas included). Well, the concept is applicable to tollbooths, coal-fired power plants, and automobile tunnels as well.

II. Introduction of the Site and Context

The Brooklyn-Battery Tunnel, which opened in 1950, connects Southwestern Brooklyn with the Wall St. area of Manhattan,its position is shown in Fig.7-3-2 .As is shown in Fig.7-3-3,it is a toll road in New York City which crosses under the East River. It consists of twin tubes, carrying four traffic lanes, and at 9,117 feet (2,779 m) is the longest continuous underwater vehicular tunnel in North America.

物燃料。某种意义而言，这将是一个一站式解决燃料、农业、生态（包括禽类和水生动植物栖息地）以及新的公共空间（包括游泳池、浮桥以及广场等）等问题的解决方案。同时，这一设计方案还适用于收费站、燃煤发电厂以及汽车隧道等项目的设计。

二、 场地及周边背景介绍

布鲁克林-炮台公园隧道开通于1950年，连接了布鲁克林西南端和曼哈顿华尔街地区，其位置如图7-3-2所示。如图7-3-3所示，它是一条位于纽约东河下的收费隧道，由两根管道构成，双向四车道，长9 117英尺（约2 779米），是北美最长的水下隧道。

本方案在城市集中产生二氧化碳的地区设计了一种新型绿色基础设施。这种新型绿色基础设施采用专业系统、产业化规模的藻类农业来封存和消耗温室气体（特别是二氧化碳），以减少其大气排放，同时通过产生氧气、生物燃料、生物塑料、营养品和农业饲料从而创造新的经济资源。如图7-3-4至图7-3-6所示，新的绿色基础设施呈码头状沿布鲁克林-炮台公园隧道链状分布。

每平方英尺的藻类每年可以消耗9.7磅（约4.4公斤）二氧化碳气体，每个藻类生物反应器面积为1英亩，沿着布鲁克林至炮台公园隧道所建立的碳技术、浮桥藻类公园中，设置28个藻类生物反应器即可消耗隧道每年所产生的26 000000磅（约11 793吨）二氧化碳气体。

三、设计理念

这一设计方案的独特之处在于，它不是单纯地在城市中大规模兴建新的绿色基础设施，而是利用已有基础设施来创造一个独特的公共休闲设施。与以往认为基础设施的形象欠佳、必须藏于地下的观点不同，设计师认为这些基础设施的改造和再设计是创造新的城市形象和社会产业的机会，这些都足以对未来美国的城市景观产生积极的影响。

设计师对新型的基础设施的类型建议是：一部

In the scenario outlined herein, a new type of green infrastructure is deployed at urban locations comprising concentrated sources of CO_2 production. This new infrastructure utilizes a proprietary system of industrial scale algal agriculture to sequester and consume greenhouse gas emissions (in particular CO_2) in order to limit their introduction into the atmosphere, while simultaneously creating a new economic resource through the production of oxygen, biofuels, bioplastics, nutraceuticals and/or agricultural feeds. In the Fig.7-3-4 to Fig.7-3-6 shown, this new infrastructure manifests itself as a series of pier-like armatures linked to the ventilation system for the Brooklyn-Battery tunnel.

One square foot of algae can consume 9.7 lbs CO_2 each year, and each algae bioreactor has 1 acre of algae. 28 algar bioreactors proposed in the Carbon T.A.P.//Tunnel Algae Park could consume the 26,000,000 CO_2 contained in the Brooklyn-Battery Tunnel.

III. Designing Conception

What is unique about this proposition is not just the introduction of large-scale green infrastructure in the context of a city, but rather the use of this infrastructure to create an exceptional public realm amenity for the

图7-3-2 布鲁克林-炮台公园隧道位置
Fig.7-3-2 The position of Brooklyn-Battery Tunnel

图7-3-3 布鲁克林-炮台公园隧道入口现状
Fig.7-3-3 The photo of Brooklyn-Battery Tunnel

分作为应对气候变化的措施，一部分作为农业生产，一部分进行生态保护，一部分承担公共休闲的作用，一部分用以推动经济。这一建议涵盖了所有对于新的城市基础设施建设的要求——从根本上提高一个城市的经济和社会发展，同时保证与之相关的当下和未来的居民生活可持续发展。

四、 生物反应器设计原理

藻类是地球上生命最强大的生物之一。它在世界各地每一块大陆上生长——只要在这里它们能够通过光合作用或有机碳的摄取来吸收能量，特别是吸收二氧化碳，它们就能够高度适应任何生存环境。

各种藻类在水生生态环境中发挥着重要作用，它们悬浮在水体中的微观形态（浮游植物）成为大多数海洋食物链的食物来源，成为鱼类和海洋哺乳动物的食物。它们每年消耗地球近50%的二氧化碳，同时产生超过45%的氧气。

单细胞藻类会消耗二氧化碳或氮氧化物，然后演变成为一种可被重复利用的形态。这种形态是一种类似于木炭的生物质能，可以像煤或液化气一样燃烧，或用于制造塑料制品、保健品或食品。

大气中二氧化碳浓度并不高，不适合作为原材料用于产业化规模生产藻类。然而，煤矿、热电厂或制造业设施可以作为高浓度二氧化碳的提供源。特别是行车隧道，每年可以产生数以亿计立方英尺的二氧化碳。2003年，世界上有6亿辆汽车，全球

隧道行车通道
以及所排放的二氧化碳资源
existing vehicular tunnel and
contained CO_2 emission
source

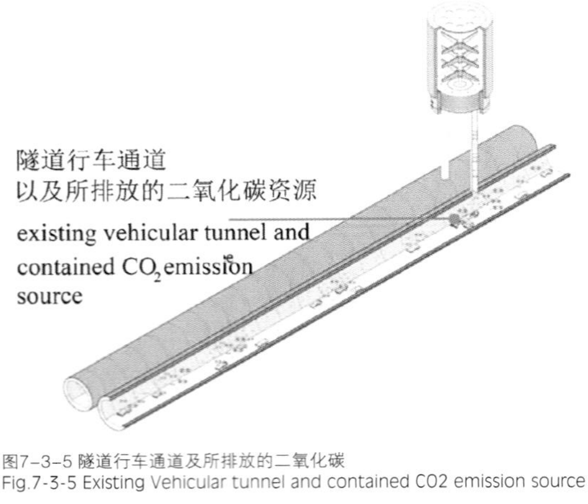

图7-3-5 隧道行车通道及所排放的二氧化碳
Fig.7-3-5 Existing Vehicular tunnel and contained CO2 emission source

city. Rather than considering urban infrastructures as a necessary evil only to be hidden or mitigated, the architects view the renovation and re-imagination of these systems as opportunities to create new forms of civic and social domain that have the capacity to positively transform the American urban landscape.

The architects' proposal for a new infrastructural typology that is one part climate action; one part agricultural production; one part ecological preserve; one part public realm; and one part economic catalyst represents what should be the aspiration for all newly deployed urban infrastructures – the ability to fundamentally improve the economic and social quality of a city, as well as the associated lives of its current and future residents.

IV. Bioreactor Design Principle

Algae are one of the most robust classifications of life on earth. Thriving on every continent, they are highly adaptive to any physical environment where they are able to derive energy from photosynthesis and the uptake of organic carbon, particularly in the form of CO_2.

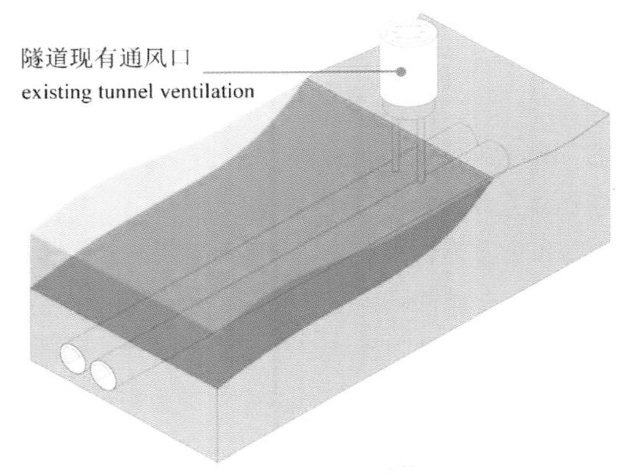

图7-3-4 隧道现有通风口
Fig.7-3-4 Existing tunnel ventilation

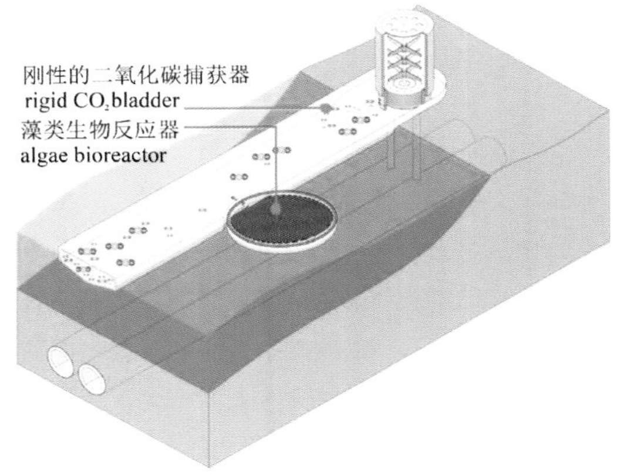

图7-3-6 刚性二氧化碳捕获器以及藻类生物反应器
Fig.7-3-6 The Rigid CO2 Bladder and algae bioreactor

14%的二氧化碳产生量来自于交通运输，特别是汽车。而通过藻类光合作用，可以将二氧化碳转化为氧气，其生化方程式为：

太阳光+$6CO_2$+$6H_2O$=$C_6H_{12}O_6$+$6O_2$（光合作用）

五、 二氧化碳二氧化碳捕获和储存设计

捕获并利用布鲁克林–炮台公园隧道二氧化碳排放量中的一小部分，就可以为一个大型藻类生产基地提供食物来源。但是，在利用这些二氧化碳之前必须攻克的一道难题是在它传递原料给藻类之前如何捕获它。

在方案设计中，设计师利用了一个分体式设计来捕获二氧化碳，并且控制其为一个工业规模的藻类反应器提供原料。在系统中，排放出的二氧化碳存储在一个刚性的罐子或气球中，可以用来为20 000平方英尺（约1 858平方米）的藻类作物提供原料。生物反应器可用于分离藻类作物的收获和加工。

这些被捕获的高浓度二氧化碳大多集中存储于水体附近，使得二氧化碳捕获器可以形成一个大的码头或者滨水填充区。由于二氧化碳的需求量和产生量都十分巨大，将这些数量巨大的二氧化碳捕获器集成设计成为其附近城市范围内的码头显然是个巧妙的设计。也就是说，设计师将这些涵养藻类的码头，设计成为一个新型的公共开放空间，兼具滨水空间和绿色基础设施的功能。

六、 意义

大气中二氧化碳浓度虽然较高，但没有高至可以工业规模化生产藻类。而在行车隧道中二氧化碳浓度极高，可以利用其捕获二氧化碳为藻类提供能量。PORT公司建筑师正是试图通过藻类利用二氧化碳生产氧气的特性来进行公共空间设计。如图7–3–13所示，遍布全美的18条长于2 000英尺（609.6米）的行车隧道，以及成千上万的煤炭和天然气电厂，都可以采用这种系统，以求在全美更广泛的范围内更好地保护环境。如图7–3–14至图7–3–29所示，PORT公司分别为这18条隧道分别设

The various sorts of algae play significant roles in aquatic ecology. Microscopic forms that live suspended in the water column (phytoplankton) provide the food base for most marine food chains that feed fish and marine mammals. They are responsible for nearly 50 percent of the earth's annual carbon-dioxide consumption and more than 45 percent of the oxygen production.

Single-cell algae can consume carbon dioxide or nitrogen oxides and then be "harvested" in a reusable form. This form, a biomass similar to an artist's soft charcoal, can be burned like coal or liquefied into oil or used to make plastics, nutraceuticals or food.

Atmospheric CO_2 concentrations are not high enough for industrial-scale production of algae. However, concentrated CO_2 sources such as Coal/Co Gen power plants or manufacturing facilities offer potential sources of high-level CO_2 concentrations. Vehicular tunnels in particular can produce hundreds of millions of cubic feet of CO_2 per year. There are 600 million automobiles in the world as of 2003. 14% of the carbon-dioxide produced globally comes from transportation emissions, particularly auto mobiles. But the CO_2 can be conversed to O_2 via photosynthesis by algae, its Biochemical equation is:

Sunlight+$6CO_2$+$6H_2O$=$C_6H_{12}O_6$+$6O_2$(Photosynthesis)

V. Design for CO_2 Capture and Storage

Capturing even a small fraction of these CO2 emissions in the Brooklyn-Battery Tunnel would offer an enormous food source for large-scale algae production. However, the challenge of these sources is holding the CO2 before it is delivered to some vessel containing the algae.

In this scenario, the architects use a two-part system for the capturing of CO_2 and providing its controlled delivery to an industrial-scale algae bioreactor. In this system, CO_2 emissions are captured and held in what is essentially a giant bladder or rigid balloon which is configured to deliver CO_2 to a series of 20,000-sq.-ft. bioreactors that can be detached for harvesting and processing of their algae crop.

Many of these concentrated CO_2 sources are sited near bodies of water, allowing the CO_2 bladder to function as essentially a large pier or an expanded waterfront. As the volume of CO_2 needed and produced is quite large, the pier into its urban context is an absolute thoughtful integration. That said, they propose that these algal piers become the sites of a new typology of public open space that bundles waterfront access with productive green infrastructures.

计了碳技术、浮桥藻类公园。其分布位置、长度见表7-3-1。

七、 总结

布鲁克林–炮台公园隧道是北美最长的水下隧道。1936年罗伯特·墨西斯曾提出要从布鲁克林建设一座桥梁连接布鲁克林的雷德胡克与曼哈顿的炮台公园。但这座桥自始至终没有建成。墨西斯梦想中的桥是“汽车世纪”的产物。虽然路线类似，但是图7-3-29所示的碳技术、藻类浮桥公园却是“生态世纪”的趋势，这代表了未来新一代基础设施的设计风向。

本案例为典型的生态设计作品，西姆·万德瑞认为：生态设计具有现场解决问题、计较生态得失、设计结合自然、每个人都是设计师以及让对象的天性可视化5个基本特征。本方案作为生态设计的获奖作品，是对自然过程有效的适应和统一，营造

VI. Significance of the Proposal

Atmospheric CO_2 concentrations are high, but not high enough for industrial scale production of algae. However, in vehicular tunnels CO_2 concentration is high, which makes it possible to harvest it and use it to feed algae. Port Architects is trying to use the ability of algae to produce oxygen when exposed to CO2.With 18 vehicular tunnels of greater than 2,000 feet across the U.S. shown in the Figure 7-3-12, and thousands of coal and natural gas driven power plants, deployment of adaptations of this system have the potential to reinvigorate a wide range of urban environments throughout the country. As is shown in Fig. 7-3-13 to 7-3-28, Port Architects propose Carbon T.A.P.//Tunnel Algae Park for all these 18 tunnels.There position and length are shown in Tab.7-3-1.

VII. Summary

The Brooklyn-Battery Tunnel is the longest underwater tunnel in North America. In 1936 Robert Moses proposed to build a bridge from Red Hook, Brooklyn to the Battery in Lower Manhattan. It was never built. Moses' bridge was the product of the automobile century. Though it traverses a similar

图名	隧道名称	连接地点	长度（英尺）
图13	布鲁克林–炮台公园隧道	纽约纽约市曼哈顿–布鲁克林	9 117
图14	荷兰隧道	纽约州纽约市–新泽西州泽西市	8 558
图15	特德·威廉斯隧道	马萨诸塞州波士顿市区–洛根机场	8 488
图16	林肯隧道	纽约州纽约市曼哈顿–新泽西州	8 216
图17	巴尔的摩–哈勃隧道 麦克亨利堡隧道	马里兰州巴尔的摩 马里兰州巴尔的摩	7 650 7 200
图18	汉普顿路桥隧道	弗吉尼亚州汉普顿–诺福克	7 479
图19	皇后区中心城隧道	纽约州皇后区–中心城	6 414
图20	桑纳隧道 卡拉汉隧道	马萨诸塞州波士顿 马萨诸塞州波士顿	5 653 5 089
图21	西姆布尔舒尔海峡隧道	弗吉尼亚州弗吉尼亚海滩–开普查尔斯	5 280
图22	切萨匹克海峡隧道	弗吉尼亚州弗吉尼亚海滩–开普查尔斯	5 280
图23	底特律温莎隧道	密歇根州底特律市–加拿大安大略省温莎市	5 160
图24	梅里麦克纪念大桥–隧道	弗吉尼亚纽波特纽斯–萨福克	4 880
图25	波西–韦伯斯特街隧道	加利福尼亚州奥克兰市	4 500
图26	城中隧道	弗吉尼亚州波茨茅斯–弗吉尼亚州诺福克	4 194
图27	市区隧道	弗吉尼亚州波茨茅斯–弗吉尼亚州诺福克	3 813
图28	乔治·华莱士隧道	阿拉巴马州莫比尔	3 300

表7-3-1 全美18条2000英尺以上隧道分布位置、长度
Tab.7-3-1 The length and position of 18 Vehicular tunnels over 2 000 feet

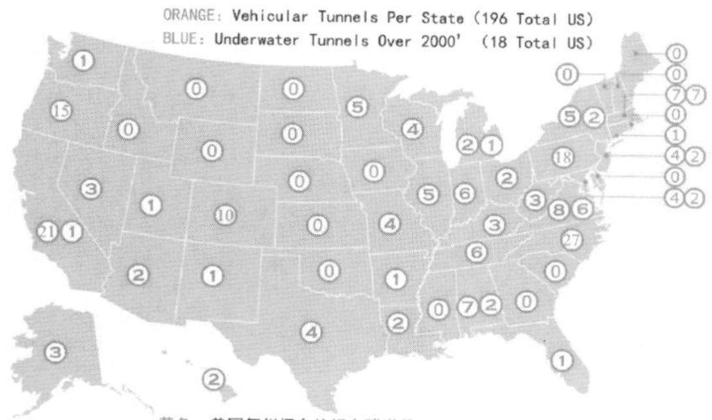

图7–3–13 全美行车隧道及水下行车隧道分布（数字核实）
Fig.7-3-13 The Vehicular tunnels and underwater Tunnels over all the USA

图7–3–14 布鲁克林–炮台公园隧道公园
Fig.7-3-14 The Carbon T.A.P.Tunnel Algae Park over the Brooklyn-Battery Tunnel

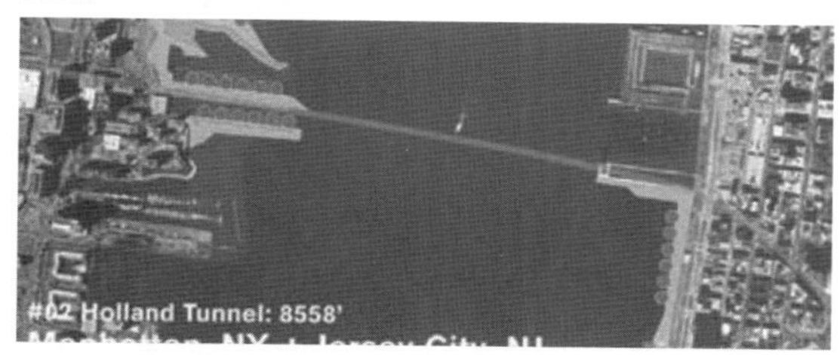

图7–3–15 荷兰隧道公园
Fig.7-3-15 The Carbon T.A.P.Tunnel Algae Park over the Holland Tunnel

图7–3–16 特德・威廉斯隧道公园
Fig.7-3-16 The Carbon T.A.P.Tunnel Algae Park over the Ted Willams Tunnel

图7–3–17 林肯隧道公园
Fig.7-3-17 The Carbon T.A.P.Tunnel Algae Park over the Lincoln Tunnel

图7–3–18 巴尔的摩哈勃隧道及麦克亨利堡隧道公园
Fig.7-3-18 The Carbon T.A.P.Tunnel Algae Park over the Baltimore Harbor Tunnel and Fort McHenry Tunnel

图7–3–19 汉普顿路桥隧道公园
Fig.7-3-19 The Carbon T.A.P.Tunnel Algae Park over the Hampton Roads Bridge Tunnel

图7–3–20 皇后–中心区隧道公园
Fig.7-3-20 The Carbon T.A.P.Tunnel Algae Park over the Queens-Midtown Tunnel

图7–3–21 桑纳隧道及卡拉汉隧道公园
Fig.7-3-21 The Carbon T.A.P.Tunnel Algae Park over the Sumner Tunnel and Callahan Tunnel

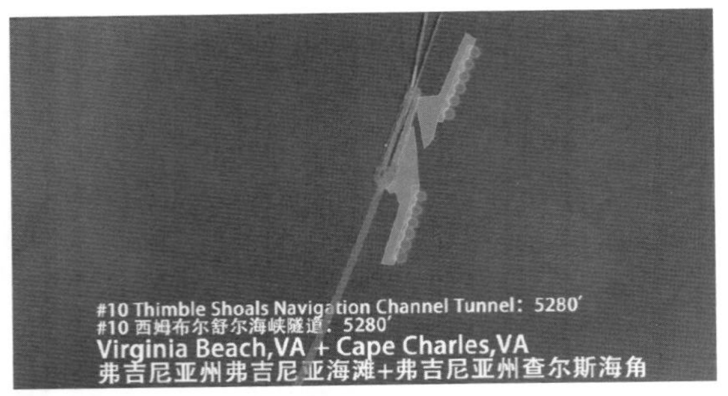

图7-3-22 西姆布尔舒尔海峡隧道公园
Fig.7-3-22 The Carbon T.A.P.Tunnel Algae Park over the Thimble Shoals Navigation Channel Tunnel

图7-3-23 切萨匹克海峡隧道公园
Fig.7-3-23 The Carbon T.A.P.Tunnel Algae Park over the Chesapeake Navigation Channel Tunnel

图7-3-24 底特律温莎隧道公园设计
Fig.7-3-24 The Carbon T.A.P.Tunnel Algae Park over the Detroit-windsor Tunnel

图7-3-25 梅里麦克纪念大桥-隧道公园
Fig.7-3-25 The Carbon T.A.P.Tunnel Algae Park over the Monitor Merrimac Memorial Bridge Tunnel

图7-3-26 波西-韦伯斯特街隧道公园
Fig.7-3-26 The Carbon T.A.P.Tunnel Algae Park over the Posey and Webster Street Tunnel

图7-3-27 城中隧道公园
Fig.7-3-27 The Carbon T.A.P.Tunnel Algae Park over the Midtown Tunnel

图7-3-28 市区隧道公园
Fig.7-3-28 The Carbon T.A.P.Tunnel Algae Park over the Downtown Tunnel

图7-3-29 乔治・华莱士隧道公园
Fig.7-3-29 The Carbon T.A.P.Tunnel Algae Park over The Sumner Tunnel and Callahan Tunnel

具有生态本质的景观与环境。从项目中众多技术措施和设计细节可见：生态设计与传统设计的区别不仅在于形态特征的差异，更涉及能源与材料、环境污染、平衡性与持续性、敏感性、多样性、整体性等各个方面。

在偌大的浮桥上建公园本身就颠覆了人们对桥的印象和理解，桥不再是简单的交通的链接，不再是金属或者是钢筋混凝土的构筑物，而变成了一个有湿地、有动植物的公共休闲场所，成为人们可以休闲漫步的城市公园。

设计团队对基础设施的理解很有创新性，他们认为基础设施可以同时承载多种功能，除了大家所熟知的公共休闲功能外，还要具备对环境的保护和调控能力，甚至对农业生产和经济都要有推动作用。藻类公园对二氧化碳有理想的捕获和存储功能，可以在滨水城市或污染集中的地块周围应用。这种多功能的设施提高了土地的利用率，在我国人口密集的东部地区有很大的推广意义。

route, this proposed Brooklyn-Battery Bridge shown in Figure 7-3-29 is a product of the ecological century, representing a new infrastructural paradigm.

This case is typical of the ecological design work. In his book "Ecological Design", Sim van der Ryn has wrote that there are 5 principle of ecological design, which are "solutions grow from place, ecological accounting informs design, design with nature, everyone is a designer, and make nature visible". As the award of the ecological design, this case is the effective adapt and unified of the natural process, which has built with ecological landscape and environment. Many of the technical measures and design details in this design shows that the ecological design and traditional design is not only the difference in characteristics of morphology, but in energy and material, environmental pollution, balance and sustainability, sensitivity, diversity, integrity and other aspects.

Design team has make a innovative design on the infrastructure , they think infrastructure can produce the desired result in variety of functions, which are not only recreational function as well known, but also the protection of the environment and control ability, even promoting agricultural production and economic. Algae park has the fuction of capture and storage for carbon dioxide. It can be used in waterfront city or seriously polluted area.The multi-function infrastructure which has improved the ratio of land utilization, has a great significance for popularization in our country's densely populated region in the east.

思考题

1. 请简述碳技术/浮桥藻类公园的项目背景，特别是其场地及周边背景。

2. 碳技术/浮桥藻类公园的设计理念是什么？

3. 本设计中的生物反应器设计原理是什么？是否能写出其反应方程式？

4. 关于二氧化碳的捕获和储存，设计师进行了哪些奇思妙想？

5. 虽然碳技术/浮桥藻类公园是一个概念性设计，其所具有的重要意义是什么？

Questions

1. Please introduce the overview of Carbon T.A.P.//Tunnel Algae Park,especially the site and context.

2. What are the designing conception of Carbon T.A.P.// Tunnel Algae Park?

3. What are the Principle of Bioreactor Design? Can you give the Biochemical equation?

4. What absolute thoughtful integration had the artist made for CO_2 Capture and Storage?

5. Though this is a concept-park design, what are the Significance of the Proposal?

注释/Note

图片来源/Image：本案例图7-3-2，图7-3-3来自http://en.wikipedia.org/wiki/Brooklyn%E2%80%93Battery _ Tunnel；其他图片引自Carbon T.A.P. // Tunnel Algae Park : PORT Architecture + Urbanism

参考文献/Reference：本案例引自"Carbon T.A.P. // Tunnel Algae Park : PORT Architecture + Urbanism"

推荐网站/Website：http://portarchitects.com/profile/carbon-tap--tunnel-algae-park/

第四节 生态城——阿科桑底
Section IV Arcology — Arcosanti

一、 项目综述

1. 项目简介

1970年，科桑底基金会组织开始着手建立阿科桑底城（图7–4–1，图7–4–2）。它是位于凤凰城以北60英里（约96.56公里）的科德斯交界处的沙漠小镇，17号州际高速262出口，它是亚利桑那州高原沙漠区的试验城镇，距离菲尼克斯市以北70英里（约112.65公里）。主体建筑建成之后可以容纳5000人居住，向世人展现了既能改善城市居住环境，可以降低对地球富饶土地破坏的设计理念的设计理念。复杂的大型结构和附带的大型温室占地仅25英亩（约10公顷），保留有4 046英亩（约16.4平方公里）的广阔土地，有利于城市居民与自然乡村景观保持亲近的关系。

阿科桑底是在城市建筑生态学理论基础之上设计的（生态建筑及城市），由保罗·索勒里提出。城市建筑生态学理论主张建筑体和生活其间的人两者之间的关系，应该像高度进化的生物内部器官一样相互影响。这意味着许多的系统共同运作，有利于人和资源高效流通或循环，建筑能够满足多用途需求，朝向有利于采光、加热和降温。在这个复合体中，创造性环境、公寓、商业、制造、技术、空地、工作室、教学和文化活动场所都是方便可达的，同时私密性也是整个设计中的重点考虑对象。温室提供了私人、公共的

I. Overview

1. Introduction

In 1970, the Cosanti Foundation began building Arcosanti (Fig.7-4-1,Fig.7-4-2). The Arcosanti site is located about 60 miles north of Phoenix, taking Interstate-17 to Exit 262 at Cordes Junction. It is an experimental town in the high desert of Arizona, 70 miles north of metropolitan Phoenix. When complete, Arcosanti will house 5000 people, demonstrating ways to improve urban conditions and lessen our destructive impact on the earth. Its large, compact structures and large-scale solar greenhouses will occupy only 25 acres of a 4060 acre land preserve, keeping the natural countryside in close proximity to urban dwellers.

Arcosanti is designed according to the concept of arcology (architecture + ecology), developed by Italian architect Paolo Soleri. In an arcology, the built and the living interact as organs would in a highly evolved being. This means many systems work together, with efficient circulation of people and resources, multi-use buildings, and solar orientation for lighting, heating and cooling. In this complex, creative environment, apartments, businesses, production, technology, open space, studios, and educational and cultural events are all accessible, while privacy is paramount in the overall design. Greenhouses provide gardening space for public and private use, and act as solar collectors for winter heat.

Arcosanti is an educational process. The four-week workshop program teaches building techniques and arcological philosophy, while continuing the city's construction. Volunteers and students come from around the world. Many are design students, and some receive university credit for the workshop. But

图7-4-1 阿科桑底鸟瞰环境
Fig.7-4-1 The overview position of Arcosanti

造园空间，也是冬天时采集太阳能的集热器。

阿科桑底生态城也是一个教育过程。四个星期的工作坊中将教授志愿者建造技术以及在建造的过程中教授生态哲学。志愿者和学生来自世界各地，有许多都是设计专业的学生，有些是为了拿到大学中的工作坊学分。然而，设计专业或者建筑学的专业背景在这里不是必须的。许多来自各种各样的兴趣和不同背景的人到此来在工程中贡献他们宝贵的时间和技术。到现在，在过去5 000名工作坊参与者的努力下，阿科桑底已经包含了各种多用途建筑和公共空间。

2. 阿科桑底基金会

1956年，保罗迁居到美国亚利桑那州的斯科茨戴尔，和夫人蔻丽，两个女儿一起在这里进行长期城市规划的试验和探索，他创建了科桑蒂基金会这个非盈利型的教育基金组织。如今，这个组织的首要工作是：教授学生、设计专业者、城市规划者和大众关于保罗·索勒里的建筑观念和哲学。定期举行集中的教学、文化集会和艺术活动。

二、 具体项目

1. 具体建成环境

科桑蒂基金会拥有860公顷土地，基地内阿瓜福利亚河流将其分开，阿科桑底坐落在海拔3 750英尺（1143米）的半干旱沙漠环境独特的区域，而与周边的凤凰城平均温度较低的情况不尽相同。下面数据大概描述了该地区气候状况（图7-4-3）。如图7-4-4所示，整个项目分为六期进行建设。

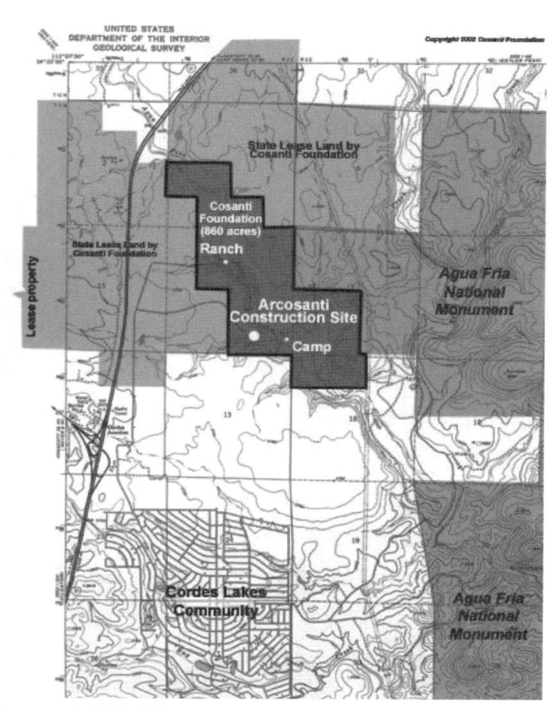

图7-4-2 阿科桑底位置
Figure.7-4-2 The position of Arcosanti

a design or architecture background is not necessary. People of many varied interests and backgrounds are all contributing their valuable time and skills to the project. At the present stage of construction, Arcosanti consists of various mixed-use buildings and public spaces constructed by 5000 past Workshop participants

2. The Cosanti Foundation

In 1956 he settled in Scottsdale, Arizona, with his late wife, Colly, and their two daughters. Dr. and Mrs. Soleri made a life-long commitment to research and experimentation in urban planning, establishing the Cosanti Foundation, a not-for-profit educational foundation. Today, the primary areas of the Foundation's work are: Educating students, design professionals, urban planners and the general public about Soleri's architectural concepts and philosophy. Hosting educational and cultural conferences, as well as performing arts events, on a regular basis.

II. Detailed Description of Project

1. Built Environment

The Cosanti Foundation Owns 860 Acres divided by the Agua Fria River. Arcosanti is located at an elevation of 3750 feet in a semi-arid desert desert climate which different from urban Phoenix with lower average temperature. The following data approximately

2. 生产性景观与农业项目城市建筑生态学原理

1) 城市建筑生态学原理

城市建筑生态学理论将以低冲击方式使城市居住与乡村空间紧密结合，并且在城市周围设置农业区，食物分发系统效率最大化。

两大人造系统（人居环境建设、世界城市建设）解答了我们该如何建造自己的家园，如何满足人类生存的需求（农业系统）。而另一个巨大的系统——能量系统，以能量形式融入其他两个系统，能量对于建造和维持这些系统极为重要。现今这些系统产生了远远超过环境及社会容忍能力范围的资源消耗和大量废弃物。阿科桑底是在一个更为可持续替代方案中的一项创新性实验。阿科桑底最终计划建设宜居、步行尺度、紧凑型的人居城市，一个控制城市无限制蔓延的城市。

2) 生产性景观

在阿科桑底的设计中，索勒里将食物生产融入城市。朝南的附属温室不仅可以生产食物、加热能量，还可以为市民提供一个生产性的环境，让人们很容易体验并享用。饭馆坐落在邻近种植床的区域里，这样食客挑选的新鲜农作物可以立刻收割、烹饪然后端上饭桌。在太阳城—简约线性城市中，北部的露台可被用作果园来生产食物，当太阳高悬空中时还能提供夏季阴凉。太阳城两个结构物中间设计的绿丝带城市公园，既有装饰性景观，又能提供粮食作物（图7-4-5）。

沿着基地北侧建了23个果园台地，在设计中融入了生产性植被。整个设计尽量减少建成环境的痕迹，如此，这块很适合农业生产的土地就不会被开发为人

describes our annual weather patterns (Fig. 7-4-3). As is shown in the Fig.7-4-4, the construction of the whole project can be divided into six phases.

2. Productive Landscape & Arcosanti Agricultural Program

1)Arcology Theory

An arcology's would provide the city dweller with constant immediate and low-impact access to rural space as well as allowing agriculture to be situated near the city, maximizing the logistical efficiency of food distribution systems.

The two largest human-built systems on this planet address how we house ourselves (our built habitat, culminating in the great cities of the world), and how we feed ourselves (our agricultural system). A third enormous system, the energy system, imbues the other two in the form of the energy necessary to build and maintain these systems. These systems are currently consuming resources, generating. These systems are currently consuming resources, generating pollutants and creating inequities that dramatically exceed the ability of the environment and society to endure. Arcosanti is an innovative experiment in a more sustainable alternative. The purpose of Arcosanti is to construct a living, walk-through demonstration model of a compact town, a proposed alterative to suburban sprawl.

2)Productive Landscape

In the designs for Arcosanti, Paolo Soleri make an effort to integrate food production into the urban design. South facing attached greenhouses not only produce food and heat energy but also provide a productive environment for the city residents to easily experience and enjoy. Restaurants could be located adjacent to planting beds so one could literally choose the living crops to be immediately harvested, cooked and served. The north patios of Solare: Lean Linear City would be planted as orchards and provide food and summer shade when the sun is high in the sky. The Green Ribbon Urban Park between the two structures of Solar: Lean Linear City would include both ornamental and food crops(Fig.7-4-5).

Along the north side, 23 orchard terraces will be

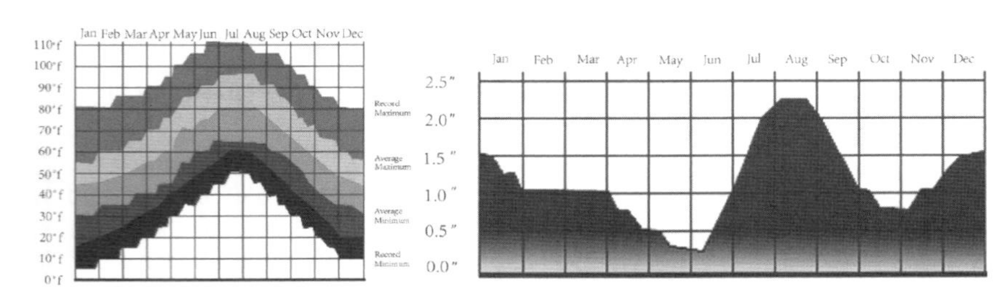

图7-4-3 阿科桑底的全年温度统计（左）及全年降雨量统计（右）
Fig.7-4-3 Annual temperature statistics (Left) & Annual rainfall of statistics (Right)

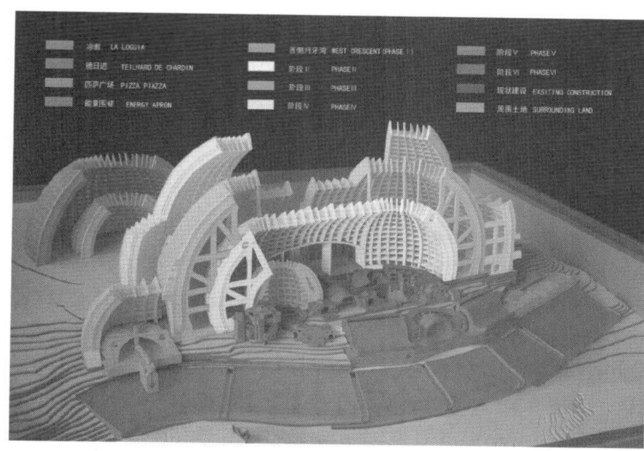
图7-4-4 整个项目六个阶段建设
Fig.7-4-4 Six phases in the construction of the whole project

类的住宅用地。保住现有的，能生产食物的农业用地，在边缘的土地上建设，是索勒里设计核心概念。

景观部门与索勒里，城市规划部门一样，首要关注的是阿科桑底（15.5英亩[约6.3公顷]）的整个城市维护工作，而且还要设计现状建筑与新建筑周围环境及花园。他们采用节水技术、永久培养技术、可食用景观技术设计高品质低能耗有机花园。

景观设计师采用有机的方式塑景、造园，拒绝使用人工合成、化学药剂、农药、化肥等手段设计花园。与建筑具有多用途一样，景观中的植物既能在炎热的夏季提供阴凉，增加湿度，又可以食用，种植的香草既可烹饪又可药用。阿科桑底旨在使可食用景观既具有极高的欣赏价值，又满足居民的食物及草药供给需求。

3）阿科桑底农业项目

阿科桑底共有四部分食物系统：开放领域、公园农业、果园生产、密集型温室农业（图7-4-6），其四大项目目标为：

教育功能：通过阿科桑底农业项目提供的科学研究、财务学习，全世界的研究者和学生能够共享信息，获得有价值资源，它也向全世界合作研究机构开放。

示范功能：阿科桑底旅游项目、食堂、农场超市、限制性社区农业支持体系展示了当地食物生产过程，能量及资源利用过程。

美学经验：城市农业所采用的创新建筑、硬质景观、能量技术能够产生独特定性经验。

built, so integrated into this design is productive vegetation. The whole design is trying to shrink the footprint of the built environment, so that land well suited for agricultural production will not be developed for human habitat. To build on marginal land in order to save current agricultural land for food production is the key idea that Paolo has.

The primary focus of the Landscaping Department is the upkeep and maintenance of the Arcosanti (15.5 acre) site while working with Paolo Soleri and the Planning Department to implement landscape designs and gardens around existing buildings and the newly constructed portions of Arcosanti. Their aim is to design high quality low maintenance organic gardens using xeriscape, permaculture, and edible landscaping methods.

Here at Arcosanti they use organic landscaping and gardening practices, meaning they use no synthetic or chemical herbicides, pesticides or fertilizers on any of our gardens. As the buildings exhibit multifunctional uses, so do the gardens. Many of the plants on site provide shade and moisture in the harsh summers and are edible. They grow herbs for both culinary and medicinal uses. They want to show edible landscaping can be both beautiful to look at as well as supplementing Arcosanti's food and herb needs.

3)Arcosanti Agricultural Program

Arcosanti is involved with four areas of food production systems: open field, garden agriculture, orchard production and intensive greenhouse horticulture(Fig.7-4-6).

The goals of the program are:

Education: Through scientific research as well as emperical learning opportunities provided by Arcosanti agriculture program, researchers and students around the globe can share information and gain experiential values. It opens up for more collaborative research opportunities with other institutions.

Demonstration: Local food productions and effective energy/resource usage can be demonstrated thoughArcosanti tours, food served at our restaurant, farmer's market, and limited community supported agriculture (CSA).

Aesthetic Experience: The innovative building/ hardscaping/ energy technology applied in various urban agriculture settings can lead to unique qualitative experience.

Economic Viability: Our systemic (holistic) approach to food/energy production can be a viable alternative to the current energy/resource intensive agricultural system that puts a burden on our environment.

Currently our agricultural program, manages 3 experimental greenhouses and 14 acres of garden and field production. Solar Greenhouse work is central to our agricultural program. Collecting data and researching growing and construction methods for sloped greenhouses is critical for the future

经济可行性：对于能够带给环境负担的集约型农业体系来说，食物、能量生产的系统方法（综合方法），实施具有可行性。

目前，整个生态城拥有三个试验型温室、14英亩（约5.67公顷）的公园、生产区域。整个农业项目中太阳能温室是核心重点。整个阿科桑底设计重要部分未来能量围裙建设，关键点在于数据收集、倾斜温室的建设研究。

3. 阿科桑底当前建设项目

1）原理

城市建筑生态学理论能够充分利用被动式太阳能建筑技术，如半球效应、温室建筑等降低城市能量损耗，尤其是热能、光能、制冷等方面。总之，城市建筑生态学理论试图寻求“简约线性城市”节俭、高效、智能化城市设计控制超级资源消耗及能源浪费。

2）阿科桑底当前建设项目

工艺品III：13 500平方英尺（约1 254平方米），于1972年动工，分阶段建造并于1977年完成，形成了一个多用途的结构体。最初，它的底层作为游客中心提供住房和休息室，第二层为咖啡厅，三层夹层为面包店，第四层为展厅。这个多用途的设施示范了保罗·索勒里城市建筑生态学的概念，即将居住、休闲娱乐和工作结合到一个结构体中去。冬天的时候，咖啡厅通过一个管道从天窗采集热量并加热下部大厅（图7–4–7）。

construction of the Energy Apron, a key design component of the Arcosanti.

3. The Current Construction Projects in Arcosanti

1)Arcology Theory

Arcology would use passive solar architectural techniques such as the apse effect, greenhouse architecture to reduce the energy usage of the city, especially in terms of heating, lighting and cooling. Overall, arcology seeks to embody a "Lean Alternative" to hyper consumption and wastefulness through more frugal, efficient and intelligent city design.

2)The Current Construction Projects in Arcosanti

Craft III: 13 500 square feet, begun in 1972, Crafts III was built in phases and completed in 1977. A multi-use structure, it is primarily the Visitor's Center and provides housing on the first level, a Cafe on the second level, a Bakery on the third floor mezzanine, and a Gallery on the fourth level. This multi-function facility demonstrates one of the characteristics of Soleri's Arcology Concept, which integrates living, recreational and working conditions within a single structure. The Cafe is warmed in the winter through the use of a warm air collected by the skylight blown through a fabric tube into the atrium(Fig.7-4-7).

Ceramics Apse: Built between 1971 and 1973, the Ceramics Apse serves in the production of ceramic wind bells and tiles at Arcosanti. Used initially as a temporary living space, it moved into production in 1975. It is constructed of both poured-in-place and precast concrete. The Apse shell was poured in place utilizing silt on shoring to form the quarter-sphere. The tempered micro-climate created by the Apse Effect, the amphitheater terracing, and the removable stage that can be erected over the slip bins combine to provide an excellent performance space(Fig. 7-4-8).

The outside shell of the ceramics apse was shaped

图7–4–5 生产性景观
Fig.7-4-5 Productive landscape

图7–4–6 阿科桑底农业项目
Fig.7-4-6 Arcosanti agricultural program

Colly Soleri音乐中心内景Interior view of Colly Soleri Music center

Colly Soleri音乐中心内景Interior view of Colly Soleri Music cente

工艺品III内景 (Interior view of Craft III)

工艺品III外景 (Exterior view of Craft III)

图7-4-7 科里索勒瑞音乐中心与工艺品III
Fig.7-4-7 Colly Soleri Music Center & Craft III

制陶圆顶：1971年到1973年间建造，是制作风铃和陶片的场所，最初用于临时的居住，自1975年开始用作生产。其结构由现场浇注的混凝土和预制混凝土板结合完成。利用半球效应来调节微气候。古罗马竞技场台地结合安装在滑轮上的可移动舞台，提供了一种完美表演空间（图7-4-8）。

制陶圆顶的外壳在由细沙雕筑的结构上造成。整个穹殿坐北朝南。当夏季来临，太阳直射地面，穹顶可以起到遮光庇荫的作用；冬天阳光从南面斜射入殿内，为空间内部提供热源。

铸造圆顶：建造于1972年到1974年间，作用是制造青铜风铃和提供额外的居住空间。最先设计成一层楼的住宅，之后改为两层，且结构上在圆顶两端各增加了一个环状物。冬天，铸造熔炉的余热会通过热力管道输送到住宅区。铸造圆顶的建造方式和制陶圆顶相似，在夏季，当遮阳屏竖立起来的时候，会达到比制陶圆顶更好的遮阳效果。

东新月建筑群：在建中的新月建筑群由10幢建筑物构成。建筑群的热能来自连排温室收集的暖气。气体通过各个建筑物外部的水泥管道输送，使得整个建筑群成为一个巨大的散热器（图7-4-9）。

over a form carved of finesilt. The apse faces south, allowing for shading in the summer from the sun straight overhead. In the winter month the sun is at a low angle, warming the interior of the apse.

The Foundry Apse: Excavation for this structure was begun in 1972, and its completion in 1974 facilitated the expansion of bronze bell production and provided additional housing. Initially designed with one level of housing, site excavation suggested a double level organization and the structure was modified to include an additional ring of housing units encircling the rear of the apse. Exhaust heat from the foundry furnace, ducted through the living areas and stored in concrete heat sinks, contributes to the heating needs during the winter. Construction of the Foundry is similar to that of the Ceramics Apse and the apse shading effect is extended during the summer months by the erection of a fabric shade screen.

The East Crescent Complex.:The East Crescent Complex under construction consist of10 buildings. The complex will be heated by warm air collecting within and rising from a greenhouse section. The air will be channeled to the crescent units through a large concrete duct running along the outer edge of the building. The mass of the structure functions as a heat sink (Fig.7-4-9).

4. Future Projects

Arcosanti construction is far from complete, they will continue to build Arcosanti5000;Energy Apron;La Loggia;Basilicas;Experimental Greenhouses;Waste Treatment;Solar cogenset;East Crescent Canopy,and road paving.

1) Arcology Theory

Hyper-Building: The Hyper Building is an Arcology (Fig7-4-10).In an Arcology, architecture and ecology come together in the design of the city. Arcology is the implosion of the flat megalopolis, the modern city of today, into a dense, complex, urban environment which rises vertically. The concept of a one-structure system is not incidental to the organization of the city, but central to it. Such an urban structure hosts life, work, education, culture, leisure, and health in a dense, compact system which also puts the untouched open countryside at the fingertips of the residents. The compactness of an Arcology gives 90

制陶圆顶 (Ceramics Apse)

铸造圆顶外景 (The Foundry Apse)

图7-4-8 制陶圆顶
Fig.7-4-8 Ceramics Apse

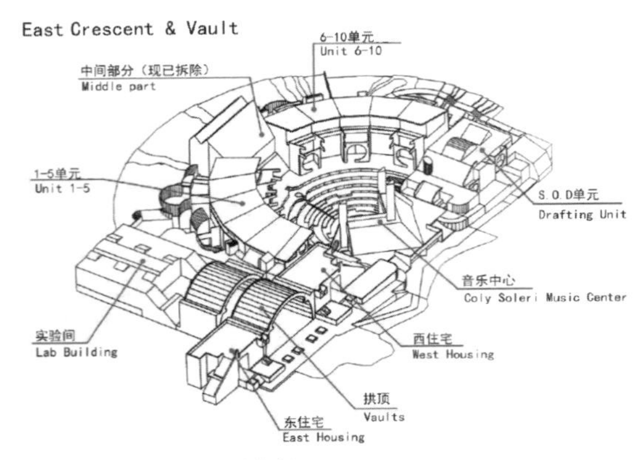

图7-4-9 东新月及拱顶组合功能介绍
Fig.7-4-9 The function of the East Crescent Complex

图7-4-10 巨构建筑
Fig.7-4-10 Hyper-Building

4. 未来项目

阿科桑底的建设还远远没有完成，计划中要继续建造有阿科桑底5000、能量围、凉廊、长方会堂、实验温室、垃圾处理场、太阳能发电机组、东新月天棚和铺设道路。

1) 原理

巨构建筑:巨构建筑是生态建筑的一个类型（图7–4–10），对于一个生态建筑而言，城市设计过程中建筑和生态应该结合起来考虑。城市生态学是平坦的大都市带，逐渐“压缩”为高密度、复杂、垂直型现代城市。单结构体系概念是城市有机体的核心，而非偶然性。整个城市结构以密集而紧凑的系统支撑生活、工作、教育、文化、休闲，健康等功能，而且可以在居住区间隙地带保留有未开发的乡村开放空间。与当今城市所面临的郊区无限蔓延的现状相比，一个紧凑的生态建筑可以节余90%甚至更多的土地用做农田及保护区。紧凑性特征使得生态建筑成为一个更为高效运作的系统。生态建筑很重视人类尺度，在步行尺度占统治地位的设计区域，以墙体和分钟为测量距离标准。机动车尺度却成为荒谬的代名词。生态建筑比传统的现代城市更能高效地利用能源。生态建筑比传统的现代城市更能高效地利用能源。污染是最直接的浪费。增加效率，降低污染意味着污染程度的降低（图7–4–11）。

轻触空间：轻触空间是指一个生态建筑能够最大限度地减少城市对周边自然环境的侵犯，同时实

percent more land to farming and conservation than today's urban and suburban sprawl. This compactness makes an Arcology a more workable system. Greater attention is given to human scale in an Arcology. In it the pedestrian reigns. Distances are measured by walks and minutes. Within it the automobile is nonsensical. In an Arcology energy is used more efficiently than in a conventional modern city. Pollution is a direct function of wastefulness, not efficiency. The increase in efficiency and reduction of wastefulness means a reduction of pollution(Fig.7-4-11).

Nudging Space: "Nudging Space" is an Arcology (Architecture / Ecology), Paolo Soleri's concept of cities that impinge minimally on their natural surroundings while trying to maximize the cultural, educational and economic benefits of urbanity. The overall form of Nudging Space is intended to create various micro-climatic environments within the building envelopes. Soleri utilizes the concept of "garment architecture" to moderate thermal and lighting conditions. The hot season garment is opaque, providing shade while the cold season garment is transparent and transmits the solar energy and traps the heated air (Fig7-4-12).

2)Future Projects

Arcosanti 5000: Developed from the Super Critical Mass in Arcosanti 2000 with the design elements of Nudging Space Arcology added, Arcosanti 5000 features seven phases of truncated super apse structures. It It will be the main building in the future,and it re-establishes the macro nature of this prototype arcology for 5,000 people,. This design is still in development, waiting on the architectural and structural resolutions (Fig.7-4-13~Fig.7-4-15).

The Pierre Teilhard de Chardin Complex. The Cosanti Foundation is planning to start construction of the Pierre Teilhard de Chardin Complex on the south slopes

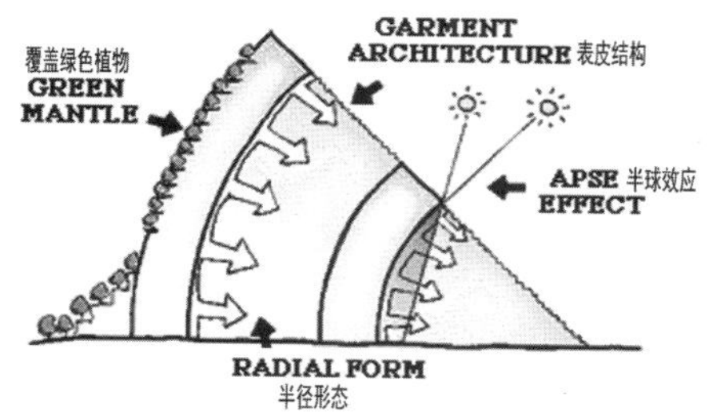

The Hyper Building Design Parameters: Form
巨构建筑设计参数:构成

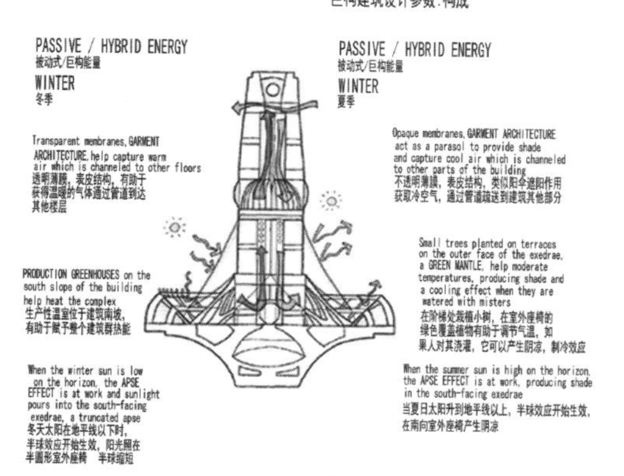

The Hyper Building Design Parameters: Energy
巨构建筑设计参数:能量

图7-4-11 巨构建筑
Fig.7-4-11 Hyper-Building

立面图 The Elevation

鸟瞰图The Bird's-eye view

立面图 The Elevation

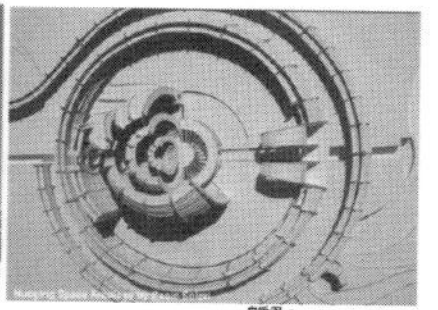

鸟瞰图 The Bird's-eye view

图7-4-12 轻触空间
Fig.7-4-12 Nudging Space

现城市文化、教育和经济利益的最大化。整个轻触空间的结构设计旨在为建筑物内部空间营造一个微型气候环境。为此，索勒里提出了“外套建筑”概念以实现空间内热量与光线的自我调节。在炎热的季节里，用不透光的“外套”来遮蔽阳光，使室内保持凉爽；而寒冷的季节则着上透明“外衣”，使空间内部得到充分的日照，保持温暖（图7-4-12）。

2)未来项目

阿科桑底5000：从阿科桑底2000的超临界质量发展而来，沿用轻触空间设计元素，阿科桑底5000采用七阶段紧缩半球结构。该建筑是未来基地上的主体建筑，它为5000居民重新定义了城市建筑生态学原型的宏观本质特征。此方案还只是目前阶段的，将来还有可能进一步修改，等待建筑与结构方面的解决措施（图7-4-13~图7-4-15）。

皮埃尔德日进建筑群：科桑底基金会计划在阿

of Arcosanti. The complex is mainly intended to be the eventual residence and work place for participants in the Paradox Program, those digital "insiders" who are at home on the Internet, working in digital media, developing their nexus withCyberspace.It also features production greenhouses placed on the slopes below providing warm air into the buildings above.

Basilicas: The Basilicasis locatedon the south of the East Crescent Complex, on the north of the Energy Apron. It takes advantage of the passive solar features that Paolo Soleri has been approve of .The BASILICAS, articulated in two wings, contain large atriums, galleries, archives, cyberspace studios and storage. CRESCENT ROOMS for different sizes of operational activities will be integrated into the APSE structure around the Amphitheater that has a seating capacity of 700 (Fig. 7-4-16).

The Amphitheater is an “extroverted” structure that takes advantage of the passive solar features. The South Facing Apse provides the sun penetration deep into the seating areas in winter and provides shade in summer.

Energy apron:An integral part of the design and coupled to this prototype town will be five to seven acres of south facing sloping greenhouses, an "energy apron" acting as a central system for producing food,collecting energy and providing leisure place to support the prototype town (Fig. 7-4-17).

The main feature of the Energy Apron is the incline of the topography. The membrane that produces the greenhouse effect is thus covering a series of large terraces on which this main layout are also developed. The incline of the whole system has two main effects: A temperature differential between the terraces with the lower terrace as the cooler, the top as the warmer. An updraft of air warming up as it moves from the bottom toward the top (Chimney Effect).

On the environmental - energy level, The better-controlled water use facilitated by the Energy Apron is also paramount in arid climates. When feasible, areas of the Energy Apron can be “leased” to individuals for their garden patch.

科桑底南坡建造皮埃尔德日进建筑群。建筑群主要功能是为是为居民、项目参与者、家居办公数字网络发展事业工作人员提供工作场所。另一个突出特征是于建筑斜坡上建造生产性温室大棚，由此可以向整个建筑中输入温暖气流。

长方会堂：该建筑位于东新月的南边及能量围裙的北边，并一如既往地运用了保罗·索勒里所主张的被动式利用太阳能。长方会堂外部形象为带有两翼的建筑，建筑内部有天井、美术馆、档案室、网络空间、工作室及储藏室。新月房间为开展不同的活动设计成多种尺寸，与圆形剧场周围拱点结构相结合，能容纳700人会议等活动（图7-4-16）。

圆形剧场开放式结构充分利用被动式太阳能资源，而因建造“南侧拱顶”，所以休息区冬暖夏凉。

能量围裙：即将要在阿科桑底南面斜坡上建立

Arcodanti are trying to find alternate ways of producing energy that are renewable and self sustaining like wind and solar. Energy is not something that comes from somewhere without effort. People have to produce it, so energy production is an investment and there are tremendous loses of energy when it is transmitted hundreds of miles to the end user. Arcodanti attempts to produce energy locally; to consume it where the energy is produced. And the dense compact design serves to reduce energy requirements for heating and cooling and the marginalization of the automobile dramatically reduces energy needs for transportation.

Experimental Greenhouses: The conceptual research of the greenhouse as a large central system for food and energy production was carried out from 1974-1976. Extensive research, carried out from 1976-1978 , resulted in the construction of a prototype greenhouse in 1979. This facility has been operating since March 1979, generating both agricultural and climactic data necessary for further greenhouse development, specifically aiding in the designs of the first full scale segment of the energy apron. Generate climatic data in order to better understand the nature of greenhouse

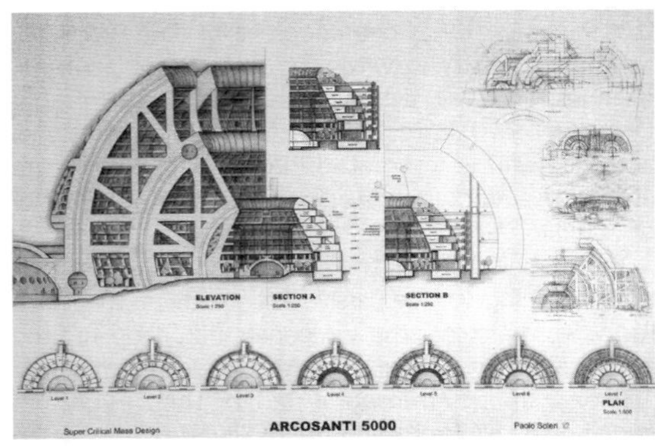

图7-4-13 阿科桑底5000剖面图
Fig.7-4-13 The section of Arcosanti 5000

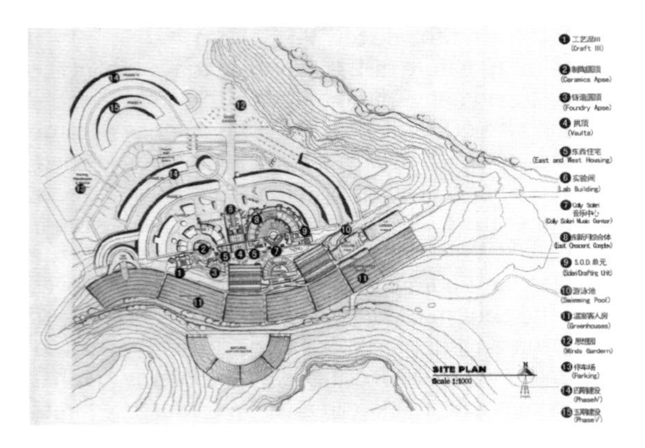

图7-4-14 阿科桑底5000
Fig.7-4-14 The master plan of Arcosanti 5000

图7-4-15 阿科桑底5000透视图
Fig.7-4-15 The perspective of Arcosanti 5000

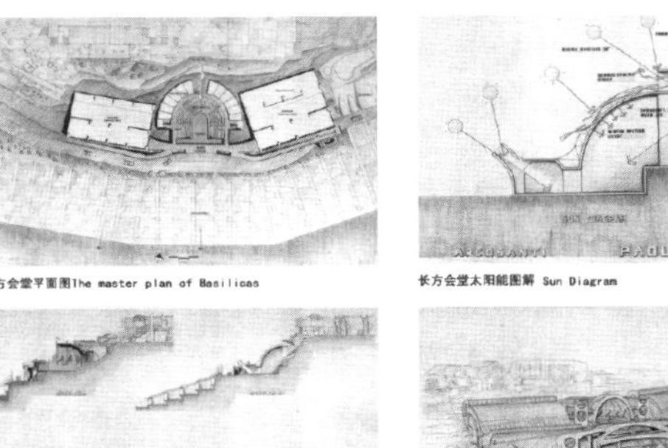

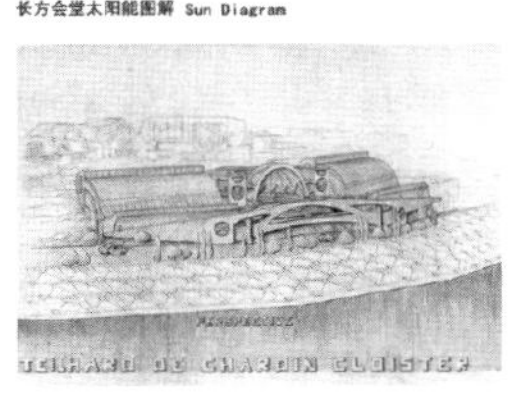

图7-4-16 长方会堂
Fig.7-4-16 Basilicas

五到七英亩的大型温室，该温室是设计中不可或缺的部分，以城镇为原型，为整个阿科桑底5000提供热能，制造新鲜空气，还为整个城市提供水果和蔬菜，聚集能量，同时也是一个给人们休闲散心的地方（图7–4–17）。

能量围裙的最大特征是倾斜的地形。产生温室效应的薄膜覆盖一系列巨大的阶梯，这样的布局逐渐形成.整个系统存在两种效应：阶梯之间温度不同，低处阶梯作为冷却区域，高处阶梯作为加热区域；通过气热运动将能量从底部传送至顶部（烟囱效应）。

环境能量方面，能量围裙所提供的优质水源在干旱的气候条件下是至关重要的。如条件允许，该区域可以向个人“租借”土地作为种植花园。

阿科桑底试图找到制造能源的替代方法，它应该是可再生的、可持续的，就像风能和太阳能。能量不是从某地毫不费力就可以白白获得的，而是需要制造它，因此能量生产是项投资。能量在经过几百英里传输到终端用户时会产生巨大的能量耗损，阿科桑底试图就地生产能量，然后在能量生产地消费能量。紧凑高密度的城市设计减少了城市中用于加热和冷却的能量需求，将汽车边缘化的做法显著地降低了交通的能耗。

实验温室：为整个阿科桑底5000提供热能、水果和蔬菜的中枢系统的温室在1974—1976年期间已经开始实施。进一步研究工作在1976—1978年期间展开，最终在1979年建立温室原型。自1979年3月设施开始启动以来，所得到的农业及气候数据为日后温室发展、能量围裙的设计奠定了基础。气候数据的研究可以更好地理解温室微气候的本质及如何与温室外部气候相比较（图7–4–17）。

线性城市景观：景观设计通过模块和节点将各个建筑结构连接起来。建筑和景观特点水平联系的加强让整个结构看起来像是一座拔地而起的山脉。垂直梯度决定了植物群落类别和建筑内部花园植物的种类。建筑的外墙通过雕刻结构形成一个三维空间，可以在建筑内部设置树木，花草。这一设计进一步将地面景观和中层景观、交通走廊整合了起来，使得巨型景观能够进入到建筑的中层和上层（图7–4–18～图7–4–22）。

microclimates and how they compare to the outside climate(Fig.7-4-17).

Landscape of Linear City: Using the modules and nodes of liked infrastructure, the landscape explores the relationship of structures to each other and their setting. The reinforcement of horizontal connections between structures and the character of their landscape result in a landscape metaphor that treats the structure like a mountain rising from the plains. The vertical gradient determines the kinds of plant communities and gardens sustained within the architecture. The façade is altered by sculpting volumes of three dimensional spaces to allow for vertical landscapes of trees and other structures within the building. This enhances the integration of the on-the-ground landscapes within the terra level, a mid-level landscape/transportation corridor, which brings large scale landscape to a midpoint in the architecture and higher(Fig7-4-19~Fig.7-4-22).

III. Summary

The Arcosanti site is located in Arizona. It is the legendary holy land of architecture,and established by Paolo Soleri, the italian architect .He made a life-long commitment to plan and design, invest and establish the Cosanti , that devoted all his painstaking efforts to operation and propaganda, in order to establish a experimental eco-community. Now Arcosanti has achieved certain experimental purposes, But, the aim of Arcosanti will house 5000 people proposed at first is very distant.

Although, it is mixed to make arcosanti construction technology popularization at home and abroad, but one is certain, it raises the efficiency of resource and space, emphasis ecology, and try to explore Arcology realistic possibility, Arcology is in the pursuit of minimize the use of energy, raw materials and land; reduce waste and environmental pollution, in the meanwhile, it designed to maximize human

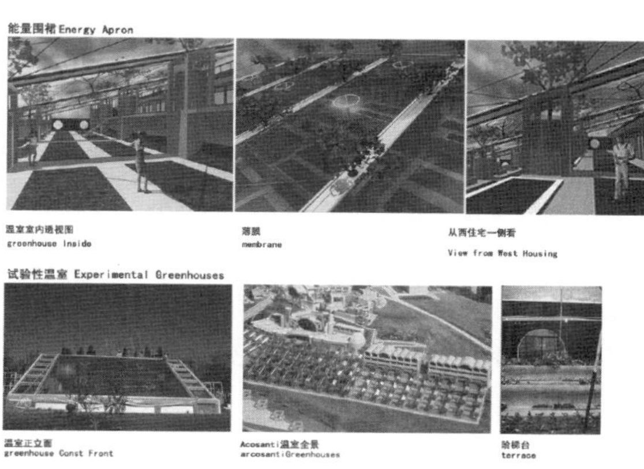

图7–4–17 能量围裙
Fig.7-4-17 Energy apron

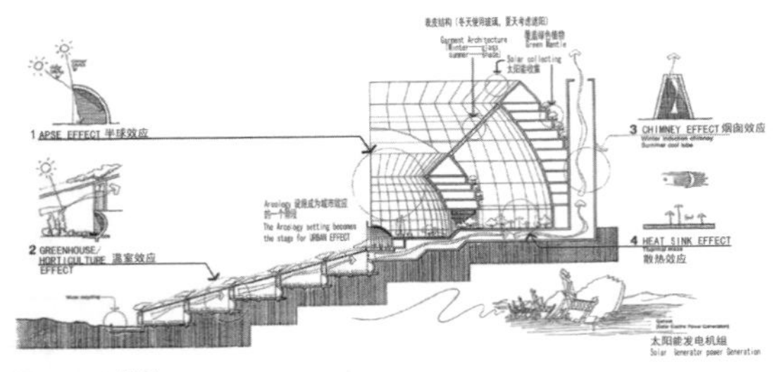

图7-4-18 能量
Fig.7-4-18 Energy

interaction and allow interaction with the surrounding natural environment.

Paolo Soleri advocates "Productive Landscape", and makes an effort to integrate food production into the daily urban life. Garden landscape are not only beautiful to look at, but also supplementing food production. The Hyper Building is Paolo Soleri's another idea, it is similar with Le Corbusier's opinions in The City of Tomorrow. Energy apron, Nudging Space, The foundry Apse, etc. in Soleri's design, all reflect the values of frugal and the emphasis of ecology.

三、 总结

阿科桑底位于亚利桑那州，是传说中的“建筑圣地”，由美籍意大利建筑师保罗·索勒里创意构想，亲自规划设计、投资兴建，并注入了毕生心血运营与宣传，力求营建一个实验性生态社区。经过40多年的建设，阿科桑底达到了一定的实验目的，但距离其最初构建5000人居住规模的目标还相距甚远。

虽然，国内外对阿科桑底建设技术的普遍推广意义褒贬不一，但有一点是肯定的，它重视资源与空间的高效利用，重视生态，试图探讨建筑生态学理论的现实可能性，“追求能耗、资源消耗、土地占用、废物排放和环境污染的最小化的同时，又能达到人际交往和人与自然接触的最大化。”

索勒里倡导“生产性景观”，将食物生产过程融入日常城市生活；提出花园景观不仅要满足观赏功能，更要发挥新鲜果蔬的生产功能；他还提出巨构建筑设计，这一点与柯布西耶《明日之城市》中提到的“现代城市”有几分相似。索勒里设计中的能量围裙、轻触空间、铸造圆顶等构想，都反映了其“节俭”的价值观和意识上对生态的重视。

索勒里一直没有引入大量资金加快阿科桑底的建设，就是为了保持生态城的理想性，为了避免商业机制的干扰，也为了坚持内心的信仰。这样的理想城近似于乌托邦，而随着时间的流逝，这样的生态精神会不会被现代商业价值观所蚕食？也许这并不重要，重要的是，阿科桑底已经影响到了世界建筑设计和城市规划的发展方向。

图7-4-19 部分线性都市的概念景观规划图
Fig.7-4-19 Conceptual landscape master plan for segment of linear city

图7-4-20 建筑凹进去的景致
Fig.7-4-20 View of building abcove

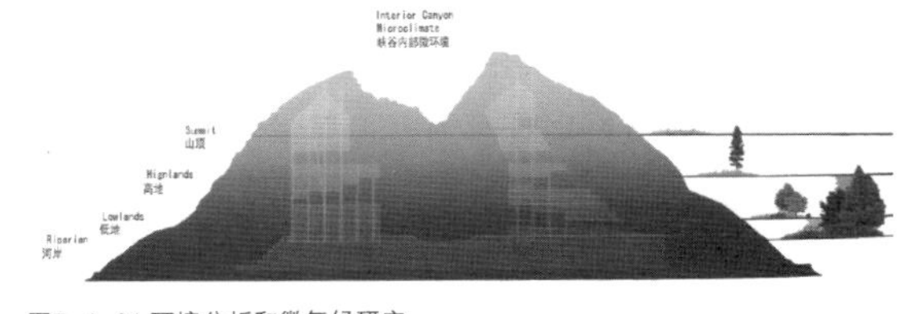

图7-4-21 环境分析和微气候研究
Fig.7-4-21 Environmental analysis and microclimate study

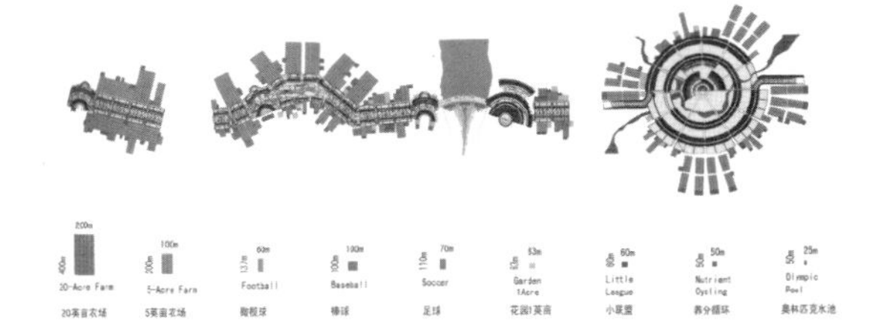

图7-4-22 土地利用提案
Fig.7-4-22 Land use proposal

注解/Comment：在高亦兰的《鲍罗•索勒里和他的阿科桑底城》中，城市名被译作阿科桑底。清华大学建筑学院的高亦兰教授早在20世纪80年代就撰文介绍过鲍罗•索勒里。为了使人更好地理解他的思想和主张，索勒里创造了一个新词“arcology”;为了营造全新的城市居住环境，建筑学与生态学应该是整体不可分割的概念，是一个问题的两个方面。高亦兰教授首次将其理论和实践介绍到国内时，将“arcology”翻译为“建筑生态学”。考虑到不存在“ecology of architecture”这样一个同“建筑生态学”更为贴切的英文词组，因此，本文暂将其改译为“城市建筑生态学”。为简化起见，文中用“城市生态学”代替。

There has been no a large number of capital to accelerate arcosanti construction, in order to keep the ideal arcosanti, to avoide interference of commercial mechanism, to keep the faith of the heart. This is similar to the utopian ideal city, but whether the ecological spirit will be nibbling by modern commercial values or not? Maybe it is not important at all, on the contrary, it is most important that arcosanti has influenced the development direction of architecture design and city planning in the world.

思考题

1. **简要介绍“Arcology”？**
2. **简要介绍阿科桑底生产性景观？**
3. **阿科桑底当前建设项目与其应用原理？**
4. **阿科桑底未来建设项目与其应用原理？**
5. **建设当今中国生态城市如何运用Arcology理念？**

Questions

1. What is “arcology”?
2. What is productive landscape in Arcosanti ?
3. What are the current construction projects in Arcosanti and their theory?
4. What are the future projects in Arcosanti and their theory?
5. How does the arcology operate in the construction of eco-city in China?

注释/Note

图片来源/Image：图7-4-1 http://www.阿科桑底.org&googlearth ；图7-4-2～图7-4-8；图7-4-10～图7-4-16 http://www.阿科桑底.org. ；图7-4-9，图7-4-17 Soleri: Architecture as human ecology，作者译；图7-4-18～图7-4-21朱莉，生态景观之旅-查尔斯•安德森景观建筑事务所

参考文献/Reference：[1]Jay Powell.Green Architecture/Arcology: The City in the Image of Man. Architecture. Washington:Nov 2000.Vol.89，Iss.11;pg.80，1 pgs ；[2]高亦兰.鲍罗•索勒里和他的阿科桑底城.世界建筑，1985，（05）；[3]朱莉•戴克编，常文心译， 生态景观之旅查尔斯•安德森景观建筑事务所Wandering Ecologies The Landscape Architecture of Charles Anderson Edited by Julie Decker，沈阳：辽宁科学技术出版社，2011:176-183 ；[4]http://www.阿科桑底.org/

第八章 景观的再生与转型

CHAPTER VIII LANDSCAPE REGENERATION & TRANSFORMATION

第一节　北杜伊斯堡景观公园 Section I North Duisburg Landscape Park

第二节　郭瓦纳斯运河海绵公园规划 Section II Gowanus Sponge Park Design

第三节　纽约东河滨水景观的转型 Section III Transforming the East River Waterfront, the City of New York

第一节 北杜伊斯堡景观公园

Section I North Duisburg Landscape Park

一、 项目概况

杜伊斯堡公园位于德国鲁尔区杜伊斯堡市的北部，是国际建筑展埃姆舍公园的一部分。昔日曾经是奥格斯特・蒂森钢铁厂。1985年钢铁厂关闭了以后，很快陷入荒废破败之中。1989年，政府决定将工厂改造为公园，德国景观师彼德・拉茨在该项目的国际竞赛中赢得了一等奖，并担任设计任务。从1990年到2001年，共完成9个子项目，之后公园的局部仍在继续建设和完善。杜伊斯堡公园作为国际建筑展埃姆舍公园的一部分，成功地尝试了在一个衰败的旧工业区，如何通过景观设计的途径，达到改善当地生活环境、为受到破坏的地区重新注入生机的目的。设计的主导思想是整合、连接、重塑、开发旧工业区业已形成的元素，并且用一种新的语汇加以重新阐释（图8–1–1）。

杜伊斯堡景观公园对工业设施处理方法可以概括为：完整保留工厂的工业结构，在工业景观框架内，作镶嵌式更新设计庞大工业景观作为工业历史的象征，大地艺术的手法，自然侵蚀和野生植被的保留，通过崭新的景观设计语言完成工业景观的转变。它具有以下的设计特点：工业遗产的多元化利

I. Overview

Duisburg Park locates at north Duisburg city, Germany, and it is part of IBA Emscher Park. Once it was August. Thyssen steel works. However, it closed in 1985 and descended into ruins soon. In 1989, government decided to transfer it into a park with Peter•Rhazes who won the first prize in the international competition for this program as its designer. The transformation lasted from 1990 to 2001 with 9 subsidiary programs completed, and certain parts of the park were still under construction for improvements from then on. As a part of IBA Emscher Park, Duisburg Park transformation program has managed to improve local living environment and rejected vitality into a deserted old industrial area through landscape design. The design principles were integrating, combining, rebuilding, and developing original industrial factors, and then representing with new factors (Fig. 8-1-1).

Means for industrial facility transformation in Duisburg Landscape Park could be summarized as integrate preservation of the works' widespread industrial framework with inlaid redesigns as a symbol for industrial age, and reservation of land art, natural erosion, and wild vegetation with transformation by brand new landscape design languages. The program absorbs following features: multi-use of industrial heritage, conflict and contrast of old and new elements, participation of various disciplines and arts, full use of ecological techniques, and perfect

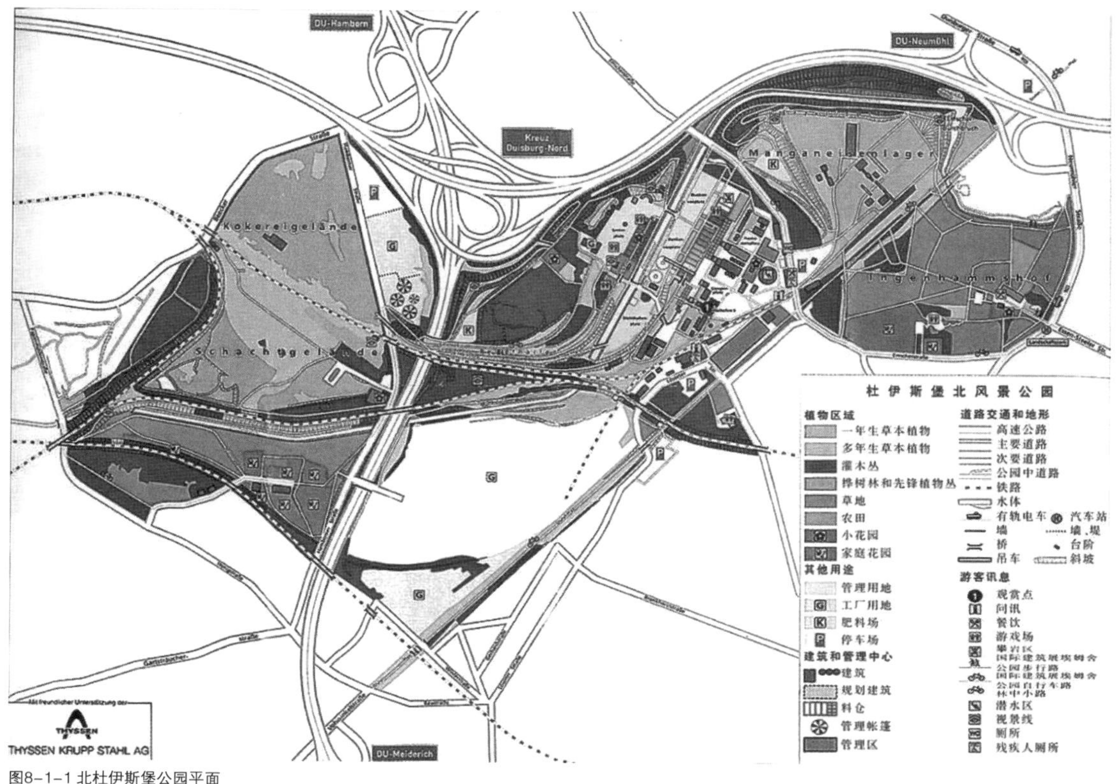

图8–1–1 北杜伊斯堡公园平面
Fig.8-1-1 The master plan of North Duisburg Landscape Park

用，新旧元素的冲撞与对比，多种艺术门类的融合多学科的参与，生态技术的充分利用，完整体现后工业景观美学，成为后工业景观设计的代表作。

二、 规划目的与具体措施

设计理念可以简要地概括为：公园发展是需要具备生态敏感度的，要能够不断创新，合理使用资金，能够为各式各样的游客提供服务，并且能够迅速地完工。设计还需要负起将工业遗址与现存开放空间格局整合的责任，以求能够在公园与基地历史之间建立起紧密的联系。

完成上述目标主要通过以下的措施：

许多现有的建筑经过改造之后直接投入使用；美化和保护现有植被与工业设施，包括巨大的高炉；提供各种休闲活动与文化活动的场地；重建与清理横穿公园的埃姆舍河也是整个埃森公园再生计划的一部分；通过设计“工业历史路线”与“项目

representation of post-industrial landscape aesthetics as a magnum opus.

II. Planning Purposes and Strategies

The design brief stated that the development must be ecologically sensitive, innovative, reasonably priced, provide for a wide range of users, and be completed rapidly. Those responsible for the landscape design also introduced the aim of integrating the industrial heritage of the site with the existing pattern of open spaces to create a park with strong links with its past.

The aims have been achieved through the following main strategies:

The renovation of some existing buildings for use in employment projects; Landscaping and preservation of existing vegetation and industrial facilities including huge blast furnaces; The provision of a variety of leisure activities, and areas for cultural events; As part of the whole Emsher Park regeneration programmers, the reconstruction and cleaning of the River Emscher where it passes through the park; An 'industrial history trail' and 'project trail' have been created as a means of exploring and understanding the site; The creation of seven 'development areas' within the site; Educational and learning farm; Wind water tower (Fig. 8-1-2).

路线”的方式来探索与了解场地；在基地内建立起7个“发展区域”；提供用于教育与学习的农场；风水塔的设计（图8–1–2）。

三、 规划设计的原则与层次

杜伊斯堡公园面积达到230公顷，面对规模庞大、情况复杂的基地条件，设计必须有重点的进行。规划对其中的一部分不作过多的干预，任其自由发展；被严重污染的部分则需要继续关闭（比如，一处煤焦油汇集区）。为此设计师设定了一系列的原则，而且这些原则在总体规划上面得到了体现。其中各个系统分别独立自主的运行。这些不同的层次只是在某些特殊点通过视觉上、功能上或者仅仅是意向上的元素联系起来。景观师通过景观分层的方法达到对破碎的工业景观结构的理性整合。拉茨在对杜伊斯堡公园进行景观分析的时候，将公园景观分为四个层次：

最底层：以水渠和储水池构成的水园；最顶层：高架铁路和高架步道组成的铁路公园；使用区域：相对独立的活动区和种植区域；散步道系统：联接公园各部分步道、自行车系统等交通线路在城市街道的层面将城市被分隔的部分联系起来。

1. 铁路公园

图8–1–2 北杜伊斯堡公园中的各种改造策略
Fig.8-1-2 The strategies of North Duisburg Landscape Park

III. Principles and Levels for Planning and Designing

Duisburg Park covers 230 hectares, so the design must have focuses in front of completed large scale base conditions. The designer planned to leave part areas into free development without intervenes, while continued to shut the heavily polluted parts such as a collecting zone of coal tar. Then he created a series of principles that were displayed in overall planning. Different systems run independently, which form various levels that are linked by certain special spots in visual, functional, or imaginary elements. The designer realizes rational integration for sporadic industrial landscapes by dividing the overall landscape into different levels. When analyzing the transformation for Duisburg Landscape Park, Rhazes divided the landscape of the park into 4 levels:

The lowest level, a Water Park consisting of water channels and water storage ponds; The highest level, a Railway Park consisting of aerial railways and elevated promenades; Areas of use, consisting of relatively independent playgrounds and planting areas; Promenades system, linking the park with footpaths, cycleways, and other transportation lines; it connects the isolated parts of the city from avenue level.

1. Railway Park

Railway Park: aerial railways run across most area of the park, with a height of 12 meters, much higher than any hypsography changes, which create a perfect view. The designer deals with the railways in various resolutions, such as viewing it as a framework for huge industrial constructions, building aerial footpaths on its basis to cross the whole base, or just leaving the criss-cross railways as works exhibition of FarhtArt. The designer named parts of the railways as rail harp. Once collected creations of engineers, now non-artist-created works of art displayed to people (Fig. 8-1-3).

2. Water Park

Water Park: the second system “Water Park” is only partially realized, with long time ahead of the completion of the rest. Given ages of industrial pollution, there is no natural body of water in the base, and the riverway is blocked in large area. “Old Emscher River” was once a pollution discharge channel, running from east to west and finally into the Rhine. The park officials spent huge investment in dealing with polluted water, and introduced a new water purification system. The rain collecting system collects rain from architectures and hard ground surface, and over flow from storage bin and cooling bay through airtight pottery pipes. The rain flows into pipe system through outdoor drainageways and then into the cooling bay, which could supply fresh water

铁路公园：高架铁路穿越了公园大部分区域，与地面的高差大约12米，高度远超过地面上任何自然的地形起伏，从而提供了绝佳的视野。设计师对高架铁路进行了不同的处理，或者将其作为庞大的工业构筑，限定空间的景框；或者在它的基础上构建空中步道，穿越基地；或者干脆将纵横交错的铁轨作为大地艺术作品展示。设计师为一部分高架铁路起了一个很好听的名字，叫做“铁路竖琴”。它们是由当年的工程师们集体创造的，而今却作为非艺术家创造的艺术品展现给世人（图8–1–3）。

2. 水园

水园：公园的第二个系统“水园”只是部分得到实现，其余的部分还需要很长时间才能完成。长期的工业污染使基地中已经不存在天然的水体，河道在很大的范围内是封闭的。“老埃姆舍河”在过去是一条开放的排污渠，由东向西穿越基地最终流入莱茵河。公园花费了巨资治理污水，安装一条新的净水系统。雨水收集系统通过密闭的陶管收集建筑、硬质地面的雨水、料仓和冷却池的溢流。雨水经过露天的排水道进入原有的管道系统。雨水在途中流入冷却池，原有的储水池经过净化可以提供清洁的水。利用风力能源系统将水从水渠中泵出，流经公园各处既增加了水域氧气接触的机会，有利于水质的净化，同时也成为游人水景（图8–1–4）。

3. 开放空间

开放空间：“考珀区”位于高炉工厂区的里面。这里原是堆砌矿渣的场地，现在土壤经过改良，种植了抗性较强的桦树，成为一处生机盎然的林荫广场。一开始没有人相信在鼓风管道和高炉的“钢铁丛林”中间，会有树木繁茂、鲜花盛开的景色。现在这里经常举办庆祝活动。在1994年（IBA）中期展示的时候，有50 000人参观了“考珀广场”。熔渣公园是一处开放空间，在废弃垃圾的地面上种植有臭椿树林。树林的后面还隐藏着一个花园和一个罗马圆形剧场，剧场由循环利用的红砖碎屑涂成明亮的红色。舞台面向500个座位，而加上周边的广阔区域可以在节日和音乐会期间容纳5 000人。一个老旧

through purification. The purified water will be piped out by wind energy system to flow around the park, which not only in favor of water purification given exposure to oxygen , but also becomes new water scenery(Fig. 8-1-4).

3. Open Space

Open space: "Edward A. Cowper Place" locates in the blast furnace area. It was slag collecting zone, but now is a exuberant shady ground with high-resistance birches after soil improvement. At the beginning, no one believed that there would be a picture of flourish trees and blooming flavors among steel woods of blast pipes and furnace, but now people often celebrate here. There once were 50,000 visitors in Cowper Place during IBA mid-exhibition in 1994. Sinter Park was an open area, with ailanthus woods on waste land. There are a garden and a Roman amphitheater shrouded behind the woods, while the amphitheater is painted bright red by cycled red brick fragments. There are 500 seats in front of the stage, yet it can hold 5000 persons during festivals and concerts with the surrounding areas included. Meanwhile an old transfer post was preserved and was rebuilt into a stage architecture after changing color and door (Fig. 8-1-5).

4. Plantation

Plantation: the local ecological system quietly recovered on its own when the works were deserted, and large scale vegetation thrived on the desolate places and abandoned pollutants. In early February, the surrounding areas along the railway are shrouded by bright yellow vegetation, with mosses and lichens perching on rocks in the slagheap. Plants spread over coal ash mixed soil, metal-forged sediment and manganese ore slag. The attitude towards wild vegetation reflects new design strategies for planting

图8–1–3 铁路公园
Fig.8-1-3 The Railway Park

图8–1–4 水园
Fig.8-1-4 The Water Park

的转运站被保留下来，在更换颜色和重新安装大门之后，被改造成为一个舞台建筑（图8–1–5）。

4. 公园植被

公园植被：工厂在废置期间，这里的生态系统悄然地进行着自我恢复的进程。大片植被在工厂荒无人烟的地方和那些废弃污染物的表面繁衍生长起来。早在二月份，炫目的黄色植被就覆盖了铁路周边的区域，青苔和地衣在矿渣堆的石块上生长。在煤灰混合的土壤上面、金属浇铸的沉积物和锰矿矿渣上面生长着类似草原上的植物。对于这些野生植被的态度体现了工业废弃地上种植设计的新策略。一些区域受到严重污染，但是只要禁止进入，仍然可以维持现状。保护这里多姿多彩的植被对设计师来说，比完全清理和覆盖这些区域更加重要。这些覆盖在原来的焦炭厂区地面上的煤矿弃土，被多环芳烃（严重污染。对于这类问题有两种不同的解决方案：一种是用黏土永久的封闭污染层，同时也就丧失了所有的野生植被；另一种是在采取适当手段降低污染，允许轻微的气体扩散，对其加以有限利用（比如自行车、步行）。拉茨及其合作者们选择了第二种方案。

公园中的植被并没有像自然风景中通常的那样均匀地分布，而是在铁路和水系之间的狭长“条带”状的空间里一块块独立存在着。相互独立的区域有着各自不同形式和颜色的植被种类。它们相互独立，但是在鲁尔区不断发现来自世界各地的植物种类（在杜伊斯堡大约有200种之多）。它们被称作“外来物种”。大量不同种类的植物需要培训专门的园艺师对其养护和管理（图8–1–6）。

四、 杜伊斯堡后工业景观改造的手法

杜伊斯堡公园2001年获德国景观设计优秀作品奖，2000年获第一届欧洲景观设计奖，杜伊斯堡景观公园被学术界称为“21世纪模式公园”。作为一个新的公园模式，杜伊斯堡公园极大拓展了后工业景观设计的内涵和方式。可以说杜伊斯堡公园代表着90年代以来德国旧工业区的改造，发展出一套独立的后工业景观设计范式。新的景观设计范式有它一套独立的

on wasted land. Some heavily polluted areas could still sustain current situations without visits. It is more important for designer to preserve the various vegetations rather than clear or cover these areas. The waste coal ash that covers the ground in coke works is heavily polluted by polyaromatic hydrocarbons, for which there are two solutions: one is to permanently seal the polluted area by clay at the price of losing all the wild vegetation; the other is to take measures to lower pollution with slight gas diffusion and take limited advantage of it such as paths for bike and hike. Rhazes and his cooperators chose the second solution.

The vegetation in the park does not distribute uniformly as one in natural scenery does. Instead, it is isolated in the strips between railway and waters, with diverse kinds in different styles and colors for various strips. Though the plants are isolated, people continue to find species with alien origins in Rhur areas, totaled more than 200 kinds, which are called neophytes. These neophytes call for specially trained gardeners for cultivation and management(Fig. 8-1-6).

IV. Methods from Post-Industrial Landscape Transformation at Duisburg

Duisburg Park won the Excellent Design Prize for landscape design in Germany in 2001, and the 1st Europe Landscape Design Prize in 2000, while it is regarded as the "21-century-mood Park" by

图8–1–5 考珀区内的广场
Fig.8-1-5 The square in Cowper Place

图8–1–6 公园内的植物
Fig.8-1-6 The species with origins in the park

思想 方法和设计手法。这种新的景观设计范式是后工业景观设计纷繁多样背后更加核心和本质的内在精神，因而也更加具有普遍价值。在1998年哈佛大学“人工场地：后工业景观的反思”景观学术会议上，与会学者在杜伊斯堡公园景观设计的讨论中间提出：“杜伊斯堡公园”开创了一种新的偶像，一种新模式，一个颂扬社区与文化的“工业公园”。并且认为它开创了一种新的公园设计范式。

1. 场地改造

在杜伊斯堡公园中，设计师彻底的保留了原炼铁厂的整体结构。而在结构体系内部进行适应性的改造和设计。工厂东部密集布置着鼓风高炉、发电车间等工业建筑，建筑师在完全保留这组建筑的前提下，对内部进行改造设计和功能置换。景观师在建筑群内部的开敞空间进行污染处理和景观设计，创造了供游人聚集活动的“金属广场”和“考珀区”林荫广场，成为“钢铁丛林”中的绿色空间。鼓风系列车间的北部是大片的料仓区域，过去用于储藏铁矿石和煤炭。景观师和园艺师将巨大的混凝土料仓改造为“料仓花园”和儿童游戏场，“料仓花园”用一种景泰蓝添色的方式镶嵌在混凝土的方格中间（图8–1–7），使新景观完全融入旧有的结构体系之内。仙游岛公园的设计同样保留了原净水厂的结构，特别是水循环框架，设计中的主题花园完全利用了原净水厂内几个容积巨大的沉淀池和净化池。

2. 控制规模

在保留工业景观结构的前提下，景观师首先需要面对工业场地普遍具有的超大规模的尺度，凌乱破碎的景观元素。景观师经常采取一些特殊的手段来达到对场地的规模和破碎的景观元素的控制，主要包括景观分区、运用格网结构和主题游线串联（图8–1–8）。

1）景观分区

景观师会根据场地的特点选择一部分区域进行重点设计，作为游人活动的主要区域，而对于其他次要区域则任凭其自由发展。这在杜伊斯堡公园的景观规划中表现得最为明显。公园的总面积230多公

academic circles. As a new park mood, Duisburg Park significantly enriches the connotation and develops the pattern for post-industrial landscape design. Represented by Duisburg Park, transformation for old industrial areas in Germany since 1990s generates a unique paradigm for post-industrial landscape design, with unique thinking patterns and design techniques. This paradigm reflects the core spirit within numerous post-industrial landscape designs, thus it holds more universal value. In the academic conference "Manufactured Sites: Rethinking the Post-Industrial Landscape" at Harvard in 1998, participants claimed in the discussion for Duisburg Park landscape design that "Duisburg Park" creates a new idol, a new mood, and a "industrial park" adoring community and culture, and it generates a new paradigm for a park.

1. Site Transformation

In Duisburg Park, the designer totally reserved the original framework in the steel works, while conducts flexible reform and design within the structure system. In the eastern part of the works stands closely the blast furnace, power workshop, etc. The designer preserves this cluster of industrial architectures, takes pollution disposure and landscape design in the open space within these buildings, and finally creates "Metal Square" and "Cowper Area" shady square as gathering zone for visitors, making them green space between "steel woods". The storage bin area locates at north blast workshops. It was once used for stocking iron ore and coal. The designer and gardener re-forge the huge concrete storage bin into "Storage Bin Garden" and children playground. "Storage Bin Garden" is inset among concrete checks by means of cloisonné add-color, which realizes a fusion of the newly created landscape into existing structure system (Fig. 8-1-7). Seonyudo Park transformation also keeps original structure in the water purification plant, especially the water cycle framework, and the theme garden in design entirely makes use of the large sedimentation ponds and purification ponds.

2. Controlling the Scale

Under the precondition of preservation of industrial landscape structure, the designer has to face typically extra large scale industrial sites with messy and fragmentized landscape elements. The designer usually adopts special means to control over the

图8–1–7 料仓花园与儿童游乐场
Fig.8-1-7 The Storage Bin Garden and children playground

顷，不可能对公园每一处都进行设计，而且由于历史原因，公园的大部分区域都处于荒芜状态，在人迹罕至的厂区，野生植被自由生长。景观师放松对这部分区域的人为干预，一方面可以集中力量改造工厂设施密集区域，另一方面可以为城市和公园提供大范围的生态保护地。这种景观分区的方法不仅使景观师面对错综复杂的基地条件时，可以有重点的展开设计，而且使景观获得对比和秩序。

2）格网结构

面对松散破碎的景观元素，景观师通过格网模式整合统一破碎的景观元素。拉茨在港口岛公园的设计中通过碎石瓦砾构筑了新的景观格网。无独有偶，这种格网模式在建筑师屈米设计的巴黎拉维莱特公园中同样能够找到。设计师通过建立120米×120米的方格网，并在格网的交汇点上阵列排布40座红色房子“狂想”，达到对55公顷的大尺度公园的控制。不管这种方法是源自拉茨宣称的“结构主义方法”，还是评论家声言的“解构主义思想”，终究格网模式有效控制了工业场地的庞大尺度和破碎散乱的景观。

3）游线组织和主题串联

除了建立格网和景观层来统摄和控制工业场地的景观元素，设计建立贯穿景区的游线和路径直接关系着游人对于空间景观的体验，同时也是设计师传达某种精神内涵的有效手段。杜伊斯堡公园通过高架步道的设置，有效的限定了游人观赏料仓花园和其他景观的特定视角和路径。公园中还设置了不同标高层面的游览道路，包括自行车游览系统，这些交通游线串联了不同的景区和景点，巧妙的限制了游人的活动范围，并且帮助游人体验并建立公园景观意象，达到设计师的意图。

3. 尺度转变

后工业景观设计根据人们对工业景观的感知方式，衍生出来的两种尺度设计——“视觉尺度”和“触觉尺度”。庞大的工业建筑和场地构筑最初并非是为人设计的，而是为机器设计的，也可以叫做“非

scale and fragmentized landscape elements, such as landscape zoning, linking by grid structure and theme tour, etc (Fig. 8-1-8).

1) Landscape Partition

The designer usually chooses some areas on site features to design for main activity zone for visitors, while leaves other minor areas to free development, which features Duisburg Park landscape planning. The park spreads over 230 hectares, thus it is impossible to design for every corner. Also, most areas are deserted into desolation given history reasons, with wild vegetation roaming over. Designer retreats from intervention in this area, because on one hand he could focus with all strength into the transformation of closely distributed industrial facilities, and on the other hand this area could become large scale ecological preservation zone for the city and parks. This zoning method enables the designer who faces complicated base conditions to take focuses on design to gain contrast and order.

2) Grid Structure

The designer integrates fragmentized landscape elements by grid structure. In the design of Buergpark Hafeninsel, Rhazes builds new grid landscape by rubble, and similarly, in the design of Paris Parc de la Villette by Tschumi we can also find gird structure. The designer builds 120 m*120m grids and arranges 40 red houses "Folie" at the intersections to control over the 55-hectares large park. This grid structure has managed the widespread industrial sites and fragmentized landscapes despite whether it belongs to Rhazes claiming "Structuralism Method" or commentators' "Deconstruction Thoughts".

3) Organizing Line and Concatenating Theme

Besides grid structure and landscape levels to govern and control landscape elements in industrial sites, a theme tour line throughout the park affects

图8-1-8 景观改造后的北杜伊斯堡公园的结构
Fig.8-1-8 The structure of North Duisburg Landscape Park after the landscape transformation

人尺度”，或者称作“机械尺度”，工业场地转变为城市公园以后，大部分工业景观对人的作用仅仅是视觉上的，因此可以作为“视觉尺度”而存在；但是当涉及游人的行走、触摸、休息和进行多种娱乐活动时，必须考虑对工业景观中的非人尺度进行调整，在人体直接接触的界面和使用的空间必须进行“触觉尺度”上的改造。杜伊斯堡公园在向人们展示尺度巨大的料仓花园的时候采用了空中俯视和底层穿梭两种参观方式。前者通过焊接在料仓墙体上面的钢结构的步行桥，为游人提供一个移动中俯视花园的视角，获得一种俯瞰“屋顶花园”的整体感受，同时在移动中浏览，获得了一种参观画廊般的感受；为了便于游人近距离接触和体味花园，设计师在料仓的底层平面，开辟了穿越“混凝土画廊”的通道，在厚达1.2米的墙壁上切割出入口。林荫广场、熔渣公园都是通过在开敞空间种植树木，为游人提供人性化的聚会和举办各种庆祝活动的场所。

4. 隐喻和象征手法运用

景观师在对工业元素进行改造设计的时候，往往利用某些特殊的工业遗迹，通过某种象征性的手法，表达设计师的思想和新景观的文化内涵。这些景观也成为公园中寓意深刻而又极具象征意义的点睛之笔。在利用绿色技术处理近代工业场地的本身就是一个强烈的隐喻。拉茨在杜伊斯堡公园中充分发挥了他的关于“物质自然”的象征手法。对自然进程进行标识，成为贯穿杜伊斯堡公园景观设计的核心理念。“梅陶利卡广场”（意为金属广场）锈蚀的铸铁板就成为自然侵蚀（erosion）的天然标尺。广场中央49块重达7～8吨的铸铁板，在清理了表面的灰渣和沉积物之后，显示出自身特有的质感。

五、 总结

北杜伊斯堡位于莱茵河与鲁尔河的交汇处，是德国19世纪重工业区的中心，该区域是世界上最大的工业区之一，有近200年的工业发展历史。20世纪60年代起，钢铁行业不断缩水，这片

tourists' experiences for the landscape, and is also an effective way of convey certain spiritual connotation. Through aerial footpath in Duisburg Park, the designer limits the views and routes for visitors in travels around Storage Garden and other landscapes. There are also travel routines with different heights, bicycle travel included, which link different areas and spots, limit visitors' activity sphere, and help them experience and forge images for landscapes in the park to reach the design intentions.

3. Changing the Dimension

According to people's sense styles to industrial landscape, post-industrial landscape design derives 2 criterions, "Visual Scale" and "Tactile Scale". The large industrial architectures and site construction were originally designed for machines rather than people, so it can be called "non-human scale" or "machine scale". After the transformation from industrial sites into city parks, most industrial landscape only acts in vision, hence the "Visual scale", but when related to people's walk, touch, rest, and entertainment, it should be adjusted according to "Tactile scale". When displays the large scale storage bin garden in Duisburg Park, the designer employs 2 visit styles, overlook from the high and shuttles at down below. The previous style provides an overlook of the garden during walks on the aerial steel-structure bridge welded on the walls in the storage bin, an integrate experience of bird's eye view over "roof garden", and visit to gallery when walking. In order to expose visitors to touch and feel the garden in short distance, the designer opens up a tunnel through the "concrete gallery" at downstairs and cuts out an entrance on the 1.2-meter-thick wall. The shady square and slag garden all turn into places for gathering and celebrities by planting trees in the open space.

4. The Strategies of Metaphor and Symbol

When tackles with industrial elements, the designer usually employs certain symbols for special industrial heritage to express his thinking and the cultural connotation for the new landscape. And these landscapes represent the profound punchline with symbolic meanings, while adopting green technology to transfer industrial site itself is a strong metaphor. Rhazes fully plays his symbolic strategy for physical nature in Duisburg Park, and the marking of natural process stands for core idea in Duisburg Landscape Park design. The rusting iron plates on Piazza Metallica become natural marks for nature erosion. After cleaning the ash and sediments on the surface, those 49 iron plates weighing 7-8 tons scattered in the middle of the Piazza display their own texture.

V. Summary

曾经繁荣的土地逐渐沦为工业废弃地。北杜伊斯堡景观公园的原址是钢铁厂和炼炉厂，设计师彼德·拉茨面临最关键的问题是如何处理工厂的各种遗留物，例如庞大的工业建筑、高耸的烟囱、铁路、桥梁、沉淀池等。这些凌乱甚至丑陋的工业遗迹在设计师眼中却饱含技术之美。彼德·拉茨用生态手段处理场所记忆，造就场地肌理，使之成为工业地更新与改造的经典设计案例，引起人们更加广泛的对逆工业化产物的思考。

公园设计所体现的最为重要的一条意义是对于生态美学的肯定。建筑物的整个生命周期都是需要能耗的，不能够只计算建造费用而不考虑运营维护和拆除重建的费用。在拆除重建的过程中各种材料的运输堆积、建筑垃圾的处理等等都是需要大量的能耗，并且这些能耗以热量的形式耗散到整个生态系统之中。因此建筑的循环利用不仅可以变废为宝，循环利用，而且避免了拆除与重建造成的不必要的能耗。在人类营建活动已经日趋饱和的今天和未来，适应性改造与循环利用将是一个切实有效的途径。后工业景观设计就是将废弃工业建筑和设施进行重新的利用，不仅节省了土建的投资，更重要的是它在毫不损伤基地“元气”的情况下，实现了工业遗产资源的循环利用，减少和降低了能耗。

在生态主义思想的影响下，人们的美学观念发生了深刻变化。原来认为是脏的、混乱的、丑陋的东西，在赋予了深刻的文化内涵以后，在生态伦理背景下去观察和认知的时候，就变成是和谐的、理性的和美的东西了。传统的美学原则是基于视觉形式的基础之上的，形式成为决定因素；但是在生态美学的原则指导下，原生、自发、荒野的景致只要是符合生态原则的，依然是美的，是可以为人们所接受的。可以说，生态美学关注的是景观的生态内涵，而非表面的形式。

拉茨演绎了一种全新的景观公园类型，即后工业景观公园。他采用反常规的设计手法，保留原有的旧工业基础骨架，保持原有大尺度空间景观特征，仅通过小尺度景观重塑，使原有景观产生质变，给人以视觉冲击。拉茨赋予旧工业景观以美学

North Duisburg locates at the interchange of River Rhine and River Ruhr and it is a center for heavy industry in Germany in 1800s, one of the largest industrial areas with nearly 200 years industrial history. But due to the decay in the steal industry from 1960s, this prosperous land fades into deserted field. The former address of North Duisburg Landscape Park is steel plant and furnace, so the most crucial problem for Peter Rhazes is how to handle with the deserted architecture, such as immense industrial architecture, soaring chimney, railway, bridge, settling pond and so on. These industrial relics, though messy and even ugly, contain numerous technological beauty in the designers' view. Peter Rhazes imposes biological means to deal with the location memory, create new geographic site, and finally makes it a classical case of industrial area update and transformation, which generates an even wider thinking on anti-industrial products.

One of the most important meanings reflected in the project is the affirmation of ecological aesthetics. The life cycle of architectures consumes energy, so the designer should consider not only building budget but also cost for maintenance and removal for reconstruction. Collection and transportation of materials, as well as handling with construction waste consume much energy which dissipates into the whole ecological system in heat. So the cycle of architecture revives the old building and avoids unnecessary energy consumption, adaptable transformation and architecture recycle will be an effective way solving the problem today and in the future with construction saturation. Post-industrial landscape design recycles deserted industrial architecture and facilities, which not only saves the construction investment but also plays more important role in lowering energy consumption with no harm to the genius loci of the site.

People's aesthetic values change profoundly under the influence of ecologism. The dirty, disordered and ugly things in previous opinions turn into harmonious, reasonable and beautiful ones when viewed from ecological ethics after being endowed with deep cultural connotation. The traditional aesthetics principles are based on visual form, thus styles become the decisive factor; with the guide of ecological aesthetics, authentic, spontaneous and wild landscapes are still beautiful if agreeing with ecological principles, as well as acceptable. We can conclude that ecological aesthetics focuses on ecological connotation of the landscape rather than superficial style.

Rhazes performs a brand new landscape park style, a post industrial landscape park. He adopts unusual design means and save all former industrial frame and large scale space landscape features, and bring qualitative changes only through small scale landscape reshape, which is quite impressive to us. Rhazes endows old industrial landscape with aesthetic implication and makes this post industrial landscape a combination of "idealized nature and

含义，这种后工业景观是一种“理想化了的自然及理想化了的工业”。实践证明，工业遗迹的再利用能够振兴衰退地区的经济发展，通过产业用地的再开发，恢复地区活力，使工业遗址在新时期以新的面貌继续运转并实现可持续发展。

idealized industry". This case illustrates that the reuse of industrial relics can vibrate the previous economic recession and restore its vigour by redevelopment of industrial land, thus the relic can run in new face in new period and realize sustainable development.

思考题

1. 纽约的一名记者亚瑟·卢伯称，彼得·拉茨的设计是反奥姆斯特德式的，为什么？
2. 后工业景观的改造手段反应了多样性的设计，试以杜伊斯堡景观公园中的任意改造案例来分析。并从景观人文的角度分析这些改造手段的产生。
3. 北杜伊斯堡景观公园被称为是“21世纪模式公园”，但是现在各式各样的改造措施层出不穷，如果可以你会选取哪些新的改造措施来取代原来的改造措施，并试着说明你的选择可以更好地诠释该公园的设计理念。
4. 分析公园规划的四个层次中的“散步道系统”是如何将“水园”、“铁路公园”以及“使用区域”在不同的空间上串联起来的，以及如何与城市街道相结合的。
5. 通过本案例的介绍与参考文献和推荐网站，来谈谈你对埃姆舍公园以及欧洲工业遗址游线的看法。

Questions

1. Why did the New York journalist Arthur Lubo call Peter Latz's design "The Anti Olmsted"?
2. The reconstruction methods used in postindustrial landscaping reflect a multi-design idea. Try to discuss it using one of the reconstruction cases in North Duisburg Landscape Park. Discuss the formation of these methods through the landscape culture aspects.
3. The modern North Duisburg Landscape Park is one of the so called "21st century style parks", however as the quick rise of various new reconstruction methods, there is still reconstruction space. Try to discuss how the new reconstruction methods can be applied in this case instead of the old methods. And explain why these methods can better explain the design idea of the park?
4. Base on the four aspects of the design of the park, discuss how the Promenades System affects on intergrading the water park, railway park and using area in space as well as combining them to the city streets.
5. Talk about your view on Emscher Park and European Industry Ruin tour line base on the information provided in this introduction, references and recommend websites of this case.

注释/Note

图片来源/Image：图8-1-1 贺旺.后工业景观浅析[D]. 清华大学，2004；图8-1-2，图8-1-7，图8-1-8 http://howardchanxx.net/?p=300；图8-1-3~图8-1-6 http://www.arch.hku.hk/teaching/cases/duisburg/Duisburg_photo.htm

参考文献/Reference：贺旺.后工业景观浅析[D]. 清华大学，2004；Udo•Weilacher, Syntax of Landscape: The Landscape Architecture of Peter Latz and Partners, 2007

推荐网站/Website：http://en.landschaftspark.de/startseite；http://www.erih.net/welcome.html；http://www.arch.hku.hk/teaching/cases/duisburg/Duisburg.htm

第二节 郭瓦纳斯运河海绵公园规划
Section II Gowanus Sponge Park Design

一、项目总述

郭瓦纳斯运河是隐藏在纽约布鲁克林区的标志性建筑，对于当地的社区是一笔宝贵的财富。原湿地的小河周围是工业建筑，还有一些当地居民。运河承载了很多社区的历史，但目前公众获得的水仅限于公有的街道。舱壁的状况一直被忽略，污水排放口定期溢出的污水和街道地表径流直接排入运河，并且该地区的工业活动污染周围的土壤和河床（图8–2–1）。

1999年当地政府机构通过重新治理隧道和运河来解决污染问题，并计划在未来加强清理和疏浚郭瓦纳斯运河的工程。最近对临近水域的规划以及房地产需求的增加，刺激了零星的私人项目发展，虽然这些规划为场地提供了到达滨水区的通道，但他们不能为公众的开放空间系统提供统一规划。为了向公众在沿运河区域提供一个潜在的休闲空间，并改善土壤和水质，海绵公园诞生了。

海绵公园的设计是一系列的城市公共滨水空间，它慢速吸收和过滤地表水径流，以净化污染运河水为预期目标，激活了运河滨水空间。同时在更高的层面上，沟通了管理中许多相互竞争的声音，议程和热点问题。

I. Overview

The Gowanus Canal is a hidden landmark in Brooklyn, New York that is a valuable asset to the local and broader community. Formerly a wetland creek, the canal is bordered by industrial buildings and surrounded by residential neighborhoods. The canal gives character and history to the surrounding neighborhoods, yet public access to the water's edge is currently limited to publicly-owned streets that terminate at the canal. The physical condition of the bulkhead has been neglected and combined sewer outfall overflows and street surface runoff regularly drain directly into the canal. Industrial activities in the area have polluted the surrounding soil and canal bed, introducing industrial toxins to the water (Fig.8-2-1).

In 1999 local government agencies addressed the pollution issues by reactivating the Gowanus Flushing Tunnel and the Gowanus Pump Station at the head of the canal and also plan for additional clean-up and dredging projects in the future. Recent plans for rezoning of areas adjacent to the water, as well as the increased demand for real estate in this part of Brooklyn have spurred sporadic private development projects, and although these plans provide access to the waterfront within their immediate sites, they do not present a unified plan for the development of a publicly accessible open space system for the canal as a whole. In order to provide public access to potential recreational spaces along the canal, and to improve the soil and water quality, the Sponge Park was invented.

The Sponge Park is designed as a series of public, urban waterfront spaces that slow, absorb and filter

图8-2-1 郭瓦纳斯运河海绵公园区位图
Fig.8-2-1 Location map of Sponge Park

二、 区域环境介绍

坐落在纽约布鲁克林的郭瓦纳斯运河海绵公园将规划为一个公园网络，沿着过去污染严重的郭瓦纳斯运河而延伸的社区绿色空间。最初的盐碱沼泽地浅滩，郭瓦纳斯河曾经是一个鱼类、野生动物和湿地植物良好的天然的栖息地。在工业革命期间，河流成为布鲁克林海事和航运的枢纽，在1849年，为了容纳较大的船只，这里建成了工业的航运码头，沼泽地的水被排干，河床经过疏浚后被水泥硬化。工厂、仓库、皮革厂、煤店和天然气厂纷纷沿河而建，充分利用河道这一交通优势与外界连通。多年来，这些工厂倾倒有毒废弃物，污染径流，将未经处理的污水直接排入河中。随着时间的推移，居民区取代工厂和仓库，工业企业搬迁到更廉价的土地，而如今，纽约市排水系统仍然继续污染郭瓦纳斯运河。暴雨期间，雨水混合着未经处理的污水直接进入运河。落后的基础设施和水污染，压低了当地的房地产价值，形成一个城市街区的毒瘤，并成为一种威胁人类和野生动物安全的有毒环境（图8-2-2）。

多年的工业垃圾排放和水污染，使郭瓦纳斯运河被定为纽约州盐渍地表水环境质量标准的SD类。SD类的水体被定为危险的污染地区，无法举行如钓

surface water runoff with the intended goal of cleaning up the contaminated canal water, activating the canal edge and communicating a larger vision for stewardship of the environment to a community with many competing voices, agendas and concerns.

II. Regional form Natural Pattern

Located in Brooklyn, New York, the Gowanus Canal Sponge Park is a planned network of public parks and neighborhood green spaces along the historic yet highly-polluted Gowanus Canal. Originally a shallow tidal inlet of saltwater marshland, Gowanus Creek was once a thriving habitat for fish, wildlife, and wetland plants. During the industrial revolution, the creek became a hub for Brooklyn's maritime and commercial shipping industries. In 1849, to accommodate larger vessels and industrial shipping docks, the marshes were drained and the creek bed was dredged and reinforced with concrete bulkheads. Factories, warehouses, tanneries,coal stores, and gas refineries popped up along its edges, taking advantage of the waterway's valuable connectivity to the region and beyond. For years, these industrial sites dumped toxic waste, polluted runoff, and raw sewage directly into the water. Over time, residential neighborhoods replaced factories and warehouses, as businesses relocated to cheaper land. Today, New York City's combined sewer system continues to pollute the Gowanus Canal. During heavy rainfalls, stormwater combines with raw sewage, overflowing directly into the canal. Its crumbling infrastructure and contaminated water drag down local real estate values, form a barrier between urban neighborhoods, and create a toxic environment dangerous for people and wildlife (Fig.8-2-2).

Years of exposure to industrial waste and water pollution has currently classified the Gowanus Canal as a New York State Saline Surface Water Quality Standard Class SD. Class SD water bodies are identified as dangerously polluted areas where activities such as fishing, swimming, and secondary contact are discouraged. This poor water quality is exacerbated by a lack of maintenance that has caused 85 percent of the canal bulkheads to erode, high cleanup costs, and minimal private investment. Recently, the U.S. Environmental Protection Agency began investigating the Gowanus Canal as a potential Superfund site (Fig.8-2-3).

New York City has a combined sewer system. During storm events, rain falling within the Bergen watershed enters the storm drains and mixes with raw sewage in the sanitary sewer system. During heavy rainfalls, the combined sewage and stormwater overflow directly into the Gowanus Canal, discharging over 1.1 million cubic meters of combined sewage. Consequently, the Gowanus Canal remains listed on the New York State

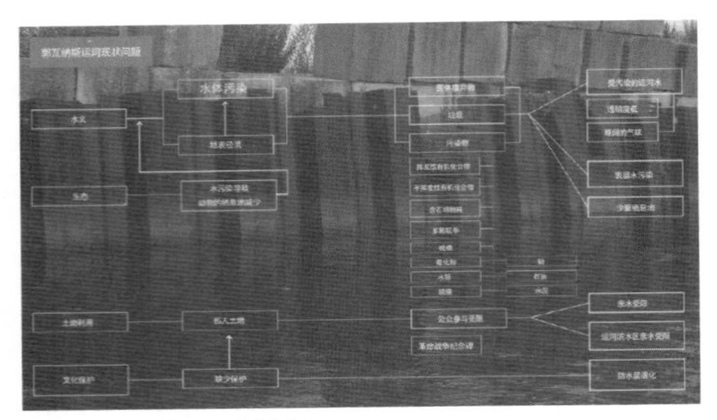

图8-2-2 郭瓦纳斯运河现状问题
Fig.8-2-1 Current issues facing the Gowanus Canal

图8-2-3 严重污染的郭瓦纳斯运河
Fig.8-2-3 Highly-polluted Gowanus Canal

鱼，游泳和与水体接触的活动。由于缺乏保养而导致的水质恶化，已经造成85%的河岸被侵蚀，清理成本高，很少有私人投资。最近，美国环境保护署开始调查郭瓦纳斯运河，并将其作为一个潜在的基金投资地点（图8-2-3）。

纽约市有一个综合的排水系统，在暴风雨时，卑尔根流域范围内的雨水和未经处理的污水混合进入清洁雨水排水系统。在强降水期间，混合后的雨水和污水及天然降水直接排入郭瓦纳斯运河雨水，总量为超过110万立方米的综合污水。因此，由纽约市环境规划局制定的纽约州303（D）污染水体清单上，郭瓦纳斯运河仍然被定为雨污合流直溢规划水体。新的海绵园计划建议将地表径流水引入到一个水体管理系统中，并将其转变成开放的公园，包括规划的城市户外空间和植被景观缓冲区，吸收和过滤污染水，减少雨水直接进入排水系统，并创建一个公共的亲水空间。

三、总体规划

郭瓦纳斯运河100英尺（30.48米）宽1.4英里（2 253米）长，起始于巴特勒街头并流向郭瓦纳斯湾，其深度从4～16英尺（约1.2～4.9米）不等。从生态系统上的考虑，无论是城市还是农村，都需要从地理的角度去理解，因此我们同时关注自然地理边界和管辖权流域的行政边界——包括2.5平方英里（647.5公顷）的布鲁克林公园、卡罗尔花园和波恩兰姆山社区。运河表面径流通过下水道排水、雨水和街道直接地表径流等手段，从1 758英亩（约

Section 303(d) list of impaired water bodies and is designated as a Track I Combined Sewage.Overflow planning water body by the New York City Department of Environmental Planning. The innovative Sponge Park plan proposes diverting surface water runoff into a water management system and publicly accessible park that includes programmed urban outdoor space and vegetated landscape buffers to slow, absorb and filter the contaminated water, reduce the input of stormwater into the sewer system, and create an accessible public waterfront.

III. Master Planning

The Canal is 100' wide and 1.4 miles long from the head at Butler Street to where it opens to the Gawanus Bay. It varies in depth from 4' to 16'. Consideration of ecological systems, whether urban or rural, involves understanding the inputs from a geographic perspective. For this reason, we looked at both the physiographic and jurisdictional watershed boundaries of the Gowanus which included 2.5 square miles of Brooklyn's Park Slope, Carroll Gardens and Boerum Hill neighborhoods. Surface water runoff from the 1758 acre watershed flows into the Gowanus canal by means of combined sewer outfalls, storm outfalls and direct street surface runoff.

Gowanus Sponge Park open space development includes a new 40' canal edge setback and conversion of street ends to park spaces, resulting in 16 acres of available land with which to work. Our design consists of 7 acres of esplanade, 4.4 acres of recreational open space and of which 3.5 acres of permeable water remediation and retention area manages the runoff from the more immediate 326 acre catchment area adjacent to the canal. Proposed back-up water storage units similar to cisterns would mediate extreme weather conditions and provide additional storage capacity.

The Sponge Park is an open space system that remediates surface water runoff while also adding

711.4公顷）流域流入郭瓦纳斯运河。

海绵公园开放空间的规划，包括一个新的40英尺（约12.2米）宽的河堤以及由街道尽端转变而成的公园空间，使得16英亩（约6.5公顷）土地发挥其作用。我们的设计包括7英亩（约2.8公顷）空地，4.4英亩（约1.8公顷）的休闲开放空间，其中3.5英亩（约1.4公顷）修复和保留区的管理直接与326英亩（约132公顷）的排水区相连，毗邻运河径流。提出后备水存储区——类似于蓄水箱——将调解极端天气条件，并提供额外的存储容量。

海绵公园是一个开放空间系统，这一系统不仅改进了地表水径流，同时也增加了城市地下水。城市逐渐成为以铺装覆盖为主的土地。许多城市仍然雨污合流，并直接排入自然水体。海绵公园设计服务，通过建筑设计，景观规划和城市景观设计，修复地表水径流，利用景观、建筑和工程战略，以减缓，保留和过滤雨水，清洁水道。

郭瓦纳斯运河海绵公园的功能主要在于恢复运河的水质，并通过一系列相互关联的公共空间使市民沿河岸活动。这种创新的理念来源于它的名字——"海绵公园"，通过一系列景观缓冲区和人工湿地降低、吸收，并过滤地表径流水。为了截流和储存污染径流，地下贮水池将截流地下排水系统的污水。该贮水池将收集到的污水缓慢释放进入人工湿地，经过过滤和净化再流入运河。而建设居民区和基础设施就是为了防止盐碱沼泽恢复到17世纪的状态。景观师的设计是，恢复植物群落，及那些过去有助于控制洪水和水质的过程（图8-2-4）。

除了改善水质，海绵公园为单调的社区创造了必要的公园和休闲场所，加强了邻里社区的联系并提升了周边的房地产价值，现在运河将成为方便当地居民散步的走廊。线性空间或散步区，使人们可以沿着运河边漫步。现有的绿地，历史遗迹，新的户外公共空间，如社区花园，遛狗场，表演场，咖啡厅，休息区，游船码头和展览空间等，共同组成了社区公园网络。风景秀丽的廊道空间构建出丰富多彩的景观，也超越了最初规划的过滤沼泽、修复性湿地及运河的恢复（图8-2-5）。

accessible urban open space to underserved neighborhoods. Cities tend to be predominantly paved land. Many cities also have combined storm and sanitary sewer systems where storm water and sewage mix and pour directly into our coastal waterways. Sponge Park design services create architectural designs that can work to remediate surface water runoff through landscape plans and designing urban landscapes. The Sponge Park design services use landscape, architectural and engineered strategies to help make waterways cleaner by slowing, retaining and filtering storm water.

The Gowanus Canal Sponge Park emerged as a vision to restore the water quality of the canal and engage citizens with its shoreline through a series of interconnected public spaces. This innovative idea gains its name "Sponge Park" from its functional ability to slow, absorb, and filter surface water runoff through a series of landscape buffers and constructed wetlands. Underground tanks will intercept street sewer drains in order to capture and store polluted runoff. The tanks will release collected water slowly into the artificial wetlands, where it will be filtered and cleaned before entering the canal. While built neighborhoods and infrastructure prevent the saltwater marsh from being restored to its 17th century state, the landscape architect plans to reintroduce plant communities and processes that historically helped control flooding and water quality in the Gowanus Bay (Fig. 8-2-4).

Beyond improving water quality, the Sponge Park would create needed parkland and recreational spaces for underserved communities, enriching neighborhood social connectivity and enhancing the value of surrounding real estate. Now the canal will become a convenient pedestrian corridor for local residents. A linear esplanade, or canal walk, will meander along the edge of the canal, weaving together a fabric of street end parks, existing parkland, historic sites, and new public outdoor spaces, such as community gardens, dog runs, performance spaces,

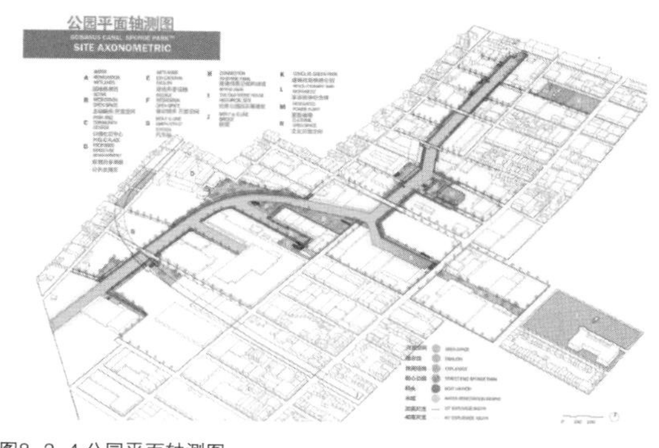

图8-2-4 公园平面轴测图
Fig.8-2-4 Site axonometric

cafes, seating areas, boat launches, and exhibition spaces. The scenic walkway will frame views into the colorful filtration basins, remediation wetlands, and restored canal beyond (Fig.8-2-5).

四、专项规划

1. 公众可达性

海绵公园要解决两个之前郭瓦纳斯运河所面临的问题：缺乏公众亲水空间；现代化的城市设施对社区造成物理隔离。

郭瓦纳斯运河河岸用地属性是私有，除了在街道两端，公众亲水受到限制。运河曾经把交通运输作为运河的主要产业，但在经济发展的变化中，目前许多行业已经很少把运河作为交通工具。运河代表一度欣欣向荣的工业大动脉，由于对水路运输的需要下降，是运河被忽视和恶化。海绵公园规划提出了连接市区的公共土地和私人土地，毗邻运河创建连续的滨海休闲空间。现有的公共街道两端将成为作为公园入口，提供亲水休憩空间。这些街心公园提供面向社区的方案，如宠物空间、花园小区、公共展览空间和临时市场空间。针对每个社区的要求，进一步创造土地和水之间的交融过渡。

目前运河城市被基础设施分开的社区将有共同聚集在海绵公园的游憩空间。带状公园将提供一个平行于运河的城市长廊，并连接波恩兰姆山，公园坡和卡罗尔花园，同时也作为一个社会的“海绵”，吸收和融合多样的社区居民（图8–2–6）。

2. 水体修复

公园最独特的功能，是其修复景观的能力，它能随着时间逐步改善运河的环境，同时支持公众积极参与运河的生态系统。目前，郭瓦纳斯流域内的

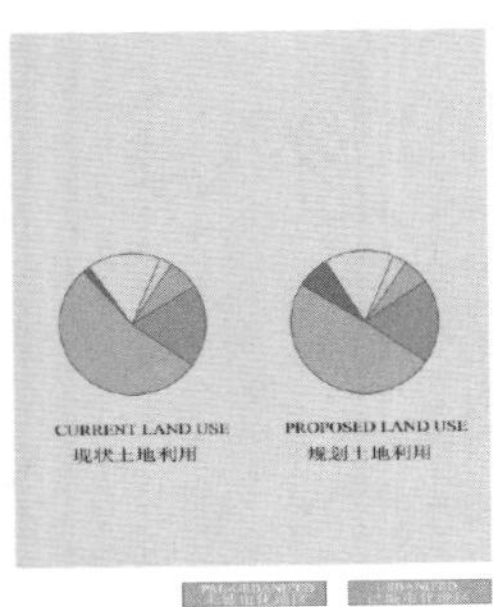

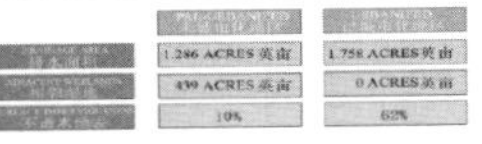

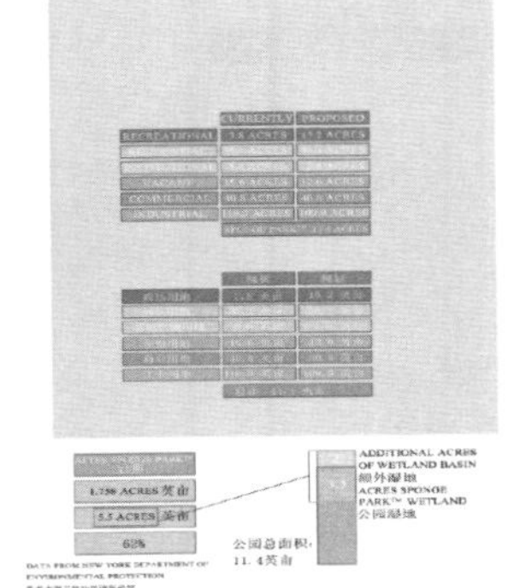

图8–2–5 土地利用
Fig.8-2-5 Land use

IV. Special Planning

1. Public Access and Urban Stitching

The Sponge Park proposes to address two key urban issues present on the Gowanus Canal: (1) Lack of public waterfront access(2)Physical division of neighborhoods resulting from modern urban infrastructures.

Properties adjacent to the Gowanus Canal are privately owned, limiting public waterfront access except at street ends. The canal once functioned as a major transportation route for industries along the canal but with changes in the economic Landscape, currently less than handful of these industries utilize the canal as means of transportation. The canal represents an artifact of a once thriving industrial corridor, neglected and deteriorating due to lack of need for water transport. The Sponge Park plan proposes a strategy of urban stitching connecting the public and private Lands adjacent to the canal to create a continuous esplanade with recreational spaces running the length of the canal. Existing public street ends would serve as entry-parks providing access to the esplanade to the water. These street end parks provide space for community oriented programs such as dog runs, community gardens, public exhibition spaces and temporary markets. Per request of the community, boat launch or ladders are located at every street, creating a further connective stitching between land and water.

Neighborhoods currently separated by the urban infrastructure of the canal will now have common space in which to congregate in the recreational spaces of the Sponge Park. The linear park will provide physical access parallel to the canal providing a waterfront urban promenade that links the neighborhoods of Boerum Hill, Park Slope and Carroll Gardens functioning as a social "sponge" that absorbs and mixes citizens of diverse neighborhoods (Fig.8-2-6).

2.Remediating Water

The most unique feature of the park is its character as a working landscape: its ability to improve the environment of the canal over time while simultaneously supporting active public engagement with the canal ecosystem New York City has a combined sewer system (see EPA diagram). Currently rain that falls within the Gowanus watershed enters the storm drains and mixes with raw sewage in the sanitary sewer system. In a heavy rainfall the combined sewage and storm water overflow directly into the Gowanus Canal. The innovative Sponge Park

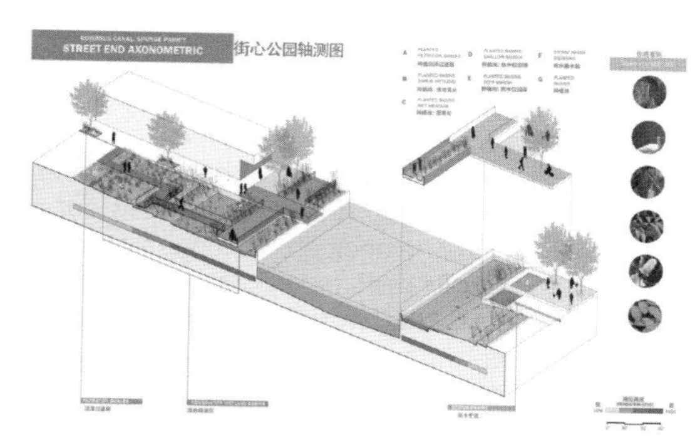

图8-2-6 街心公园轴测图
Fig.8-2-6 Street end axonometric

雨水，进入雨水渠，并与未经处理的污水混合流入下水道系统，并直接进入联合雨污管道溢出。创新海绵公园设计将分流地表径流与污水进入水管理系统，并开放公园，包括系统的城市户外空间和植被景观缓冲区，以减缓吸收和过滤污水，减少雨水下水道系统的负担，并创建宜人的海滨（图8-2-7）。

针对收集的城市供水基础设施的研究和数据，规划对于城市水管理的研究包括：现有的城市污水处理系统，流域边界，水生植物群落，水体污染物，水体修复方法，透水材料，土壤透气性及与水有关的公共项目。对水的分析分为以下几点：

（1）了解现有的城市雨水径流和排水机制。

（2）指导，收集和吸收多余的雨水径流。

（3）确定如何通过植物修复净化雨水。

（4）制定恢复湿地栖息地的策略。

3. 文物古迹保护

设计也认识到公园目前和未来的文化资源，包括一些历史遗址，休闲区和社区设施。重要的包括：目前不开放位于运河端头的革命战争纪念馆，位于拜恩公园内滨水的旧石屋，托马斯格林游乐场，郭瓦纳斯挖泥船船屋，和历史悠久的海滨本身。该计划要求宏伟的建筑风格和历史价值，坐落于地理空间中心，作为文化展览，表演空间和社会团体的办事处等功能。

在设计修复性湿地的使用中，也认识到郭瓦纳斯流域的自然历史。两个转折盆地位于运河东侧，将作为修复湿地和湿地教育中心——以促进郭瓦纳斯运河治理景观和栖息地的研究。沼泽将不能恢复

plan proposes diverting surface water runoff into a water management system and publicly accessible park that includes programmed urban outdoor space and vegetated Landscape buffers to slow absorb and filter the dirty water, reduce the input of storm water into the sewer system and create an accessible waterfront (Fig.8-2-7).

Research and data collection focused on urban water infrastructure. For urban water management, we researched existing urban sewer systems, watershed boundaries, aquatic plants plant communities, water contaminants, water remediation strategies, permeable materials, soil permeability and water related public programs. The water analysis was divided into the following foci:

(1)Understanding existing urban rainwater runoff and sewer mechanics.

(2)Developing strategies to direct, collect and absorb excess storm water runoff.

(3)Determining how to clean stormwater runoff through phytoremediation.

(4)Strategies for restoring wetland habitat.

3.Historic Preservation

The design also recognizes the cultural context of current and future resources by linking important historic sites, recreation areas and neighborhood facilities. Important sites include the currently inaccessible Revolutionary War Memorial at the head of the canal, the Old Stone House in JJ Byrne Park, Thomas Green Playground, Gowanus Dredgers boathouse & launch, and the historic waterfront itself. The plan calls for a derelict industrial structure with tremendous architectural character and historic value located at the geographic heart of the space to be revisioned as cultural amenity with exhibition, performance space and offices for community groups.

The use of remediation wetlands in the design also recognizes the natural history of the Gowanus watershed as a wetland creek. Two turning basins located on the Eastern side of the canal will be restored as remediation wetlands and a wetland educational center is programmed to promote studies of remediation landscape and habitat of Gowanus Canal. The former marsh will not be restored to its 17th century state but the plant communities and processes that helped control flooding and kept the Gowanus Bay clean will be reintroduced.

4. Plant Design

The Sponge Park utilizes plants that remediate contaminants and thrive with different periodic water inundation levels to control and clean surface water inputs. After the water is treated through the phytoremediative qualities of the plant communities it would be slowly released in to the canal. In addition,

到其17世纪的状态，但植物群落和生态过程将帮助控制洪水和保持郭瓦纳斯湾洁净。

4. 植物规划

海绵公园利用植物修复污染物和不同的周期性淹没水平，以控制和净化地表水的输入。通过植物群落对水进行修复之后，水会缓慢流入运河。此外，浮动的修复性湿地也纳入了协同工作的水生生物中，用来吸收或分解污水中的有机毒素，重金属和生物污染物的混合物。

在设计中包含了在不同淹没程度下生长的植物，并吸收污染水中的重金属。此外，浮动湿地岛和富含水生生物的牡蛎养殖场，可以用来吸收和分解有毒的有机物质、重金属及在水中发现的21种细菌。利用自然生态系统修复高度污染的水体，这样的设计展示了可行的低冲击技术如何替代昂贵的工程化解决方案（图8-2-8）。

5. 规划实施：政府与私人关系处理

客户和设计师联合当地政府机构、社区组织和利益相关者，都开始积极地参与海绵公园合作规划中。我们的目标是通过一个涉及社会各个成员的动

floating remediation wetlands would incorporate a mixture of aquatic organisms that work in concert to absorb or break down organic toxins, heavy metals, and biological contaminants from sewage.

Plants included in the design thrive at different levels of inundation, and draw heavy metals out of contaminated water. In addition, floating remediation wetlands and oyster beds incorporate a mixture of aquatic organisms that work in concert to absorb and break down organic toxins, heavy metals, and 21 species of bacteria found in the water. By using natural systems to remediate highly polluted water, this design demonstrates how low impact techniques can offer viable alternatives to expensive, highly engineered solutions (Fig.8-2-8).

5. Process & Implementation: Public and Private Partnership

The client and designers are currently engaging local government agencies, community organizations, and local stakeholders to begin the process of collaborative implementation of the SPONGE PARK. Our goal is to ensure that the banks of the canal are fully open to the public through a dynamic process that involves all members of the community. The Sponge Park conceptual framework will then be developed by private developers organically over time. Concurrent with this development are plans by public agencies such as USACE, NYSDEC and NYCDEP to clean the water and adjacent toxic sites. The site may well be designated a Superfund site in the future

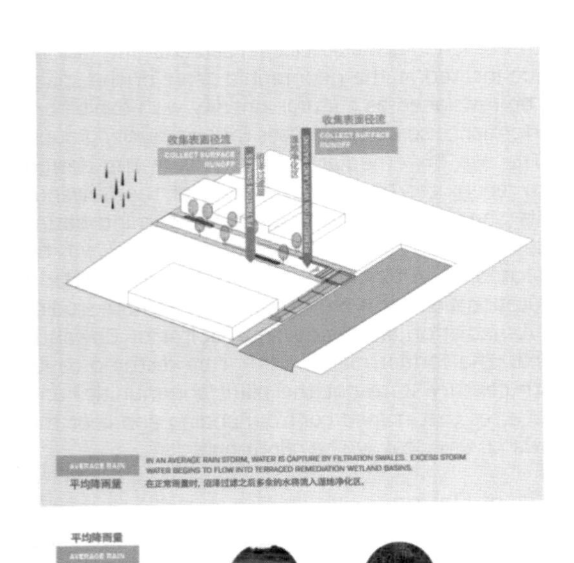

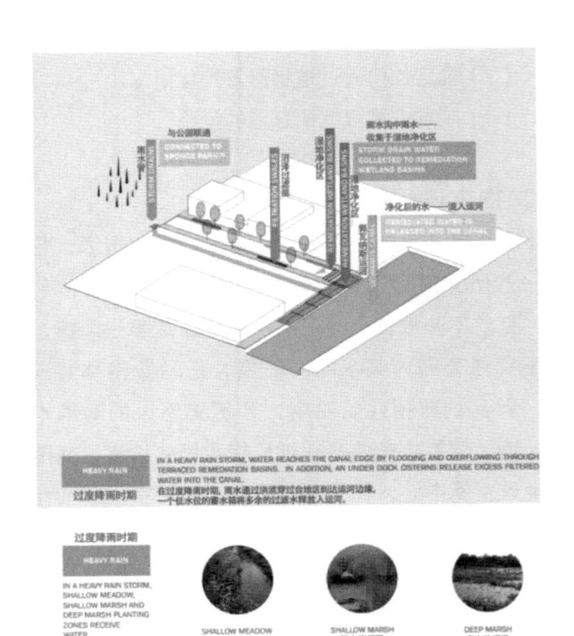

图8-2-7 水体修复系统
Fig.8-2-7 Remediating water system

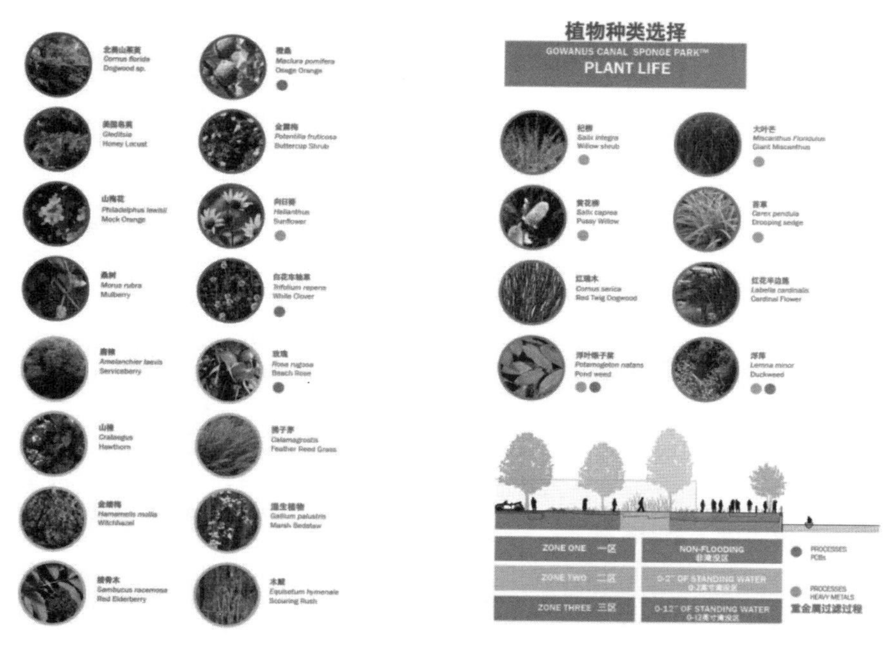

图8-2-8 植物种类选择
Fig.8-2-8 Plants selection

态参与过程，确保运河的河滨向公众全面开放。然后海绵公园概念框架将由私人开发商提出发展。这方面的发展的同时，公共机构，如美国陆军工程兵团，纽约州环境保护局和纽约市环境保护部着手清洁水体和周边污染地区。这里在未来可能设立一个基金，专门用于修复受损景观建。海绵公园的诞生是基于在城市错落有致的策略。异质文化的保存，运河的美学价值和环境价值，在开放空间网络中都是重要的目标。我们认识到环境的质量不在于建立一个地标，而在于一个进程的创建。

五、 总结

西方滨水区的开发建设，主要经历了繁荣发展期——污染衰败期——更新与复兴期这样的历程。自上世纪中期开始，由于城市滨水区工业的衰退和水上交通运输业的弱化，人们开始对滨水地带有了新的认识。郭瓦纳斯运河滨水区的复兴不仅能满足发达国家的城市产业升级，而且能更好地为城市创造出富有地方特色与时代魅力的公共空间环境，实现经济、社会、生态效益的共赢。本规划的重点是通过整治水体环境，营造滨水景观绿地，提升滨水

which would help with funding the renewal of the damaged landscape. Implementation of the Sponge Park is based on an urban patchwork strategy to retain and compliment the character of the neighborhood. Preservation of the idiosyncratic culture, aesthetic and context of the canal is an important goal in realizing the open space network. We recognize that quality of place isn't built with a one-stroke gesture but is created within a process.

V. Summary

The development and construction waterfront mainly has three stages: Prosperous period, development period, pollution and decline period, regeneration and redevelopment period. Since the middle of the twentieth century, people began to rediscover the water transportation, because of the recession of inclusions in urban waterfront area, and the weakeness of The regeneration of the Gowanus Canal Sponge will not only prompt urban industrial upgrading of developed countries, but does create public space with land feature and the charm of the times. The plan will achieve the ultimate good of win-win of economic, social and ecological benefits. Water environment management, the construction of waterfront green space and improvement of waterfront function and vigor are the key strategies which co-create characteristics of the city.

The function of the Sponge park, as the same implies, is to filter suface runoff water and improve

地区功能与活力，从而塑造城市个性及特色。

顾名思义，海绵公园的主要功能是过滤地表径流水，改善运河水质。规划设计了一系列景观缓冲区和人工湿地，用以吸收、净化地表水，在满足净水功能的基础上，营造了丰富的景观形态。设计秉承“以人为本”的理念，留出足够的开敞空间，为城市居民提供了亲水的乐趣。公众可达性研究是海绵公园专项规划之一，滨水空间的活力与其可达性密切相关，人们从各个方向到达滨水区的机会越多，滨水区的吸引力就越大。公园规划中强调滨水空间可达性和交通组织顺畅性的有机结合，并注重城市功能向滨水空间的有机渗透和重组，使之成为城市的“门户”和“窗口”。

该项目最重要的是认识到海绵公园是未来城市基础设施的发展方向。将雨洪管理与文化方面的一些重要历史遗迹、娱乐设施和社区设施相结合，可以减少昂贵的下水道设施清洁用水，同时也提供了系统的城市开放空间。这是一个可以推广下去的综合概念。它适用于城市基础设施成熟，并且在地区产业发展中留下了不良景观的城市。对海绵公园来说，环境是系统功能不可分割的一部分，设计提出了现实的战略，打破固定的基础设施模式，支撑一个更清洁的未来。

water quality. The plan designs a series of landscape buffer areas and man- made wetland to absorb, and purificate suface water. Also, it provides abundant landscape forms. Design adhering to the principe of "people oriented", the programme provide enough open space and the chance to enjoy activities to water. Accessibility is one special plan, which stay closely with the waterfront vigor. The more opportunities people go to the waterfront area, the more attraction it is.

It is important to recognize how significant development of the Gowanus Canal Sponge Park is to the future of urban infrastructure development. Integrating storm water management with cultural context of important historic sites, recreation areas and neighborhood facilities can reduce pressure on expensive sewer infrastructure while also cleaning the water and providing programmable urban open space. This environmental urbanism is a synthetic idea that can be implemented across the country. It is applicable to mature cities whose infrastructure is taxed by age and growth as well as in areas where industrial development has left behind inhospitable toxic landscapes. With the Sponge Park, consideration of the environment becomes an integral part of the way the system functions. The design proposes realistic strategies for fixing broken infrastructure in a manner that supports the promise of a cleaner future.

思考题

1. 海绵公园规划是如何激活运河滨水空间的?

2. 总体规划是如何处理运河与周边环境的关系?

3. 海绵公园采取了什么措施来恢复运河水质，同时提供了一系列相互关联的公共空间?

4. 规划采取什么措施来满足公众可达性的需求?

5. 植物规划采用了哪些植物? 它们有什么样的作用?

Questions

1. How does the Sponge Park plan activate the canal edge?
2. How does the master plan deal with the relationship between the canal and the context?
3. What measures are taken to restore the water quality of the canal and provide a series of interconnected spaces?
4. What measures are taken to meet the needs of public access?
5. What plants are selected in plant plan? What are the functions of them?

注释/Note

图片来源/Image：本案例图片均引自http://www.asla.org/2010awards/064.html

参考文献/Reference：本案例引自http://www.asla.org/2010awards/064.html

推荐网站/Website：http://www.asla.org/2010awards/064.html

第三节 纽约东河滨水景观的转型

Section III Transforming the East River Waterfront, the City of New York

一、 介绍

1. 项目背景

纵观全球，伟大的城市都在它们具有历史意义的滨水区寻求发展、再生和转变。在不断动态变化的城市背景下，在动态变化的城市环境中提高城市质量和加强可持续性于任何一个城市社区都是挑战。在巴萨罗那、阿姆斯特丹、悉尼和伦敦都可以看到成功的例子。这些方案是将码头港口重新打造成全球性的城市滨水区，让每个人都能去体验一下。当然纽约也不例外的关注曼哈顿下区东河唯一的滨水区。

东河滨水区，是贸易和商业的发源地，是纽约市350年来的中心区。尽管它现在没被充分利用也没和周边其他的社区很好的连接起来，曼哈顿下区朝南的东河滨水区是非常特殊的未开发资产。正是因为享誉世界的市中心背景、历史性桥梁的展望及布鲁克林的城市天际线使得将此滨水区打造为世界级的滨水区成为可能。纽约市为这个曼哈顿再开发的重要组成部分发起了一年的规划研究。

2. 设计团队研究方法

从一开始，设计团队即城市规划部门和经济发展公司就致力于恢复地区的发展。这个团队认为一个成

I. Introduction

1. Background

Across the world, great cities are regenerating, transforming and seizing new urban opportunities on their historic waterfronts. Improving urban quality and reinforcing sustainability in dynamic and changing city contexts is a challenge for urban communities everywhere. Successful results can be seen in Barcelona, Amsterdam, Sydney and London. These initiatives are reclaiming and opening waterfront ports globally for everyone to access and experience. The City of New York is no exception as it focuses on the unique waterfront of the East River in Lower Manhattan.

The East River waterfront, birthplace of trade and business for New York, has been a centerpiece of New York City for three hundred and fifty years. Though currently underutilized and poorly connected to the surrounding community, the south-facing East River waterfront of Lower Manhattan is a spectacular untapped asset. Its world renowned downtown backdrop and great vistas of historic bridges and the Brooklyn skyline provide a unique opportunity for waterfront revitalization in a world class setting. City initiated a one-year planning study for this crucial component of the redevelopment of the Manhattan Waterfront.

2. Design Team Methodology

From the outset, the design team, the Department of City Planning and the Economic Development Corporation worked to recover this area for the community. The team involved as many opinions as possible by consulting with the local community and

功的项目是通过全面的集体性研究方法才能实现的，所以他们通过与地方社区和行政人员的接洽来了解尽可能多的意见。经过十二个月的时间，一份作为项目计划基础的共同协议达成了，现在称为“基础项目”。这个基础工程通过提供滨海中心基础设施而为广泛的设施需求提供了一个助跳板。它由大量包括软质和硬质景观的设施，步行区、绿道连接、与现存社区的连接区、与公园连接的水域及公共区。

这个项目进程包含了一份出自城市机构的深入概要和一些从前期接洽中得到的社区团体的愿景和心声。一个全面的方法会继续考虑以下几点：滨水区基础设施、交通及运输限制方面的资料收集；城市联动机会、滨水屏障的移除，河畔两岸现用的桥墩结构的限制及海运业对两英里长的滨水区生态方面的限制状况方面的调查。

3. 公共进程

东河滨水区研究的公共参与项目于2004年初期开始，这个项目的目标是尽快与研究区域中的利益相关者建立良好的关系。有关公众参与进程的会议、访谈、简报、研讨会使得研究团队从相关部门得到了有价值的投入，这些部门包括社区委员会、居民区、行政官员、都市滨水联盟等。这个研究不仅帮助了最初示意图的选择也促进了规划设计的形成。

二、 项目研究

1. 历史背景

东河滨水区作为纽约史上一个中心区已经有350年的发展历程。在1613年，荷兰人在曼哈顿岛建立了贸易站，随后开发了其在美的殖民地新阿姆斯特丹。现在曼哈顿下区的珍珠街是岛上东边区的历史海岸线。

在1600年的后期，纽约市开始掩埋垃圾以扩展岛屿至东河内。随着岛屿的扩张，海岸线越来越硬质化并且用作城市提高填埋区以扩张岛屿面积的滨水区。这些填埋使得现在的海岸线比原始海岸长3个城市街区的距离。在20世纪50年代这个2英里（约3.2公里）的滨水区有40个码头，但是今天东河滨水区只留下不到10个码头。在1954年，为了岛屿上的

City officials with the intent that a successful project is created from an informed collective approach to the problem. Over a twelve month period, a consensus of agreement was reached which has formed the basis of the scheme now called the 'The Foundation Projects'. The foundation projects provide a springboard for a wide range of facilities by delivering the basic amenity of a new esplanade. This is supported by a range of facilities including hard and soft landscape, new pedestrian realm and greenway connection, linkage back into the existing communities and new water related parks and public realm.

The project process has involved an in-depth brief from City agencies and the visions and aspirations of community groups from early consultations. An extensive process proceeded taking the following into account: research and data collection on physical infrastructure constraints of the waterfront, the extensive traffic and transportation constraints and imperatives, the identification of urban linkage opportunities, removal of waterfront barriers, the constraints of pier structures an existing riverside uses and the marine an ecological constraints of the two mile long waterfront.

3. Public Process

The East River Waterfront Study public involvement program began in early 2004 with the goal of quickly establishing good relationships with stakeholders in the study area. Meetings about the public involvement process, interviews, briefings, and workshops allowed the study team to get valuable input from interested parties. Such as Community Boards, neighborhood groups , elected officials , Metropolitan Waterfront Alliance (MWA),and so on .This helped shape the design and planning process as well as aided in the development of concepts and preliminary schematic options.

II. Study Area

1. History Context

The East River waterfront has developed over the past 350 years as a central place in the history of New York. In 1613 the Dutch established trading posts on Manhattan Island and subsequently developed the colony of New Amsterdam. The present location of Pearl Street in Lower Manhattan is the historic shoreline of the east side of the island.

By the late 1600's the City began a process of landfill to extend the island into the East River.As the island was expanded, the shoreline "hardened" and was used as waterfront lots where the city promoted their infill in order to expand the island. This infill has resulted in the current location of the shoreline more than3 city blocks from the original shore. In the 1950's there were over 40 piers in this 2-mile stretch

交通更加便捷，罗斯福路高架桥沿着东河在南大街上面建设起来。

作为一个城市日常生活发生的场所，由于商业区将搬至深水集装箱港口，东河滨水区很快被人遗忘。纽约市在接下来45年的大量总体规划中注重城市重建及改造工程，东河滨水区也成为主要的一个重建区。

2. 水岸类型

纽约市的高密度展现了各种各样的风格独特的滨水类型。科尼岛让纽约拥有了木板路和沙滩，炮台公园城拥有滨海中心，哈德逊河公园有滨海中心及再利用的码头（图8–3–1）。由于罗斯福路高架桥及滨水的宽度，东河滨水区因其独特之处而不同于所举的例子。这是一处城市基础设施、海事活动及滨水景观的复合区。罗斯福高架桥下现有堵塞的解决将会使这个地区开放并且转变为另一个新旧混合型的滨水区。

3. 存在问题

1）断点连接

研究区域范围由其南北的两个地方为界即在曼哈顿下区边界的炮台海事大楼及东北部的东河公园。现在的东河滨海中心有一处较窄并且受损很严重的地方，使得行人、骑车的人及其他人不能容易的穿过，也不能安全的环绕岛屿及有名的地方。东河公园的入口通道也较窄，设计不合理使得车辆、行人及骑车的人不能安全的进入到公园中（图8–3–2）。

2）设施的缺乏

目前，大部分的开放空间都用来停小汽车及公交车。17号码头是在2英里长研究区域内唯一吸引人们到滨水区的场地。由于河畔不再是人们消磨时光的地方，在河畔也就没有太多供休闲停留的设施了。现在，东河滨水区很大程度上是城市工业的遗产，沿着滨水区有很多荒废的码头。

3）堵塞问题

现在沿着东河的土地利用主要是停车场。公交车、汽车和卡车都停在罗斯福高架桥的下面，阻碍了

of waterfront, today there are less than 10 remaining piers on the East River Waterfront. In 1954, The FDR Drive viaduct was constructed along the East River, over South Street in order to more efficiently move traffic around the island.

The East River Waterfront was soon forgotten as a place of everyday life in the City as moved to containerized ports in the deeper waters. The City focused on redevelopment and urban renewal projects and the East River Waterfront in Lower Manhattan became a major redevelopment site for numerous master plans over the next 45 years.

2. Waterfront Types

The density of New York provides a wide variety of waterfront types each with it shown unique characteristics. Coney Island gives New York a boardwalk and beach, Battery Park City provides an esplanade along the water, Hudson River Park provides an esplanade along the water and reuses existing piers for park land (Fig.8-3-1). The East River Waterfront has its own unique set of criteria that separate it from the given examples due to the FDR Drive and width of the esplanade. It is a mixture of urban infrastructure, maritime activities and waterfront views. Removing some of the current blockages under the FDR Drive will open this site up and turn into another exciting waterfront type mixing old with the new.

3. Existing Problems

1)Broken Links

The study area is defined by two distinct locations at its north and south – The Battery Maritime Building at the tip of Lower Manhattan and East River Park to the northeast. The current East River Esplanade has a narrow and severely compromised connection for pedestrians, cyclists and others to move easily and safely around the island and between these significant locations. The entrance to East River Park is similarly narrow and inadequately designed to allow for the safe movement of vehicles, pedestrians and cyclists into the Park (Fig.8-3-2).

2)Lack of Amenities

Currently, much of the open space is used for car and bus parking. Pier 17 serves as the only attraction along the 2-mile study zone to draw people to the City's edge. Since the river was never a place for people to linger or spend time, there are no amenities which would allow for a more leisurely stay at the river's edge. Today, the East River waterfront is largely a legacy of the City's industrial past. There are many dilapidated, defunct piers along this waterfront.

3)Blockages

Currently the predominant land use along this

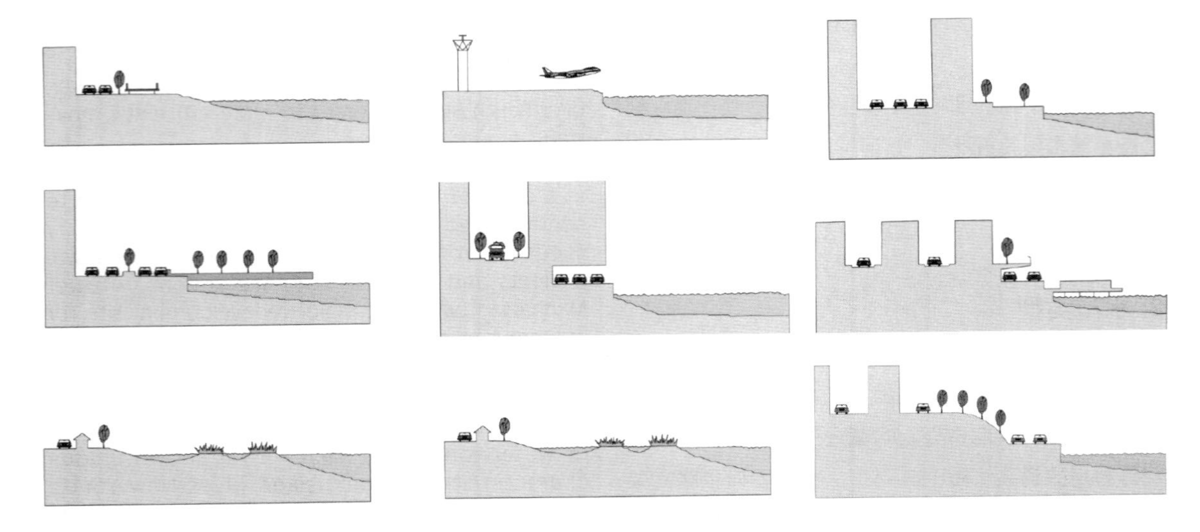

图8-3-1
Fig.8-3-1

行人们去滨水区。从高地到滨水区的可达性主要取决于滨水区的可用性、安全性及为社区提供的设施多少。由于滨水区的大量工业遗产，行人通道并没有得到改善。20世纪80年代出现的严重的无家可归问题，迫使纽约市将罗斯福高架桥下面的那块滨水区用栅栏和混凝土墙隔离开。南大街也被用作通勤巴士和旅游巴士的停车处，使得地方社区感到很沮丧。

4) 未利用的水岸

现在的停车区、栅栏及废弃物占据了大部分滨水区。同样地，城市服务机构也设在滨水区，环境卫生部门、交通部门、紧急医疗服务及消防部门占据了35号、36号、42号码头的大部分地方。这个地方由其过去的工业发展显示出了它的使用痕迹，但是大部分区域对公众不开放并且没被使用。因为这曾经是一个贸易、商业及其他海事活动的区域，所以步行通道和开放空间在它的开发中一直都不是优先考虑的事。

三、 相关工程介绍

1. 基础工程

1) 基础工程设计理念

东河滨水区将会维持丰富并具创造性的环境，在此地区中如此，在纽约市也同样。这个区域将会

portion of the East River is parking. Buses, cars and trucks all park under the FDR Drive and block access to the waterfront for pedestrians. Access to the waterfront from the upland is directly tied to its usability, safety and degree of amenity provided to the community. Given the industrial legacy of much of this waterfront, access for pedestrians was not promoted. A severe homeless problem in the 1980's prompted the city to cordon off portions of the waterfront beneath the FDR Drive with fences and concrete barriers. Outside of the fences, South Street is also used as a parking area during the day for commuter and tour buses much to the dismay of the local community.

(4)Under-utilized Waterfront

Parking, barriers and trash today occupy large parts of the waterfront. Similarly, City services were placed here at the water's edge, the Department of Sanitation, Department of Transportation, Emergency Medical Services and the Fire Department of New York's facilities now occupy large areas of piers 35, 36 and 42. The site shows traces of use through its industrial past but most of these areas are not open to the public and remain un-used. Since this was an area of trade, commerce and other maritime activities, pedestrian access and open space has not been a priority in its development.

III. Projects Introduction

1. Foundation Project

1)Foundation Project Design Philosophy

The East River Waterfront will help to sustain the

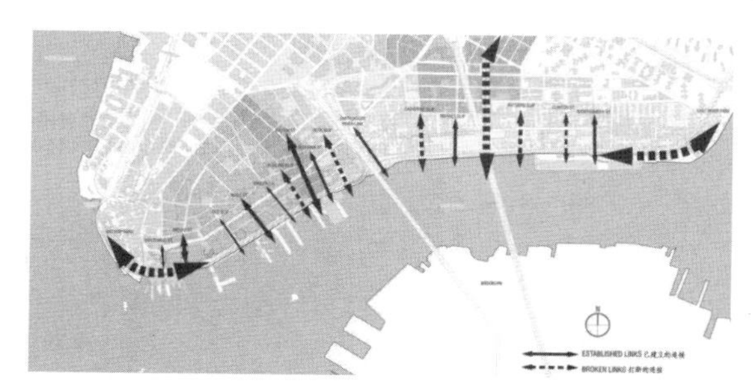

图8-3-2 损坏及已建立的城市与滨水区的连接通道
Fig.8-3-2 Broken and established links between city and waterfront

提供更多样化的社区、更多的就业机会及企业。罗斯福高架桥的实际限制，让水边及南大街为创造完全连续的视觉景观的提供了基础。通过一系列适当的城市干预去复兴东河，人们能够更加充分的享受这个公共领域的乐趣。首先是提供为滨水区及社区活动服务的基础设施，在罗斯福高架桥下为公众提供场馆，并为公众提供开放的码头。这规划是用循序渐进的方法而不是革命性的方法。将新的东西添加到旧的东西中去，将小公众干预融入大的滨水环境规划中去，积极的改变将会为邻近地区带来新的项目及利用途径，也会带来更大的城市效益。

2）需求及机遇的了解

规划的其中一个目标就是通过引进公共开放空间使全年滨水区的利用率达到最大。2英里的延伸的重点是增长滨水空间的长度，同时也连接了城市与滨水区。新的设施如罗斯福高架下的场馆、码头的公共通道、行人专用道及自行车道将会为居民及观光者提供一系列全年的活动。这些新地点也会加强已建立的与邻近地区的连接，并在滨水区创造更多的商业、文化及社会发展机会（图8-3-3，图8-3-4）。

3）改善现状环境的途径

生态方法：与曼哈顿下区紧邻的东河水生环境因现状的物理生境及水生生物而独具特色。这些生境可能会被沿岸的与现状资源密切相关的重建活动所影响。现有的水生生物方面资料表明大部分鱼类活动都是短暂的，与季节性的生境需求有关，反映出潮汐状况。

深海测探及潮流调查数据记录了研究区域的自

rich, creative environment that is enjoyed within the area and is synonymous with New York. This is an area that has pride in its diverse communities, job opportunities and entrepreneurial settings. The physical constraints of FDR Drive, the waterside and South Street provide a base to create a clear and coherent vision for the area. Through a set of appropriate urban interventions it is intended to revitalize the East River for a fuller enjoyment of this extensive public realm. This begins by providing a base infrastructure that can support both waterfront and community activities and providing pavilions under the FDR Drive and opening piers to the public. This is planning by evolutionary rather than revolutionary measures. By adding new among the old and smaller public interventions into the larger waterfront context, positive change is in place to bring new programs and uses for neighborhood and city wide benefit.

2)Understanding Needs and Opportunities

One of the planning goals is to maximize year round waterfront use by introducing layer of public open space along the existing esplanade. The 2-mile stretch establishes focal points that enhance the length of the esplanade space and at the same time link the City to the waterfront. New amenities such as pavilions under the FDR Drive, public access to piers, designated pedestrian and bike lanes will provide a mix of activities for residents and visitors through out the year. These new locations also enhance established links into adjoining neighborhoods and create further commercial, cultural and social opportunities along the length of the waterfront (Fig.8-3-3,Fig.8-3-4).

3)The way to improve the existing environment

Ecology:the aquatic environment of the East River adjacent to the Lower Manhattan shoreline is characterized by existing physical habitats and aquatic life. These habitats may be effected by shoreline redevelopment activity that would interact with the existing resources. Existing information on aquatic life indicates that the majority of fish species are transient, moving in relation to seasonal habitat needs, response to tidal flow conditions.

Bathymetric (depths)and tidal current surveys describe the physical conditions of the study area(Fig.8-3-5).The tidal current survey provides baseline information for calibrating a two dimensional computational fluid dynamics model. This model of the downstream portion of the study area shows flow and sedimentation patterns of alternative in water redevelopment options.

Aquatic environmental conditions are a prominent factor in the evaluation of redevelopment options and form a set of base criteria for the Foundation Projects preferred options.

Sustainable Design:Sustainable design for the East

图8-3-3 活动意向
Fig.8-3-3 Activities intentions

图8-3-4 活动地图
fig.8-3-4 Activities

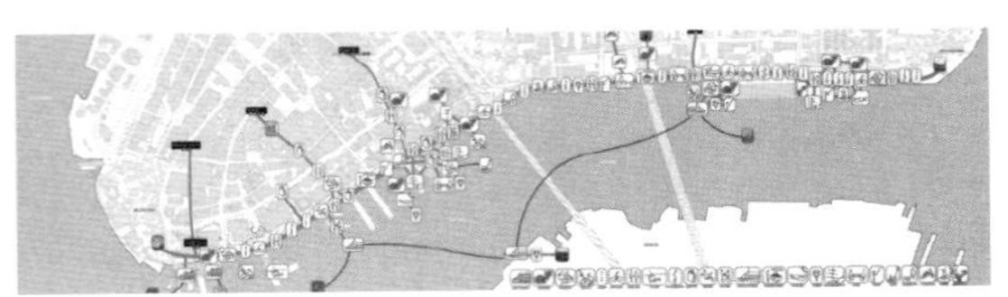

然状况（图8-3-5）。潮流调查提供了二维的流体动力学模型的基准信息，研究区域下游部分的模型说明了流动和沉淀的交替模式。水生环境状况是重建选项评价中的主要影响因素，也形成了基础工程优先选择的基础标准。

可持续性设计：东河滨水区的可持续性设计意味着对环境问题的考虑，对社会责任及经济发展的同样关注。广义上说，这个项目代表着可持续发展，因为它提高城市生活，鼓励人们在纽约市这样高密度的都市区生活。纽约市在美国有最低人均能源使用率，最广泛公共交通系统，最多样化的种族文化，最强大的经济动因及最密集的土地利用。所有这些因素使得纽约毫无争论地成为“最绿”城市。

但是纽约市缺少一些环境特征。城市生物多样性低，产生大量的固体废弃物，并且从大气到水到声音污染都有严重的集中污染问题。东河滨水区研究在基础工程规划上注重可持续发展问题。15号码头重建工程包含了一个新的生境敏感设计，这个将会使水生生物在东河中大量繁衍。一些小的附加物如在现有的码头敦上放礁球将会提高生物多样性且不需要大的花费或大的干预（图8-3-6）。重新计划及再利用罗斯福高架下的地方，打造成商业、娱乐或者社区项目用地，并促进新材料及潜在文化点

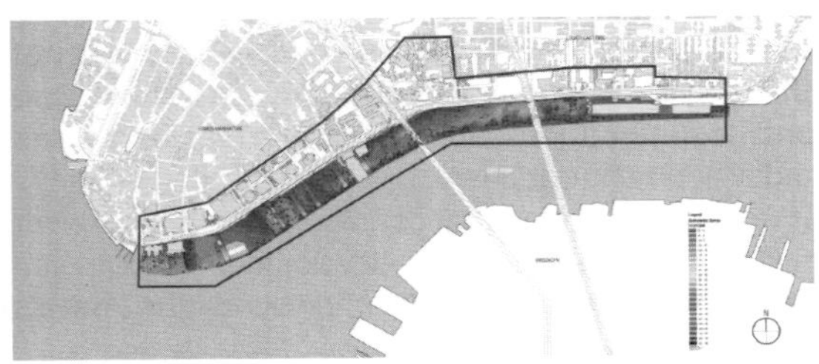

图8-3-5 东河深海测探——河床深度
Fig.8-3-5 East river bathymetric survey—rivered depth

River Waterfront means responding to environmental issues, social responsibility and economic development all with equal concern. In a broad sense, the project represents sustainable development as it enhances city life and encourages people to live in the dense urban area of New York City. New York has the lowest per capita energy use, the most extensive public transportation system, the most ethnic and cultural diversity, the strongest economic engine and densest land use in the United States.All of these factors make it arguably the 'greenest' city in the country.

But New York lacks some environmental features. It has low biodiversity, it generates large amounts of solid waste and has concentrated pollution issues from air to water to noise pollution.The East River Waterfront Study addresses issues of sustainability throughout the Foundation Plan. The rebuilding of Pier 15 with a new habitat sensitive design will allow aquatic life to flourish in the East River. Small additions such as placing reef balls on existing pier promote biodiversity without great cost or intervention(Fig.8-3-6). Reprogramming and reusing the space under the FDR with commercial, recreational and community programs promotes the use of new materials and potential cultural destinations. By implementing the Foundation Plan a base framework will be put in place as an example for future sustainable projects. New York is an exemplary model of economic, social and sustainable development. This waterfront development project will encourage and improve the unique qualities that are already in place.

Transportation:In recent history, the waterfront has been the location for the placement of major elements of the City's transportation infrastructure; local streets provide major connections within the Lower Manhattan street network. Access to the waterfront from the north and south is constrained by existing infrastructure such as the Battery Park Underpass and the current entrance into East River Park. To enhance access and egress to these facilities for both vehicular and pedestrian traffic, the study team reviewed a series of alternatives for the FDR Drive, South Street, the Battery Park Underpass, and so on.

To evaluate the potential benefits and impacts of the proposed foundation plan alternatives, the study

的利用。通过实施基础工程计划，一个基础的框架将会作为未来可持续性工程的例子。纽约是一个经济、社会可持续发展的典型例子，滨水区开发项目将会鼓励和提高地方已有的独特品质。

交通：在近期历史中，滨水区已成为城市公共交通设施中主要元素的所在地，当地街道为曼哈顿街道网络提供了主要的连接。从北部到南部到达滨水区被现有的基础设施所限制，如炮台公园地下通道、东河公园的入口。为了加强车辆及行人交通进出这些设施的可能性，研究团队评估了罗斯福高架、南大街、炮台地下通道等一系列的替代选择方案。

为了评估各个已拟定基础工程的备选方案的潜在效益和影响，研究团队制定出一个现存的2004年交通网络来预测2025年的交通网络，2025年是作为备选方案分析的基准年。2025基准年将次要的和主要的基础设施改善合并在一起，这些改善包括总督岛的复兴、富尔顿鱼市场的再定位及布鲁克林大桥引坡道的重建（图8-3-7，图8-3-8）。

2. 滨海工程

1) 滨海中心

一个成功的滨水区最重要的方面就是对边界的处理。一个充满生气的滨水边界将会激发全年的活动并为居民、工作者和观光者提供新的康乐设施。东河滨水区一系列新的工程展示了形式上及材质上的特点，为这个新滨水区创造了独特标识（图8-3-9，图8-3-10）。

滨海中心有两种基本形式，一种是主要在罗斯福高架下的典型代表，一种是有水上减载平台结构的宽大的滨海中心。在这两种形式中，连续的环绕曼哈顿的绿道被保护并延伸和改善。滨海中心的布局包含了一个沿水边并有坐椅和种植物的娱乐区，一个在罗斯福高架下提供馆内和户外活动的场所以及一个沿着南大街的专用自行车道。在罗斯福高架下规划的场馆及各项活动将活跃的街头生活由城市带到了水边。这些场馆沿着滨水区分布，与现有的街道平行，这样视觉通廊就不会被遮住。

在东河和布鲁克林大桥之间的滨海中心一边紧邻堤岸，另一边紧邻南大街。在罗斯福高架的边缘之外

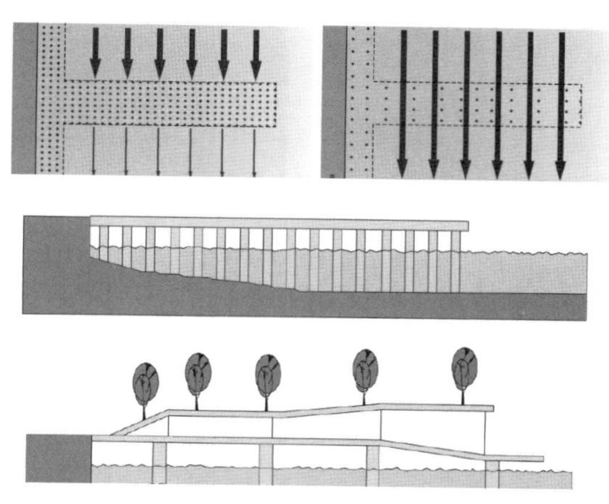

图8-3-6 新旧码头对比
Fig.8-3-6 The comparison between new and old

team established an existing 2004 transportation network that was projected to reflect the 2025 horizon year transportation network, which was established as the base year for alternatives analysis. The 2025 horizon year incorporated minor and major infrastructure improvements that include the reactivation of Governors Island; the relocation of Fulton Fish Market; and, the reconstruction of the Brooklyn Bridge approach ramps (Fig.8-3-7, Fig.8-3-8).

2. The esplanade projects

1)The Esplanade

The most vital aspect of a successful waterfront project is the treatment of the edge

condition. A vibrant edge will generate year-round activity and provide new amenities for residents, workers and visitors alike. The new East River Waterfront a series of 'projects' share formal and material traits, creating a unique identity that signals this new waterfront(Fig.8-3-9;Fig.8-3-10).

The esplanade has two basic configurations along its length, a typical condition which exists primarily below the FDR Drive and a wide esplanade where an over-water relieving platform structure already exists. In both configurations the continuous Greenway around Manhattan is maintained, expanded and enhanced. The layout of the esplanade consists of a recreation zone along the edge with seating and plantings, a program zone under the FDR Drive for the pavilions and outdoor activities, and a dedicated bikeway along South Street. The pavilions and activities planned underneath the FDR Drive bring the active street life from the city to the water's edge. These pavilions are located along the waterfront, parallel to the existing street, so that view corridors are not obstructed. The esplanade between East River and Brookly Bridge is bordered on the side by the bulkhead and the other by South Street. There is little room beyond the edge of the elevated FDR which is located over

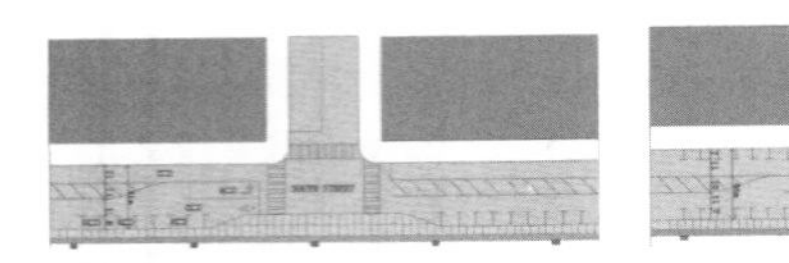

图8-3-7 布鲁克林大桥平面图
Fig.8-3-7 Brooklyn bridge-plan

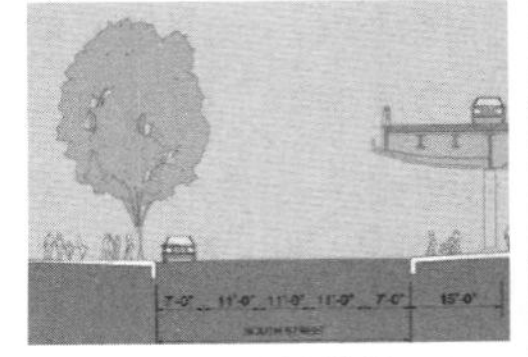
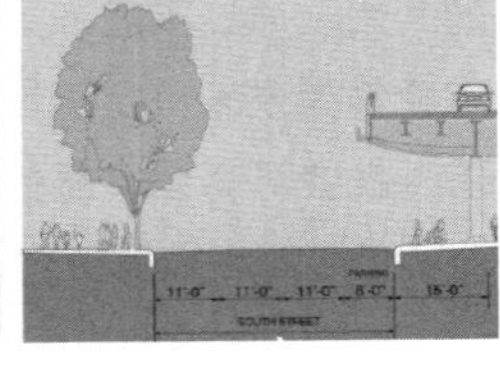

图8-3-8 布鲁克林大桥剖面图
Fig.8-3-8 Brooklyn bridge-section

图8-3-9 规划总图
Fig.9-3-9Planning overview

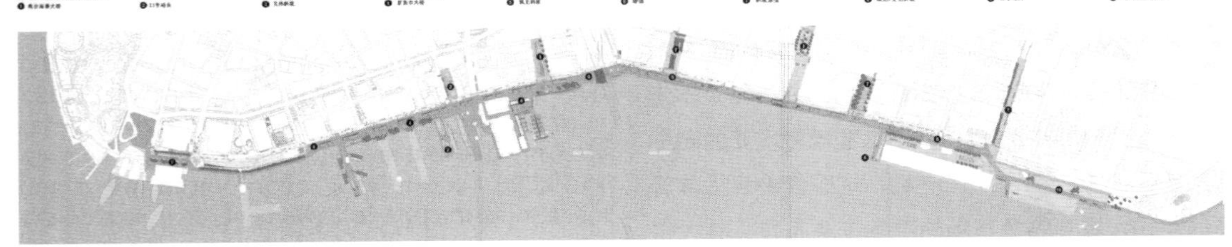

图8-3-10 项目索引
Fig.8-3-10 Project index

仅有很少的坐落在堤岸线上的空间。宽度一致的人行道沿着南大街铺开来，自行车道和场馆也一样。

the bulkhead line. A consistent sidewalk width along South Street is provided, as well as the bike lanes and pavilions underneath.

2)滨海中心组成要素

东河滨海区规划成一系列组件，这些组件创造了一致性但又是唯一性的特征，它代表着一个新的滨水环境。这些组件显示出材质和形式特点，可以用来满足社区的需求并可利用地理位置的优势。这系列的组件包含了长椅、栏杆、种植池及凉亭（图8-3-11）。

2)Esplanade components

The East River Esplanade is planned as a system of components which create a consistent, yet unique identity that signals a new waterfront environment. These components share material and form ,can be placed to best meet community needs and to take advantage of local conditions. The system's components consist of benches, railing, planters, and arbors(Fig.8-3-11).

3. 场馆工程

为了达到将城市活动和城市活力带到滨水区的目的，而在罗斯福高架下规划了一些场馆。这些场馆是根据位置及项目内容而定的面积在1500～8000平方英尺（约139～732平方米）间的小型建筑。它的一个重要作用就是将城市构造延伸至滨水区。场馆可能规划用于社区、文化及商业方面。

利用罗斯福高架天然的屋顶，规划建议在罗斯福高架的下表面覆盖一层天花板及声音衰减材料并提高照明。除明显减少噪音外，还将滨海中心变为一个更加令人愉快的休闲区（图8-3-12）。

3. Pavilion Programs

To meet the goal of bringing the activity and vitality of the city to the waterfront, pavilions are planned underneath the FDR Drive viaduct. These pavilions would be small buildings between 1,500 – 8,000 square feet depending on location and program. It performs a key function of continuing the city fabric to the water's edge. The pavilions would be programmed for community, cultural, and commercial uses.

Taking advantage of the natural "roof" that is the FDR Drive viaduct, the plan proposes that the underside of the FDR be clad with a new ceiling and sound attenuating material with enhanced lighting. In addition to significantly reducing the noise making the esplanade a more pleasant place to spend time(Fig.8-3-12).

图8-3-11 滨海中心组成要素
Fig.8-3-11 Esplanade components

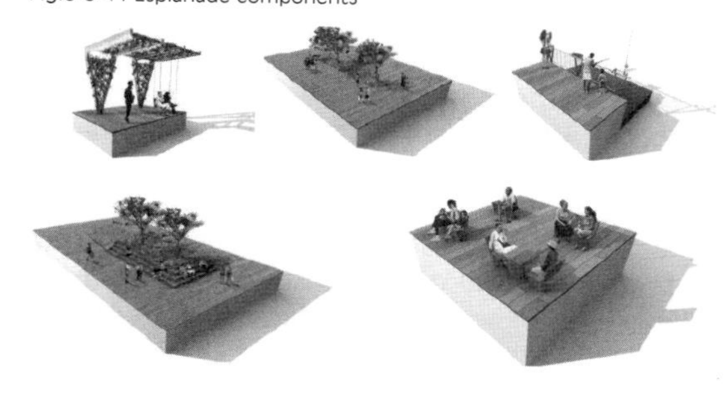

4. 斜坡工程

斜坡一般都是水拍打栅格的区域，并将水边活动带到城市中去。现在，这些斜坡已经被填充并铺设上材料。通过在这些地方重建活动和公共区，这些斜坡又重新变成水边充满生气的部分了。贝林斜坡目前是鱼市的大型停车场，未来将会变成孩子们的游戏场。佩克斜坡目前也是停车场，有成为曼哈顿大型公共空间的潜力。已经为卡萨林、罗格斯和蒙哥马利斜坡做好了景观原型，将会把这些斜坡作为新滨水区的入口（图8-3-13，图8-3-14）。

5. 入口工程

1）目标

创造一个环绕曼哈顿岛的连续绿道是设计师们的目标，也同样是跑步者、骑自行车者及居民的愿望。东河滨水规划不仅为东河下游创造一个世界级的滨水区，也延伸和加强了与周边岛屿的联系。这个2英里的延伸区将会补充另外一个30英里（约48.3公里）长的滨水区，它环绕岛屿并用美妙的滨水区将纽约市民重新联系在一起。

2）东河公园滨水连接处

4. The Slip Project

Historically the slips were an area where the water 'slipped' into the grid, bringing waterfront activity into the city. Today, these slips have been filled in and paved. By reestablishing activity and public open space in these areas, the slips once again become a vibrant part of the waterfront. Burling Slips , currently a large parking lot for the Fish Market will become a children's playground. Peck Slips , also used as a parking lot and potentially one of the great civic spaces in Lower Manhattan. A 'landscape prototype' has been developed for Catherine, Rutgers and Montgomery Slips that will establish these slips as gateway to the new waterfront(Fig.8-3-13,Fig.8-3-14).

5. The Gateways Projects

1)Goal

A continuous Greenway around the island of Manhattan has long been a goal of planners, runners, joggers, cyclists and residents alike. The East River Waterfront Plan not only creates a world-class waterfront for the Lower East River, but expands and enhances the connections all around the island. This two-mile stretch will complement the other 30 miles of waterfront which encircles the island and reconnects New Yorkers with fantastic waterfront which surrounds them.

2)East River Park Connection

Pier 42 is a crucial link to East Park and the Manhattan greenway north of the study area. Currently access to the park is granted via a 20 foot path which needs to accommodate vehicular, service, pedestrian and bike traffic. This access road is squeezed between the FDR Drive and the vacant Pier 42 shed which constrains the entrance to the park. The plan calls for the removal of the fence and the creation of a wider, safer connection to East River Park. In the future, both the pier and shed will be rebuilt to make way for a new large public gathering space (Fig.8-3-15).

图8-3-12 场馆意向
Fig.8-3-12 Pavilion intentions

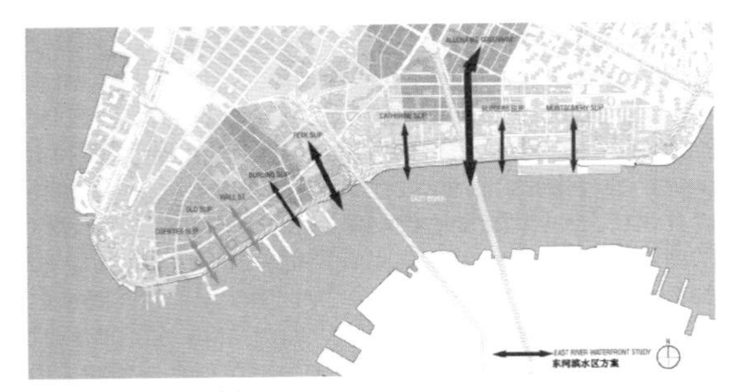

图8-3-13 目前的斜坡方案
Fig.8-3-13 Current slip initiatives

42号码头是东河公园和研究区北部曼哈顿绿道的一条重要的连接。目前到达公园的一条20英尺（约6米）的通道还是需要协调车辆、服务设施及行人和自行车才能通过的。这个入口通道被挤在罗斯福高架和废弃的42号码头棚之间，从而限制了公园入口。规划需要除去栅栏并创造一条与东河公园连接的更宽、更安全的通道。在将来， 码头和棚子都会被重建，为一个新的大公共集聚空间让路（图8-3-15）。

四、 总结

东河滨水区是纽约市的中心区，是城市最宝贵的财富。东河滨水区具有丰富的景观要素、深厚的历史文化积淀和浓郁的生活气息，是纽约城市人居环境的缩影。本规划尊重现状、延续文脉、保护地域特色，强调滨水空间可达性、满足居民亲水性要求，展示了滨水区演进过程中的“历史足迹”。

图8-3-14 斜坡方案
Fig.8-3-14 Slip plan

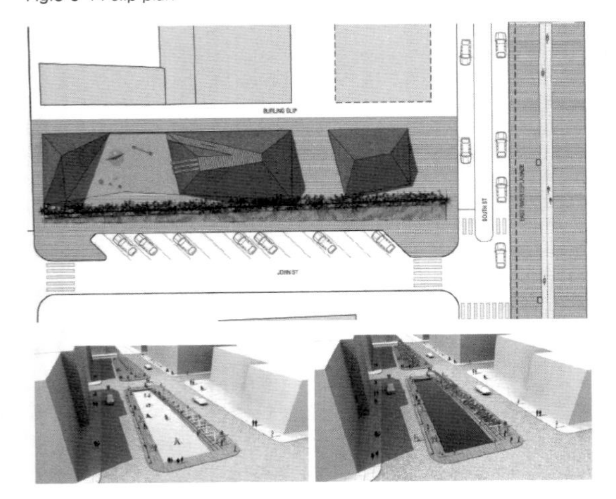

IV. Summary

The East River waterfront is the central zone of New York city, it's the most valuable wealth in the city. The East River waterfront has rich landscape elements, long history culture and life vitality,is a microcosm of the living environment in New York city.The planning respected current situation.continue cultural context, protected regional characteristic,emphasized the accessibility of waterfront space,meet residents hydrophilic requirements,showed the "historical footprint"in evolution process of the waterfront .

The new park set an pedestrian area for the East River in Lower Manhattan ,,and a vitality public open space,tourists can see the charming scenery of East River and New York harbor.The new waterfront way is the unique scenery line of New York city.It linked 500 miles long waterfront region again,integrated every corner of the People's Daily life in New York.For improving the environment,,the planning used the ecological method to record and study natural condition in the area, have done some new explorations in waterfront development pattern,space structure, operation mechanism,trying to realize the sustainable development of the East River waterfront .

New York is the world's most important seaport cities, the planning and design of the East River waterfront in Lower Manhattan through the recall of history and culture and the revival of vitality of the

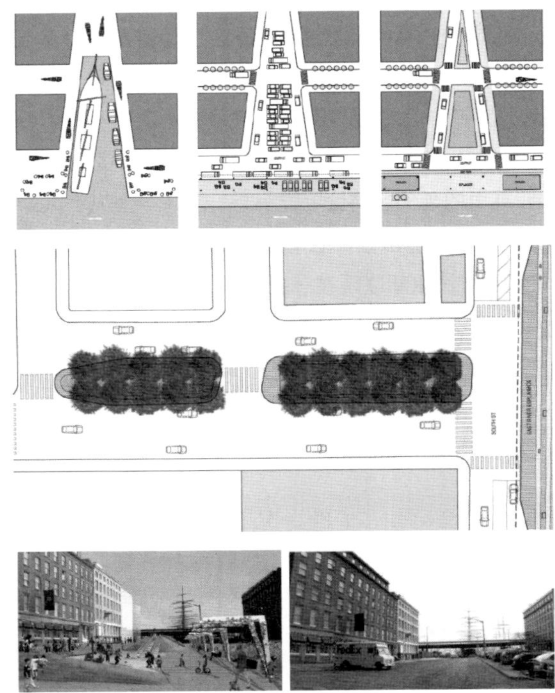

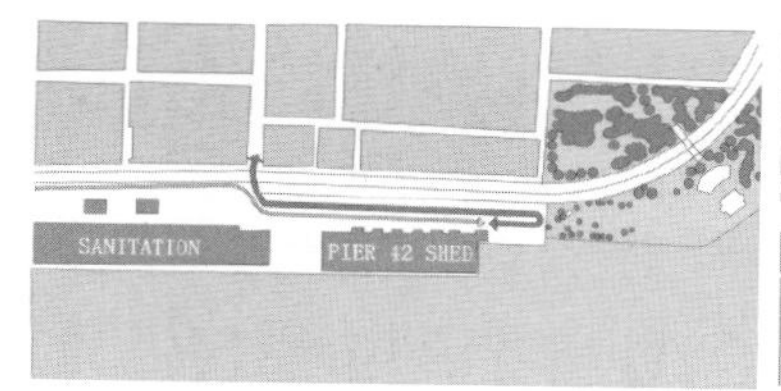

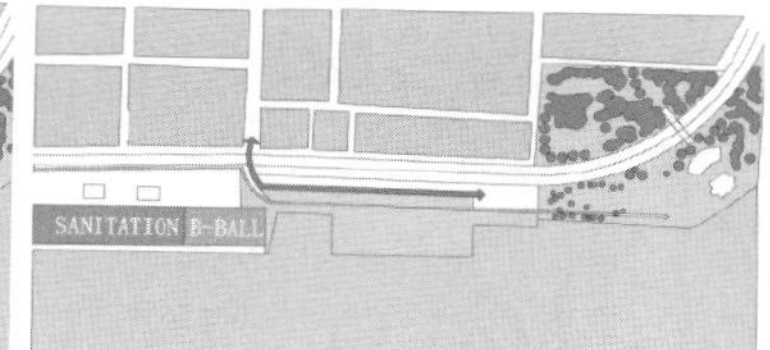

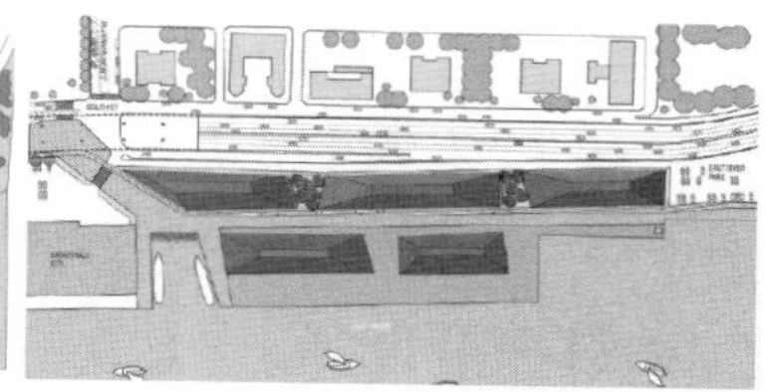

新公园设置了下曼哈顿东河的步行区，以及富有活力的公共开放空间，游者可一睹东河与纽约港的迷人景色。新的东河滨水散步道是纽约市的独特风景线，它把500英里（约805公里）长的滨水地带重新连接起来，融入纽约市民日常生活的每个角落。为改善环境，规划采用生态方法记录并研究区域的自然状况，对滨水区开发模式、空间结构、运作机制等方面进行新的探索，努力实现东河滨水区的可持续发展。

纽约是世界上最重要的海港城市之一，下曼哈顿东河滨水区的规划设计通过对历史文化的追忆和滨水区活力的复兴，营造了一处集城市基础设施、海事活动和滨水景观于一体的复合区，其独特的地理位置和历史文化氛围，使其成为纽约最具吸引力的地区之一。

图8-3-15 东河公园滨水连接处
Fig.8-3-15 East river park connection

waterfront,built a composite area which set the city infrastructure, maritime activities and waterfront landscape together.Its unique geographical position and history culture atmosphere, will make it become one of New York's most attractive regions.

思考题

1. 历史滨水区复兴的背景是什么?

2. 城市滨水区开发建设的趋势及要求是什么?

3. 怎样在历史滨水区的开发与保护之间最大限度的满足人们需求，体现场地特色?

4. 我国城市滨水区开发建设与西方在背景、目标和方式方面有哪些异同? 对我们有哪些可借鉴之处?

5. 怎样在规划过程中真正实现社区参与制?

Questions

1. What's the background of historical waterfront revival ?

2. What's the trend and requirements of development and construction of urban waterfront ?

3. How to satisfy the people's demand and reflect site feature in maximum between the development and protection of the historical waterfront ?

4. What are the similarities and differences in the background ,goals,and the way of the urban waterfront development and construction between China and the west?

5. How to achieve the community participatory truly in the planning process?

注释/Note

图片来源/Image：本案例图片均引自纽约东河滨水景观的转型规划文本，Transforming the East River Waterfront, The City of New York

参考文献/Reference：本案例引自纽约东河滨水景观的转型规划文本，Transforming the East River Waterfront, The City of New York

推荐网站/Website： http://www.nyc.gov/html/dcp/html/erw/index.shtml; ; http://www.lowermanhattan.info/construction/project _ updates/east _ river _ waterfront _ 78637.aspx

第九章　适应全球气候变化的景观生态规划设计

CHAPTER IX　LANDSCAPE ECOLOGICAL PLANNING & DESIGN ADAPTIVE TO CLIMATE CHANGE

第一节　格罗宁根：适应气候变化的设计 Section I Groningen: Adaptation to Climate Change
第二节　重庆龙湖项目 Section II The Longhu Project, Chongqing

第一节 格罗宁根：适应气候变化的设计

Section I Groningen: Adaptation to Climate Change

一、背景介绍

这个世纪最急迫的问题就是地球上将要经历的气候变化。因此减少人类对于这些出现的变化的干扰是非常重要的。然而，除了尽力让气候变化程度最小化，我们也应该去适应将要来临的各种变化。可以实现适应变化的最重要的一方面就是地区的空间布局。在适应气候变化的过程中，区域空间规划的作用是非常重要的但也是有限的。它通过一系列的与区域尺度相关的主题为未来发展指引方向。但是，当今的空间规划实践关注于经济发展的促进，如农业、基础设施、生活区和企业，也关注于为水管理和自然预留所需空间。

对于气候变化的适应可以在策略、方针、媒体中很好地表现出来，也可以在各部门措施中体现。但是适应性的空间规划的综合设计是很难实现的。景观应该变得更加适应气候，为了预期并更加适应气候变化，城市规划中的景观必须要在设计中实施适应措施，或者规划需要将适应措施包含在内。一个新的景观应是人们在一起工作并以一种网络化方式预期变化的地方。特别是，在空间规划和设计中

I. Background Introduction

Probably the most urgent problem of this century is the change the climate on earth will undergo. Therefore it is of the highest importance to reduce human impact on these emerging changes. However, despite all efforts to minimize the changes in climate, the world needs to adapt to the upcoming changes as well. The most important field where adaptation can be realized is the spatial lay out of the area. In the process of adaptation to climate change the role of regional spatial planning is important but limited. It gives directions for the future on a number of themes that are relevant at a regional scale. However, in spatial planning practice of today the focus lies on facilitating economic developments like agriculture, infrastructure, living areas and enterprises as well as creating the necessary space for water management and nature.

The adaptation to climate change is well represented in strategies, policies and the media. It can even be found in sectoral measures, but integrated designs for adapted spatial plans are hardly available. Landscapes need to become more resilient for climate. In order to anticipate on climate change and to adapt better and better to it the landscape en urban plans will have to implement adaptation measures in the designs or the planning will need to be adaptation-inclusive. A new landscape evolves in which people work together and anticipate on changes in a network-based way. Especially, in spatial planning and design the changes

变化和适应措施应该合成一体。因为长期的发展是不可预测的，空间规划不再明确未来的最终情形，但是可以致力于战略干预，这些敢于可以控制和激发未来往更适应的方向发展。

海平面升高和饮用水的短缺是荷兰城市面临的难题，格罗宁根位于荷兰的最北部，是荷兰重要的贸易及商品转运中心。在格罗宁根案例中的设计方法说明了空间规划中的适应集成方法是可行和便捷的，并为其他国家提供了新思路。

and adaptation measures will be incorporated. Because of the fact that long-term developments are unpredictable, the spatial planning no longer defines end images of the future but aims to define strategic interventions, which steer and initiate future developments in a more resilient direction.

The Rising of sea level and shortage of drinking water are the serious urban problems the Dutch meets. Groningen located in the north of the Netherlands,is the important trade and transit center of goods,The design approach in the Groningen case proves that integration of adaptation in spatial planning is possible, easy and offers new ideas for other countries.

二、 方法介绍

1. 气候方案

荷兰气象机构开发了四种气候方案。为2050年预测的方案基于两个特别影响荷兰气候的变量即大气模式的改变和温度的上升。四个方案被定义为：G（大气模式没有改变，温度上升1℃），G+（大气模式改变，温度上升1℃），W（大气模式没有改变，温度上升2 ℃），W+（大气模式改变，温度上升2 ℃）。大气模式改变仅指由于主导东风而引起的一个更温和、更潮湿的冬季。并在规划过程中确立了适应气候变化的目标（图9–1–1）。

当格罗根林省开始新的区域规划时，它就使用了这个预测的方法。除了常规的经济和人口规划主题，气候变化是规划过程中的第三根支柱。区域规划通过三个阶段来进行：分析、相互作用和思考。在每个阶段中气候变化都与规划过程的每一步连接起来。

1) 格罗根林的出发点：两种情景

气候变化政府小组收集最新的科学知识，在2007年公布了第四次评估内容。在这次评估中，没有预测气候变化，但是最有可能的预测和情形有所发展。这些情形包含了大部分未来的不确定性。皇家荷兰气象研究协会建立了四种荷兰气候方案，这四个方案描述未来气候变化带宽，并为方针决策提供了基础。在KNMI方案之外的一种极端元素被用来建立另一种新的方案。系统性和方法性不是最纯粹的形式，但一个方案的使用会更加容易理解复杂的方案。除了这个联合的方案，也建立了另一个以荷

II. Methods Introduction

1. Climate Scenarios

The Dutch Meteorological Institute developed four climate scenarios for the Netherlands (KNMI,2006).The scenarios for 2050 are based on two variables, which influence the Dutch weather in particular: changing air patterns and temperature rise. Thus, four scenarios are defined(Fig.9-1-1):G(no change in air patterns and 1℃ rise of temperature),G+(changed air patterns and 1℃rise),W(no change in air patterns and 2 ℃ rise) and W+(changed air patterns and 2℃temperature rise). The changed air pattern simply a more moderate and wet winter caused by a dominant eastern wind.And established the goals of adapt ion to climate change in the planning(fig. 9-1-1).

The province of Groningen has used this anticipative approach, while it designs its new regional plan. Beside the more regular themes of economy and demography, climate change is the third meaningful pillar in the planning process. The Regional plan is carried out in three phases: analysis, interaction and reflection. In each of these phases climate change is connected to the steps in the planning process.

1)Starting Point Groningen: Two Scenarios

The Intergovernmental Panel on Climate Change (IPCC) brings together the latest scientific knowledge. In 2007 the fourth assessment was published. In the assessment the climate is not predicted, but the best possible projections and scenarios are developed. Ideally these scenarios contain the largest part of the uncertainties of the future. The KNMI developed four Dutch climate scenarios, which describe the bandwidth for future climate change and offers the basis for policy decisions. For the Groningen situation the extreme elements out of the KNMI scenarios were used to build one new scenario. Scientifically and methodologically not the most pure form, but the use of one scenario make it possible to understand the complexity of the scenarios better. Beside this combined scenario, an extreme scenario is developed, which uses the starting point of an accelerated melt

兰及南极洲西部冰山的加速融化为出发点的方案。这个方案被认为可能性较小，但是北极研究人员预测北极的极端发展模式是有可能实现的，而且皇家荷兰气象研究协会的方案是非常适中的。在这个方案中，海平面上升的速度将会比过去的10−15年快十倍多，这看起来是不可能的。然而，通过一个适中、一个极端方案的渐渐发展，在整个带宽中都发起了可能性方法的讨论。 两种方案的出发点在表1中总结出来。

2）气候知识

目前气候变化方面的资料在国家级尺度范围内是可靠的，这些知识在以气候地图形式的区域尺度上也会是可靠的。格罗宁根走在这方面的前端，主要通过观察空间功能的三个重要因素：温度、降水、海平面上升。在这里我们讨论分析降水和海平面上升。

降水：降水量和降水形式的未来变化预测在表9−1−1中列出。这些图显示以皇家荷兰气象研究协会2006年冬天和夏天的方案来预测格罗宁根省和德伦特省在2050年的变化（图9−1−2）。

预测—分析的主要结论为：

of the ice-caps of Greenland and western Antarctica. This scenario is seen as less probable, but arctic researchers expect that extreme developments on the poles might be realistic and that the KNMI scenarios are too moderate. In this scenario the sea level rise is predicted to rise ten times faster than the last 10-15 years has happened. This seems very unlikely. However, by developing two scenarios, a moderate and an extreme one, the debate on possible measures took place in full bandwidth. The starting point of the two scenarios are summarized in Table1. 9-1-1

2) Knowledge of Climate

Information on climate change is available on a national scale so far. The knowledge will be available on a regional scale in the form of climate atlases. Groningen runs ahead of this by visualizing three important factors for spatial functions: temperature, precipitation and sea level rise. Precipitation and sea level rise are discussed here.

Precipitation:The possible changes in future precipitation amounts and patterns are shown in figure. The maps show the possible changes for Groningen and Drenthe provinces (Alterraetal.2008) in 2050 in two KNMI '06 scenarios (KNMI,2006) for the winter and summer period(Fig. 9-1-2).

The main conclusions of the precipitation-analysis are:

In the summer period drought will most likely increase. In the 'dry' scenario of the KNMI (W+) in the eastern part of the provinces (the Peat Colonies) drought becomes a serious problem. Nowadays,

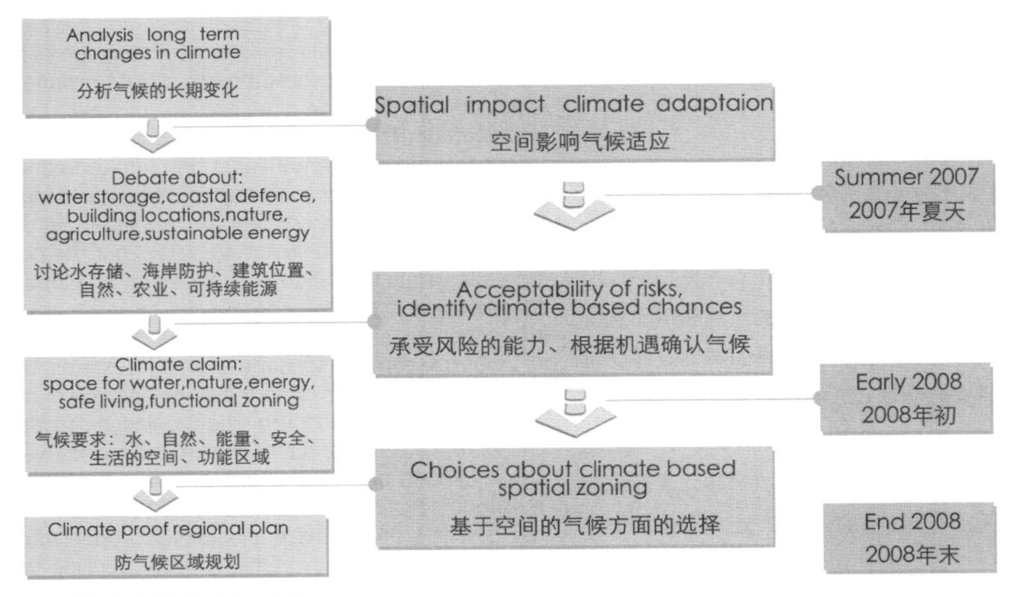

图9−1−1在规划过程中适应气候变化的目标
Fig.9-1-1 Anchoring of adaptation to climate change in the planning process

Elements from the KNMI scenarios 来自KNMI方案的数据		Extreme accelerated melting of land-ice 冰川极速融化方案
Precipitation spring and autumn 春季和秋季的降水量	+20%	+30%
Precipitation summer 夏季的降水量	-20%	-40%
Precipitation winter 冬季的降水量	+15%	+30%
Temperature 温度	+1.5	+3.0
Sea level rise 海平面上升	+35cm	+150cm

表9-1-1 格罗宁根2050年预测方案
Table. 9-1-1 The Groningen scenarios for 2050

夏季的干旱期将可能会增长。在皇家荷兰气象研究协会的“干旱”方案中，东部各省的干旱成为一个严重的问题。现在，这些地区的水资源短缺已经是一个问题了。水资源短缺问题和供应的不确定问题将会越来越大，主要由于天气变化导致夏天的干旱期增长。另一个降水方面的变化就是阵雨强度的增强。这个结果在图中不能直接看出来，但是对于水管理有很大的影响，并意味着城市地区洪涝风险的增大。

根据皇家荷兰气象研究协会2006年方案预测，秋、冬和春三季将会变得更加潮湿。尽管干旱夏季会在秋季造成一定的干旱影响，但是整个冬天的降水总量将会增大。这个增强的问题关乎水管理的方式。加大排放量（伴随增加泵的容量）是不是最好的解决方案？或者增大蓄水量的方法比较好？除了降水变化的主要影响，还有些由于降水引起的次要影响。

所以，将多雨期的多余的水储存起来或者将水用大型抽水机抽入海里的必要性越来越大；在更干旱的时候对于水的需求越来越大，特别是在干旱地区，需要更多的供给;城市地区需要处理多余水量不能排放的问题，在夏季将会陷入严重的阵雨问题，导致周期性的洪涝。

海平面上升：海平面在接下来的几十年，甚至几个世纪内都将会继续上升。海平面上升的速度和程度依赖于格陵兰和南极西部陆冰的融化速度。尽管CO_2的排放量将会控制在今天的排放水平上，陆冰的融化过程在接下来的几十年也会持续下去。因此，适应海平面的上升是不可避免的。下面的地图以现在的海平面高度对比两种方案（＋50厘米和＋

water shortage in this area is already a problem. The problem of water shortage and uncertainty of supply will increase in the future, due to longer dry periods in summer, which are caused by climate change. Another change in precipitation is the increased intensity of severe rain showers. This effect is not visible in the maps, but it has a huge impact on water management and implies an increased risk of foods in urban areas;

According to the KNMI'06 scenarios , autumn, winter and spring become(much) wetter. Although a dry summer will have its 'drying' effects in autumn, which leads to average dryer autumns, the total amount of precipitation in the winter period increases. This raises questions as to how the water management must be arranged. Is an increased discharge towards the sea (with accompanying increase of pump capacity) the best solution or is an increased amount of water storage preferred? Beside the primary effects of changes in precipitation, there are also secondary effects, which are driven by these changes in precipitation.

So,there is an increasing necessity to store the extra water in wet periods or pump it into the sea with heavier pumping engines;In dryer periods an increasing demand for water, especially in dry areas, asks for extra supply;Urban areas increasingly have to cope with the impossibility to discharge the extra water, falling in the form of severe showers in the summer period, leading to periodical foods.

Sea Level Rise:The sea level will continue to rise for the next decades and most probably for the next centuries. The speed and degree of sea level rise is dependent on the pace of melting processes of land ice on Greenland and Western Antarctica. Even if the emissions of CO_2 could be frozen at today's emissions level, the melting process would continue for the next decades. Therefore, adaptation to the rise of the sea level is inevitable. The maps show the possible impact of the rise of the sea level for Groningen, compared with current altitudes, in two scenarios (+50and +150cm)(fig. 9-1-3).The maps show the

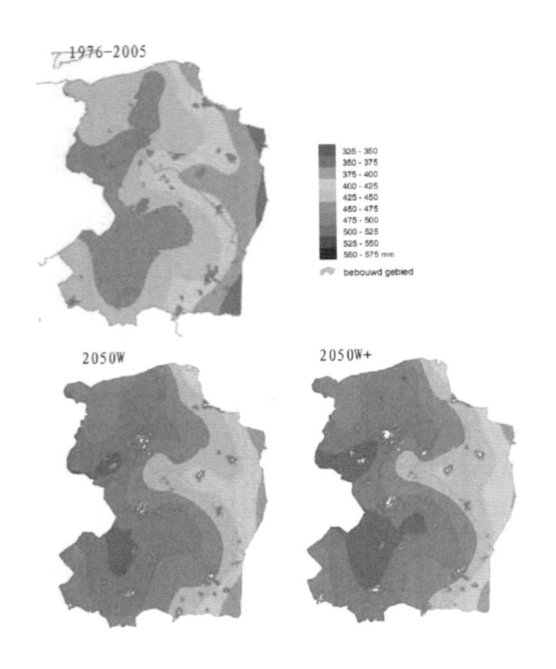

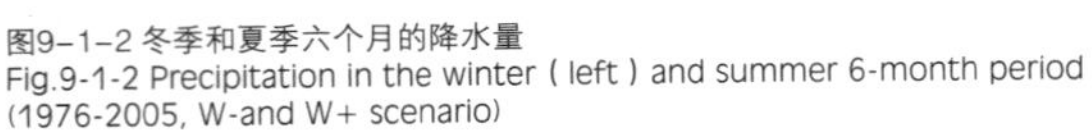

图9-1-2 冬季和夏季六个月的降水量
Fig.9-1-2 Precipitation in the winter (left) and summer 6-month period (1976-2005, W-and W+ scenario)

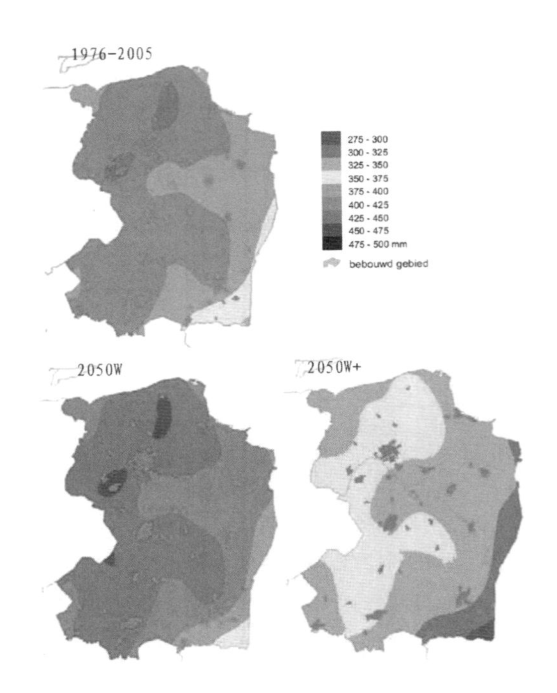

图9-1-2 冬季和夏季六个月的降水量
Fig.9-1-2 Precipitation in the winter(left) and summer 6-month period (1976-2005, W-and W+ scenario)

150厘米）来显示出海平面上升对于格罗宁根的可能影响（图9-1-3）。这些图显示出的是最大影响，因为它们是以堤坝决口后的洪水泛滥的情况来建模的。实际上，景观上的一些障碍物（道路、小堤坝）可以防止海水不受障碍直接侵袭土地。

图纸表明了在两种海平面情景下将会发生的一些现象。图像主要根据高度线并且不考虑会出现突发状况的真实环境条件，也就是说海平面比平常（春潮）要高，也会有严重的雨水和大风。图中也不会显示出堤坝维护的积极作用，而堤坝维护将减少突发状况的发生。另一方面，陆冰快速融化的影响会导致海平面比现在上升至3米或是10米以上，这些在图上不能直接看出。

通过这些图可以得到以下几个结论：

省内的南部地区及格罗宁根市是省内最高的地区，这些地方洪涝风险比较低；工业地区位于人工建造的高地，这些地区相对来说比较安全，即使在海平面上升更极端的情况下；尽管图中不能直接观察到，沿着海岸的农地的盐度增大了，因为海平面上升导致渗流度增大；省内的低地和湿地显示了劳

maximum impact, because they were modeled with an undisturbed flooding by the sea after a dike breach. In reality several obstacles (roads, little dikes) in the landscape prevent the sea from entering the land without barriers.

The maps show an indication of what might happen in the two sea level scenarios. The images are based on altitude lines and do not take into account the real circumstances in which a breakthrough takes place, namely when the sea level is much higher than normal (spring tide) and heavy rain and wind are present. Also, the map does not show the positive effect of good maintenance of dikes in order to keep them strong, which makes a breakthrough less likely to happen. On the other hand, the impact of a much faster melting land ice, leading to possible sea level rises of 3m(Gore,2006)or up to 10m(Carlson,2006)above the current level, is not visualized in the maps either.

Based on these maps a couple of conclusions can be drawn:

The southern parts of the province and the city of Groningen are the highest parts of the province. In these areas flood risk is low;The industrial areas are located at artificially constructed higher levels. These areas are relatively safe, even in a more extreme sea level rise scenario;Though not visible in the maps, the salinity of agricultural land along the coast increases, due to the sea level rise, which results in increased seepage;The lower and wetter parts of the province

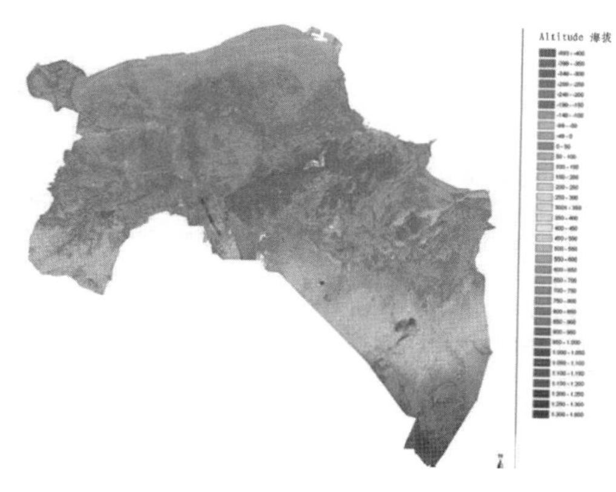

图9–1–3 由海拔高度决定的洪水范围（两种方案）
Fig.9-1-3 Flooding according to altitude lines(two scenarios)1

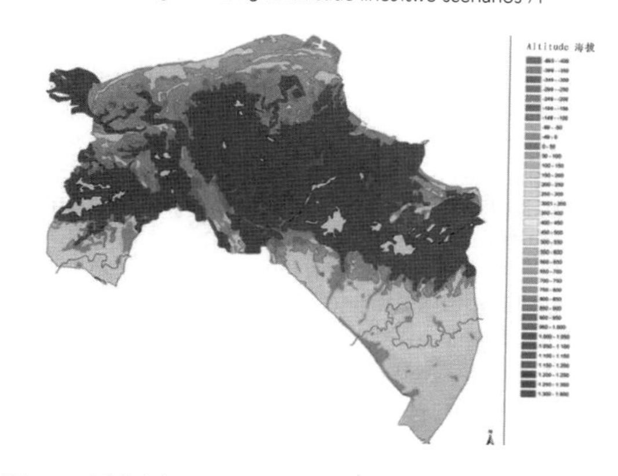

图9–1–3 由海拔高度决定的洪水范围（两种方案）
Fig.9-1-3 Flooding according to altitude lines(two scenarios)2

尔斯河和多拉德之间的空间连接，这个连接与德伦特的溪流系统都是将来最重要的湿生态结构。这里，大自然可能要经受更长时间的干旱。

2. 对不同功能的影响

1) 自然和农业

自然和农业依赖于土壤类型，这对于格罗宁根来说，皮特殖民地因多沙干旱的土壤而和格罗宁根高地的黏土地区或者格罗宁根周边地区发展的出发点不一样。皮特殖民地的问题是在越来越潮湿的冬天里，水能不能被储存以供干旱夏天水荒时使用（图9–1–4）。

格罗宁根黏土区主产小麦，因为黏土保持水分的能力较强，使得这个地区夏天水分的可利用率较高。格罗宁根的南部地区即洪兹吕赫，沙土现有的林地（橡树和山毛榉）的扩大可以防止这个地区土地干旱发生。格罗宁根的生态结构应该与省内其他

show a spatial connection between Lauwers Lake and Dollard. This connection, together with the brook system in Drenthe is potentially the most important wet ecological structure for the future. Here, nature is able to survive the longer periods of drought.

2. Consequences for Different Functions

1) Nature and Agriculture

Nature and agriculture is dependent on the soil type. This means for the Groningen situation that the Peat Colonies, with a sandy dry soil, have a different starting point for development than the Groningen Highland, with its clay soil, or the area around the city of Groningen. The question in the Peat Colonies is whether water can be stored in the increasing wet winter time in order to use it in the dry summers with water scarcity(fig. 9-1-4).

In the clay areas of the Groningen Highlands wheat production has an advantage because the availability of water in dry summers in this area is better, because clay is better capable of keeping water in the soil. In the area southern of the city of Groningen, the Hondsrug, extension of existing forests of sandy soils (oak and beech), can be applied to prevent the area from drying out. Ecological structures in the province of Groningen need ideally be connected to the lowest and wettest parts of the province. These areas contain enough water in the dry summer periods to allow the ecosystems to survive.

2)Urban Developments

The urban functions in the province of Groningen are under threat of floods. Existing artificial hills-the so-called Wierden-have historically a higher altitude, which makes them easy to protect against water. The safest way to house the people in the province is to choose locations for concentrated living areas at higher altitudes in the landscape. Existing lower areas with a high historic value, may be transferred into new islands in the landscape, which will be surrounded by water from time to time. A firm bastion needs to protect these valuable villages (Fig. 9-1-5).

The highest part of the province, the utter most piece of the Hondsrug in the middle of the city of Groningen densities may be risen tremendously. A new Groningen emerges, where nature, landscape and recreation is dominant in the outskirts and on the islands and where the space for working and living efficient, multifunctional and in high densities takes place at higher altitudes. This new Groningen is capable of moving along with climate change.

3. Improving Resilience

1) Understanding the system: mapping climate and energy potentials

In order to understand the existing system and

最低洼最潮湿的部分相连接，这些地区在干旱的夏天有足够的水分使得生态系统能够生存下去。

2）城市发展

格罗宁根省的城市功能受到洪水的威胁，现有的人工丘陵即威尔登，历来就在高海拔区而使得它们能够比较容易地防止水的侵袭。省内人们安家的最安全的方法就是将集中生活区选址于高海拔处。现存的历史价值较高的低地，可能会转移至有时被水包围的新岛屿上，需要一个坚固的堡垒来保护这些价值的村庄（图9-1-5）。

省内最高的地方，即格罗宁根中部地区洪兹吕赫的大部分地方的人居密度将会急速上升。一个新的格罗宁根会出现，那里的郊区和岛上的自然、景观和娱乐条件优良，高海拔区将是工作效益高、功能多样化的高密度地区。这个新的格罗宁根可以随着气候变化而转移。

3. 增强适应力

1）了解系统：气候地图和能源潜力

为了了解目前的系统，并探索引起问题的原因如气候变化和能源供应功能，而发展了一种绘制潜力的地图。这种绘制的方法不仅用于能源还用于气候变化可能性。对于空间规划最重要的气候变化因素和能源一样都被明确定义出来，并直接在图上表示出来。

2）集群规划范式使用

规划过程中的下一步就是将区域气候和能源系

explore the way turbulence causing factors like climate change and energy-supply function a method of mapping the potentials was developed. This mapping method was used for energy as well as for climate potentials. The key-factors, which are crucial for spatial planning were defined for climate change as well as for energy and put subsequently on maps.

2) Use of Swarm Planning Paradigm

The next step in the planning process was the transition of the knowledge about the regional climate and energy system into interventions and ideas, which might improve the resilience of the region. Two ways were explored. The first way aimed at integrating the potential maps into a climate proof map of Groningen: the idea map. This map shows an end-image of a climate proof province. The map functioned as a source for debate and gave direction to the planning process. The second way focused on the definition of strategic interventions, which should be introduced today in order to start the change of the region towards higher resilience. These strategic intervention stimulate the desired changes on the long term

3)Strategic Interventions: the Groningen Impulses

These interventions are not meant to define exactly an end state of the area, but they mark the start of processes, which emerge from that point on by themselves and influence a larger area. The objective of this approach is to increase resilience. This is realized by loosening the fixed state of the existing situation. These strategic interventions are the impulses, which are added to the area and adjusting the area without changing its function. The impulses make use of the capacity the complex adaptive spatial system has (if enough space can be created odoso) to adjust itself to new circumstances and developments. In the Groningen case several of these interventions are proposed. They have in common that a single intervention opens the way to an indirect effect in a

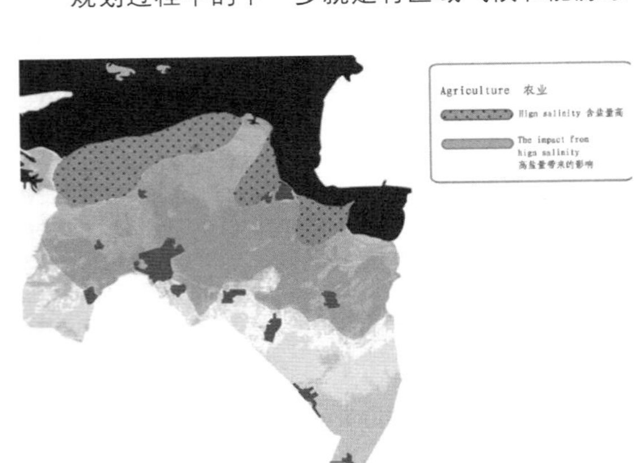

图9-1-4 对格罗宁根农业的影响
Fig.9-1-4 Consequences for agriculture in Groningen

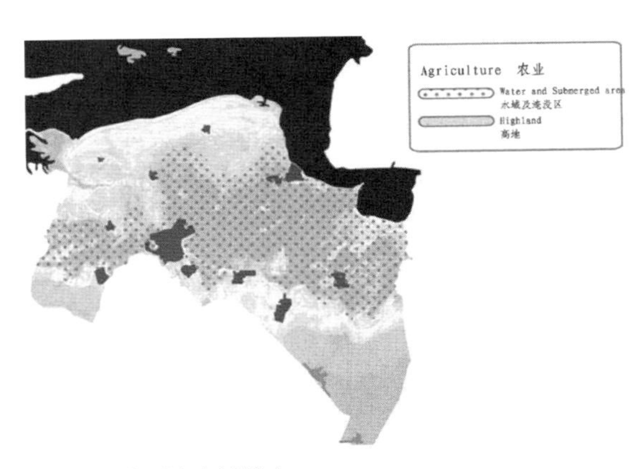

图9-1-4 对格罗宁根农业的影响
Fig.9-1-4 Consequences for agriculture in Groningen

图9-1-5 对城市发展的影响
Fig.9-1-5 Consequences for urban developments

图9-1-5 对城市发展的影响
Fig.9-1-5 Consequences for urban developments

统方面的知识融入措施和想法中去，这可能会提高区域的适应力。研究了两种方法，第一种关注于整合潜力地图到格罗宁根气候验证地图中去，即概念地图。这个地图显示了一个省的最终气候图像，作为讨论的源头并为规划过程指出方向。第二种方法关注于战略干预的定义，为了将区域朝着更强的适应力改变而应该在今天引入。这些战略干预措施在长时间中会激发所期望的变化。

3）战略干预：格罗宁根的推动力

这些干预不是准确地定义地方的最终状态，而是为过程的开始做标记，这个开始于场地自身并影响更大的地区。这个方法的目的是增强适应力。通过放松现存的情况来实现。这些战略干预措施是地区不需要改变自身功能而进行调整的推动力。这些推动力利用复杂的适应性空间系统的能力来调整它们自己去适应新的环境和发展。在格罗宁根案例中，有几个干预措施被提议，它们都是一个简单的干预措施在更大地区发挥间接效益。

在劳尔斯湖中储存淡水：战略干预措施的一个例子就是处理上升的海平面和饮用水的即将短缺。在设计中，淡水将通过升高水平面被储存在劳尔斯湖中。升高的水面会导致格罗宁根的整个水系统被迫自我调节。储存雨水的能力也会增强，这会帮助处理大阵雨和乡镇中潜在的洪水（图9-1-7）。

引进老式的盐沼湿地：第二个战略干预就是在格罗宁根海岸前重新引进老式的盐沼湿地，特别是在伊姆斯港口附近最脆弱最易受冲击的部分。盐沼湿

larger area.

Fresh Water Storage in Lauwers Lake:The first example of a strategic intervention is dealing with the rising sea level and the upcoming shortage of drinking water. In the design fresh water will be stored in the Lauwers Lake.by heightening the level of the water in the Lake. As a result of the risen water level the entire water system of Groningen is forced to adjust it self. The capacity to store rainwater is increased by this simple intervention, which helps to deal with heavy rain showers and potential flooding in villages and towns (Fig. 9-1-7).

Re-introduce old-fashioned kwelderworks:The second strategic intervention is to re-introduce old-fashioned kwelderworks in front of the Groningen coast and especially in front of the weakest and most vulnerable part near the Eems harbour. The kwelderworks start natural processes, which fixate sand and mud. The introduction of the kwelderworks implies a better protection of the coast and makes it possible to postpone the choice what to do with the reclaimed land until it is needed(Fig. 9-1-6).

Improve the condition in Blauwe Stad:The Blauwe Stad area is one of the lowest places in the province and in fact, has led to a third strategic intervention: the introduction of a new luxurious village around a lake, the Blauwe Stad. This intervention resulted in the upgrading of the entire area (the swarm) around the village, where economic development improved, the amount and quality of amenities increased, the infrastructure is improved, the possibilities to deal with large amounts of water is increased and unemployment is decreased(Fig. 9-1-9).

Build Dynamic Coast Fivelboezem:The coast line between Eems harbour and Delfzijl is the most vulnerable in the province. This second dike is absent in this area. This strategic intervention consists of the perforation of the existing sea dike and the creation of an extra dike in the hinterland. The area between the old and new dike will be flooded semi-permanent.

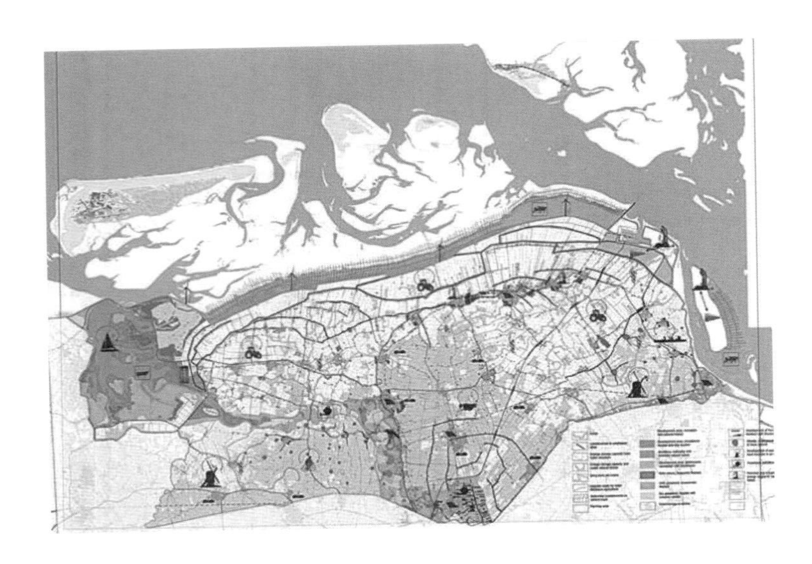

图9–1–6 在伊姆斯港湾附近的盐沼湿地
Fig.9-1-6 Kwelderworks near Eems harbour

地开始自然的过程，固定沙和泥土，它的引进意味着一个更好的海岸保护，并且可以延期决定开垦的土地的用途，直到确实需要这块地前（图9–1–8）。

改善布洛韦斯泰德的状况：布洛韦斯泰德地区是省内最低洼的地方之一，制定了第三个战略干预：将一个新的豪华村引进到布洛韦斯泰德河周边。这个干预导致村子周边的所有地升级，这些地方的经济水平将会提高，便利设施的数量和质量也会增加，基础设施也会被改善，处理的水量也可能会增大，失业也会减少（图9–1–9）。

构建动态海岸费佛布森：在伊姆斯海港和代尔夫宰尔之间的海岸线是省内最易受冲击的地区，在这个区域没有第二堤坝。战略干预包括现存海堤的穿孔，并在内地贸易区建一个新的堤坝。在新老堤坝之间的地方将会半永久性地被淹没。水进入，然后离开，这个地方的水将会很容易地回到原来的位置（图9–1–8）。

4）控制集群

目前的空间规划体系要创造这些有效的干预是有困难的，这些干预能够处理越来越不稳定的环境（如能源供应的不确定性、上涨的能源价格、气候的变化）。这些方面不稳定是因为它们的长期性、不确定性，会在遥远的未来发挥作用。如果地区有机会和空间随着突变而变化，那么处理这些地区的不确定性并增强适应力是有可能的，同时也可以在现在为将来的风险和挑战经验累积。

After water enters the area, it leaves and the area can turn back into its old position very easily (Fig. 9-1-8).

4)Steer the Swarm

The existing spatial planning system has difficulties creating those effective interventions, which are capable of dealing with increasing turbulent circumstances (like the uncertainty of energy supply and rising energy prices and changes in climate). These aspects are turbulent because they are long-term, uncertain and play a role in the far future. It is possible to deal with these uncertainties and create resilient areas if areas are given the opportunity and space to change along with sudden changes and at the same time build experience today which is required for future threats and challenges.

The planning system of the future has to include steering principles, which enable areas to adapt more easily and change its spatial patterns as required by future threats and challenges. These steering principles include, according to the Groningen example two elements: space to change and strategic interventions. These elements differ from place to place, depending on the characteristics of the natural system and spatial identity. Thus, the swarm in the area can be steered in a desired direction of higher resilience.

5)Discussion

The constructed idea-map for an adaptive Groningen functioned as a trigger for debate. The idea-map is an interesting starter for discussion, but it contains the risk of stepping into the same pitfalls as the existing planning practice: a far to fixed image of the future is translated into strict spatial measures, while the future becomes increasingly unpredictable.

Although these innovations were developed during the planning process of the new regional

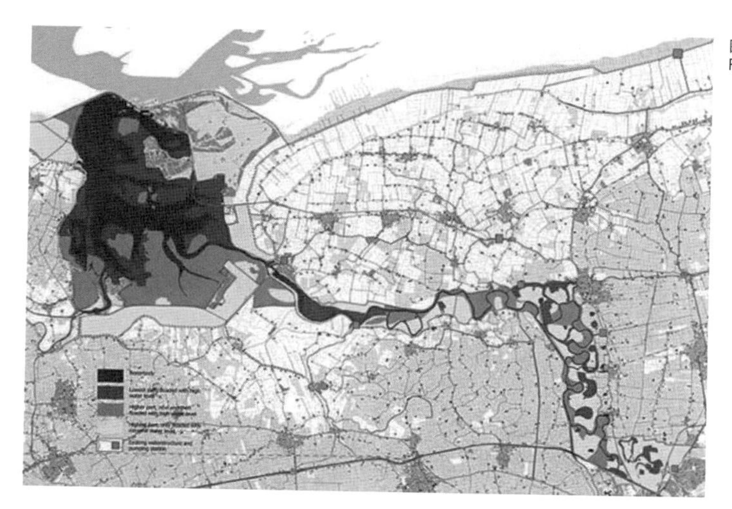

图9-1-7 在劳尔斯湖中储存淡水的提议
Fig.9-1-7 Proposal for fresh water storage in Lauwers Lake

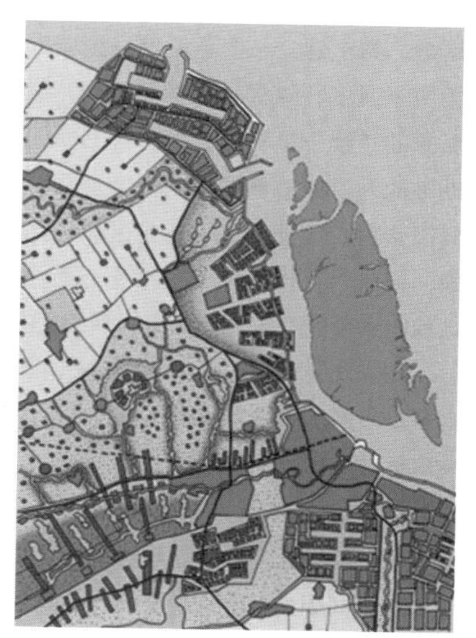

图9-1-8 费佛布森的动态沿海开发
Fig.9-1-8 Dynamic coastal development Fivelbozem

未来的规划体系必须包括控制原则，这可以让地区能够更容易地适应变化，并根据未来风险和挑战的需要而改变它的空间格局。控制原则根据格罗宁根例子的两个元素主要包括：变化空间和战略干预。这些元素各个地方不尽相同，主要根据自然系统的特点和空间特性。因此，地区的群规划可以控制在高适应力的发展方向。

5) 讨论

为格罗宁根适应性建立的概念地图成为争论的导火线，这张地图是讨论的一个有趣的开端，但是它也含有与现存规划实践一样的风险：当未来变得越来越不可预料，确定未来的图像就会被转化成绝对的空间措施。

尽管在格罗宁根新的区域规划过程中发展了这些创新，但问题是已提出的干预是否恰当并怎样实现。现存的规划仍是较平淡的过程，它的目标是问题分析及解决方案，而不是增强空间体系的适应能力。真正的问题是当今决策者为未来的不稳定状况做准备的紧迫感是不是足够高。

三、 总结

随着全球变暖和极端气候事件频繁发生，气候

plan for Groningen, the question is if and how the developed interventions will be realised. The existing planning process is still a tame process, where the problem analysis and the derived solutions are the objective instead of aiming to increase the adaptive capacity of spatial systems. The question is if the sense of urgency to prepare for future turbulence is high enough among decision makers of today.

III. Summary

As global warming and extreme weather events occur frequently, climate change and its adverse effects become increasingly human issues of common concern.The planning and design of Groningen has done some new explorations to deal with the Climate change.The project adhering to the city planning idea of slow down and adapt to climate change ,then strive to establish an environment friendly city.A lot of research has focused on temperature, precipitation and rising sea levels,these factors can cause the last drought in summer and more moisture in autumn ,winter or spring,and the periodic flood are threating the safety of city.The Planning analyzed the different effects of these climate factors for agricultural areas, urban areas and the natural areas,then pointed proposed intervention measures.And steering the swarm can guarantee the effectiveness of intervention,and it has the high flexibility.

"China 's agenda in twenty-first century - The White

变化及其不利影响日益成为人类共同关心的问题。格罗宁根市规划设计在应对气候变化方面做了新的探索。该方案秉承减缓和适应气候变化的城市规划理念，力求建立一个环境友好型城市。大量的研究集中于温度、降水和海平面上升，这些因素会引起夏季的持续干旱和秋、冬、春三季的更加潮湿，周期性的洪涝威胁着城市的安全。规划分析了这些气候因素对农业地区、城市地区和自然区域的不同影响，有针对性地提出干预措施，而控制集群方法可以保障干预的有效性，具有较高的适应力。

1994年《中国21世纪议程——中国21世纪人口、环境与发展白皮书》发表，随后，中国政府主要从温室气体减排和增强碳汇能力入手，应对气候变化。国内外经验表明，在城市规划与设计时，应考虑气候影响，重视规划之间、规划与气候之间的协调整合。通过城市空间规划、城市规模控制、城市交通规划等多规融合，来缓解气候变化给城市带来的种种问题。

图9–1–9 代尔夫宰尔在格罗宁根东部风景中已实施的战略
Fig.9-1-9 Blauwe Stad implemented in the landscape of eastern Groningen

paper of twenty-first century Chinese population, environment and development"was published in 1994.Then, the Chinese government mainly from the reduction of greenhouse gas emission and the enhancement of carbon absorbing ability to cope with climate change.The experiences from domestic and overseas shows that in urban planning and design should considered the climate impacts,and pay attention to the coordination and integration between plannings or between planning and climate. Through the many rules fusion such as city planning, city space control, urban traffic planning and so on to mitigate the various problems in the city which caused by climate change.

思考题

1. 由于预测的不确定性，怎么能保证现阶段所做的规划和措施的有效实施?
2. 不同尺度上的气候适应性设计应该具体到哪些方面?
3. 控制集群方法在战略层面上的主导因素是什么?
4. 气候适应性设计与生态规划设计的异同?
5. 动态海岸线的构建还有哪些典型模式?

Questions

1. Because of the uncertainty of the forecast, how to ensure the present plannings and measures could implement effectively?
2. What are the aspects should be specific in different scales during the adaptation to climate change design?
3. What's the dominant factor of the way of steering the swarm in strategic level?
4. What's the similarities and differences between the adaptation to climate change design and ecological planning and design?
5. Does the construction of the dynamic coastline has any other typical model?

注释/Note

图片来源/Image：本案例及图片均引自"Adaptation to Climate Change: A Spatial Challenge", Rob Roggema, November 18， 2009

参考文献/Reference：本案例引自"Adaptation to Climate Change: A Spatial Challenge", Rob Roggema, November 18， 2009

推荐网站/Website：http://www.knmi.nl/index _ en.html; ；http://www.climateresearchnetherlands.nl/climatechangesspatialplanning; ；http://www.cittaideale.eu/01 _ CV _ RobRog.html; ；http://www.epa.gov/climatechange/index.html.

第二节 重庆龙湖项目

Section II The Longhu Project, Chongqing

一、项目综述

1. 项目简介

位于中国重庆市北部新区礼嘉组团，项目场地处于丘陵地区并邻近嘉陵江（图9-2-1）。该场地属于广大开发区中的一部分，居住区中结合了福利设施、基础设施以及游憩设施的布置。

场地南向、西向和北向视野开阔，往东甚至可以看到嘉陵江。场地总面积达201 444平方米，规划的地上建筑面积约为296 800平方米，地下停车面积约为57 000平方米。

2. 规划目标

为了应对气候变化导致的种种困难，本规划整合了一系列综合措施。旱情的加剧、夏季的热岛效应以及生物多样性的减少被视作综合问题，这些问题的理想解决办法需要通过自然环境来实现：该地区现存的水体尽量长期地用于蓄积雨水，同时对城市空间起到降温效果，山丘及平原用来创造自然通风以降低气温的攀升，而山坡的不同朝向正是丰富生态性的基础，同时也能对城市起到降温效果。

本项目的目标是实现“绿色建筑评价标准”的要求，包括六个方面：土地效用及户外环境；能源效率及利用；水源效率及利用；材料效率及利用；

I. Overview

1. Introduction

The Longhu project is located at the north new district of Lijia Group in Chongqing, China (Fig.9-2-1). The project site is at a hilly region and close to the Jialing River. It is part of a widespread development zone, where living areas are combined with amenities, infrastructure and recreation.

It is an open vision to the south, west and north orientation and even possible to watch the Jialing River to the east orientation. The total area of the complete site is 201,444m2. The total building area above ground is planned to be approximate 296,800m2 and the underground car parking is 57,000m2.

2. Goals

In order to meet the difficulties caused by climate change a combination of measures are integrated in the plan. The increasing droughts and the heat island effect in summer as well as the decrease of biodiversity are seen as combined problems, which are solved by using the natural circumstances optimally: the existing waterbodies in the area are used to keep rain water as long as possible in the project area and have a cooling effect on the urban spaces, the hills and plains are used to create natural ventilation in order to minimize temperature rises and the different expositions of slopes are the base for an enriched ecology, which has a cooling effect on the city also.

The target of the project is to fulfill the requirement of the Assessment Standard for Green buildings. This includes six aspects: Land efficiency and outdoor environment; Energy efficiency and utilization; Water

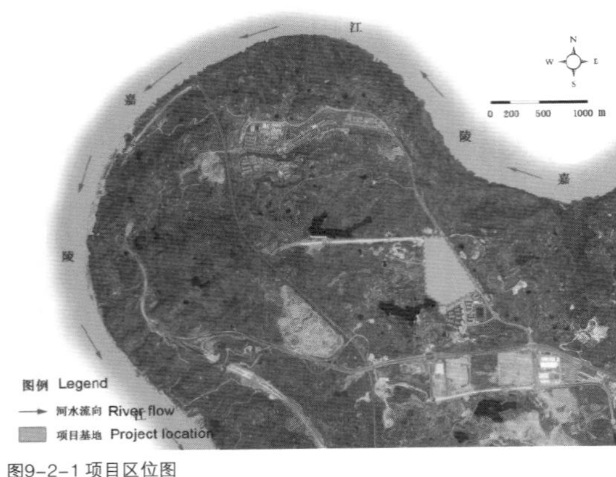

图9-2-1 项目区位图
Fig.9-2-1 The location of the project

室内环境质量；运营和管理。

为了强调绿色建筑设计的各项条款，本项目着重探讨了一些话题，即：对目前的概念性城市设计的建议；能源效率及技术途径；水资源效率及技术建议；材料效率及技术建议。

二、项目背景

1. 气候变化影响

重庆市坐落于青藏高原与长江中下游平原之间的过渡地区，属亚热带季风气候，年平均温度为18至20摄氏度，冬季最低温度4摄氏度，夏季最高温度40摄氏度。重庆市同样也是一座雾城，一年中约有100天为雾天，通常集中于春夏两季。重庆市降水量丰富，年均降水量约1 000～1 400毫米。全年多夜雨，但多集中于夏季。此区域的气候变化影响可能表现为夏季炎热天数的增多，而这最终会导致更强的城市热岛效应。在夏季，伴随着西藏山区降雪量的减少，重庆市降水量的减少将导致水位下降和干旱。大量农田因干涸而难以耕种，饮用水的获取成为更多人面临的问题。由于干旱，难以在夏季的炎热干燥中维持其生态性，生物多样性也减少。

2. 场地分析

项目场地位于丘陵区，山坡以约30米的高差自北向南下降，由于过去的农业利用，自然山坡已转变为带有小水池的绿色梯田。一条蜿蜒的陡坡沿东西向从该区域中间穿过，绿色的梯田和错综复杂的水系是相

efficiency and utilization; Material efficiency and utilization; Indoor environment quality; Operation and management.

In order to focus on the terms for the Green Building design, several topics are focused in this project, namely: Advices on the exist conceptual urban design; Energy efficiency and technical solutions; Water efficiency and technical proposal; Material efficiency and technical proposal.

II. Context

1. Climate Change Effects

Chongqing is located in the transition area between the Qinghai-Tibet Plateau and the Middle-lower Yangtze Plain. It is part of the humid sub-tropical monsoon climate belt. The annual average temperature is 18–20℃ with a low temperature of 4℃ in winter and a high temperature of 40℃ in summer. Chongqing is also a fog city that has about 100 foggy days a year usually in spring and summer. Chongqing receives abundant rainfall, averaging about 1000–1400 mm annually. It has plenty of evening rain all year round, but most of it falls in summer. The effects of climate change in the region are expected to show an increase in number of hot days in summer. This will eventually lead to higher impact of heat island effects in urban areas. In the summer period the rainfall is predicted to decrease, which will, in combination with lesser snowfall in the Tibetan mountains, leading to lower water levels and droughts in summer. More farmland becomes too dry to cultivate and the access to drinking water becomes for more people a problem. Due to droughts ecology will meet difficulties in surviving these hot and dry summers and biodiversity decreases.

2. Analysis of the Site

The project is at a hilly area, with slopes declining from the north to the south with a height difference of ca. 30 meters. Because of the past agricultural use, the natural slopes have been turned into green terraces with small water ponds. A curved steep slope is running in east-west direction through the middle of the area. The green terraces and the fine amazed water structure are valuable agricultural heritages. To the northeast of the site there is a panoramic view to the river which lies in the depth. To the south, near the boundary of the site, there is a big lake, which provides an attractive recreational environment (Fig.9-2-2&Fig.9-2-3).

The landscape pattern is very rich: A diffuse network of hills, slopes, flat areas and curving routes can be distinguished in the site. There is a large difference in altitude between the northern and southern part of the site. The steepness of the slopes varies largely. A highly differentiated and fine mazed water system

当有价值的农业遗产。场地东北向能看到位于深处的河流全景，靠近场地南边有一个大湖，它提供了一个充满吸引力的游憩环境（图9-2-2，图9-2-3）。

场地的景观格局相当丰富：可以辨识出山丘、斜坡、平原以及曲折的道路所构成的扩散状网络。场地北部与南部高差巨大，斜坡的陡度也变化多端。该地区中的水系高度分化并错综复杂。水流流经多层梯田而放缓流速，通过打开或关闭小型堤坝（障碍物）或石头来蓄水或流通至下一块农田，最终水被反复地用于多种不同的目的，如农作物生产、洗衣服。

三、 规划设计介绍

1. 总体概念性规划

1）以景观为核心的新开发

现有的梯田景观可以转变为这一新开发区中极具吸引力的核心。这里存在着一条潜在的南北向“蓝绿轴线”，中央由弯曲的陡坡形成了一个半封闭的山谷，从而具备了特殊的自然吸引力。通过利用现有的山坡、梯田和水系，可以建立与场地的（农业）文化遗产和自然品质之间的联系。减少不必要的土地平整工程，同时节省成本。

在城市规划的调整和进一步的深入中，应当注意以下有关景观的事项：充分并创造性地利用现有的山坡和梯田，利用并突出地面高差，尤其要充分发挥该区域中弯曲陡坡的优势；利用南北向的绿色廊道来连接场地的高低两部分，使公园面积最大化，与南边的大湖保持开放性关系；保持并改变绿色梯田和山坡的多样性，有助于给该居住区带来与众不同的吸引力；保持并促进动植物多样性。

exists in the area. The water is slowed down in many steps. By putting open or close down little dikes or stones the water is kept or transported to the next field. Finally, the water is used over and over again for many different purposes, production of crops, washing clothes, etc.

III. Planning and Designing

1. Overall Conceptual Planning

1) Landscape as Heart of New Development

The existing terraced landscape could be transformed into an attractive heart of the new development. This has the potential of a long green and blue axis, stretched in south-north direction, with a special natural attraction in the middle where the curved steep slope forms a half-enclosed valley. By using the existing slopes, terraces and water structure, a connection is made with the cultural heritage and the natural qualities of the site. Unnecessary leveling of the ground could be minimized, which also can save costs.

The following items regarding the landscape should be regarded in the adjustment and further deepening phases of the urban plan: Make sufficient and creative use of existing slopes & terraces, use and accentuate differences of ground levels, especially take advantage of the curved steep slope in the middle of the area; Use the north-south green corridor to connect the upper and lower part and maximize park area, keep an open relationship to the big lake in the south; Keep and transform the diversity of the green terraces and slopes, which contribute to unique attractiveness of the residential area; Keep and develop the fauna and flora diversity.

2) Conceptual Remarks: Living in the Landscape

Make use of the different qualities in the landscape, preferably by accentuation. For example: the differences in altitudes can be accentuated by putting high-rise buildings on the higher spots and lower buildings on the lower areas.

Strive for a poly-culture of buildings, instead of putting over the existing landscape like one big layer that reduces gradients and diversity. Organise and allocate

图9-2-2 场地鸟瞰图
Fig.9-2-2 An aerial view of the site

图9-2-3 场地的景观
Fig.9-2-3 The Landscape of the site

2）概念性意见：居住于景观之中

更好地通过强调和突出来利用景观的不同品质。例如：通过在高处布置高层建筑而在低处布置较低建筑来突出高差。

努力实现建筑的多元化，而非将现有景观视为一个巨大的缺少梯度和多样性的层面来布置建筑。将不同的建筑类型组织和分配到具体的位置上。

努力实现水上和水中居住、坡上和坡顶居住、水花园居住和风景式居住相互间的积极作用。

2. 具体措施

1）水体处理

项目区中现存的大量水体形成了一个相当灵活的系统。如果大量水体可利用的话，则能将其导向多个地方并保留在该区域中。如果该区域转变为市区，可对这些水体进行调整，但不能缩小或移除。如果将它们保留，该地区的恢复力则能保持较高水平。这些水体作为一个系统能够汇集大量的水并长期的保留于项目区之中，从而在干燥的夏季派上用场。水体同样也能减轻因该区域的城市化和炎热天气的增多导致的城市热岛效应。

基于自然水系，提出了如下的技术性建议。下图表现了水系的不同水流（图9-2-4）。

这些技术特性被移植到了该场地的一个整合水系中。针对不同的水质而推出了有细微差异的处理措施。建筑和道路雨水经预处理后排放到水池中，洗涤用水经预处理后也排放到水池中。而厨房、厕所和洗衣房废水则直接排放到下排污系统中。最终，植被和

different building typologies on specific locations.

Strive for positive interaction On&In the Water-living, At&On the Slope-living, Water-garden-living and Living with a view.

2. Specific Recommendations

1) Water-bodies

The large amount of existing water-bodies in the project area causes a very flexible system. If much water is available it can be directed to many different places and kept in the area. If the landscape is transformed to an urban area, the water-bodies can be adjusted, but not minimised or removed. If they stay in the area the resilience can be kept at a high level. The water-bodies function as a system that is able to incorporate large amounts of water and the water can be kept for a long time in the project area, thus being available in dry summers. The water-bodies are also capable of mitigating the urban heat island effect, which is due to increase if the area is urbanised and the number of hot days is increasing

Based on the natural water system a technical suggestion has been made. In the next scheme the different flows of the water-system are represented.

These technical features are transplanted in an integrated possible water-system for the site. For each different water-quality slightly different treatment is suggested. Rainwater from buildings and roads is pre-treated and discharged towards the pond. Bathwater is pre-treated and also transported towards the ponds. Water from the kitchen, toilet and laundry is directed towards the sewage system. Finally the green and surface water is maintained as an eco-aquatic balanced water-system (Fig.9-2-4).

2)Natural Ventilation

During summer the site can get warm and sticky and the number of hot days is increasing in the future. The urban heat island effect will increase if the site is urbanised. The most natural way of cooling and ventilation

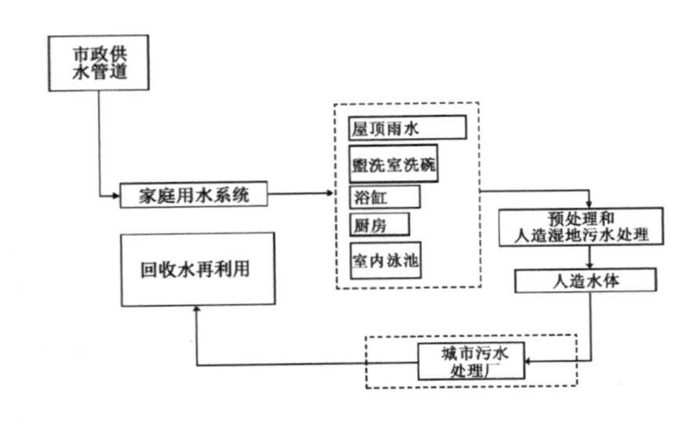

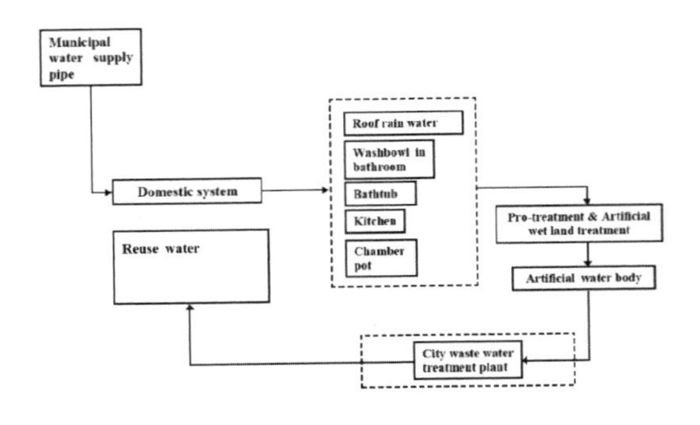

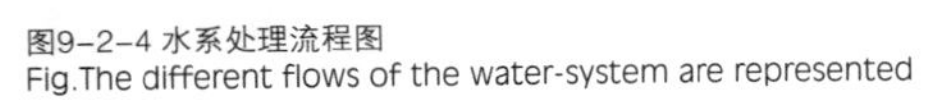
图9-2-4 水系处理流程图
Fig.The different flows of the water-system are represented

地表水被保留下来而成为一个平衡的水生生态系统。

2）自然通风

夏季，场地将变得温暖而湿润，炎热天数也将增多。此地一旦城市化城市热岛效应则会增强。最自然的降温和通风办法则是在该区域创造自然风和空气流动，这可以通过建筑布局或墙体绿化来实现（建筑间可产生更强的自然风，而这能促使空气流通）。

常风向为西北风，在场地的这一侧布置高层建筑（图9-2-5）将会给场地北部带来更强的通风。在场地（刚好位于最陡山坡的坡顶边缘）边缘布置建筑则会给该区域的较低部分带来同样的效果。如果场地边缘布置了一条绿带（图9-2-5），也能实现同样的效果并增加了空气湿度。如果将高层建筑布置于底部台地上，则能给该场地最湿润的部分带来一股凉爽的微风。采用这种自然通风办法不会增加能耗却能减轻城市热岛效应。

为改善自然通风而进行的建筑布局同样也会增加景观的可读性。这一做法的结果则是最高的建筑位于场地的最高位置而较低的建筑位于场地的最低部分。从而呈现出不同的居住环境：水上和水中、

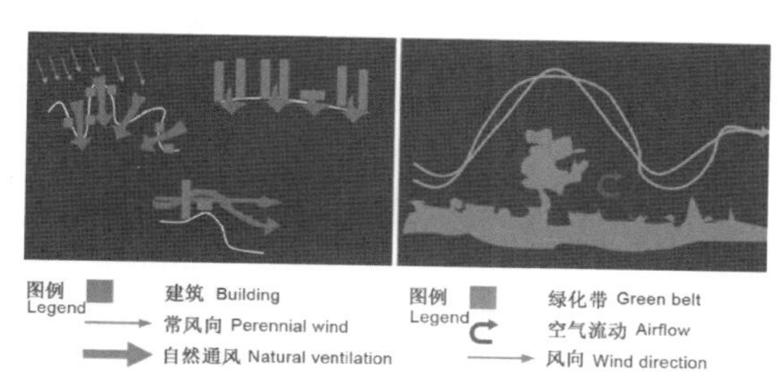

图9-2-5 通过建筑布局（左）和采用绿带（右）来增强自然通风
Fig.9-2-5 Increasing natural ventilation by positioning of buildings

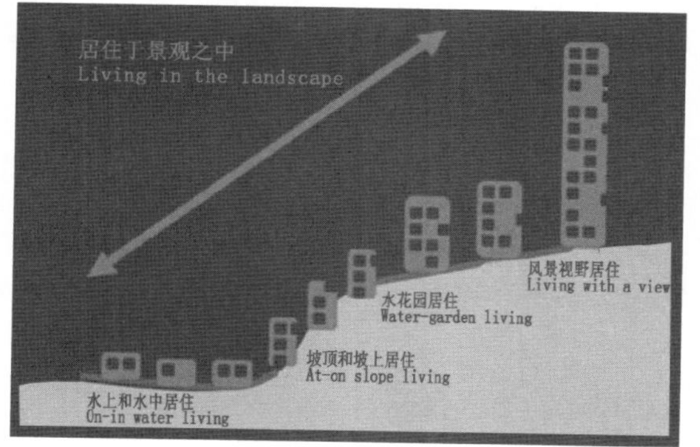

图9-2-6 景观中几种不同居住类型
Fig.9-2-6 Several living typologies in the landscape

is to create winds and ventilation in the area. This can be done by the positioning of buildings (in between the building stronger winds occur there occur, which create ventilation) or by implementing green walls.

The most common wind direction is from the Northwest. Positioning of high-rise buildings (Fig.9-2-5) at this side of the area will lead to stronger ventilation in the northern part. Positioning of buildings at the edge of the site (located just at the top edges of the steepest slope) results in the same effect for the lower parts of the area. If a green belt is positioned at the edge (Fig.9-2-5), the same effect is reached and it also adds more humidity to the air. If high-rise buildings are added at the bottom plateaus a cool breeze will occur in the stickiest parts of the site. Introduction of this natural ventilation does not increase energy use and mitigates the urban heat island effect.

Positioning of buildings to improve the natural ventilation increases the readability of the landscape also. As a result of this the highest buildings are located at the highest altitudes, and lower buildings at the lowest places of the site. Different living environments emerge: On and In the Water-living, At and On the Slope-living, Water-garden-living and Living with a view 3)Biodiversity

In order to withstand a decrease in biodiversity the main element is to deal with droughts. Therefore, the ecological qualities and the water elements show a strong connection. If there is not enough water the ecological qualities will be different. Based on the existing landscape four different qualities are defined (Fig.9-2-7): A balanced eco-aquatic ecosystem: Clean water provides fish and water-plants with enough air and water at flat plains. Water is kept here as much as possible to provide wet circumstances in dry periods; The ecology of slopes: these gradients are positioned in the middle of residential areas and buildings. Park-like gardens with water-basins form an ecology, which can be used by inhabitants; Steep-slope ecology: at the steep slopes the water is running

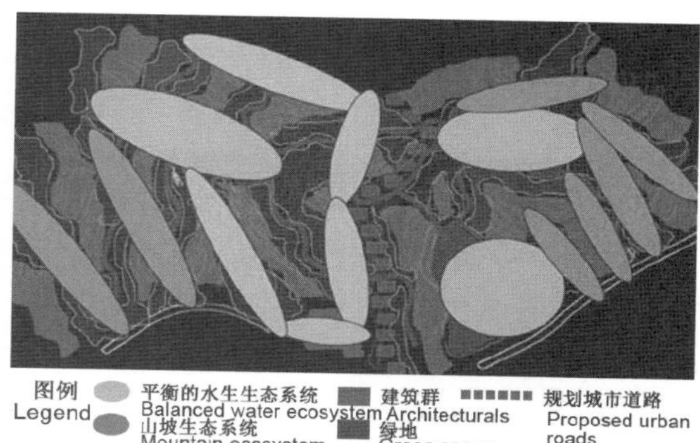

图9-2-7 生态差异
Fig.9-2-7 Ecological differences

坡上和坡顶、水花园和风景视野（图9–2–6）。

3) 生物多样性

为了防止生物多样性的减少，首要任务是解决干旱。因此，生态质量和水之间有着紧密的联系，如果水分不够则生态质量将大打折扣。基于目前的景观，这里确定了四种不同的类型（图9–2–7）：平衡的水生生态系统：清洁水为平原地区的鱼类和水生植物提供足够的空气和水源；山坡生态系统：这些斜坡布置于住宅区和建筑中间，拥有水池的类似于公园的花园形成一个生态系统，并能为居民所使用；陡坡生态系统：在陡坡上水流以瀑布的形式飞流直下，坡上长满草和苔藓植物，湿润又清新的环境能够吸引特定的鸟类、爬行动物以及昆虫；立面生态系统：墙体绿化为亚热带鸟类和昆虫提供了适宜的生存环境。

4) 各要素的布局方式及位置

水体、生态要素以及绿化带和建筑群相互联系，并基于场地现状所呈现的差异而布局。阶梯和斜坡决定了创造水流、自然通风或生态多样性的可能性。等高线的紧密程度反映了斜坡陡度的四种类型（平坦、缓坡、较陡和极陡），而斜坡则引导了水系（图9–2–8）。基于不同的坡型可以确定相应的水系类型：平坦地区的水静止不动，只有当水流充沛或将堤坝打开时，水才会流向周边较低地区。2类和3类斜坡中的水缓缓流淌，而最陡斜坡上的水流如瀑布一样飞流直下。每一种坡型都拥有其自身特有的水系。

3. 规划模型

1) 两种基本模型

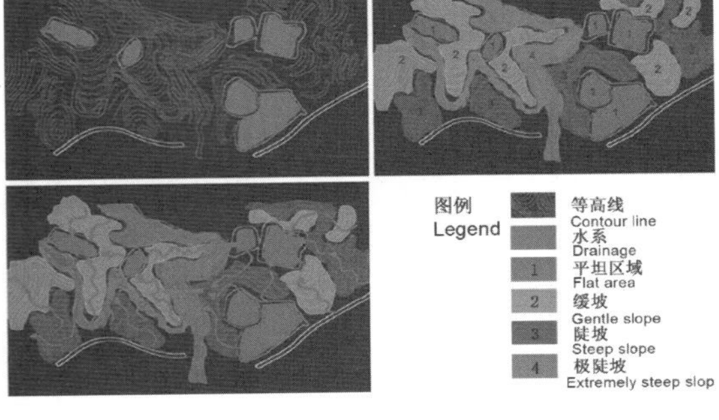

down quickly in the form of waterfalls. Grasses and mosses fill the slopes with vegetation. Humid and fresh circumstances attract specific birds, reptiles and insects; Façdes-ecology: green building walls provide subtropical birds and insects with appropriate habitats.

3)The Elements How and Where Positioned

The water bodies, ecological elements and green belts and building blocks are positioned in relation with each other and are based on the differences the existing landscape has to offer. The gradients and slopes direct the possibilities for the creation of water, natural ventilation or biodiversity. The intensity of altitude lines gives an estimate on the steepness of the slopes in four categories (flat, slow, steep and very steep) and the slopes direct the water system (Fig.9-2-8). Based on the different slope types the water-typology can be defined: At the flat areas the water stand still and only when there is enough water or the dikes are opened the water is transported to the neighboring and lower area. The water at slope categories 2 & 3 is gently running down and the water at the steepest slopes is running down like waterfalls. Every slope-type has its own water-system.

3. Planning Models

1)Two Basic Models

Based on the landscape and water system different models can be developed (Fig.9-2-9). The first model reserves building sites next to the brooks and water bodies. The buildings are surrounded by green, which is planned at the edges of the water. The second model combines building zones with the water system. Building structures are positioned on and across the water system. Both models are able to react and anticipate on future changes in precipitation, natural conditions and rises in temperature.

The models are not meant to create a design, but function as pathways in which the different possibilities of the natural landscape can be explored and taken advantage of.

2)An Integrated Model

The models and ideas are integrated in one integrated vision on the site (Fig.9-2-10).The building

图例 Legend
等高线 Contour line
水系 Drainage
1 平坦区域 Flat area
2 缓坡 Gentle slope
3 陡坡 Steep slope
4 极陡坡 Extremely steep slop

◀ 图9–2–8 等高线密集度，坡型和水系
Fig.9-2-8 Intensity of altitude lines, slope types and the water system

▼ 图9–2–9 两种气候适应模型
Fig.9-2-9 Two climate adaptive models

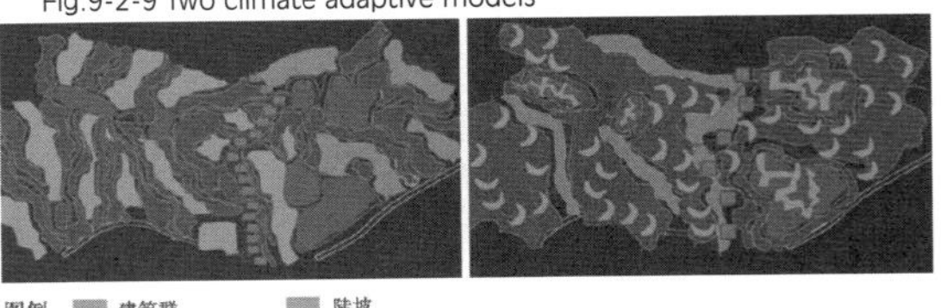

基于景观和水系能建立不同的规划模型（图9-2-9）。第一种模型将建筑基地保留在溪流和水体旁边，建筑被绿色植被所环绕，这些植被被规划于水体的边缘。第二种模型将建筑区与水系结合在一起，建筑被布置于水系上和水系间。两种模型都能反映并预测降水、自然条件和升温在将来的变化。

这两种模型不是为了创造一种设计，而是作为一个途径，在其中可以探索和利用自然景观的不同可能性。

2）整合模型

模型与概念被整合到了场地的整体构想中（图9-2-10）。建设区被规划于自然水体结构之间，居住区位于斜坡之上而高层建筑被规划在平原上。两条车行道将各建设区连接起来：其中一条穿过较高的北部地区，而另一条则通往较低的南部地区之中。主要的福利设施被布置在该区中央并与一条绿带结合在一起，而绿带则为较低地区提供自然通风。场地能够将尽可能多的水保留在错综复杂的水系之中，而斜坡的差异则能被用于创造多种生态类型。自然通风、生态空间以及水体都能对该地区产生降温效果。这样，预计的干旱和炎热天数的增多便通过综合方式得以解决。

四、 总结

该项目作为本书中唯一一个国内项目，其规划亮点同样是以积极的姿态应对气候变化。重庆是我国典型的山地城市，因其地理环境和地貌特征的特殊性，形成了与平原城市截然不同的城市布局和空间形态。这里处于亚热带季风区，具有温暖湿润的区域气候特点，受地形影响，形成了南北风向、温润多雾的微气候特点。本项目位于城北新区，规划目标是达到“绿色建筑评价标准”，营建环境友好型住区。大量的研究集中于城市绿地对温度的调节，以及不同建筑环境对地表和空气温度关系的影响。这些研究表明，可以通改变绿地形态、景观异质性，以及地表粗糙度来环减轻“热岛效应”，从而降低人工环境对气候变化的影响。

设计首先从重庆的地貌形态、气候因素、植

zones are projected between the natural water-structures. The residential areas are located at the slopes and the high-rise buildings are projected at the flat plains. Two car traffic routes connect the different building areas: one through the higher northern part and one cul-de-sac in the lower southern area. In the center of the area the major amenities are placed and are combined with a green belt that provides the lower parts with natural ventilation. The site is able to store as much water as possible in the fine mazed water system and the differences in slopes are used to create different ecological typologies. Both natural ventilation, the ecological spaces as well as the water-bodies provide the area with a cooling effect. Thus, the expected droughts and increase of the number of hot days are dealt with in an integrated manner.

IV. Summary

As the only case located in China of this book, the bright spot of Longhu Project is its positive attitude towards tackling climate change. Chongqing is a typical mountain city in China with special geographical conditions and geomorphic features, so it differs from plain cities in urban layout and spatial pattern. Chongqing is located in the subtropical monsoon zone which has warm and humid regional climate characteristics, and because of the terrain, it forms a north-south wind and moist and foggy microclimate characteristics. This project is located in the north new district in Chongqing, Its planning target is to achieve the requirements of the Assessment Standard for Green Buildings, and construct an environment-friendly residential community. Lots of the studies focuses on the temperature regulation by the urban green space, and the impact from different constructing environment to the relations between the earth's surface temperature and the air temperature.

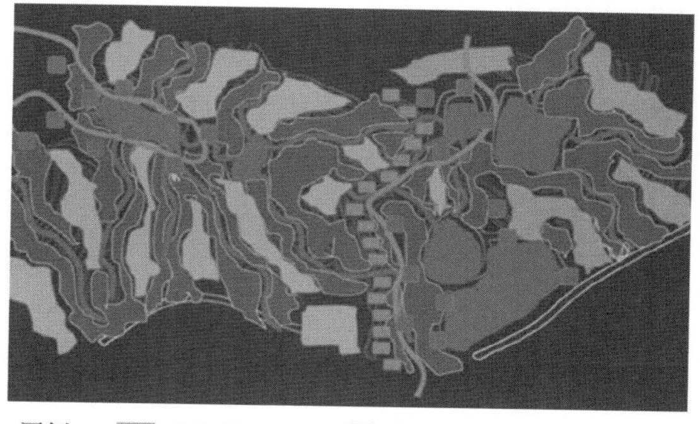

图例 Legend
建筑群 Architectural complex
绿地 Green space
水系 Drainage
高层建筑 High-rise buildings
规划城市道路 Proposed urban roads
居住区车行道 Residential roads

图9-2-10 整合模型
Fig.9-2-10 Integrated model

被、工程措施、风俗文化等要素入手进行分析，注重坡地的功能空间组织和视线的控制，推敲坡地生态环境的营造以及对环境审美的追求，最终融合为理想的山地城市坡地景观设计秩序。

规划因地制宜、依山就势布置空间，充分利用环境中地形及空间属性，处理好坡地环境和项目的关系，这种环境秩序是生态的、可持续发展和自我循环的环境系统，在满足活动功能和生态环境的基础之上，蕴涵了地域之美、文化之美，塑造了不可复制的城市性格和品质。

These studies show that we can change the form of green belt, the heterogeneity of landscape, and the roughness of the earth's surface to alleviate the "heat island" effect, thereby reducing the impacts of climate change from the artificial environment.

The design starts from the analysis of geomorphology, climatic factors, vegetation, engineering measures, customs and culture and other factors of Chongqing, focuses on the organization of the functional spaces on the slope and the control of sight, and then deliberates the construction of slope ecological environment and the pursuit of environmental aesthetics. Finally, it integrates the order of the ideal sloping landscape design in mountain city.

This plan arranges the space based on local conditions and the terrain of mountains, take full advantage of the topography and spatial properties of the environment, settles down the relation of the slope environment and projects, it is an ecologic, sustainable and circulative environmental system. On the basis of fulfilling the function of activity and ecology, it implicates the beauty of geography and culture, and shapes an urban character and quality that cannot be copied.

思考题

1.本案例在应对气候问题时从哪些方面进行了考虑？除此以外，你认为还有哪些因素应纳入考虑？

2.本规划是如何因地制宜的利用场地现状条件的？

3.如何理解建立的两种基本模型“不是为了创造一种设计，而是作为一个途径”的含义？它对我们进行景观规划设计有何借鉴意义？

4.试考虑，当场地位于其他地形和环境条件（如滨海城市、水网城市等）时，景观规划设计在应对气候问题时应着重考虑哪些方面？

5.请列举出你所了解的任何与应对气候问题有关的景观规划设计措施或理念，与大家分享交流。

Questions

1. What factors have been taken into consideration to cope with the climate issues in this case? In addition, what other factors should be taken into account in your opinion?

2. How does this plan make use of the current conditions of the site to adjust to the local situation?

3. How to understand that the establishment of the two basic models "are not meant to create a design, but function as pathways", what can we learn from it to improve landscape architecture planning and design?

4. What factors should be taken into consideration primarily if the project site is in other terrain and environmental conditions (such as coastal cities, water network city, etc.)?

5. Please list any landscape planning and design measures or concept to address climate issues that you know, share them with others and make a communication.

注释/Note

图片来源/Image：本案例图片均引自R. Roggema， Adaptation to Climate Change: A Spatial Challenge, 2009.

参考文献/Reference：本案例引自R. Roggema， Adaptation to Climate Change: A Spatial Challenge, 2009.

后记

EPILOGUE

作为同济大学建筑与城市规划学院景观学专业的核心框架“风景园林——环境生态——游憩旅游”、“资源与生态——规划与设计——技术与管理”的重要构成部分，生态规划设计一直是重点培育和发展的学科方向之一。在相继出版《景观生态规划原理》（王云才，2007；2013）、《园林植物与应用》（李文敏，2006）、《现代生态规划设计的基本理论与方法》（骆天庆、王敏、戴代新）、《群落生态设计》（王云才，2009）等多部生态规划设计教材的同时，承担了国家“十一五”科技支撑计划“城镇绿地生态构建和管控关键技术研究与示范”（项目编号2008BAJ10BO2，刘滨谊，2008）和国家自然科学基金委工程与材料科学部建筑、环境与结构工程学科风景园林“十二五”科学研究发展战略研究，将城乡景观的生态化设计理论与方法研究作为重要的学科建设方向（刘滨谊、王云才、刘晖、徐坚）。在开展自然生态与风景、城乡绿地生态研究的同时，开展了以国家自然科学基金项目“传统地域文化景观破碎和孤岛化现象及形成机理”（项目编号50878162，王云才，2008）为切入点的人文生态系统研究，试图建立融合自然生态和人文生态为一体的整体人文生态系统的规划设计体系，成为风景园林生态

As parts of the core framework of landscape Architecture, College of Architecture and Urban planning, Tongji University, which are composed of "landscape Architecture, ecological environment, recreation & tourism", "resources & ecology, planning & design, technology & management," ecological planning and design is always one important direction of the subjects have been cultivating and developing . During the past years, some books relatived to this field have been published ,such as Landscape Ecological Planning Principle (WangYuncai, 2007; 2013), The Garden Plants and Application(Li Wenmin, 2006), The Basic Theory and Method of Modern Ecological Planning and Design (Luo Tianqing, Wang ming, Dai Daixin), Community Ecological Design (WangYuncai, 2009), at the same time, we finished some important scientific research projects, such as the national eleventh- five-year science and technology plan "Urban Green Space Ecological Construction and The Control Key Technology Research and Demonstration" (No. 2008 BAJ10B02, Liu Binyi, 2008) and The Scientific Research Development Strategy of Architecture, Environment & Structure Engineering Discipline and Landscape Architecture of Engineering and Material department, national natural science foundation, in which the theory and method of urban and rural landscape ecological design was ranked as an important construction direction of landscape Architecture (Liu Binyi, WangYuncai, Liu hui, Xujan). With the developing of natural ecology and landscape, rural and urban green space ecological research, we carried out the national natural science fund project "The Fragmentation and Islanding Phenomenon of Traditional Regional Culture Landscape and Its Mechanism" (No. 50878162, WangYuncai, 2008) , as the breakthrough point of the human ecosystem research, the research tries to build a comprehensive system of total human ecosystem which

规划设计的重要理论和方法。

新时代和新格局需要有新的研究思路，才能摸索出新的教学体系和方法。从2003年开始就试图建立这样一个贯穿本科、硕士、博士三个阶段并富有同济特色的生态规划设计教学体系，构建了包括生态学原理、景观生态学、植物生态学、园林植物与应用、景观生态规划原理、植物景观规划原理、种植设计与群落生态设计、生态规划设计、生态规划实践等课程组成的生态规划设计课群，成为同济景观学专业人才培养的重要理论与技能平台。《景观生态规划设计案例评析》作为同济大学景观规划设计专业研究生教学的系列教材之一，担负着专业教育的试点改革的重任。为了开展这项研究，投入了大量的人力，在这个领域作了诸多有益的尝试和拓展。我的博士研究生崔莹、吕东，硕士研究生赵岩、李洋、张英、瞿奇、邹琴、傅文历经2年时间完成案例的选择、翻译和整理与初稿的编写；我的博士研究生杨眉担负了初稿的校对和评论的起草。在此感谢所有给予帮助的各位前辈、同事和同学。可以预见到将有更多的同学在此领域进行辛苦的工作和拓展，成为推动风景园林生态规划设计发展新生力量。

integrate natural ecology and human ecology system in planning & Design , and it becomes the important theory and method of landscape ecological planning and design.

It needs to have new research ideas in the new era and new situation in order to find out a new system and method for teaching. Beginning from 2003 we have tried to build such a teaching system of ecological planning and design to integrate bachelor, master, doctor together and of which has the rich characteristics of Tongji University, the courses cluster mainly include ecology principle, landscape ecology, plant ecology, landscape plant and application, principle of landscape ecological planning , principle of the plant landscape planning , planting design and community ecological design, ecological planning and design, ecological planning practice etc., at last it becomes the theory and skills professional platform for landscape learning of Tongji University. As one of landscape planning & design professional teaching book series of Tongji University, Reviews on Landscape Ecological Planning & Design Cases plays an important role and takes the responsibility of the pilot reform in professional education. In order to carry out the research, we have spent a great deal of labor in this field and have done lots of beneficial attempts and expansion. My doctor candidates Cui ying, Lv Dong and master candidates LiYang, Zhao yan, ZhangYing, Qu qi, Zou Qin, Fu wen, they almost spent 2 years time in case selection, translation and writing the draft, My doctor candidate YANG Mei, a lecturer of Xi'an JiaoTong University, checked the first draft and prepared the review for each case of the draft. So I want to say thanks to each elder, colleagues and students for their helping in the course of writing. We can foresee more and more students will have hard work and make much progress in the fields, and becomes new blood and power to push the development of landscape ecological planning and design .